Foundations of Early Childhood Education

TEACHING CHILDREN IN A DIVERSE SOCIETY

Seventh Edition

Janet Gonzalez-Mena

mheonline.com/advancedplacement

Send all inquiries to:
McGraw Hill
8787 Orion Place
Columbus, OH 43240

ISBN: 978-1-26-475590-5
MHID: 1-26-475590-2

Printed in the United States of America.

1 2 3 4 5 6 7 8 9 LKV 30 29 28 27 26 25 24 23 22

To the memory of my dear friend Patricia
Monighan Nourot (1947 to 2006)—
mother, wife, and educator—
who dedicated her life to teaching children
and teaching others about *children.*

Brief Contents

Part 1

Foundations of the Teaching–Learning Process: The Role of the Early Childhood Educator 2

Part 2

Foundations in Supporting Development and Learning 208

Part 3

Planning for Learning and Development by Enhancing Children's Curiosity, Joy, and Sense of Wonder 356

Contents

Thinkstock Stockbyte/Getty Images

Corbis/VCG/Getty Images

Steve Debenport/Getty Images

IT Stock Free/Alamy Stock Images

Rob Van Petten/Photodisc/Getty Images

SW Productions/Photodisc/Getty Images

Part 2

Foundations in Supporting Development and Learning 208

Rob Hainer/Shutterstock

Ariel Skelley/Getty Images

DGLimages/Shutterstock

Ingram Publishing

Hero/Corbis/Glow Images

Preface

Society is waking up to the fact that the early years last forever as the saying goes. Whereas families once were the only ones responsible for their children and their early education until they went to school, now the responsibility is more shared. That puts a good deal of pressure on us in early care, and education to do the best we can to make sure each program is of high quality.

In *quality* early care and education programs, children not only gain the foundations they need for school success and beyond, but they also learn to interact in cooperative ways with others, the basis for gaining a sense of community. Rugged individualism was badly needed in the frontier days of the United States, but today we face huge challenges in creating unity through diversity and keeping the economy healthy. A good beginning in a *high quality* early care and education program can lead to both social and economic benefits and is a great investment for the society to make. It is the kind of investment that will grow from generation to generation. Individuals reap the benefits of this investment and so does society.

This textbook is designed to help increase the quality in early care and education programs through training teachers. It features skill building with a solid theoretical base. Many students taking early childhood classes are already working with young children–in practicum placements, as volunteers or staff in centers or as family child care providers. This book addresses their needs as well as those of beginning students who have no hands-on experience.

An advantage of using this book is that it provides students with an overview of what goes on in early childhood programs through the use of examples, anecdotes, and scenarios. Some students may have opportunities to see master teachers at work, but others will not. To address this reality, the text finds ways to transport readers to early childhood classrooms and family child care homes so they can "watch" how effective educators facilitate the teaching-learning process. These examples are designed to help readers put themselves in the educator's shoes, examine their own reactions, and anticipate how they might handle similar experiences.

What is New to this Edition?

1. *Reexamining Professionalism.* What does it mean to be a professional Early Childhood Educator? Is ECE a true profession? There is no clearcut answer to that question. Students will be engaged in being part of this current discussion as they prepare for a career in ECE. In this edition, Stephanie Feeney and Nancy Freeman examine the various sides of this issue.

2. *Using Reflective Practice* integrated throughout to help students fine tune observation skills and think about what they observe. Reflective practice helps students pay close attention and become sensitive to what children might need at any given time and what next step might further their learning. Observing, stepping back, reflecting, and figuring out what's needed is more demanding than instructing, teaching, and showing children how to do things, but the benefits are much greater.
3. *Encouraging Problem-Solving in Children.* Research by Carol Dweck, Angela Duckworth, and others provides students with a perspective on effective adult responses to help children learn and explore their interests. The science strongly suggests that the outcomes are more positive and enduring when adults praise effort and process over success because it cultivates grit and perseverance in the face of failure or struggle. It is becoming clear that children who hear "Good job!" often tend to avoid activities that bring an enthusiastic adult response.
4. *Recognizing the Impact of the Environment and Culture on Learning.* As numbers of children under three years old enter into out-of-home programs, students need knowledge of what is in an environment that serves this age group. This edition will expand on the differences between infant-toddler and preschool physical environments, illustrating how children learn through their interactions with developmentally appropriate environments. Moreover, this edition will also examine how creating an inclusive social-emotional environment can support a sense of belonging by acknowledging and accepting both cultural and developmental differences.
5. *Teaching Children How to Play.* Not all children need to be taught to play, but some do. This new material helps students know how to observe the child and comment on what he or she is doing. Speculating on what the child may do next can serve as a simple suggestion that encourages the child to go further.
6. *Respecting Infants and Toddlers as Capable Self-Learners.* Respect is not always a word used to describe attitudes toward the youngest children. How we perceive the capabilities of infants and toddlers affects how we teach. If we underestimate the potential of both infants and toddlers, we may hamper their learning by doing for them what they can learn on their own through their own natural instincts to explore and experiment. Students will discover respectful behaviors by contrasting them with the more common disrespectful ones.
7. *Putting an Emphasis on Supporting and Augmenting Learning and Development.* Curriculum is a word that comes down from training teachers for olders students and is used more and more in Early Childhood Education—even in infant-toddler programs. Misconceptions about curriculum abound. As an introductory text, this new edition changes the emphasis on *curriculum* and instead puts it where it more rightly belongs for the beginner by focusing fully on learning and development. From the many examples throughout the

book, students will become more sensitive to each child's developmental pathway, what the child is learning and what the next steps are. When students can move beyond individuals to patterns in the group, they will be more prepared to be early childhood teachers. The last three chapters of this book transition the student into the realm of curriculum, but still put a major emphasis on learning and development.

8. Updated scholarship appears in each chapter and is reflected in References and F or Further Reading.

THEMES OF THIS BOOK

Critical Thinking Skills

This text explains theory in such practical ways that students can take sophisticated information in stride and understand its usefulness right away. The book talks directly *to* the student *from* the author, person-to-person. The text makes it clear that there is no formula for "correct" behavior in every situation. Students are encouraged to use critical thinking along with reflective practice rather than look for right answers. To do this, the text sometimes provides a particular viewpoint but then asks students to use it as a backboard off of which they can bounce their own ideas. Following the National Association for the Education of Young Children's (NAEYC) advice in the *Developmentally Appropriate Practice* book, students are urged to make decisions about what is best for each child and family based on child development principles as well as the child's and family's individual and cultural background.

Integrated Subject Matter

As you become familiar with the text, you will notice that the book is unique in many ways. Part of the uniqueness shows in the way important themes and subjects are integrated throughout. For example, diversity is not a separate chapter but a main theme throughout the book. Working with children with special needs is also integrated throughout–in chapters on guiding behavior (Chapter 5), modeling adult relationships (Chapter 7), and setting up the physical environment (Chapter 8) among others. This book is about the care and education of *all* children, including those with special needs. Observation and reflective practice are also strands that weave through the book. Guidance is another strand. Besides being set apart in a separate chapter, guidance strategies also appear wherever students might need them for helping children solve problems and resolve conflicts. Learning to resolve conflicts equitably is an important part of moral development so one could say that moral development is embedded throughout the book, though it is not pointed out as such.

The Link Between Care and Education

A quick look through the chapter headings makes evident some of the unconventional approaches the book takes. Why is there a chapter on routines–those essential activities of daily living that involve physical care such as eating, toileting, and even diaper changing? That emphasis reflects the fact that care and education cannot be separated in the early years. Also routines are featured because this book addresses the needs of infants and toddlers as well as children ages three through eight. The younger the child, the more important (and educational) are the care-focused activities. The chapter on routines also reflects the fact that even in preschool and primary classrooms, some children with special needs may still need the kind of physical care that falls under the classification of caregiving routines.

Developmental Information: Birth Through Age Eight

Why does the information about ages and stages appear in Chapter 11 instead of Chapter 1, and where are the familiar chapter headings of physical, cognitive, and social-emotional development? Those have all been reconstructed into new forms and folded into the text in innovative ways. These changes may challenge traditionally minded people. But with an open mind and a little readjustment, the organization makes sense. Developmental information is presented differently but never minimized. One of the basic glues that holds our profession together is the general agreement about the value of developmental research. The Program Standards for the NAEYC Accreditation uses the word *development* 19 times in their 10 standards.

Modeling as Teaching

Another innovation is reflected in the two chapters on modeling as a way of teaching children. I do not mean showing children how to create a picture, make a craft, or solve a problem, but rather, more subtle unconscious modeling through interactions and everyday actions and manners that children pick up on. The point is that when adults are focused outwardly on the children, they often ignore their own behavior and the unspoken messages they are giving. Children pick up adult attitudes and accompanying behaviors. Since diversity and equity are important subjects of this book, adults must model acceptance of diversity and behaviors that lead to equity.

Family-Centered Approach

In early care and education programs, the relationship with families and those who work with their children is vital. The strand weaves throughout and is also featured in a chapter in Part I—not left until the end of the book. Although early educators play a prominent role in the lives of children, they cannot ignore that families play a much greater and more long-term role. Combined with a focus on diversity, this strand makes clear that professionals cannot ignore what families want for their children, even when they do not see eye-to-eye with all families. What is at stake are children's identity formation and connections to their family. This emphasis on parents' goals and values reflects the vision of a pluralistic, democratic society. This viewpoint is quite different from a "let's study diversity" approach to multicultural curriculum. This book teaches students to use an antibias, activist approach.

Features

- **Part-Opening Introductions** provide readers with an overview of the chapters to follow, pointing out how each informs the larger message of the part.
- A **Chapter Outline** begins each chapter to lay out the key topics.
- The **In This Chapter You Will Discover** sections provide readers with a listing of what they should learn by reading the chapters.
- Marginal links **provide** key content to the **NAEYC Early Childhood Program Standards.**
- *Focus on Diversity* boxes allow readers to understand differences in new ways.
- **Marginal definitions** explain key terms at the point where they appear in the text.
- *Point of View* boxes provide two sides of an argument or idea.
- *The Theory Behind the Practice* boxes link content to the theory supporting it.
- *Voices of Experience* boxes present real-life stories from real-life practitioners.
- **A Story to End With** section concludes each chapter with a brief scenario related to the chapter's topic.
- A **summary** provides a conclusion to each chapter.
- **Reflection Questions** encourage students to consider and apply the chapter's topics.
- **Terms to Know** lists key terms discussed in this chapter.
- **For Further Reading** presents a listing of suggested related readings.

Supplements

The seventh edition of *Foundations of Early Childhood Education: Teaching Children in a Diverse Society* is now available online with Connect, McGraw-Hill Education's integrated assignment and assessment platform. Connect also offers SmartBook© for the new edition, which is the first adaptive reading experience proven to improve grades and help students study more effectively. All of the title's website and ancillary content is also available through Connect, including:

A full Test Bank of multiple choice questions that test students on central concepts and ideas in each chapter.

An Instructor's Manual for each chapter with full chapter outlines, sample test questions, and discussion topics.

Lecture Slides for instructor use in class.

FOR INSTRUCTORS

You're in the driver's seat.

Want to build your own course? No problem. Prefer to use our turnkey, prebuilt course? Easy. Want to make changes throughout the semester? Sure. And you'll save time with Connect's auto-grading too.

65%

Less Time Grading

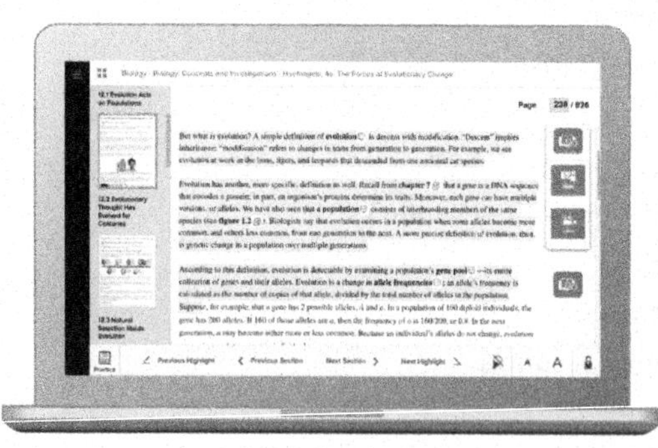

Laptop: McGraw-Hill Education

They'll thank you for it.

Adaptive study resources like SmartBook® help your students be better prepared in less time. You can transform your class time from dull definitions to dynamic debates. Hear from your peers about the benefits of Connect at **www.mheducation.com/highered/connect**

Make it simple, make it affordable.

Connect makes it easy with seamless integration using any of the major Learning Management Systems—Blackboard®, Canvas, and D2L, among others—to let you organize your course in one convenient location. Give your students access to digital materials at a discount with our inclusive access program. Ask your McGraw-Hill representative for more information.

Padlock: Jobalou/Getty Images

Solutions for your challenges.

A product isn't a solution. Real solutions are affordable, reliable, and come with training and ongoing support when you need it and how you want it. Our Customer Experience Group can also help you troubleshoot tech problems—although Connect's 99% uptime means you might not need to call them. See for yourself at **status.mheducation.com**

Checkmark: Jobalou/Getty Images

FOR STUDENTS

Effective, efficient studying.

Connect helps you be more productive with your study time and get better grades using tools like SmartBook, which highlights key concepts and creates a personalized study plan. Connect sets you up for success, so you walk into class with confidence and walk out with better grades.

Study anytime, anywhere.

Download the free ReadAnywhere app and access your online eBook when it's convenient, even if you're offline. And since the app automatically syncs with your eBook in Connect, all of your notes are available every time you open it. Find out more at www.mheducation.com/readanywhere

"I really liked this app—it made it easy to study when you don't have your textbook in front of you."

- Jordan Cunningham, Eastern Washington University

Calendar: owattaphotos/Getty Images

No surprises.

The Connect Calendar and Reports tools keep you on track with the work you need to get done and your assignment scores. Life gets busy; Connect tools help you keep learning through it all.

Learning for everyone.

McGraw-Hill works directly with Accessibility Services Departments and faculty to meet the learning needs of all students. Please contact your Accessibility Services office and ask them to email accessibility@mheducation.com, or visit **www.mheducation.com/about/accessibility** for more information.

Top: Jenner Images/Getty Images, Left: Hero Images/Getty Images, Right: Hero Images/Getty Images

ACKNOWLEDGMENTS

I would especially like to thank Ani Shabazian, of Loyola Marymount University, for all of her editing and support, which was essential to complete this seventh edition. Ani's contributions, insight, and wisdom were essential for the revision of this seventh edition. I would also like to thank the following reviewers:

Zana Wilkie: *De Anza College*

Sue Langwell: *Moraine Valley Community College*

Diane Smith: *Ivy Tech Community College*

Greg Mullaney: *Quinsigamond Community College*

Lucia Obregon: *Miami Dade College*

Peggy Mitchell: *Ozarks Community Technical College*

Heather Logan: *Central Piedmont Community College*

About the Author

Janet Gonzalez-Mena was on the faculty of the Child and Family Studies Program at Napa Valley College for 15 years where she taught academic classes and also worked in partnership with campus child care as a practicum teacher. Since 1991 she has also worked for WestEd Program for Infant-Toddler Care PITC as a "trainer of trainers." Janet has experience as a preschool teacher, home visitor, child care director, family child care coordinator, and supervisor of a pilot program of therapeutic child care for abused infants, toddlers, and preschoolers. She has a special interest in diversity, equity, and partnering with parents. Gaining more experience with full inclusion programs came when Janet joined the faculty of another WestEd training project called *Beginning Together*, a project designed to promote inclusion of children with disabilities and other special needs. Janet occasionally works for WestEd's Center for Child and Family Studies as a writer and has contributed a chapter called "Culture and Communication in the Child Care Setting" to WestEd's *Guide to Language Development and Communication* (2011). She also worked on a project called *Bridging Cultures in Early Care and Education*, which resulted in coauthoring a training guide published by WestEd and Erlbaum. Janet has written a number of other books and articles about early childhood education and parenting, including *Dragon Mom.* Janet's writing appears in *Child Care Information Exchange* and *Young Children.* For example, her article "On the Way to Friendship: Growing Peer Relationships Among Infants and Toddlers" appeared in *Exchange* in May 2012. Her coauthored article "Self-Regulation: Taking a Broader Perspective" appeared in the January 2011 issue of *Young Children*. Janet also coauthored an article for *The Signal*, which is the newsletter of the World Association for Infant Mental Health (2011). Her book *Diversity in Early Care and Education: Honoring Differences* was distributed by the National Association for the Education of Young Children as a comprehensive member benefit for 2008. A book coauthored with Anne Stonehouse called *Making Links* (2008) reflects Janet's interest in collaboration and is about planning and practice in partnership with children and parents. Reflecting that interest is her book 50 *Early Childhood Strategies for Working and Communicating with Diverse Families* (2013) now in a third edition. Although she has wide experience in early childhood education, major interests focuses on the care and education of infants and toddlers. Janet studied with Magda Gerber, the Los Angeles infant expert, and also with Anna Tardos, daughter of Emmi Pikler, founder of the Pikler Institute in Budapest, Hungary. Janet has a Master in Arts Degree in Human Development from Pacific Oaks College.

Foundations of Early Childhood Education

PART

1

Foundations of the Teaching-Learning Process

The Role of the Early Childhood Educator

Part 1 was written for two purposes: to give background and theory to students going into early childhood education and to act as a mini-crash course on how to work with children when you're in an introductory class and taking a practicum or field experience course at the same time. Because this is an introductory text and presumably the first class students take, the professors who reviewed the manuscript wanted to see background material in the first chapter. So if you're reading this sequentially, you start out with some history, theory, and an idea about early childhood education as a special kind of profession. That's fine if you have the luxury of studying the history and theories first before entering the field. Many coming into the field, however, find themselves faced with a group of children from day one. Others are already working in the field before they take their first academic class. If you're in that situation, you can't dig into which theorist said what while two children are fighting over a toy in the corner. You need to know what to do!

Chapter 2 takes a big leap from the background material and theory in Chapter 1 dives straight into practice. The next five chapters focus on the vital skills that adults need to work effectively with young children. This is not the ordinary sequence of most introductory texts because it focuses so heavily on practice. Watch though how in each chapter practice is tied into the theory behind it. Theory may not be explicitly stated except where highlighted, but rest assured that theory lies behind practice on every page.

The chapters of Part 1 focus on the adult's role in the teaching-learning process in a holistic way; in contrast, Parts 2 and 3 focus more specifically on supporting and promoting development and learning in a more traditional way. Part 1 emphasizes the adult's role in intentionally planning or thinking things out and also in spontaneously taking advantage of opportunities that emerge. The goal in both cases is to enhance *all* children's learning in the various areas of development. When *all* is stressed in such a way, it assumes that the group includes diversity–either obvious or invisible. It also assumes that the diversity is not just in family background, but also in ability. Children with disabilities and other special needs, identified or not, belong with their typically developing peers. This book recognizes that reality and makes suggestions about how to respond individually and appropriately to each and every child.

The idea of unwritten plans as an approach to learning relates to the fact that children learn every minute of the day, whether adults are aware of it or not. When the adults don't pay attention to what the children are learning, the lessons may be detrimental. For example, if you're not paying attention, the children in the block corner are learning that those who grab get the blocks they want. By paying attention, you can help them learn to share and still get what they want to play with. Furthermore, if you are not careful, other children may learn that hitting works when other children will not give in as long as no adult notices. Sometimes children learn that loud crying is the only behavior that gets adult attention. Careful observation, reflection, and awareness can help you see that when adults notice, encourage, and intervene in ways that support positive behavior they make a difference in what children learn.

Most people know they should set good examples for children, but some are not fully aware of the power of the modeling effect. Much of what children learn–desirable or not–comes from imitating the powerful adults around them. Adults are aware of their own behaviors and consciously try to model ones they want the children to learn, which is an example of an unwritten curriculum. When you grab a child, roughly sit her down, and speak to her in a threatening voice, you model aggression. When you make eye contact with every adult you speak to except the one you dislike, you model differential behavior. If that adult happens to have darker skin than you, or speaks a foreign language, children will notice. Even if your feelings may be completely unrelated to race or culture, these differences that children notice can eventually become racist behavior on their part. Most textbooks devote little to modeling or setting examples for children, but in this book, two entire chapters, Chapters 6 and 7, are devoted to what children learn from the attitudes and behaviors–implicit and explicit–of the adults around them. This is not only unwritten curriculum, but it's practically invisible. Yet modeling is a powerful teaching tool when used with awareness.

Unwritten curriculum also includes planning. Planning and preparation are part of everything. Being responsive and spontaneous is important, but so are observing and reflecting in order to arrange the environment, supervise groups, interact with individuals, and be ready to make appropriate interventions when needed. This is why observation and also reflective practice play such an important part in the chapters of this first section. Even though the chapter on more formal records and assessment comes later, you are always observing and doing informal assessment whenever you are working with children. This is how you know when and how to interact with them.

At this point you are ready to increase your knowledge of the foundations of early childhood education.

1 Early Childhood Education as a Career

Thinkstock Stockbyte/Getty Images

Four Themes in Early Childhood Teacher Training

The Value of Reflective Practice

A Multicultural Perspective

A Holistic Approach

Professionalism

Child-Development History

Historical Trends and Figures

Child-Development Theorists and Their Theories

Pioneer Educators

Brain Research

What It Means to Be an Early Childhood Educator

Legal Responsibilities

Code of Ethics

A Story to End With

In This Chapter You Will Discover

- what reflective thinking has to do with knowing yourself.
- how a multicultural perspective relates to a pluralistic goal.
- what is meant by the term *whole child.*
- what it means to be an early childhood professional and the legal responsibilities involved.
- why early childhood educators need to know about "ages and stages."
- what is developmentally appropriate practice.
- which standard-setting early childhood organization is the world's largest.
- the history of early childhood education and why you can't separate care and education.
- whether there is an answer to the "nature-nurture question."

Some say, "All you can ever teach is yourself, but if this is true, why would anyone need a text like this? Indeed, early childhood education (ECE) is a career and can also be considered a special kind of profession. Those who enter it must learn to speak the language of their chosen field because it binds early childhood educators together. The goal of this book is to give you the concepts and the vocabulary shared by the ECE community and to introduce you to the reality of the early childhood culture.

FOUR THEMES IN EARLY CHILDHOOD TEACHER TRAINING

This book carries four themes throughout that are important to the early childhood educator. One is **reflective practice:** we must examine our experience, both past and immediate, in order to understand it, learn from it, and grow. The second theme is **multiculturalism:** we must recognize, respect, and value the diversity that encompass the "American people." The third theme is **holism:** early childhood education focuses on the "whole child" and fashions its curriculum accordingly to facilitate the teaching-learning process. The last theme is **how we define ourselves:** we are educators, not glorified babysitters.

The Value of Reflective Practice

If the statement "All you can ever teach is yourself" is true, ask yourself honestly, "How well do I know myself, particularly in relation to children unrelated to you?" Working with other people's children is an experience that teaches you a lot about yourself. Tucked-away feelings, forgotten experiences, and buried treasures may surface as you embark on this career. This book is designed to help you deal with the negatives as well as rejoice in the positives that come from working with young children.

As you read, be mindful of who you are–your gender, your race, your ethnicity, your culture, your family circumstances, and your background. You are a real person, bringing with you the sum total of your experiences. The children, too, have their own gender, race, ethnicity, culture, and family background that influence their interactions with you. The interface between an adult and child is where the greatest learning takes place and constitutes what is often referred to as the "unwritten curriculum." The challenge is to be reflective about what you perceive during those interactions–the ones that involve you and the ones you observe that do not necessarily involve you.

As you read this book, you will learn more about yourself and about early childhood education as a career.

A Multicultural Perspective

This book takes a pluralistic view of the United States; it recognizes, respects, and values the many cultures that constitute the "American people." This book recognizes that the "predominant culture" is only a culture, not a universal reality to

FOCUS ON DIVERSITY

Aspects of Culture

Culture is much more than mere ethnicity or national origin. The term *culture* includes the way lives are influenced by race, gender, age, abilities and disabilities, language, socioeconomic class, education, religion and/or spiritual practice, geographical roots of the family, and present location as well. Sexuality, including sexual orientation, is also included. We are a sum of numerous influences and so are the children and families served by early care and education programs.

avoid the insinuation (intended or not) that the predominant culture is "normal" and that other cultures (Asian American, African American, Latin American, and so on) are deviations from the norm. As shown in the *Focus on Diversity* box, culture is a complex composite of attributes, beyond food, music, and customs, that influence every facet of our lives.

I, Janet, author of this book, am an Anglo American married to a Mexican. I live in a multicultural, multiracial family through marriage and adoption. Members of my family have been challenged with a variety of disabilities. Although I have had plenty of exposure to diversity, I still have to work at seeing the world through various perspectives. Sometimes I fail to do so.

In this text, I often speak from my own experience in the first person to make this book as authentic as possible. We all know our own truth best, and although we must try to see other perspectives, we sometimes miss the mark. Yet the more we speak from an honest place within ourselves, the more we can share our perspectives with others and invite them to share theirs with us.

A Holistic Approach

The "whole child" is an important concept behind this text. Although we may sometimes focus on a child's mind, body, or feelings, we cannot separate one part from the other. The child operates as a whole. We may plan an activity with intellectual objectives in mind, but we can never ignore how the child responds physically or emotionally.

Furthermore, we cannot work with children without considering the context of their family and home life, which are influential to a child's individual, cultural, ethnic, linguistic, and socioeconomic makeup. We welcome not only the individual child into our early childhood classroom, but also his or her family. Also even when the family is not present, we must remember that the family, which represents a larger context, is always part of the child's individual makeup.

In order to teach the whole child, early childhood education must offer a holistic **curriculum.** Rather than offer separate subjects taught at separate times, the teaching-learning process occurs in a holistic way throughout the day. Curriculum is a broad

curriculum A plan for learning.

An integrated approach allows both the teaching and learning processes to occur in a holistic way.
Ariel Skelley/Getty Images

concept and can be thought of as a plan for learning. For instance, a bread-making project, an example of a holistic curriculum, that starts early in the day can weave in and out of the day's activities and may continue the next day. This activity can encompass a variety of concepts and skills related to math, science, culture, feelings, eye-hand coordination, sensory development, social relationships, language, symbolic representation, and emergent literacy; it also teaches self-help skills and can be incorporated into the daily food plan.

emergent curriculum
An approach to teaching that focuses on "bottom up" rather than "top down" curriculum. It emphasizes being responsive to children's interests and creating meaningful learning opportunities that organically emerge from their interests.

Projects can flow from one to another—an approach called **emergent curriculum.** For example, a simple water-play activity involving a hose and a sandbox could lead to a variety of projects, depending on the interest of the children and adults. In one program, the children were frustrated when water disappeared instead of pooling in the sandbox. They ended up with an extended project of creating a "beach." It took days to dig out the sandbox, lay down a plastic tarp, and put the sand back in. When the children filled the sandbox with water this time, the water pooled at one end where the sand level was lower, and the other end became the beach. Taking the project a step further, the teacher documented the process through photos and words and encouraged the children to draw, dictate, and write about what they did, what happened, and how they felt about it. The documentation was displayed, which gave the project a past and helped the children conceive of future remodelings and related projects. The display also gave them much to talk about and informed the parents of their children's projects, thus bridging what has been called the "home-school gap."

This is one example of how teachers can encourage a holistic approach by continuity in learning—emphasizing depth over breadth of learning. The goal of this approach is not to merely watch the children, keep them safe, and allow them to play with whatever toys happen to be present; nor is it aimed at presenting children with a string of isolated, unconnected learning activities. Instead, curriculum is developed to provide continuity.

This book is about planning for learning, but it is not a curriculum book. It is a text that introduces the reader to the idea that learning is always a part of every child's life, and the teacher must be attuned to that fact.

Another aspect of this text's holistic approach is that it is a prosocial approach. Although there is no chapter devoted specifically to character and value education, suggestions about how to guide children toward prosocial skills, attitudes, and behavior are woven throughout the book.

Professionalism

professionalism A set of attitudes, theories, and standards that guides the early childhood professional.

Perhaps it is clear by now that early childhood education is both a career and also a special kind of profession, not just glorified babysitting. Early childhood education is a branch of education that deals with children from birth to age eight. What children in this age category need is different from what older students need.

What is a profession and what makes early childhood education a "special kind of profession"? According to Stephanie Feeney and Nancy Freeman,[1] a profession has nine attributes as requirements, which include:

1. Requirements for entry.
2. Specialized knowledge and expertise.
3. Prolonged training in using professional judgment.
4. Standards of practice.
5. Distance from clients.
6. Commitment to significant social value.
7. Recognition as the only group that can perform its societal function.
8. Autonomy.
9. A code of ethics.

Early childhood education does not have all of these attributes.

- We do not have requirements for entrance in the same way that other professions do.
- We do not have professional autonomy because the field is governed by laws and licensing agencies rather than early childhood educators.
- We do not have prolonged training, except for those who choose it, because training requirements vary by state.

What we do have is specialized knowledge and expertise that is supported by the science of child development. We also have professional organizations that set standards and a code of ethics.

Early childhood educators combine care and education in many different kinds of programs, but they share common goals. They agree that early childhood is to be appreciated as a unique stage in the life cycle. They strive to educate the whole child, taking into consideration mind, body, and feelings. They create educational goals designed to help each child achieve his or her individual potential in the context of relationships. In addition, early childhood professionals recognize that the child cannot be separated from the social context, which includes family, culture, and society. They not only strive to understand and relate to children in context but also appreciate and support the ties that bind the child to his or her family.

Early childhood educators look to the science of child development for their knowledge base about what children need and how they learn and develop; they use research to distinguish science from myth. Those untrained in early childhood education may rely more on their own assumptions, background, experience, and bits of research. For example, many people still believe that spanking is effective in teaching a child to behave properly. The early childhood educator, however, knows that research indicates that harsh physical punishment models violence, creates hostile feelings, and doesn't improve behavior.

Without a background in child development, some adults might expect a child to act much older than he or she is, so they might say to a very young child, "Don't cry! You're acting like a baby." Or they might expect a slightly older child to sit still and behave like mature students who have the ability to learn by listening. Early childhood professionals, however, are familiar with scientific evidence that shows what appropriate behavior expectations are for each developmental stage.

ages and stages
A catch phrase that relates to childhood developmental features and behaviors that tend to correlate with specific ages. Each stage describes a particular period of development that differs qualitatively from the stages that precede and follow it. The sequence of stages never varies.

Ages and Stages. "Ages and stages" is a catch phrase that refers to particular sets of tasks and behaviors that are specific to different periods of child development. Usually, the stages correspond to particular ages, but not always. Developmental variation within an age group can be great. Some children take longer to get to and pass through each stage and others move more quickly, yet, the stages tend to occur in an unvarying sequence. Cultural differences also play a role in defining expected behaviors at any given age.

Physical milestones of development were introduced by Arnold Gesell (1880–1961) based on his research of children's behavior. Following Gesell's tradition, Benjamin Spock and T. Barry Brazelton have brought the concepts and the specifics of stage development to the attention of the general public. Many others have continued to research and standardize **ages and stages** norms, expanding on Gesell's rather narrow sample.

Stage norms are useful, but remember, children are individuals. Also, keep in mind that, research does not always have the final answer; it may not even be asking the right questions. Early childhood education addresses cultural and value differences that is sometimes overlooked by research. Nevertheless, as the cultural diversity of researchers more closely reflects the demographics of the population, we will move closer to solving these kinds of problems.

Professional Organizations. Early childhood educators have professional organizations that guide and support them, help them make professional ties, and keep them abreast of current issues through the publication of journals. Two of the oldest respected organizations in the United States are the Association for Childhood Education International (ACEI) and the National Association for the Education of Young Children (NAEYC). Both organizations have long histories as advocates for children, their families, and education, and they continue to have substantial influence on improving the field. **ACEI** started in the late 1800s as a kindergarten organization but changed its name and expanded its focus to preschool and elementary school in the 1930s. Today, ACEI's work includes publishing a journal and books focusing on children from birth to early adolescence and hosting annual international study conferences. **NAEYC,** established in 1926 (originally named the National Association for Nursery Education), began the same year that the federal Head Start program was launched, which brought for the first time ever national public attention and funding to preschool education. The NAEYC began to rapidly grow and by its 50th anniversary in 1976, it had a membership of 31,000. As more and more women joined the workforce and dual-income households began to rise as well as single parents, concern about both the need and quality of child care led NAEYC to launch a national voluntary accreditation system for early childhood programs. As a result, during the years 1985–1990, NAEYC membership doubled from 45,000 to 90,000 members. Also, during this time, NAEYC begin issuing position statements, publishing journals, and hosting annual conferences, which until this day remains one of the largest conferences of early childhood educators.

The NAEYC is by far the largest and best-known early childhood education organization and has set standards for the field by: creating an accreditation process, which is administered through its National Academy of Early Childhood Programs, working to improve pay and working standards for teachers, and creating a code of ethics to guide early childhood educators in their work and decision making. The NAEYC advocates for young children and their families through its position papers, designed to influence government policies and early childhood education practices.[2]

Another momentous organization, the Children's Defense Fund (CDF), was started by Marion Wright Edelman in 1982. The **CDF** is a lobby based in Washington, DC, and its primary purpose is to advocate for children, particularly those in poverty and those of color. In 1996, the CDF drew national attention with a demonstration called "I Stand for Children"; people came from across the continent to stand in the nation's capital to shine a spotlight on children and their needs. The "Stand for Children" campaign went local after this event. Today communities across the nation each create their own versions of "Stand for Children."

An organization called Zero to Three has established a National Center for Infants, Toddlers, and Families that emphasizes the care and education of the first three years of life. With welfare reform in the 1990s, growing numbers of infants and toddlers are being cared for outside their families. In addition to some compelling research on brain development, which indicates that, as the slogan goes, "the first three years last forever," Zero to Three has become an important supportive agency to the early childhood field. The organization provides resources such as conferences, training institutes, a journal, and a book publishing press to advocates of children under the age of three.

WestEd's Center for Child and Families Studies houses the Program for Infant-Toddler Care (PITC), the largest training organization in the United States devoted to improving the quality of infant-toddler care.

Developmentally Appropriate Practice (DAP). An important document to come out of the NAEYC is entitled *Developmentally Appropriate Practice in Early Childhood Education Programs.* DAP is a statement about what the NAEYC and its members believe constitutes quality care and education for young children. The document is designed to guide professional decision making using the following three knowledge bases:

1. Knowledge about how children develop and learn, including information about ages and stages and what the appropriate experiences, materials, activities, and interactions for age and stage are
2. Knowledge about each individual child in the group
3. Knowledge about the social and cultural context in which each child is growing up

For instance, either-or decisions, such as those that compromise cultural appropriateness for age appropriateness, prioritizes one knowledge base over the others. In contrast, both-and thinking entails decision making that balances conflicting elements to ensure developmental, individual, and cultural appropriateness in all situations. The *Voices of Experience* story by Lynne Doherty Lyle a former preschool teacher, takes both-and thinking out of the realm of the abstract and brings it into an adult-child interaction that she experienced. She realized that both-and thinking validates that two different realities can exist at the same time. Sometimes such decision making requires great creativity on the part of professionals to look beyond their own perspective. The NAEYC document makes it clear that early childhood educators must themselves be learners as they work with children and families.

both-and thinking An approach to decision making in which the early-childhood educator considers what is developmentally, individually, and culturally appropriate in all situations; it involves coming up with a solution that may incorporate all the conflicting elements. Both-and thinking contrasts with either-or thinking, in which the choice is between one solution or the other.

What kinds of conflicts might require both-and thinking? One common conflict is how to respond to both group needs and individual needs simultaneously. For instance, one child might need a morning nap at a time when the rest of the group is noisy and lively. An example of creative problem solving would be to find a quiet corner where the child can rest with minimum disruption.

This document encourages moving away from "either-or thinking" to "both-and thinking" and stresses that each of the three knowledge bases is dynamic and changing. Another kind of conflict involves diverse views of what children need such as when the child's home stresses interdependence instead of independence. Again, both-and thinking is called for. Sometimes what is age appropriate is not what is individually or culturally appropriate. Instead of telling the family they are wrong, the early childhood educator should try to understand their point of view. The family's cultural value may be to prolong dependence and to "baby" the child rather than to stress independence and teach self-help skills. The family may be teaching their child that it is more important to accept help from adults than to try to do

Accepting Dual Realities

It happened again; tearful, three-year-old Ben refused to get out of the car. We were in the parking lot at the grocery store, with Ben screaming, "No store!" People glanced our way as we battled in our parked car.

All my skills from working with preschoolers disappeared, and I could not distance my own feelings of frustration, fear, and shame from my son's passionate declaration of what he wanted. We were both in an emotional stew until I summoned up my teacher self and said, "I know you don't want to go to the store, but we are going anyway." Once I said this, I felt in control again and in charge. I often used this technique with children that I had worked with. It was clear, set limits, and yet validated the child's wants. Well, I thought it did.

While struggling to pull Ben from his car seat that day, I remembered someone saying that using the word "and" instead of "but" is more effective in these situations. She explained that using "and" allows for two realities to exist at the same time. These types of challenging and trying situations do not have to be either/or situations with the adult's needs being more important than a child's. The experiences and feelings could coexist.

After remembering this, I said to Ben, "I know you don't want to go into the store, *and* we are going to go." What happened? Ben kept screaming, and I realized that I too did not want to go into the store any more than he did; and we went anyway. Sometimes you must do things you don't want to do, and Ben and I could connect with each other in this shared experience. We both could feel our own feelings while knowing that there was room for both: his and mine, while in the grocery store buying something for dinner.

–Lynne Doherty Lyle

things on his or her own. Such teaching represents a cultural goal that opposes a push for independence.

Under the *Developmentally Appropriate Practice* guidelines, the early childhood educator cannot simply discount this family's approach as being "developmentally inappropriate" and ignore their goals. This situation calls for discussion–lots of discussion–until the professional and family can see each other's point of view and come to some sort of agreement. (More will be said about goal conflicts as the book progresses.)

Both editions of *Developmentally Appropriate Practice* were designed to help practitioners use what is known about how children develop and learn to plan experiences that help each child move forward.

Though defining and implementing *Developmentally Appropriate Practice* started in the 1980s and 1990s, it remains important today. The twenty-first century brought a new area of focus–the creation of learning and program standards. Although the field supports standards, often the pressure to formulate them comes from governments and other funding bodies that value teaching subject matter and skill over child development and the individual construction of knowledge. Measuring outcomes has become a huge subject of controversy in the profession as children and the programs they attend are being judged on test performance.

Types of Early Childhood Programs. There are so many different early childhood education programs that they do not always fit neatly into clear categories. Table 1.1 is an attempt to categorize, but as you read through this section you will notice that the picture is even more complex than this sampling of programs depicts.

One way to classify programs is to distinguish between full-day programs and half-day programs; whereas full-day programs usually focus primarily on child care, half-day programs usually focus more on education. The true difference between these two types of programs, however, is the length of the day. Both types care for and educate children.

A common misconception in the field is the distinction between educational programs and child care programs. However, you cannot separate care from education nor can you separate education from care. Nell Noddings has made a good case for care *always* being a part of education in her book *The Challenge to Care in Schools.* Well-trained ECE professionals who work in quality programs for young children combine education and care. The challenge for professionals, funders, and policy makers is to increase the quality in programs, mandate training for the people who staff them, and find ways to pay trained professionals a living wage. These groups work together to advocate for increased investments in professional development and compensation for early learning educators. When we invest in professional development, as this book aims to do, we are creating a qualified workforce. When we create a qualified workforce, we are cultivating the quality learning environment that sets the foundation for our children's futures–and, really, all of our futures.

Early childhood education programs can also be classified by locale–in homes or in centers. There are **center-based programs** and **family child care programs.** Family child care, though less regulated, is in the process of professionalizing itself.

A second type of home-based program is a home-based visitor program, in which a professional or paraprofessional works with families of young children in their own homes. Home-based Head Start and Early Head Start are the best-known programs of this type, but other models exist.

Another type of early childhood education program is one that primarily serves children with special needs. But because enrollment of children with special needs is increasing in other types of programs, we are starting to see fewer programs of this category. There is a growing realization that segregation robs our society of the unity it so desperately needs. Special education is the area of desegregation that the early childhood field needs to work on most, but changes are happening. Today, for example, programs in the United States are bound by law to accept any child who applies unless they can prove that they are not equipped to handle the special needs of a particular child. A program risks being

TABLE 1.1 SAMPLING OF PROGRAM TYPES

Early Care and Education Program	Full Day Half Day	Center Based Home Based	Sponsorship Funding	Age Range	Mixed Age Range	Designated Population
Family Child Care	Usually full day	Home based	Parent fees	Could be all ages	Usually	General
Child Care	Full day	Center based	Parent fees	Could be all ages	Sometimes	General
Child Development	Full day	Center based	Government and/or parent	Could be all ages	Not usually	May be limted to low income
Centers Early Learning Centers Children's Centers	Full day	Center based	Employer and/or parent	Could be all ages	Not usually	May be limited to employees
Campus Children Centers	Variable	Center based	Government or student or college or parent fees	Could be all ages	Not usually	May be general or limited to students or low-income students or to staff and faculty
Child Care for School-Age Parents	Full day	Center based	Government or school district	Usually infants and toddlers	Sometimes	Teen parents
Preschool Pre-K	Half day	Center based	Parent fees or government	Mostly 3 to 4	Sometimes	General
Head Start and Other Compensatory Education Programs	Usually half day	Center based	Government	Mostly 3 to 4	Sometimes	Low income
Surround Care	Before and after preschool, kindergarten, or primary	Center based or family child care	Parent fees or government	Could be all ages	Sometimes	General or low income
Transitional Kindergarten (signed into California Law in 2010)	Usually half day	Center based	Parent fees or government	Primarily meant for a 4-year-old who doesn't turn 5 by September 1 (not eligible for Kindergarten)	Not usually	General
Kindergarten	Usually half day	Center based	Government or private (parent fees)	5	Not usually	General
Primary	Full day until 2 or 3 p.m.	Center based	Government or private	6 to 8	Not usually	General

POINTS OF VIEW

Parent Involvement or Parents as Partners?

The parent involvement and education approach requires parents to put in so many hours doing various things for the center, either in the classroom during the week or on weekends, such as making repairs or doing yard work. They also attend classes designed to expand their knowledge of child development and improve their parenting skills through the use of positive parenting strategies.

The other program considers parents as partners and focuses on looking for ways to create equality in the relationship. This means the staff or provider of such a program involves parents in both major and minor decisions about their child's care and education, including decisions about what and how their child will learn and how to best discipline and guide their child. Any parent education is designed to meet the specific needs of the particular parents enrolled.

held liable if it arbitrarily refuses to work with a particular family or if it rejects a child with special needs because it fears that working with the child would entail too much work.

Sponsorship is another way of categorizing early childhood programs. There are public-supported programs, private-nonprofit programs, and private-for-profit programs. An example of a public-supported program is **Head Start,** a comprehensive, federally funded program that since the mid-1960s has provided education, health screening, and social services to help low-income families give their children–from birth to five years of age–the start they need to succeed in public school. Some Early Head Start programs and other types of early intervention programs are examples of home visitor programs. There are also some state-supported versions of Head Start, as well as early childhood programs run by public schools. Numerous nonprofit programs–some sponsored by religious organizations–serve a variety of needs, but all are designed to provide either half-day or full-day care and education for young children. For-profit programs include both chains and independent businesses.

Employer-supported child care comes in many shapes and forms. Such programs may be run in a center built, owned, and operated by a company for its employees. Many employer-supported programs are run by child care management corporations that assume the organizational, supervisory, and liability aspects of child care.

Programs for teen parents are often housed in high schools so the infants and their mothers are on the same campus. The infants are in child care, while the mothers attend classes. The program includes a parent education component along with the child care and other supports needed for young parents to finish high school while rearing young infants. Judy is a young mother whose story Ethel Seiderman

Coming Back to Child Care

When Judy returned to the child care center that she had attended just five years ago, she enrolled her own baby Mika. No one, including Judy, expected her to be back so soon.

As a young mother, Judy says that she could have easily felt defeated by her daughter's birth. It was a hard time and she now recalls how she was "ready to give up my own goals." But as Judy returned home to the place where she and her family spent much of her childhood, she realized that she was not alone. The same people who had cared for her as a child cared for her as an adult. They welcomed her and Mika back to the center.

With the security of child care and support of others, Judy found her way. She looks forward to graduating from college. Her daughter Mika is thriving too. She feels fortunate to have found the way back. She credits the center as being a special place:

> What makes this program a special place is their focus on children *and* their families. The staff is like family. They check in with every parent so that I have a strong sense of involvement in my daughter's education and care. As parents, we choose activities for our children and ourselves. By participating, being with other parents, I found the strength, support, and courage to do what I needed to do.

Today, Judy is an advocate for quality child care. She recalled testifying at the state capitol. "I was scared but I learned that my voice advocates easily; it all came out and it has a type of power similar to the power of my smile I give my daughter when she looks up at me. By the time I finished speaking, I felt as if I could cry because for the first time I heard what it was that makes me able to do what I do. *It's knowing that my daughter is well cared for . . . but so am I.*"

–Ethel Seiderman

tells in the *Voices of Experience* box above. Seiderman is the founder of Parent Services Project, an organization dedicated to helping early care and education centers understand the benefits of family-centered programs.

A category that deserves special mention is the **parent cooperative preschool,** also known as a "parent-participation nursery school." This type of program is designed primarily for parent education; parents are educated through the program but they also serve as "co-teachers." Parent cooperatives, however, are not the only programs with a goal of parent education. Most programs, in keeping with one of the principles of the early childhood profession, regard themselves as serving families, not just children. Parent education and/or involvement is almost always part of the philosophical statement or goals of most early childhood programs. Although today programs are increasingly aiming at a "parents-as-partners" approach. See *Points of View* for different approaches to parental inclusion.

To become an early childhood educator, you will need education and training to provide you with the necessary skills, vocabulary, and concepts. But to enter the field of early childhood education, you will also need to know some of its history because history is an important part of being socialized into a profession, we will take a look at early childhood education's past in the following section.

CHILD-DEVELOPMENT HISTORY

child development The study of how children change as they grow from a qualitative rather than a merely quantitative standpoint.

The field of science that studies how children change as they grow is called **child development.** Child-development researchers study all aspects of children, but most focus on specific areas, such as how children develop thinking or social skills. The extensive study of children in the past has yielded much information that has been useful in designing a variety of early childhood programs. It is a very important field of study because (1) from the time of conception to kindergarten, development from conception through the first five years of life proceeds at a pace that exceeds that of any other stage in life, and (2) what happens during this period of development sets the stage.

What happens during the first months and years of life matters a lot, not because this period of development provides an indelible blueprint for adult well-being, but because it sets either a sturdy or fragile stage for what follows.

Historical Trends and Figures

Child development is a relatively modern area of study and *childhood* as a concept was virtually unexplored until the eighteenth century. For example, in seventeenth-century Europe, children were treated as miniature adults. Over time, however, child study has grown into a legitimate academic discipline.

nature–nurture question The question that asks, "What causes children to turn out the way they do?" In other words, is a child's development influenced more by his or her heredity (nature) or by his or her environment (nurture)? This question can be controversial because nature proponents insist that genetics plays a stronger role in influencing development, while nurture proponents make the same claim about environmental experiences.

The Nature-Nurture Question. One of the key questions child-development researchers explore is: What causes children to turn out the way they do? The possible answers often fall in between or on either side of the **nature-nurture question:** Do children turn out the way they do because of their heredity, their genetic makeup (nature), or because of how they are raised, their environment (nurture)? Today, virtually all researchers conclude that the development of a child is a highly complex process that is influenced by the interplay of nature (genes) and nurture (the environment). While each theorist/theory emphasizes a different proportion of these two forces, development is a dynamic process in which nature influences nurture and vice versa.

The Question of the Basic Nature of the Child. The nature–nurture question is a recurrent theme in child-development history. Another theme also involves a question that has been posed by philosophers and, more recently, researchers: what is the basic nature of the child?

The Church View: The Child Is Basically Evil. Down through the ages, beliefs about the basic nature of the child have influenced how people understood and treated children. The church, the highest authority in Western society before the Renaissance, had its own theory about a child's basic nature. According to early church philosophy, each child carried the seed of evil as a result of being born in original sin, and only the strictest discipline could keep a child from becoming more sinful.

Few child-development experts adopt such an extreme view today, but the idea still exists. Some believe the child's wild nature must be tamed, shaped, and molded.

Experts have long debated whether heredity or environment determines the ways in which a child develops. Today, most agree that the interaction of both genetics and environment influences a child's development.
Kali9/Getty Images

Locke and the Blank-Slate View. The English philosopher John Locke (1632-1704) was the first to write about the newborn as a blank slate, a *tabula rasa*. He was the first to state that the child had no inborn ability to influence his or her own development, and only the environment could determine the outcome. Locke saw the child as a passive recipient of experience rather than someone with specific tendencies to think or behave one way or another. (A whole line of researchers called "behaviorists" have descended from this point of view. A little later in this chapter, we'll take a look at John Watson and B. F. Skinner, two of the best-known behaviorists.)

According to the blank-slate theory, the child is open to all forms of learning that will eventually mold that child into an adult capable of functioning successfully in society. Development comes from the home and parents, but outside early experiences influence development too. The concept of the child as a blank slate places tremendous responsibility on those who are in a position to influence the child's development. Proponents of this view see both parents and educators as having the power to determine the character, talents, and inclinations of the individual–even his or her happiness. Some still believe in this view–that the way a child turns out is purely a result of his or her environment, with innate ability having nothing to do with it. If the children come out fine, parents and educators are lauded; if they do not, then they are blamed.

Rousseau and the Little-Angel View. Some see children as pure and innocent beings with great individual potential that only needs to be unlocked. In this view, an infant is like a seed. Give a seed good soil, nutrients, water, sunshine, and fresh air, and nature will do the rest. So it is with a child. If his or her needs are met, the child will bloom without training, supervision, punishments, or rewards.

Jean-Jacques Rousseau (1712–1778), a French philosopher, held this view. He believed infants were born with an inherent drive toward goodness that is vulnerable to corruption by adults. His view was very different from those who saw the child as evil or as a blank slate. Rather than constantly guiding or correcting children, Rousseau advocated for allowing them to develop naturally, with a minimum of adult supervision. Rousseau believed that nature took care of development. The very word "development" means to unfold, which is how Rousseau saw the childhood process: a child unfolds the way a rosebud does. You cannot force the petals open; they will naturally open on their own.

Some preschools in the first half of the twentieth century and beyond were guided by this view of the child as being inherently good; they initiated a movement of open, or free, schools, which peaked in the 1960s. Children were gathered into groups in natural settings and encouraged to explore and experiment with mud, water, sand, clay, and each other as media for education. Children were free to be outdoors, free to follow their own inclinations and interests without worries about objectives, accountability, productivity, or future academic demands. Society today has moved in the opposite direction to such an extreme that a movement has restarted to encourage children to play outside in nature.

So three historical trends underpin early childhood theory today: the church's idea of original sin, Locke's emphasis on the environment, and Rousseau's belief in the natural process of development. These basic views, as well as scientific research, still influence child-development theory today. To see an example of how these views influenced a group of early childhood educators, read the *Focus on Diversity* box on page 22. However, this does not mean that everyone agrees. Disagreement occurs from time to time among early childhood professionals who look to child-development theory to guide them in program design. In fact, current debate today centers around the inclusion of culturally diverse views in the study of child development–a field that historically has been dominated by research from white male figures.

Two other theorists have also contributed to the notions about the nature of the child and human nature in more recent years. Abraham Maslow (1908–1970), a leader in the humanistic psychology movement, was born in Brooklyn, New York, to Russian Jewish immigrant parents. He was interested in what motivates people, not studying either the mentally ill or altering behavior through reward systems like his other peers. He was interested in human potential and uncovering what makes people successful. Personal growth and fulfillment were his goal. With that in mind, he created a hierarchy of needs, and at the top was what he called self-actualization. Maslow's theory of self-actualization asserts that regardless of age, gender, or cultural background, every human being is motivated by a set of basic needs. The most critical needs form the base from which the other needs can be met. The top level can only be reached once the more basic needs are met. When these basic needs are met, the person is free to seek self-fulfillment. "Peak experiences," which have a spiritual flavor, intrigued him. Those kinds of experiences bring a person a sense of purpose, a feeling of integration and personal fulfillment. He wrote *Toward a Psychology of Being,* an influential book of its time that is still relevant today. This book incorporates Maslow's positive outlook on human nature and human development.

Uri Bronfenbrenner, born in 1917 in Moscow, came to the United States at the age of six. A professor at Cornell, he applied developmental research to both policy and practice. Whereas Maslow focused on individuals, Bronfenbrenner taught us to see individuals within a context, warning early childhood practitioners to expand their focus from the child alone to the family and community, and to regard the context as an important factor in child growth and development. Children come to our programs nested in ever larger contexts, which influences them and upon which they also have some influence. Bronfenbrenner's work has contributed much to the themes of this book, especially those that relate to diversity and families. As a result of Bronfenbrenner's work, is some programs are beginning to move from calling themselves child-centered to calling themselves family-centered programs. Bronfenbrenner has also played an active role in the design of developmental programs in the United States and elsewhere; he was one of the founders of Head Start. His book *The Ecology of Human Development: Experiments by Nature and Design* contains information about looking at children within the context of their families, communities, and the greater society.

Child-Development Theorists and Their Theories

A succession of scientists has contributed to the study of the young child and to the theories that influence professionals today. The way we currently see, understand, and deal with children is greatly influenced by historical theoretical frameworks.

Charles Darwin's (1809–1882) journals of his own children marked the beginning of a scientific approach to child study. Later, G. Stanley Hall (1844–1924) took another scientific step forward by focusing on groups of children rather than on the individual child. Hall also kept anecdotal records on the developmental stages of young children.

Hall's student Arnold Gesell, who was mentioned earlier, continued with a systematic scientific study of developmental stages by filming a number of infants month after month to record the average age at which they rolled over, sat up, started to walk, and so on. From his careful studies, he came up with norms to define the physical milestones of development.

FOCUS ON DIVERSITY

Differing Views on the "Nature of the Child"

At a staff retreat for small, church-affiliated preschool programs, a facilitator was brought in to do a workshop on discipline. She started by talking about how each person has a personal view of "the nature of the child." She suggested that these views lie along a continuum. At one end is the seed image, which she related to Rousseau's view of the child as a "little angel." At the other end is the "little devil," which she related to the theory that a child is born in sin and only by the use of a "firm hand" will a child move toward a path of good behavior. Then she asked participants to arrange themselves along an imaginary line, placing themselves in the position that corresponded to their personal view of the nature of the child. Most of the staff arranged themselves at various points on the angel half of the continuum. Two people were the exception, however; one stood on the far end of the devil side, and one refused to place herself on the line at all. When asked to explain why they chose to stand where they did, most participants spoke of their varying degrees of faith in the innocence of children. The person on the "little devil" end of the continuum turned out to be a minister of the church, and he spoke eloquently about the power of temptation. The facilitator noted that many of the staff were members of his congregation and wondered if their early childhood training had overridden their church's emphasis on people as sinners. The person who stood outside the line had a degree in psychology and spoke equally eloquently about how there was no "basic instinct" or "human nature." She believed that children were a product of their environment. After presenting the different views so visually, the facilitator was able to talk about how the ways in which we discipline relate to what we believe about children, their nature, and their needs.

Emmi Pikler. Dr. Emmi Pikler (1902–1984) was a pediatrician who developed her own theory of development and developmental stages in the 1930s in Budapest, Hungary. Because of World War II and the Iron Curtain that separated Eastern Europe from the West, most Americans had not heard of Pikler until recently. However, Pikler, a devoted researcher deserves the same recognition as Western-stage theorists. She is now getting attention for the practices she established that go along with her stage theory. See more about Pikler in the next section on Pioneer Educators.

Jean Piaget: Cognitive Theory. Jean Piaget (1896–1980) was less interested in the physical milestones, like sitting up and walking, than in the cognitive milestones that mark developmental stages of intelligence. Piaget is best known for his concept of **cognitive stages.** He is considered one of the giants of child development.

Piaget spent his life studying how children think. He observed children and conducted clinical interviews for years, to determine how their minds develop rationality. According to Piaget's theory, children construct knowledge and develop their reasoning abilities through interactions with people and the environment as they seek to understand the world and how it works. Initially, children explore only on very concrete levels, but eventually children begin to understand and explain things without physically trying them out each time. Piaget's final stage of cognitive development occurs when adolescents can use logic to talk about ideas.

Piaget was a **stage theorist.** He believed in the literal meaning of the term *development,* or that stages *unfold* through maturation. He described development as occurring in distinct steps that always fall in the same order. (Piaget's stages that are most pertinent to the early childhood educator–those covering children from birth through age eight–are laid out in Table 1.2.)

stage theorist A theorist who believes that children develop according to specific, sequential stages of development.

Piaget believed in putting children together in a rich environment and letting them interact in an exploratory way. Piaget believed that thinking and learning is a process of interaction between a child and their environment. Piaget also believed that all species inherit a basic tendency to organize their lives and adapt to the world around them. Piaget believed that children actively construct knowledge on an ongoing basis. Thus, he believed that children learn best when they are actually doing the work themselves, rather than being shown, or explained to. Like Pikler, he did not stress right answers, nor did he believe in molding and shaping through a system of rewards. You'll see Piaget's and Pikler's theories in practice throughout the book, but especially in Chapter 4, which focuses on the benefits of play and connections to learning and development.

Thanks to Piaget, Pikler, and others, young children are generally viewed as active learners. Those who follow Piaget's theory are adamant that children be allowed to have exploratory, firsthand experiences. Hands-on learning is more important than sitting and listening to a teacher.

According to Piaget, imaginative and pretend playing are also important to cognitive development; children create mental images through this kind of play, thus taking an early step in symbolic development. As a result, most early childhood programs have a dramatic play area, where children are encouraged to dress up and play pretend.

Sigmund Freud and Erik Erikson: Psychoanalytic and Psychosocial Theories. Rather than focus on the mind and its development, psychoanalytic theory focuses on

TABLE 1.2 JEAN PIAGET: STAGES OF COGNITIVE DEVELOPMENT[3]

Age	Stage	Description
0–2	Sensorimotor	Children use their bodies and senses to understand the world. By the end of this stage, the infant has begun to use mental as well as physical activity to learn.
2–6	Preoperational	Children engage in pretend play and talk, which shows they are able to use symbolic thinking. Children's thinking still has limitations: it is egocentric and not always logical but based more on intuition and perception. They are still working out the difference between reality and fantasy.
7–11	Concrete operational	Children think in concrete terms. They can understand their world more objectively and rationally. They are able to classify and conserve.

TABLE 1.3 SIGMUND FREUD: PSYCHOSEXUAL STAGES[4]

Age	Stage	Description
0-1	Oral	The child's focus is on the pleasures and sensations of the mouth and the area surrounding it. Feeding is a major source of pleasure and satisfaction.
1-3	Anal	The anus is the major focus of pleasures and sensations. Toilet training is a primary task of this stage.
3-6	Phallic	The genitals and their stimulation are the focus of pleasure. The Oedipus and Electra complexes are part of this stage.
7-11	Latency	Sexual needs are on hold, and pleasure is derived from a variety of activities.

feelings. The leaders in this field were Sigmund Freud (1856-1939) and Erik Erikson (1902-1994), both of whom were concerned with the subconscious, what is hidden deep in the psyche.

Sigmund Freud, the "Father of Psychology," studied troubled adults and came up with theories about how children develop in **psychosexual stages** (see Table 1.3, which shows the stages that cover the early childhood years). He believed that early experiences determine personality development and create specific outcomes. He also described how early experiences and feelings come out symbolically in children's play. In most early childhood programs, you can find children working out emotional issues through pretend play in psychoanalytic fashion. Little girls in the housekeeping corner who are playing doctor by giving each other shots with kitchen utensils are exploring their own experiences–trying on roles and playing out fears and anxieties.

Piaget would have looked at the same scene and said it demonstrates that the children are developing symbolic representation; they use one object to stand for another, showing that they have an image of the real object in their minds. Cognitive theory and psychoanalytic theory are two different ways of looking at the same behavior. But like Piaget, Freud was a stage theorist; that is, he believed that the psyche unfolds in successive stages.

Watch a newborn's mouth. It's constantly busy. The newborn period–the first year of life–is what Freud termed the "oral stage." Oral satisfaction is what newborns seek. It's not just the food in the stomach but the feelings in the mouth. Following the oral stage is what Freud termed the "anal stage"–a period in which toilet training is the main event and must be handled sensitively.

Erik Erikson was Freud's student, but he became an important early childhood theorist in his own right. Erikson rethought Freud's stages, calling his own viewpoint "psychosocial theory." Erikson saw the first year of life as the time infants develop a sense of basic trust. During this first **psychosocial stage,** they come to see the world as a safe and secure place when their needs are met, when someone consistently heeds their cries and feeds them; however, they come to see the world as cold and cruel when the opposite happens and they do not receive responsive caregiving and nobody seems to care. It is not just what is done to them but also how it's done. Infants who seldom or inconsistently get loving treatment learn mistrust. Infants who

TABLE 1.4 ERIK ERIKSON: PSYCHOSOCIAL STAGES[5]

Age	Stage	Description	Implications for Practice
0–1	Trust versus mistrust	Children come to trust the world if their needs are met and they are cared for in sensitive ways. Otherwise, they see the world as a cold and hostile place and learn to mistrust it.	• Caregivers nurture infants during routines (i.e., eating, sleeping, changes, etc.). • Caregivers should be responsive to distress. • Primary caregiving is a good practice to support healthy secure attachments at this stage.
1–3	Autonomy versus shame and doubt	Children work at becoming independent in areas such as feeding and toileting. They can talk and assert themselves. If they do not learn some degree of self-sufficiency, they come to doubt their own abilities and feel shame.	• Caregivers should offer children simple choices. • Caregivers should eliminate false choices. • Caregivers should set clear limits. • Caregivers should accept alternating needs for independence and dependence.
3–6	Initiative versus guilt	Children thrust themselves into the world, trying new activities, exploring new directions. If their boundaries are too tight and they continually overstep them, they experience a sense of guilt about these inner urges that keep leading them into trouble.	• Caregivers should encourage independence. • Caregivers should focus on gains, not mistakes. • Caregivers should consider individual differences. • Curriculum should focus on "real-life" things.
7–10	Industry versus inferiority	Children learn competency and strive to be productive in a variety of areas. If they fail to learn new skills and feel unproductive, they are left with a sense of inferiority.	• Caregivers should set children up for success not failure. • Caregivers should offer support and encouragement. • Children should be taught to explore, embrace new experiences, and enjoy challenges. • Caregivers should try to instill a growth mindset in children at this stage, helping them believe that abilities can be developed and strengthened by persistence and effort.

develop a sense of trust develop an external believe system that adults will meet their needs and an internal belief system that affirms their own power to effect change and cope with a variety of circumstances. The opposite occurs in infants who do not develop a sense of mistrust. In Erikson's view, the newborn period is followed by a stage of autonomy, when children learn to say no and protest. Children who successfully emerge from this stage acquire a strong sense of self and a sense of independence without suffering the extremes of shame and self-doubt. (Table 1.4 lists Erikson's stages of development that cover the early childhood years.)

John B. Watson and B. F. Skinner: Behaviorism and Learning Theory. John B. Watson (1878–1958), an American psychologist and the "Father of Modern **Behaviorism**," had another view of children. He believed that all behavior was learned and that training was the way to change it. In his theoretical view, waiting out a stage didn't make any sense.

behaviorism The scientific study of behaviors that can be seen and measured. Behaviorism, also called "learning theory" attributes all developmental change to environmental influences.

A star chart is a classic example of a behaviorist tool for reinforcing desired behaviors–in this case the performance of chores at home.
Monkey Business Images/ Shutterstock

Both Watson and B. F. Skinner (1904–1990), another famous behaviorist, believed that only that which can be seen and measured–outward behavior–was worth studying. Unobservable concepts like "mind" and "emotions" were unimportant. They rejected the idea of development unfolding in ages and stages, and they did not believe in innate behavior or instincts. According to Watson and Skinner, all behavior is learned through the consequences of the individual's actions: a child will repeat behavior that is reinforced and cease behavior that is not reinforced. Adults who use rewards (reinforcement)–simple ones like acknowledgment or praise or complex ones like token economies–are influenced by **learning theory** based on behavioristic principles. Nevertheless, parents and teachers sometimes misunderstand the principles of behavioristic learning theory: simply by paying attention to negative behavior, even in the name of correcting it, adults in fact reinforce that behavior. According to behaviorists, it is better to pay attention to the behavior you *want* and ignore behavior you don't want.

Albert Bandura: Social Learning Theory. Albert Bandura (1925–) is one of the researchers associated with a branch of behavioristic learning theory called **social learning theory,** which focuses on the significance of modeling and imitation in a child's development. In this view, children do not only learn by being reinforced; they also learn by observing others. Children tend to behave like the people in their lives, identifying with some and failing to identify with others.

Early childhood educators need to be constantly aware of their role as models for children. Taking this theory into practice, Chapters 6 and 7 directly focus on the implications of social learning theory for early childhood educators by discussing in detail how they are role models for children. It is quite a responsibility for adults, teachers, and parents alike to set good examples and also counteract the models

Hands Off or Hands On?

One teacher and her co-teacher, both well versed in Piaget, set up an interesting, challenging, developmentally appropriate classroom and then sit back to observe how the children use it. The two adults expect the children to explore, experiment, and discover for themselves what they can learn from all the rich materials, toys, books, and equipment available to them. Although they keep track of the children's learning and development, they don't direct it. They facilitate rather than teach. They intervene periodically when children need help and guidance.

Another teacher and his co-teacher in a different program introduce each material and piece of equipment to the children and teach them exactly how to use it. They work with the children individually and in small groups, observing what the children have learned and what they are ready to learn next. Based on this information they present activities and materials designed to always keep the children moving forward in their learning and development. They both teach and facilitate. They also encourage children who know how to do something to help children who don't have the same skills.

children see in the media. Adults can tell little boys not to be aggressive and little girls not to act helpless, but children are more apt to copy the models around them than to pay attention to words. According to social learning theorists, "Do as I say, not as I do" does not work, no matter how often you say it.

Lev Vygotsky: Social Context and the Construction of Meaning. Because social context is a significant concern of contemporary early childhood professionals, the **sociocultural theory** of Russian researcher Lev Vygotsky (1896–1934) has experienced recent popularity. Vygotsky and Erik Erikson were both interested in the effect of culture on development. Like Erikson, Vygotsky believed in the influence of culture on childhood development, but Vygotsky was not a Freudian. His interest, like Piaget's, was cognitive development. Vygotsky believed, as did Piaget, that children *construct* knowledge–they don't just take it in. They both agreed that learning is active and constructed. Vygotsky believed in the power of language and in social interaction as a vital ingredient in learning and development.

Unlike Piaget, however, Vygotsky didn't believe in letting children explore and experiment without adult help. Vygotsky was an advocate of what's called **scaffolding;** that is, providing learners with support and assistance. According to Vygotsky, assisted performance is fine–even desired. Vygotsky believed that interaction and instruction were critical factors of a child's cognitive development and that a child's level of thinking could be advanced by just such interaction. According to Piaget, children should be left alone to explore the environment and discover what they can do in it without adult help. Piaget insisted that while children needed to interact with people and objects to learn, the stages of thinking were still bound by maturation. See the *Points of View* box above for a look at how two different

scaffolding A form of assistance that supports and furthers understanding and performance in a learner.

teachers put Piaget's and Vygotsky's theories into practice. Although the examples may seem extreme, they are based on real teachers, both of whom regard relationships with the children as a primary goal. They just have different ways of approaching the teaching/learning process. Despite these two contrasting approaches, it is important to acknowledge that many teachers have integrated Piaget's and Vygotsky's theories.

Here we have an example of how two theories–two different ideas about development and what children need–conflict with each other. Piaget was critical of educators who stressed right answers and hurried children toward a goal. In response to such educators, he might have said, "You're placing too much stress on getting things right and pushing children when they aren't ready." In contrast, Vygotsky might have argued, "Why wait when you can help a little?"

There are all kinds of theories. Whose theories are right? Whose theories are wrong? There is no clear-cut right or wrong with theories of early childhood education. Even though one theory may conflict with another, each has something to contribute to our understanding of children. Educated professionals have their favorite theories and lean toward some more than others, but no early childhood educator can afford to completely dismiss specific theories. Being an eclectic educator, one who is selective, is not a weakness but a strength.

PIONEER EDUCATORS

The theorists just discussed contributed much to the field of early childhood education, but they were not primarily educators. This section will look at some well-known early childhood educators and the institutions they created.

One such educator was J. H. Pestalozzi (1746-1827), who started a school in Switzerland based on the principle that education should follow the child's nature. He believed that children learn through activity and sensory experiences, and he stressed an integrated curriculum.

Pestalozzi influenced Friedrich Froebel (1782-1852), a German educator who became known as the "Father of Kindergarten" by creating that institution. Froebel brought play into education. He thought of young children as seeds and saw the educator's role as gardener. Hence, the name *kindergarten,* which in German means "garden of children."

Maria Montessori (1870-1952), who was the first woman physician in Italy, is best known as an educator. She created her own brand of education, which still survives under her name to this day. She emphasized the active involvement of children in the learning process and promoted the concept of a prepared environment. Child-sized furniture and specific kinds of self-correcting learning materials were two contributions of the **Montessori** program.

John Dewey (1858-1952), an American, created the progressive education movement. Like his forerunners, he also advocated experiential learning. He believed that curriculum should be built on the interests of the children and that subject matter should be integrated into those interests. From Dewey comes the term **child-centered curriculum.**

FOCUS ON DIVERSITY

Culturally Diverse Perspectives on Childhood

To honor diversity, this book should ideally focus equally on childhood perspectives, histories, and trends without European roots. By observing different cultures, we see that there are many ways to care for and educate children; there is no one right way. It is important for all early childhood educators to understand the value of learning about diverse perspectives.

Unfortunately, information about non-Western childhood histories and methods is not always easily accessible. Indeed, there are numerous historical descriptions of childhood, education, and family from various cultures, but it is beyond the scope of this book to cover every one. The *Focus on Diversity* box on page 33 contains a reading list to allow you to further explore on your own, and here is a brief overview of some non-Western views of childhood and early education.

Historically, attitudes toward childhood in China and Japan were influenced by Confucius's writings (551–479 B.C.), which stressed harmony. Children were seen as good and worthy of respect, a view not held in Europe until more recently.

Native American writings show close ties and interconnectedness, not only among families and within tribes but also between people and nature. Teaching children about relationships and interconnections are historical themes of early education among many indigenous peoples. Closeness is a theme in Latino families, some of whom also have indigenous roots in North, Central, and South American soils. This theme is expressed by the title of a framework that Costanza Eggers-Pierola and others developed called *Connections and Commitments: A Latino-Based Framework for Early Childhood Educators*.[6]

Strong kinship networks are a theme among both Africans and African Americans; people bond together and pool resources for the common good. Whether these contemporary tendencies come from ancient roots, historic and modern oppression, or all three remains unclear.

Not all variations in attitudes and child-rearing practices reflect cultural differences; they may also arise from family, societal, or historical circumstances. Poverty, oppression, and other kinds of adversity affect child-rearing practices and attitudes toward education. Among populations that experience high infant-mortality rates, child rearing has certain characteristics in common: Infants are on or near a caretaker's body day and night so they can be watched for signs of illness. Crying is quickly attended to because it may signal the onset of illness. In such societies, early education is of less concern than survival.[7]

Dewey's ideas influenced educators in the United States and other countries. In fact, his approach influenced Loris Malaguzzi (1920–1994), who founded the internationally acclaimed Reggio Emilia early-education system in Italy. Many components of this program are notable: cooperation and collaboration at all levels, an emergent curriculum combined with a project approach that picks up on and enhances children's interests, and the documentation of the learning process and the work the children do.

Another pioneer educator was the German Austrian philosopher and writer Rudolf Steiner (1861–1925), who was the leader of what was called an anthroposophical movement (literally "the wisdom or knowledge of man"), which is defined as the science of the spirit that bridges the physical and spiritual world. In 1919, Steiner started the first school based on anthroposophy at the request of the owner of the Waldorf-Astoria cigarette factory in Stuttgart, Germany, which became the model for a worldwide movement of Waldorf education. Waldorf education is based on Steiner's theories of child development that are being used today in Waldorf schools around the world. Some distinctive features of Waldorf education is the emphasis on spirituality and eurythmy, which is the art of movement. Waldorf education is a unique form of education starting with preschool and going through high school and is a fast-growing independent education movement.

Emmi Pikler (1902–1984) (both a theorist and a pioneer educator who focused on infants and toddlers) was a pediatrician and researcher in Budapest, Hungary, who came up with her own theories of what children need for optimum development. Working first with families in Budapest in the 1930s she taught about autonomous motor development and the importance of self-initiated activity. She also stressed the need for building relationships during caregiving activities. In 1946, after World War II, the Hungarian government asked Pikler to create a residential nursery for children whose families couldn't care for them. Challenged by the fact that institutionalizing babies historically resulted in unhealthy outcomes, Pikler created a unique approach to infant-toddler group care and education that is still influencing programs and parents in Europe, Central and South America, and the United States. One of the keys to her success was the special kind of relationship she trained caregivers to create with the infants. The Pikler Institute, although no longer serving as a residential nursery, continues to have a child care program and conduct research and training. It is known as a world-class model program. The Pikler approach was first introduced to the United States by Magda Gerber, who came from Hungary in 1956. Gerber modified Pikler's approach to reflect the new context and Gerber's own ideas and experiences gained while working in the United States.

We have reviewed the European roots of early childhood education, but history and trends from other cultural roots are also important. See the *Focus on Diversity* box on page 29 for a discussion of culturally diverse perspectives on childhood and the *Focus on Diversity* box on page 33, which lists readings that explore the early childhood practices of various cultures.

Brain Research

One thing all theorists and pioneer educators have in common is the idea that the early years matter. What happens in the first years of life can impact later outcomes. As Sigmund Freud often quoted from a poem by William Wordsworth, "The child is the father of the man." (He also meant to include women in that statement.) Until fairly recently studying the human brain yielded little information about its development, but new research using sophisticated technology lets us actually peek into living brains to see what's going on. Finally, the current research that ties actual learning and behavior to neurological development has validated the statement "the early years matter." All of a sudden the political and financial key players who paid little attention to early childhood education are beginning to take notice. A massive research project captured in a book called *From Neurons to Neighborhoods: The*

Science of Early Childhood Development[8] relates the current science to its potential impact on children, families, and communities.

Most people born in United States who read this book went to kindergarten–that is because the kindergarten movement in the last century grew out of the idea that first grade is too late to begin a child's learning and development. By the 1960s, the Head Start movement as well as programs for children with special needs grew out of the realization that for some children, kindergarten was too late to begin addressing learning and development. Modern brain research makes it very clear that optimum development of the brain and therefore the whole child (including the cognitive part) depends on the care that is given as early as gestation in the womb and the first three years after birth. As a result we now have Early Head Start for infants, toddlers, and their families as well as a number of programs and resources for babies with special needs. This is also one of the reasons that infants and toddlers get so much attention in this book. Though many of you who plan to work with prekindergarten or older children may find it pointless to read about babies, you should understand that the early care and education in the first year significantly impact the development of children as far as age seven. Even if you do not work with the youngest children, you may one day become an advocate for quality programs that promote healthy development and learning for infants, toddlers, and thier families.

So what does the brain development research tell us? One of the tasks of the early years is for the brain cells to make connections with each other and create pathways by growing branches called dendrites, which form every time infants have experiences in their environment. For example, neural connections form and strengthen as they are spoken to or sung to by their primary caregiver. Infants are born with many more brain cells than needed because the first years involve "pruning." Brain cells that are not used or connected disappear. "Use it or lose it" is a common expression stated by those who teach about the implications of the brain research. This does not mean to sit infants down with flash cards in front of them! The message here is that infants and toddlers need the kinds of experiences that foster the optimum number and right kinds of connections and pathways that result in healthy, wholesome development.

One of the most notable findings is how vitally connected social and emotional development is to learning and cognition. The approach must be infinitely broader and more developmentally appropriate than merely stimulating the infant or focusing on academics. In other words, cognitive development builds on a foundation of social-emotional stability and security, which is why this text focuses on relationships, connections, feelings, and the social-emotional environment. See the *Theory Behind the Practice* box on page 32 for more on how to support optimum brain development.

What It Means to Be an Early Childhood Educator

This introductory chapter began with an explanation of why the early childhood education student must learn the language of the ECE profession and understand how this special profession is organized and regulated. This final section takes a different slant on what it means to be part of the profession.

THE THEORY BEHIND THE PRACTICE

Brain Development

Optimal brain development is not difficult to achieve. For example, **attachment** matters. Infants and toddlers who have warm relationships that result in feeling closely connected to someone or to several people are more likely to have optimal brain development than those who are shuffled around and feel constantly insecure. Even when there is attachment and especially if there is not, shielding babies from stress is vital. When violence is a part of family life, babies to whom violence is done can be stunted in their brain development. Even when babies witness violence done to others, the experience can have a negative effect. Infants who are nurtured in safe and secure environments tend to become children who are curious, interested, motivated, and competent learners.

attachment A warm relationship that results in feeling closely connected to one or several people helps infants or toddlers to have optimal brain development.

Early childhood educators know the meaning of behaving in a professional manner. They understand the importance of confidentiality. They never talk about one family to another or spread gossip. They have an attitude that shows they take the work seriously. They are dedicated to working with children and families, using the skills and knowledge they have gained through preparation and training.

Early childhood educators are lifelong learners. They continually pursue professional development and create professional goals for themselves, using on-the-job evaluations and feedback, as well as self-assessment, to determine future directions for learning.

They understand and follow the requirements set by regulating agencies. They adhere to the adult-child ratios, group size, and space requirements determined to be minimum standards and realize that optimum standards are what they should strive for.

Legal Responsibilities

Early childhood educators are aware of their legal responsibilities. For example, they know that including children with disabilities, particular challenges, and other special needs is a legal mandate. The original idea for special education was to separate children with special needs from their typically developing peers. Now the Individuals with Disabilities Education Act (IDEA) of 1991 makes it clear that children with disabilities must be in natural environments–such as child care and other early care and education programs. It is illegal to exclude any children just because you do not want them or do not feel you are knowledgeable enough to meet their special needs.

Early childhood educators are also aware of the seriousness of child abuse and know they have a legal mandate to report any suspected cases. Here is a situation

Books and Articles That Explore Culturally Diverse Roots of Early Childhood Practices

Basso, K. (2007). To give up on words: Silence in Western Apache culture. In L. Monaghan and J. E. Goodman (Eds.), *A Cultural Approach to Interpersonal Communication* (pp. 77–87). Malden, MA: Blackwell.

Brody, H. (2001). *The Other Side of Eden: Hunters, Farmers, and the Shaping of the World.* New York: North Point Press, 2001.

Eggers-Pierola, C. (2005). *Connections and Commitments: A Latino-based Framework for Early Childhood Educators*. Portsmouth, NH: Heinemann.

Feeney, S., Galper, A., and Seefeldt, C. (Eds.). (2009). *Continuing Issues in Early Childhood Education.* Upper Saddle River, NJ: Pearson/Merrill.

Hooks, B. (2003). *Rock My Soul: Black People and Self-Esteem.* New York: Atria.

Hudson, R. A. (2007). Speech communities. In L. Monaghan and J. E. Goodman (Eds.), *A Cultural Approach to Interpersonal Communication* (pp. 212-217). Malden, MA: Blackwell.

Lopez, E. J., Salas, L., and Flores, J. P. (2005, November). Hispanic preschool children: What about asessment and intervention? *Young Children, 60*(6), 48–54.

Maschinot, B. (2008). *The Changing Face of the United States: The Influence of Culture on Child Development.* Washington, DC: Zero to Three.

Ramirez, A. Y. (2008). Immigrant families and schools: The need for a better relationship. In A. Pelo (Ed.), *Rethinking Early Childhood Education* (pp. 171–174). Milwaukee, WI: Rethinking Schools.

Tan, A. L. (2004). *Chinese American Children and Families.* Otney, MD: Association for Childhood Education International.

Tannen, D. (2007). Conversational signals and devices. In L. Monaghan and J. E. Goodman (Eds.), *A Cultural Approach to Interpersonal Communication* (pp. 150–160). Malden, MA: Blackwell.

Villegas, M., Neugebauer, S. R., and Venegas, K. R. (2008). *Indigenous Knowledge and Education: Sites of Struggle, Strength, and Survivance.* Cambridge, MA: Harvard Education Press.

an early childhood professional might face: A four-year-old child arrives at school with an ugly bruise on his cheek. His baby sister has several small burn marks on one arm. "What happened to you?" the teacher asks the four-year-old. She then looks up at the mother, a timid woman with a new boyfriend, who gives an unlikely explanation and leaves quickly. The child later tells the teacher that "Uncle Bob" burned his sister with a cigarette and that when he tried to stop him the man shoved him hard against a wall. The teacher is very upset, but she feels sorry for the mother. She knows that this boyfriend is important to her for emotional and financial reasons. She is afraid of what might happen if she reports the suspected abuse: the mother could lose her children, or they might all have to go back to living in her car as they used to. What can the teacher do? *She has no choice.* She cannot ignore the incident and hope things will get better. She cannot just talk to the mother and hope that she will get some help. *She must report the suspected abuse to the authorities.*

Teachers, aides, assistants, family care providers, caregivers, or any other adults who work with children and families are "mandated reporters"; that is, they are required by law to report suspected abuse. There are penalties for not doing so. The purpose of this law is to stop the violence committed against children–violence that every day results in injury, permanent disability, and even death.

CODE OF ETHICS

NAEYC
Code of Ethics

Early childhood educators have legal responsibilities to guide them in some of their decision making, but they aren't entirely on their own in handling non-legal matters. Because early childhood education is a profession, a code of ethics guides its members in decision making. The NAEYC publishes a document that outlines a set of shared values and commitments based on the collective wisdom of the profession.[9] The code of ethics is based on six core principles outlined in its preamble: "We have committed ourselves to

> appreciating childhood as a unique and valuable stage of the human life cycle,
>
> basing our work with children on knowledge of child development,
>
> appreciating and supporting the close ties between the child and family,
>
> recognizing that children are best understood in the context of family, culture, and society,
>
> respecting the dignity, worth, and uniqueness of each individual (child, family member, and colleague), and
>
> helping children and adults achieve their full potential in the context of relationships that are based on trust, respect, and positive regard."

The NAEYC code of ethics is designed to offer professional guidelines for working out ethical dilemmas. It lays out four areas of responsibility: to children, families, colleagues, and community and society.[10]

By now you can see that if you go into early childhood education, you will become part of a large community of educators dedicated to the care and education of the whole child. By pursuing a career in early childhood education, you will join a special kind profession that has a past, a present, and a future. Welcome!

A STORY TO END WITH

It was on a January day long ago that I decided to go back to school. I had four children at the time, and the idea of college was far from my mind–until a flyer ended up in my hand. My life changed that day.

I was a busy woman. My oldest child was in kindergarten, my two middle children were three and four, and I had a new baby. I decided to enroll my tots three mornings a week in a parent co-op preschool. I was obligated to help out at the preschool two mornings a week, but on Fridays I was free to do what I wanted while they were in school. On that first Friday, I was glad for the day off. On the second Friday, I realized I didn't want to leave the preschool world. And by the third Friday, I had discovered that a new Head Start preschool that was opening in the same location was looking for volunteers. I signed up, and my career as a preschool volunteer was launched.

I became totally engrossed in early childhood education. While I was busy learning about the similarities and differences in the two programs I was involved in, I received a flyer announcing a community college class in early childhood education. At first, the idea of taking a class seemed ridiculous to me. I was a grown woman with four young children. What place did college have in my life? I hadn't been to school in years and couldn't even remember how to study. Besides, wouldn't I look out of place in a class with 18-year-olds?

Nevertheless, I picked up the phone and registered. I arrived at the first class practically shaking in my shoes, but it didn't take long to feel right at home. There were other students my age, and the younger ones were plenty friendly.

Well that was the beginning of my career in early childhood education. My course work allowed me to move up the ladder from volunteer to assistant teacher. The more classes I took, the higher I went until, finally, I ended up as a community college teacher in the school where I started as a student.

SUMMARY

Early childhood education is a special branch of education that deals with children from birth to eight years of age. There are four key themes in the training of early childhood educators: reflective thinking, multiculturalism, holism, and professionalism. Early childhood educators look to the science of child development for their knowledge base about what children need and how they learn and develop. As well, numerous professional organizations guide and support early childhood educators in the various types of programs they work in.

Child development is the study of how children change as they get older. Over the years, researchers in this field have devised theories that explore the physical, cognitive, emotional, social, and behavioral development of children—theories that maintain their relevance in the field of early childhood education today. Taking the cue from these theorists, many educators have made contributions to early childhood education through the development of creative and innovative programs. Early childhood education is a profession—one that entails legal as well as ethical responsibilities—that is dedicated to children, families, colleagues, and society.

REFLECTION QUESTIONS

1. Reflect on the question, "Who are you?" Write down a list of 10 words that you can use to define yourself. Look over that list and see if you can see any patterns. Does this list explain your identity? What would it take to explain your identity?
2. Did you attend any sort of early childhood care and education program before the age of five? If yes, what memories do you have of your experiences there (if any)? If no, what are one or two outstanding memories (if any) you have of those first five years?
3. Explain how nature and nurture interacted to create you as the person you are.
4. What draws you to a class in early childhood education?

TERMS TO KNOW

How many of the following words and acronyms can you use in a sentence? Do you know what they mean?

FOR FURTHER READING

Bradley, J., and Kibera, P. (2006). Closing the gap: Culture and the promotion of inclusion in child care. *Young Children, 61* (2), 34–41.

Copple, C., and Brededamp, S. with Gonzalez-Mena, J. (2011). *Basics of Developmentally Appropriate Practice: An Introduction for Teachers of Infants and Toddlers.* Washington, DC: National Association for the Education of Young Children.

Copple, C., and Bredekamp, S. (2006). *Basics of Developmentally Appropriate Practice: An Introduction for Teachers of Children 3-6.* Washington, DC: National Association for the Education of Young Children.

Copple, C., and Bredekamp, S. (2009). *Developmentally Appropriate Practice in Early Childhood Programs Serving Children from Birth through Age 8.* Washington, DC: National Association for the Education of Young Children.

Feeney, S. (2012). *Professionalism in Early Childhood Education: Doing our Best for Young Children.* Upper Saddle River, NJ: Pearson.

Feeney, S., and Freeman, N. K. (2018). *Ethics and the Early Childhood Educator: Using the NAEYC Code,* 3rd ed. Washington, DC: National Association for the Education of Young Children.

Kagan, S. L., and Kauerz, K. (2009). Governing American early care and education: Shifting from government to governance and from form to function. In S. Feeney, A. Galper, and C. Seefeldt (Eds.), *Continuing Issues in Early Childhood Education* (pp. 12–32). Upper Saddle River, NJ: Pearson/Merrill.

Lutton, A., Ed. (2012). *Advancing the Early Childhood Profession. NAEYC Standards and Guidelines for Professional Development.* Washington, DC: National Association for the Education of Young Children.

National Association for the Education of Young Children. (2018). *NAEYC Early Learning Program Accreditation Standards and Assessment Items.* Washington, DC: National Association for the Education of Young Children.

Neugebauer, B. (Ed.). (2008). *Professionalism.* Redmond, WA: Exchange Press.

2 First Things First: Health and Safety Through Observation and Supervision

Corbis/VCG/Getty Images

Observation, Supervision, and Guidance

Observation Skills for Beginners

Supervision Skills for Beginners

Focusing on Individuals and the Group

A Crash Course in Guidance

Conflict as a Safety Issue

Risk Taking as a Safety Measure

Helping Children Learn from Their Experiences

A Safe Physical Environment

Developmental Appropriateness

Maintenance as Prevention

Sanitation Procedures

Program Policies and Procedures for Health and Safety

Stress and Frustration as Health and Safety Issues

A Story to End With

In This Chapter You Will Discover

- what you can learn about children by observing them.
- how supervising young children properly keeps them safe and helps them learn at the same time.
- what you can do to stop children from doing something they shouldn't.
- why you shouldn't "blow up" at children.
- how children in conflict need to be kept safe.
- how to make an argument a learning experience.
- what part risk taking plays in teaching safety skills.
- how to create a safe and healthy environment.
- why reducing stress and frustration is a safety measure.

I remember my first day working with young children. I smiled so much my face hurt. I wanted so desperately to be liked by the children. I was the picture of warmth and friendliness–except when a child would do something he or she was not supposed to. Then I was helpless because I did not know how to be friendly and at the same time still handle the unacceptable behavior. Sometimes, I am embarassed to say, I even pretended not to see the incident if it was not too dangerous; other times I tried rather pathetically to stop the problem. But most often, I either found myself floundering or standing by paralyzed until another adult rescued me.

Fortunately, I was not by myself. I had plenty of experienced teachers and assistant teachers right there to help me out. I did not have to worry about keeping the children safe because I was not on my own. I was free to learn how things worked without having more responsibility than I was ready for.

All beginners are not so lucky. Necessity dictates that some beginners have to take more responsibility than they feel comfortable with or are ready for. And one category of beginners, family child care providers, are alone from day one. This chapter offers a quick survival course on how to keep children safe while you're learning to become a more proficient caregiver, provider, and teacher.

There are three key elements in keeping children safe: supervision, guidance, and a safe environment. For you, the beginner on your first day, supervision and perhaps guidance will be your main focus, and the amount of supervision and guidance you provide will depend on the level of responsibility you are assigned. The safe environment, most likely, will already be established.

OBSERVATION, SUPERVISION, AND GUIDANCE

NAEYC Program Standards
Program Standard 2: Curriculum

This chapter is designed to help students develop the skills they need to observe and interact with children. Teacher education and training programs differ in their approach to preparing students to work in the early childhood field. In some, the beginning student's experience is like mine (above). He or she is suddenly faced with a group of children to supervise, and the first priority is to keep them safe. This chapter will be helpful to those students as it explains how to deal with such a situation. It also gives them ideas of how to facilitate the development of each child. In some programs, students have the luxury of observing before they end up in a teacher or aide role. This chapter will also help those students. First, it will give them an overview of observation skills. Second, it will help them learn how to watch teachers and student teachers resolve conflicts and help children learn from their experiences. Third, it will provide them with a framework of the basic supervision skills they will need when they move out of a pure observer role and into an interactive one. For observation guidelines see the *Tips and Techniques* box on page 42.

Theories and Observation

Observation is part of the theories discussed in Chapter 1. Learning theory or behaviorism depends on close observation to analyze the behaviors of the learner. Skinner advocated matching the environment and rewards to the individual in order to create changes of behavior. Making those matches depends on paying attention. The use of behaviorism has made a world of difference to the lives and education of many young children, including those with disabilities. Piaget's research depended a great deal on observation of children's behavior and explanations in order to make visible what was going on in their heads. Teachers today can use Piaget's ideas along with their own observations to understand what children are thinking.

OBSERVATION SKILLS FOR BEGINNERS

What can you learn by observing? Through observation you get to know a child and what is behind his or her behavior. You discover the many unique ways children express their needs, desires, and motivations. Observing gives you information that can help you foster relationships with each individual child and can assist you in helping children relate to each other. **Observation** is an important means to making connections and building relationships. By knowing each individual child, you can better promote learning. You can also discover unique ways to support development in each child that you cannot learn from books alone. Sometimes, observation gives you information about how the environment needs to change to promote learning. Sometimes, it tells you when intervention is needed or when the adult should just step back and let things play out. See *The Theory Behind the Practice* box to learn how theory relates to observation.

NAEYC Program Standards

Program Standard 1: Relationships

observation The act of watching carefully and objectively. The usual goal involves paying attention to details for the purpose of understanding behavior.

This observation process asks you to authentically assess children by carefully observing them in their everyday interactions in the classroom. This observation happens in the natural context of teaching and providing care. Observing children's development in a structured way can make teachers' jobs easier. Classrooms run more smoothly when children's needs are met. This process allows for a guide to develop to help individualize programs to meet each child's unique needs. By systematically observing children's interests, playmates, and developmental accomplishments, you will get to know the children better. By closely watching everything children do in the classroom, you can see development in action clearly. This deeper level of understanding helps meet children's needs and makes the task of working in early childhood programs more satisfying and interesting.

Interestingly enough, adults sometimes learn as much about themselves as they do about children during observation. This happens through a process called self-reflection. Self-reflection can be thought of as observation turned inward. As the observer tunes in on his or her own reactions and examines what lies behind them,

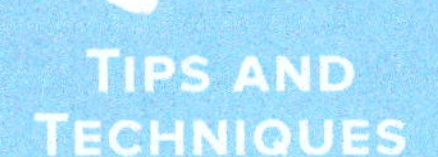

TIPS AND TECHNIQUES

Guidelines for Observing

- Make arrangements to observe ahead of time.
- Find out what the procedures and rules are for observers in the place where you will do your observation. Be cooperative about following them.
- Check in with the person in charge when you arrive.
- If you are in the same environment as the children (instead of an observation room), select a place for observing and become as invisible as you can. Get down low. Become part of the woodwork. If you move around, do so unobtrusively.
- Keep from interrupting the program or getting into conversations with other adults.
- Unless your assignment requires you to interact with children, stick to your observer role. If children talk to you or try to interact with you, respond, but then do what you can to sensitively discourage further engagement.
- Keep yourself in an open, receptive mode. Try to be objective, and at the same time note your own inner responses and reactions to what you are observing.
- If you are writing an observation, try to remain neutral and factual. Note what is actually happening. Write down what you can actually see, describing behavior in as much detail as possible. Do not second guess children's motives or meanings.

self-awareness grows. From self-awareness can come self-acceptance, and from self-acceptance can come remarkable growth and change.

Related to change in the observer is one of the most surprising and interesting parts of observation—change in the child. Sometimes, this "change" in the child is really only the observer coming to see the child from a different perspective and learning to appreciate that child for who he is. Judy Jablon, in a book called *The Power of Observation,* relates a story of her second year as a third-grade teacher. In that story, she confesses that she found one little girl defiant and provoking. It was hard for her to find anything positive about the child until a student teacher arrived in her classroom. The student teacher saw this same child very differently. She described the girl as enthusiastic and commented on what a great sense of humor she had. When Jablon looked at the child through the eyes of the student teacher, she discovered things that she had not seen before.

Through observation, we sometimes grow fond of a child whom at first we might have thought was difficult. Yes, *did not like*. As educators, we know that we are supposed to like all children and not have favorites. The reality, however, is that we each have different temperaments and, thus, are drawn more to some temperament types over others. If we are honest with ourselves (and it is important to *be* honest with ourselves), we have to acknowledge the particular reaction we have to each child, as Judy Jablon did in the example above. Of course, we are pleased when the feelings are positive. If, however, a child strikes a negative chord in us, we should never act on those feelings, but should admit, at least to ourselves, that we have them. Just by recognizing, acknowledging, and accepting our negative feelings we can make the feelings less emotionally charged. Sometimes getting to know the child better helps develop a stronger relationship with the child, which may help develop more of a positive attitude toward the child. As a result, sometimes, the child's behavior also changes.

A student of mine once conducted a child study on a boy in her class who bothered her. She could find nothing redeeming about his behavior when she first started observing him. After a few weeks of observation, however, her reactions changed. He still had behaviors that bothered her, but she had grown to appreciate him as a person. The more positive she felt about him, the less annoying his behaviors became. As she began to appreciate him more, her interactions with him changed, which in turn triggered change in him. Slowly, the annoying behaviors began to disappear and were replaced by other, more positive, behaviors. Much of their interaction was nonverbal, but it nevertheless resulted in a significant change in both of them.

Observation works best when you are open to a situation. When you observe with a set of expectations of what you think you will see, you may miss what is actually happening. Magda Gerber, an internationally known infant-toddler expert, taught her students at Resources for Infant Educarers (**RIE**)[1] that they should come to the child with an open mind. She asks them to put aside what they know and be open to letting the child teach them what they do not know. When observation is used as the learning mode, the child becomes the teacher. There is no better way to learn about a child in a particular situation than through observation.

RIE stands for Resources for Infant Educarers a group dedicated to improving the quality of infant care and education through teaching, mentoring, and supporting parents and professionals. RIE was started by Magda Gerber, A Hungarian infant expert.

Observation is not easy. Even skilled observers sometimes miss the obvious. J. Ronald Lally, who heads WestEd's Center for Child and Family Studies,[2] tells the story of some researchers who failed to see what was right under their noses. These researchers observed newborns and reported that the boys cried much more than the girls. There did not seem to be a reason. What they missed was the fact that many of the newborn boys had just been circumcised. At that time, circumcision was so routine that no one even thought about it. A lesson for us all: as good observers, we have to strive to see all possibilities, even those we tend to overlook.

Supervision Skills for Beginners

When you are supervising a group of children, a slightly different set of observation skills is called for. Although adequate supervision depends to a large extent on the adult's ability to see everything going on at once, supervising children is more complicated than that. If it were so simple, supervision would simply be a matter of stationing "guards" at the corners of the room or play yard, or monitoring the area with closed-circuit cameras.

As an early childhood educator, your job entails much more than just watching children to keep them safe. It involves being an active part of the children's education as well as building relationships with them. You have to learn to interact with

Learning to interact with individual children and small groups while at the same time constantly scanning the room or yard is very important.
Blend Images/John Lund/Marc Romanelli/Getty Images

dual focus A method of supervision that allows the adult to focus on a child or small group of children while still being aware of what else is going on in the environment at large.

individual children and small groups while relating to the larger group; and all this interacting and relating must be done in a healthy, positive, and educational way.

As you will discover, prevention is a form of guidance and plays a major role in safety. Although it is not easy to focus on individuals while paying attention to the whole group, it is vital to learn to do so. You must learn to constantly scan the room or yard so you can detect "trouble spots." Indeed, it may seem that you have to have eyes in the back of your head to supervise effectively, but the secret is in developing **dual focus.**

Focusing on Individuals and the Group

The tendency of early childhood beginners is to focus too broadly or too narrowly. Compare the experiences of the two student teachers in the following two examples and contrast them with the dual-focus approach described in the third example.

The Narrowly Focused Student Teacher. Erin is sitting outside at a table where collage materials are attractively displayed. Two children are busily gluing, one on each side of her. She talks first to one and then the other in a give-and-take fashion, being appropriately responsive and not too directive. The scene looks fine until you notice that she has seated herself with her back to the rest of the children, who are riding tricycles, playing in the sand, and climbing on a wooden structure. Another adult stands in a corner of the yard watching the entire group.

For a while, everything is going smoothly, until a child falls off a tricycle and starts to cry. The other adult repositions herself and bends down to see to the crying child. At this point, Erin should be alert to the group at large because the other teacher is now focused on one child. Unfortunately, Erin is deep in a conversation with one of the children at the collage table. As a result, Erin misses the squabble beginning under the climbing structure. By the time she becomes aware of it, the two children are both screaming loudly and the other adult is calling her to check it out.

The Broadly Focused Student Teacher. While Erin is outside, Jamie is indoors with the younger group. She stands off to one side of the room, carefully watching what is going on. Her coworker is busy at the diapering table and has her back to the rest of the children. Jamie supervises just fine, but she never bends down to talk to the children. She maintains her supervisory role without becoming any more interactive than to call out to the children when she wants their attention.

A Dual-Focus Approach. Contrast Erin and Jamie with Chantal, who works in the infant room. She is busy diapering a squirming nine-month-old, and to watch her you would think this child has 100 percent of her attention. But you would be wrong. Although she is focused on the child she is diapering, talking to him about what she is doing, involving him in the process, she still manages to sense a conflict brewing off to one side. "Watch Taylor," she says quietly to the student teacher, who, now alerted, moves to where a toddler is approaching a child rolling around on a miniature plastic car. Right before Taylor reaches the child, the student teacher takes her by the hand gently. She says calmly, "I see you really want to ride, but Brian

This teacher or student teacher is focused on one child but remains aware of what's going on in the rest of the classroom.
Wavebreakmedia/ Shutterstock

has the car now." She sighs with relief when Taylor runs off to find a book and brings it to her. "We stopped that bite before it happened," she whispers to Chantal, who gives her the thumbs-up sign.

Other experts at the dual-focus technique include family child care providers. These professionals work mostly by themselves and are often forced to develop this skill early on if they are to keep the children in their care safe and still meet all their needs, including physical care, education, socialization, attachment, relationships, and interpersonal interactions.

One of the ways you can learn to split your focus is to periodically "step out of yourself"–to observe the scene from an objective point of view. Question yourself: Do I have tunnel vision? Is my focus too narrow? or Is my focus too broad? Am I just supervising without interacting? To perform your job well, you must learn how to supervise *and* interact at the same time.

With experience, you will learn how to position yourself so you can scan an entire area. You will also learn how to focus on one child and remain alert to changes in the rest of the room or play yard. You will learn how to keep your eyes and ears open!

But just knowing what is going on is not enough. When threatening situations arise, you will need to know what to do. That is why guidance is another important skill for the early childhood educator to acquire.

A Crash Course in Guidance

A common complaint of the beginner is, "The kids don't listen to me!" In fact, it really is not a listening problem. There's nothing wrong with the children's ears, just their motivation to do what the adult wants. When someone new arrives in a program, someone who has no relationship with the children and hasn't yet earned their respect, children often ignore this person's requests or orders. When safety is an issue, the new adult must know how to respond effectively from day one.

What do adults usually mean when they say, "The kids don't listen to me"? They mean, "When I tell children to do something, they don't do it." The adults expect obedience, and they do not get it.

That is when *guidance* is needed. *Guidance* is a word that has replaced *discipline* in many early childhood settings. Discipline can have negative connotations, though the word itself comes from disciple–and can mean following a wise leader. But too often discipline is associated with punishment in many minds. Guidance moves behavior management into a positive realm.

guidance This word has replaced the word *discipline* in many early childhood settings, as it moves behavior management into a positive realm.

Cooperation as a Goal. Although some people advocate teaching strict obedience, there are reasons to focus instead on teaching cooperation. See the *Points of View* box on page 47 for contrasting perspectives on teaching children to be obedient. When you aim for cooperation, you help children develop respect for others plus an understanding of behavioral limits. One way children learn about cooperation is by watching adults. As such, you must remember to always model cooperation–with other adults and with children.

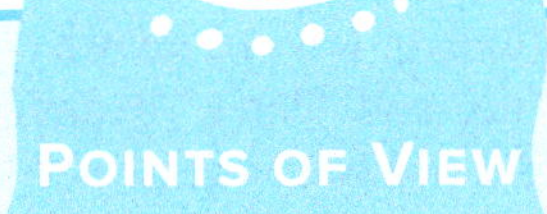

POINTS OF VIEW

Contrasting Perspectives on Teaching Children to Be Obedient

On the one hand: Obedience is overrated and usually results in adults issuing orders or giving commands, which triggers instant rebellion in children who have been taught to think for themselves. Some families do not want their children to automatically respond to what an authority tells them for a number of reasons including their culture, a different guidance system, their own rebellious nature, or even the fear of child abuse, of which the strictly obedient child is an easy victim.

On the other hand: Children have to learn to respect their elders, and part of that respectful behavior is doing what they are told to do. Children aren't old enough to think for themselves. They can make serious, even fatal mistakes. They need an older, wiser hand guiding them and telling them what to do. Children need to be taught obedience from an early age.

Developing a relationship with the children is the most important factor in building cooperation and getting them to listen. When children accept that you have their best interest at heart, they are more apt to pay attention to what you say. You cannot just tell them that you care what happens to them, *you have to prove it*–over and over. As you work at developing a relationship with each and every child, you gain their trust. This will not happen overnight; it takes time.

The number-one program accreditation standard of the National Association for the Education of Young Children (NAEYC) involves relationships. Here, in this chapter the relationship focus relates to supervision and guidance, but important to acknowledge that the goal of building relationships pervades every area of the program.

NAEYC Program Standards
Program Standard 1: Relationships

Stopping Unacceptable Behavior. "I'd appreciate it if you wouldn't climb on the fence," says a student teacher timidly. This statement is so polite that it's hard to tell whether she means what she says or not. "Please don't hit him, okay?" is not assertive enough to get the point across. "Can you please not throw the little cars around?" does not address the hazards of the behavior.

When safety or health issues are involved, you need to be firm and state your message clearly–in words all the children can understand. Your tone of voice and posture should convey "That's the way it is" and "No nonsense." It is good to be polite to children, but don't be wishy-washy. Also, be sure they understand when they have a choice about something and when they do not.

NAEYC Program Standards
Program Standard 5: Health

Be firm: "No climbing on the fence." Also, if possible, explain why you are being so firm or the reason for the rule: "It's dangerous to climb on the fence." Tell the child point-blank what you will or will not allow and why: "I won't let you hit. Hitting hurts." You may avoid some resistance by finding ways to say what you mean without using direct orders or prohibitives like "no" and "don't." Redirecting is a useful guidance technique for conveying your intent firmly but gently: "Climb on the jungle gym instead of the fence." "Tell Jaime why you're so angry rather than hurt him." The most effective **redirection** of all is to give the child more than one alternative: "I won't let you throw the little cars, but you can throw the bean bags or these foam balls instead." When the child has a choice, he or she feels **empowerment** instead of helplessness, and an empowered child has less need to defy you.

redirection A form of early childhood guidance that diverts a child from unacceptable behavior to acceptable behavior without stopping the energy flow. Ideally, redirection involves giving the child a choice to lead him or her toward a constructive behavior or activity.

empowerment Helping someone experience his or her sense of personal power. For example, an adult can empower a child by giving him or her the opportunity to make some decisions rather than being told what to do.

The Problem with Using Anger to Control Children's Behavior. When early childhood beginners do not know how to stop unacceptable behavior, they usually start by asking nicely and then turn to anger when being polite does not work. Here is a fairly common sequence:

> I'd appreciate it if you would get down off the fence. Will you please get down? Pretty please? If you don't come down off that fence I'm going to get mad! Okay, now I'm mad!

Unless the adult's anger frightens the child, the adult will have to resort to threats or else get stuck. What started as a simple request has become a power struggle. Unless the adult is willing and able to haul the child off the fence, he or she may either start to threaten dire consequences or go back to begging.

Using anger to get children to do what you want only works if they are truly afraid of you or concerned about displeasing you. If you are a stranger to them, they may have no reason to please you unless they fear you. However, if you depend on fear to control their behavior, you will have a hard time developing a relationship with the children–a relationship that is a primary goal in guidance and the teaching-learning process.

Another problem with using anger to control children's behavior is that they are likely to take your cue. Think about it. If you do not want children to use their anger to try to control other children, you must not model that approach yourself.

Self-Expression Versus Manipulation. Keep in mind that attempting to guide children's behavior through anger is different from honestly expressing feelings of frustration or anger–feelings that are perfectly normal when you are trying to control children but cannot. At this point, you want to start thinking about controlling your own behavior. Rather than use your anger to try to manipulate the children's behavior, show them how well you can control your behavior when you are angry. Do not, under any circumstances, explode because you will scare the children. Instead, find less dramatic ways to express your anger. Put your anger into words: "It bothers me when you keep dumping the toys off the shelves like that."

If you find you are getting angry regularly, do some self-searching to discover what is going on. Obviously some changes are needed–changes in you or in the

When talking to a child, position yourself within a close range and at the child's level. Rather than sit directly across from them try to sit adjacent to them (shoulder to shoulder).
Brand X Pictures/Stockbyte/Getty Images

situation. Often you can prevent the things from happening that are triggering your anger. Maybe there is a child who refuses to hang up her coat when she arrives. Is this the child's problem or *your* problem? Maybe it is something from your past, your values, this period of your life, or even an old feeling about little girls with curly red hair. Some reflective thinking and creative problem solving can shed light on why you feel so strongly and what can be done about it. Maybe you can simply find a way to change the child's behavior. Or maybe you need to work more on your perspective and your relationship with the child.

A theme running throughout this section has been communication. Learning to communicate effectively with young children is a key to guiding their behavior and keeping them safe. Let us look further at communication skills.

FOCUS ON DIVERSITY

Eye Contact and Its Meaning in Different Cultures

It is important for early childhood educators to find out about how each family teaches its children to show respect to adults. Do parents want their children to look adults in the eye or not? For European Americans, eye contact is a sign of attentiveness, but in most Asian and Native American cultures, it is a sign of disrespect. Within their family, African Americans frequently use nonverbal communication, including deep eye contact. But when speaking with white people, they may avoid eye contact.

Anthropologist Virginia Young suggests that "perhaps the eye contact that goes with the words in white cultures brings too strong a communication to the black who regards it as more important than words." In some African American families, as well as some Puerto Rican and Mexican American families, however, children do not look at adults directly, especially if they are being scolded. This lack of eye contact may suggest to some European American adults that a child is not listening.[3]

Communicating Effectively. One mistake early childhood beginners make is to stand over children while talking to them or to shout from a distance. Anytime you talk to children, you should be close to them and *down at their level.* And when safety is involved, being right on the spot is vital. A rule of thumb for talking to children is "one foot for each year": put no more than one foot between you and the child for each year of the child's life. If you are talking to a baby, you need to be right next to her–less than a foot away; a two-year-old should be within arm's distance. Squat down so you are at eye level, but note that making eye contact is a cultural matter. Accepted early childhood practice dictates that adults look children in the eye when talking to them. Making the children look back, however, should not be mandatory since eye contact has a vast range of cultural meanings. See the *Focus on Diversity* box for some of those meanings.

Try to refrain from shouting at children; raising one's voice should be saved for true emergencies only. If you make a habit of talking in a soft or moderate voice, a loud yell will get the attention it needs when the situation calls for it. Say you are not close enough to stop a child with an upraised wooden block. "Look out, Morgan!" or "Stop, Shawn!" may be a lifesaver in that case.

When Words Alone Aren't Enough. Often you cannot depend on words alone–shouted or not–especially when working with younger children. Telling a toddler not to touch things on the counter probably will not work after she has discovered she can push over the little wagon, climb up, and reach the forbidden objects. If you apply the foot-per-year rule of thumb, you will be close enough (while you tell her why it is important to stay down) to physically redirect her to something equally interesting. Of course, if the explanation works and she climbs right down, you won't

Defusing a Conflict

I had the chance to learn a new way of approaching conflict in an early childhood setting when I did my assignment for an observation class. After observing a parent/infant guidance class facilitated by my professor, I experienced a totally different way of intervening with children at play. I observed two toddlers wanting the same toy while the adults observing within the space did nothing to help them resolve the problem. I thought to myself, why isn't anyone (parents or professional) doing something because they are not sharing? After a couple of minutes of "mine" and holding on, the two toddlers lost interest in the toy and moved on to something else.

It wasn't until I saw that demonstration that I got a better understanding of why there was no need to intervene. What I observed was the group facilitator acknowledging both of the infants' need to have the same toy. In order to observe and truly be present in the situation, all that was necessary was for her to move closer to the children and reflect on (also assessing for herself) what she saw. Since she (and the rest of us) could see the children were not at risk of getting hurt, there was no need to intervene more.

–A Community College Student

have to touch her at all. But if simply talking does not take care of the problem, a gentle guiding hand or firm grip will tell her you mean business. Also, make sure your affect matches your tone. Often times adults inadvertently send mixed messages to children when they ask them to do something in a wishy-washy tone or with a smile when they are actually quite angry. This often confuses a child and does not bring about the desired behavior or outcome. Also, now would be a good time to remove the wagon as well.

Conflict as a Safety Issue

A conflict over a toy can be one of the most dangerous situations you will encounter in an early childhood program. Some young children are unable to stop themselves from pushing, hitting, biting, or lashing out when they are determined to win an argument. To avoid the injuries that can result from even a minor disagreement, you should always stay near the action, regardless of the children's age. If two three-year-olds are about to get into a biting contest, you may need to be closer than three feet to prevent it from happening. If five- or seven-year-olds are about to come to blows, talking to them from a distance will not be good enough. Sometimes by being close by, you can just observe and not have to do anything. Other times you need to intervene. The *Voices of Experience* comes from Elizabeth Memel, who is an infant-toddler expert trained in RIE and a community college teacher. Her student reports on a conflict that resolved itself. In contrast to this story, the following scenario is an example of how to handle a situation when two children are fighting over a toy and seem to be about to hurt each other.

An Empowering Way to Handle Conflict. Kyle and Cody are digging with plastic scoops in the sand next to Ashley, who is using a red-handled shovel. Suddenly, Ashley runs off to the swings and abandons her shovel. Kyle and Cody both start for it at once, Cody still holding his plastic scoop in one hand. The teacher, Danielle, who was standing at a distance from the two boys, moves quickly to the scene. At first Kyle and Cody struggle, both holding on to the red handle. Danielle says quietly from a squatting position next to them, "You both want that shovel." They ignore her, and the struggle continues. She remains quiet until Cody raises his hand with the plastic scoop in it to hit Kyle. "I won't let you hurt him," Danielle assures both boys. She looks Cody in the eyes as she holds his arm firmly but not aggressively. Cody looks away from her gaze, lowers his scoop, and drops it. Danielle lets go of him. The two boys stare at each other, both still with a firm grip on the red shovel handle.

Finally, Cody reluctantly releases his grip, and Kyle grabs the shovel and runs out of Cody's reach triumphantly. Cody stays rooted to the spot, crying quietly. Danielle puts an arm around him and says, "You really wanted that shovel!" When Kyle comes back to flaunt his victory in Cody's face, Cody screams at him angrily. "Tell him how you feel," says Danielle. Cody screams again. Danielle says to Kyle, who is dancing up and down, "He's mad that you have the shovel." She puts Cody's screams into words. Kyle considers this message briefly, then, seeing another shovel abandoned nearby, runs to get it. He tries to give the second shovel–with an unpainted handle–to Cody, who rejects it and reaches for the one in Kyle's other hand. "I want the red one!" he insists. Kyle shrugs, drops the second shovel, and walks off.

Kyle starts digging a hole a short distance away. Cody stands crying loudly over the shovel lying at his feet, which is exactly like the one Kyle is digging with except for the color of the handle. "You wanted the other one," says Danielle. She is still down at his level talking to him. At this point, Ashley returns. Seeing that the shovel she had before is in use now, she starts for the one lying at Cody's feet. He sees her coming and reaches down and grabs it up. She frowns briefly and then turns and runs back to the swings. Cody is last seen digging reluctantly, and Danielle has moved over by the climbing structure to talk to a child about his new baby sister.

Contrast the previous scene with the following one, in which the student teacher, Laura, uses a very different approach to solve the same kind of conflict.

A Disempowering Way to Handle Conflict. When Laura sees Briana and Taylor both tugging on the same plastic lawn mower, she comes over and says, "Who had it first?" They both say, "I did!" Laura makes a snap decision. "I think Briana had it first," she says and takes it away from Taylor. She walks away, and Taylor comes right back and grabs the lawn mower away from Briana, whose squeals bring Laura back. The two are tugging frantically, and it looks as if they might come to blows. Laura removes the toy, prying their fingers off the handle. She walks over and locks the lawn mower in the shed, saying, "If you two can't play nicely, then neither of you get to play with it." The two children sit listlessly in the sandbox for a while with sullen looks on their faces. Then, when Briana

Practice in risk taking is an important safety measure and should be part of any early childhood program.
Maskot/Getty Images

picks up a plastic tub and starts throwing sand in the air, Laura removes the tub as well.

Both Danielle and Laura kept the children safe, but there is a difference in how they handled the conflicts. By acknowledging both boys' feelings, Danielle prevented violence and allowed Kyle and Cody to solve the problem themselves. She did not try to talk them into any solutions, nor did she impose adult judgments on them about fairness, even though she may have had some strong opinions. Instead, she was calm and impartial and kept them from hurting each other until the problem worked itself out.

Laura, on the other hand, kept the children safe but without allowing Briana or Taylor a chance to work it out themselves. She short-circuited the conflict process and gave them the message that they needed an adult to solve the problem. Whether she meant to or not, she punished both, leaving resentment in her wake.

Like Danielle, Laura did not expect the children to obey her authority based on words alone; she conveyed her authority with the appropriate amount of words and action. Like Danielle, Laura also avoided silent action and, the opposite approach, talking the situation to death–both ineffective conflict-resolving measures. Nevertheless, the way Danielle handled the situation–by allowing the opportunity for discussion and a chance to explore feelings and choices–was more empowering to both boys. If the two boys had been more verbal, they might have argued it out, explored trade-offs, and perhaps come to some verbal agreement. As it was, the problem eventually solved itself.

People who feel empowered are often more willing to cooperate than those who feel powerless.

Risk Taking as a Safety Measure

When children learn to judge risks accurately, they take a big step toward contributing to their own safety. Children need to be exposed to a variety of experiences in which they learn to evaluate the degree of risk involved in their many activities. Without being allowed to take reasonable risks, children are unable to learn important safety skills. As such, practice in risk taking is an important safety measure and should be a part of any early childhood program.

Learning to judge the degree of risk has its roots in infancy, as demonstrated by Dr. Emmi Pikler (mentioned in Chapter 1) whose research on movement development started in the 1930s. From her studies she discovered that when children are free to move, from the early months of life, they develop marvelous control of their bodies and are able to take on amazing physical challenges. Observations at the Pikler Institute in Budapest and videos produced at the institute show children who know how to take risks without getting hurt. The accident rate at the institute is remarkably low. Only one child has broken a bone in 60 years and that was an older child the first week she arrived at the institute.[4]

Allowing children to experience the consequences of their acts helps them understand what is okay to do and what is not. Of course, you would not let a child fall out of the top of a tree—that consequence would be too severe. But you should let her climb on structures that are appropriate to her size and developmental level. That is one of the big advantages of being in a child care program—children can experience making choices in a rich, protected environment. These are early lessons that will serve them well all their lives.

Helping Children Learn from Their Experiences

Sometimes the early childhood educator must *set up* consequences that respond to a child's inappropriate or unsafe behavior. For example, the child who insists on climbing the fence is brought inside for a while because he cannot be trusted to stay in the play yard. He stays inside until he can reassure the adult that he will control his fence-climbing tendencies. This method is most effective when the child understands the connection between his own action and the consequence. In other words, it is not an arbitrary punishment imposed by an adult but a direct result of the child's own behavior. The approach used in this example becomes an empowering way of handling the situation if the child is allowed to decide when he is ready to go outside and stay off the fence.

In another example, a child is throwing sand. The teacher warns the child about the danger and then removes her from the sandbox when she threatens to continue throwing sand. This approach is more effective than lecturing, scolding, or putting the child in time out because there is a logical connection between the action and the consequence. The child comes to see that it is not the adult's whim that has affected her, but her own actions.

A Safe Physical Environment

Setting up and maintaining a safe physical environment for exploration (which includes learning while playing) and for caregiving are major factors in guiding behavior and in providing for the health and safety of children. This chapter will deal with the safety aspects of the environment that are of greatest concern to novice teachers. NAEYC's Program Standard 9 relates to the physical environment and requires that it be safe and healthful. The following sections expand on Program Standard 9.

NAEYC Program Standards

Program Standard 9: Physical Environment

Developmental Appropriateness

Although you may not have chosen how the environment is set up, it is up to you to understand the relationship of developmental appropriateness to safety. You have to think about simple things, like how to set out large, hollow wood blocks. If stacked, a child could be hurt by toppling blocks. If a wagon with a tricycle in it might invite a child to climb on the storage shed roof, the equipment needs to be moved.

Sometimes the equipment is too big for a given age group. For example, one play yard, designed for school-age children, had a tall slide with a steep ladder. At the platform on top was a hole with an eight-foot firepole down the middle. Preschool children learned to manage the equipment, though some teachers worried. But when toddlers were introduced to the play yard, it was immediately obvious that this climbing structure was inappropriate for the new age group. For a week, the teachers and aides took turns blocking the ladder to keep the toddlers from climbing it, but because this was a waste of teacher energy, the bottom rung was removed. The lowest rung was now too tall a step for any toddler to reach, and some of the shorter preschoolers could not manage it either. Everyone breathed a sigh of relief when that problem was solved.

Just as there is equipment that is too big for younger children, there may be small objects that pose a choking hazard to infants and toddlers. One way to tell is to have what's called a "choke tube"–a commercially sold device used to test which objects are dangerous. If something is small enough to fit through the tube, it has to be out of reach of infants and toddlers. Of course, older children need to have small objects available to them. One solution to this problem might be to block off an area that allows older children to move in and out but prevents younger ones from entering. Both examples demonstrate the use of creative problem solving to produce developmentally appropriate and safe environments for children of various ages.

Maintenance as Prevention

An important requirement for a safe environment is that it be orderly and well maintained. Watch out for broken toys and equipment, which should be repaired right away or removed. Store dangerous substances in their original containers (with labels intact) in locked cabinets. Whenever you need to use cleansers or other toxic items, be very careful you keep them out of children's reach at all times; never walk off and leave them, even for a moment, without locking them up.

Be constantly on the lookout for obvious safety hazards–loose nuts and bolts, splinters, electrical cords, and sockets children can get to, toys with small parts that can go into mouths, adult purses and backpacks left in reach, hot beverages, unsecured doorways, slippery rugs.

Also look for hidden safety hazards. Anything painted before 1978 may contain dangerous levels of lead, which can poison children who like to taste their environment. Toxicity lurks everywhere–in the innocent plant that came up on its own in the play yard or in the tube of paint that just showed up in the room one day. Do not take chances; check to be sure everything in the environment is safe.

Cribs must meet specifications. The old drop-side ones must be replaced as they are dangerous for babies. Also, walkers are particularly hazardous, causing more injuries than any other piece of baby equipment. (They also cause problems in children with cerebral palsy and can delay walking.) And look out for curtain cords that may be dangling near a crib or play area and can get wrapped around a child's neck.

Sanitation Procedures

NAEYC Program Standards
Program Standard 5: Health

Early childhood educators should have a good understanding of how to keep the environment, the children, and themselves clean and safe. Because consistent hand washing is the single best approach to sanitation, child care providers should wash their hands thoroughly with soap and warm water before and after diapering, before preparing food, before and after meals, after feeding children, and any time they come into contact with bodily fluids including breast milk, blood, vomit, or runny noses. Make sure that the children also wash their hands thoroughly with liquid soap (bar soap spreads germs) and warm water before preparing food, setting the table, and eating and also after toileting, coughing, nose blowing, and eating. Wash infants' hands for them. Hand-washing routines should become second nature in any early childhood program, including family child care homes. (Refer to the *Tips and Techniques* box on page 57 for more information.)

If any children are in diapers, a sanitary diapering process must be in place and prominently posted by the diapering area. (A sample diapering procedure is outlined in the *Tips and Techniques* box on page 58.) Moreover, all bathroom surfaces must be sanitized daily. And any program catering to infants and toddlers should have a policy of wiping all toys and surfaces daily with a fresh, safe, sanitizing solution. Finally, keep children's personal possessions (combs, toothbrushes, clothing, bottles,

TIPS AND TECHNIQUES

Hand-Washing Guidelines

Adults and children should wash their hands thoroughly with liquid soap and warm water at the following times:

- Whenever hands are contaminated with bodily fluids.
- Before preparing, handling, or serving meals or snacks (including setting the table).
- After toileting or, in the case of adults, changing diapers or assisting a child with toilet use.
- After eating meals or snacks.
- After handling pets or other animals.[5]

washcloths, towels, and bedding) labeled and separate from each other, and limit their use to the respective child only.

Program Policies and Procedures for Health and Safety

In addition to national standards set by the National Association of the Education of Young Children, there are other standards and regulations on health and safety, put out by such organizations as the Academy of Pediatrics, state departments of education, and other state and local agencies concerned with early care and education. There is a difference between an accreditation procedure and a licensing one. The main difference is choice. Being licensed is mandated by the state, while being accredited is a voluntary assessment process. In addition, each early childhood program establishes its own set of health and safety procedures. Find out what they are. If you are a family child care provider, you will have to develop your own systems in accordance with your local licensing regulations and/or professional standards. For example, a policy on proper food handling might include the following: put food away promptly; know how long something has been stored in the refrigerator; sanitize bottles, dishes, and flatware; and keep the food-preparation area separate from toileting and hand-washing facilities.

At the same time you are following proper health and sanitary procedures, you can be teaching the children about many of them as well. Of course, you cannot involve young children in cleaning the bathroom, but you can teach them proper hand-washing skills and how to use and care for their personal possessions. You can also teach them, say, not to share their toothbrushes, not to taste the mashed potatoes in the serving bowl before they reach their plates, or not to eat with a fork that has fallen on the floor.

Covering coughs, properly disposing of tissues, and hand washing are also important teaching goals to prevent the spread of disease. If a child does get sick, or exhibits some contagious condition, separate him or her from the group. You may be asked to keep the child quiet, warm, and comfortable while somone contacts

Tips and Techniques

Sample Diapering Procedure

1. Be sure the diapering area was sanitized following its last use. If not, discard the used liner paper, wipe down the diapering surface with soap and water and then a diluted disinfectant solution that contains bleach and allow it to air dry before placing fresh sanitary paper down.
2. Wash your hands thoroughly with soap and warm water before changing a diaper.
3. Use disposable gloves, especially when handling diarrhea or bloody stool or if you have cuts on your hands.
4. Dispose of used diapers in the designated container that has a lid.
5. Wipe the child with a clean, moist cloth or baby wipe. Dispose of the used cloth or wipe in the container provided.
6. If wearing gloves, remove them carefully without contaminating your hands. Then, put on the clean diaper.
7. Wash the child's hands thoroughly with soap and warm water.
8. Clean and sanitize the diapering area: discard the used sanitary paper in the container provided, wipe down the diapering surface with a bleach solution, and put down fresh liner paper.
9. Wash your hands thoroughly with soap and warm water.
10. If a diapering chart is provided, note the time and any information that should be shared with parents or other caregivers.[6]

the family. Although there may be a long waiting period, remain sensitive to the child's emotional state until the family arrives. Ill children sometimes feel insecure, and separation from the group may add to this feeling.

As part of its health and safety policy, every program should have on file medical-consent forms and up-to-date emergency cards to locate family members or a designated substitute. Each early childhood program should also have clearly outlined emergency procedures (including a reporting procedure). If you have not read these procedures, ask to see them. You should also ask about fire drills (and earthquake or tornado drills where applicable) and what provisions are made for infants and nonambulatory children with special needs. These drills should be carried out regularly, and every adult should be aware of his or her specific responsibilities.

Stress and Frustration as Health and Safety Issues

A final word about health and safety: A child's stress and frustration can turn a safe situation into an unsafe one. Although you cannot eliminate everything that could make a child upset and angry (and you wouldn't want to), you can examine the physical environment for ways to reduce unnecessary frustration. For example, activities that are developmentally appropriate to the age of the group create less frustration than those that are not. And just as overly frustrated children pose safety risks, so, too, do bored children who are left to their own devices to find things to do. Make available equipment and materials that are interesting, relevant, and developmentally appropriate.

Finally, a child whose needs go unmet may also exhibit stress that can contribute to an unsafe situation. When a child becomes overtired, her frustration may increase. She may be less capable of getting along with others and of making good decisions. Remember to be sensitive to individual differences, including children with varying abilities special needs. Just because the whole group is exhibiting high energy does not mean that a particular child does not need to rest. Likewise, if a child arrives for the day with unmet needs, try to determine what his needs are. If he needs to eat before the next meal is scheduled, try to accommodate him. In general, it is a good policy to keep children well rested and well fed and to meet their needs in a timely manner for their sake and for safety's sake. NAEYC's Standard 5 on health relates to this chapter in that it requires that the program promote health and protect children from illness and injury. If you are a beginner, your role is confined to careful supervision. You also can help children make wise choices and allow them to take enough age- or stage-appropriate risks for learning, all the while keeping children within safe boundaries by not permitting hazardous behavior or environmental conditions. Through your careful supervision you can anticipate and avoid accidents or problems before these occur. The trick is to guard children's safety, while also encouraging them to do what they are capable of doing for themselves.

NAEYC Program Standards

Program Standard 5: Health

A Story to End With

I remember the day I learned an important safety lesson. I was an assistant teacher of four-year-olds and had not been in charge of a group of children before. We were getting ready to go on a field trip, and the teacher was outside talking to the drivers. I had been told to get the children ready to leave at 10:00. I was standing far from the door when I glanced at the clock, saw it was 9:50, and announced to the group, "It's time to go now." What I really meant was, "It will be time to go soon." Three children took me literally and headed straight for the door. I was right on their heels, but not in time to stop them from going out the door. Luckily, there was another teacher in the room to supervise the rest of the group, so I took off after the escapees. The three split up, each going in a different direction. Even though I was young and spry, there was no way I could catch them all—and it was obvious that chase was the name of the game. I do not know if I realized that they were playing with me or if I just gave up, but for whatever reason, I stopped cold. It was the best thing I could have done. With the game over, the children were easy to round up. I learned that day that teacher positioning is everything; after that, I never made any announcements about leaving until I was in front of the door and in control of it. I also learned that I must *say what I really mean.*

SUMMARY

Developing observation and supervision skills is the most important goal for the beginning early childhood educator. In order to ensure the safety and health of the children, the early childhood educator needs to develop dual focus–a technique that allows the adult to supervise the group and, at the same time, build relationships with individual children. Early childhood educators also need to develop skills in guiding children's behavior and communicating effectively to gain children's cooperation. The first step in building cooperation is to develop a relationship with the children.

To stop unacceptable behavior, the early childhood educator must be firm and assertive. Redirecting behavior is more effective than a command or angry threat because it offers the child a choice; he or she feels empowered rather than helpless. Conflicts should also be handled in ways that empower children; the early childhood educator should acknowledge the children's feelings and allow them to solve the problem themselves. Helping children learn to take reasonable risks and experience the consequences of their actions helps teach safety and sound decision-making skills.

Early childhood educators must also know how to set up and maintain a safe environment for exploration and for caregiving. A developmentally appropriate environment not only ensures safety but also cuts down on unnecessary frustration for the children. The environment must also be clean, orderly, and well maintained.

REFLECTION QUESTIONS

1. If you have already worked with children, reflect on your first day. What was it like? How did you feel? How would you have wanted it to be different? What did you learn that first day?
2. If you still have your first day of working with children in front of you, think about what it will be like. Do you have any feelings when you consider stepping into an early childhood care and education setting for the first time?
3. How do you feel about the *Points of View* box on page 47 on teaching children to be obedient. Do you relate to one view more than the other?
4. Has anyone ever tried to control you by getting angry at you? What effect did that have on you? How do you feel about it? What do you think about it? Have you ever used anger to try to control someone else? Has a child ever made you angry? What did you do? What were the effects?
5. When did you learn something from experiencing the consequences of your own actions? What happened? Was that an effective way to learn? How much do you believe in allowing children to experience the consequences of their choices?

Terms to Know

How many of the following words and acronyms can you use in a sentence? Do you know what they mean?

observation 41
RIE 43
dual focus 44
guidance 46
redirection 48
empowerment 48

For Further Reading

Copple, C., and Bredekamp, S. (2009). *Developmentally Appropriate Practice in Early Childhood Programs Serving Children Birth Through Age 8* (3rd ed.). Washington, DC: National Association for the Education of Young Children.

Jacobson, T. (2008). *Don't Get So Upset! Help Young Children Manage Their Feelings by Understanding Your Own.* St. Paul, MN: Redleaf.

Kaiser, B., and Rasminsky, J. S. (2016). *Challenging Behavior in Young Children: Understanding, Preventing, and Responding Effectively* (4th ed.). Upper Saddle River, NJ: Pearson.

Moore, L. O. (2009). *Inclusion Strategies for Young Children.* Thousand Oaks, CA: Corwin.

Reynolds, G. (2008). Observations are essential in supporting children's play. In B. Neugebauer (Ed.), *Professionalism.* Redmond, WA: Exchange Press.

Wylie, S., and Fenning, K. (2015). *Observing Young Children: Transforming Early Learning Through Reflective Practice* (5th ed.). Toronto, Ontario: Nelson.

Design credits: Tips and Techniques: ©Ingram Publishing; Focus on Diversity: ©Pixelic/Getty Images

3 Communicating with Young Children

Steve Debenport/Getty Images

Communication, Relationships, and the Cognitive Connection

Listening: An Important Skill

Listening and Giving Feedback Are Valuable to Communication

Listening and Responding to Different Situations

How to Communicate Clearly

Ask Real Questions, Not Rhetorical Ones

Validate Feelings and Perceptions Instead of Discounting Them

Address Uncomfortable Situations Instead of Ignoring the Obvious

Be Congruent; Avoid Incongruence

Watch Out for Double-Bind Messages

Use Redirection Instead of Distraction

Be Sensitive About Questioning Children

Using Observation and Reflection to Improve Communication

A Story to End With

In This Chapter You Will Discover

- how communication skills can build positive relationships with children.
- what types of responses indicate "holistic listening."
- how warm, caring, communicative relationships contribute to healthy brain development.
- what to do when children discriminate against a playmate.
- how to give children real choices.
- how to validate feelings and perceptions.
- why it's important to address awkward and uncomfortable situations.
- what mixed messages are and how to avoid them.
- how redirection is different from distraction.
- why asking a child questions can be intrusive.
- how to keep anecdotal records.

This chapter is about mental and emotional health and a feeling of security, which are the foundations for learning and development. The underlying goal is building a relationship with children as stated in NAEYC's first Standard for Accreditation. This standard is about programs promoting positive relationships because they are essential for the development of:

NAEYC Program Standards

Program Standard 1: Relationships

- personal responsibility
- self-regulation
- positive sense of self
- feeling of security

Relationships grow when teachers are emotionally available and responsive to children. Check out the *Voices of Experience* story on page 65 where Jean Monroe, a nationally known early childhood consultant, describes how she has observed relationship building between teachers and children.

One of the ways that you can meet Standard 1 is by developing appropriate communication skills. This chapter is about those skills. You will explore how to listen and respond to children in emotionally charged situations in ways that build relationships, promote self-esteem, and empower children. You will also learn how to communicate clearly and avoid mixed and detrimental messages. Finally, this chapter will cover a number of alternatives to issuing direct orders, which are often counterproductive for dealing with young children.

The information in this chapter, and indeed in every chapter in Part 1, is the "unwritten curriculum." The unwritten curriculum refers to the learning that occurs, often unconsciously, through the relationships that children develop with significant adults in their lives. How adults communicate with children builds such relationships, and those relationships are the basis of a good deal of learning in the early years. The learning of the early years is then the foundation for learning in later years. But any activities and lessons designed to stimulate cognitive development and promote academic skills are secondary to the development of emotional health and relationships in those early years. You can't separate thinking from feeling according to Dr. Stanley Greenspan, a clinical professor of psychiatry who has written a number of books on emotional development. Intelligence doesn't come from cognitive stimulation but rather has its origins in specific emotional experiences. Children experience a series of distinctive social-emotional stages as they grow.

According to Dr. Greenspan, infants experience six stages of emotional development during infancy. These stages are as follows: Stage 1 (0+ months) is self-regulation of basic emotions; Stage 2 (4+ months) is falling in love; Stage 3 (8+ months) is developing purposeful communication with others; Stage 4 (10+ months) is beginning of an organized and complex sense of self; Stage 5 (18+ months) is creating emotional ideas; and Stage 6 (30+ months) is using emotional thinking (fantasy). By understanding these stages, caregivers can better support each child's developing sense of self, relationships with others, understanding of social rules, and ability to work out feelings through communication and creative play.

When Teachers Are Emotionally Available to Children

When I observe in ECE settings, I am very interested in the relationship that has formed between adults and children. I want to know how emotionally available the teachers are to the children. When teachers are emotionally available they are warm, accepting, empathetic, *genuine,* and show caring concern. The following is what I observed in one classroom.

I arrived at the center early because I believe that hellos and goodbyes are crucially important for preschool children separating from loved ones. The teacher was near the entryway greeting the children and their family member. There seemed to be a secret password because she asked each child what the password was for the day. Each child eagerly told her and she responded with great bravado. Then the children signed in on their own book while the family member signed them in on the roll book. Sign-ins over, children said good-bye to their family member and went to wash their hands for breakfast.

One boy whose mom had left was hanging very close to the teacher. She reached down and put her arm around his shoulder, drawing him closer to her side as she talked to another parent. When the parent left, the teacher stooped down to his eye level and said, "Carlos, are you missing your mom today?" He looked down and did not respond. She said, "I know your mom was a long time getting here yesterday to get you but she promised to be here today so you won't be the last child here. Your mom will be back to pick you up when the clock looks like this." She got a wooden clock and turned the hands to the time the mom would return. "But you know, I like having time to play with you all by ourselves sometimes." He smiled and went to wash his hands. I noticed there were several wooden clocks in the room with pick-up times on them. I liked this concrete way of reassuring children.

The teacher surveyed the room to make sure that each child had found a place in the room that was engaging and comfortable. She noticed that two boys were playing roughly with a couple of books in the library. She went over, bent down, and asked them which book they would like her to read to them while they waited for circle time. Both boys wanted their book read. She said, "Since we only have five minutes before it is circle time, I'll read the book with the fewest pages now and the longer one later!" They settled down to read the book.

–Jean Monroe

Communication, Relationships, and the Cognitive Connection

This chapter may not seem to be about cognitive matters. You may be thinking that it is so focused on emotional security that it couldn't possibly relate to cognition. If you are thinking that, it's because you don't understand that the child is a holistic being. What happens in one area of development affects the others and the brain as a central processing unit is connected to everything!

NAEYC Program Standards

Program Standard 2: Curriculum

The relationship of warm, responsive care to healthy brain development is one of the important findings of the brain research. The emotional environment deeply affects how children develop and impacts their ability to learn. Children need warm responsive care leading to close relationships, say the brain researchers.

The brain starts out as an unfinished piece of architecture. It has unlimited possibilities. What happens to it in the early years produces the actual structures that become part of the final building. The creation of the structures is influenced by emotions. Research by Megan R. Gunnar of the University of Minnesota shows how responsive the brain is to feelings.[1] Gunnar measured a steroid hormone called *cortisol* that shows up in saliva. When a child is in stress, the levels of cortisol rise. High levels affect the child in many ways, including working on metabolism, the immune system, and the brain. Cortisol levels increased by stress actually change the brain by allowing neurons to be destroyed. Further, continued stress reduces the number of synapses in certain parts of the brain. Stressful experiences over a period of time can have a negative impact on brain function, even to the point of developmental delays.

The first year of life makes a difference in how children handle stress in later years as measured by cortisol levels. The baby who feels secure and safe is more able to handle stress and regulate feelings so they don't overwhelm. Close relationships serve a protective function. Such babies grow up to be more resilient and they are more likely to thrive.[2] So the information in this chapter may be seen to be only about getting along with children and supporting their social and emotional development, but it is much more than that. Imagine all the way through as you read that the information also relates to the development of children's thinking and learning. Check out the *Voices of Experience* story on page 67 where Lynne Doherty Lyle uses her ability to pick up a child's unique communication to turn what seems to be an obsession into a chance to promote exploratory skills leading to cognitive development.

Listening: An Important Skill

One of the most beneficial things you can do for a child is to listen–really listen. Don't just listen with your ears; practice what we will call **holistic listening,** that is, listening with all your senses and a splash of intuition. Listen to the verbal message (for children who are old enough to talk), and look closely at the visual signals, such as facial expressions and body language. A child feels validated when someone cares enough to listen. The simple acts of *listening* and indicating that you *hear* both help the child develop **self-esteem**–an important factor in psychological well-being.

holistic listening A form of listening that goes beyond merely hearing. Holistic listening involves the whole body and uses all the senses in order to pick up subtle cues that aren't put into words or otherwise readily apparent.

self-esteem A realistic assessment of one's worth that results in feelings of confidence and satisfaction.

Listening and Giving Feedback Are Valuable to Communication

When a child is upset, listening is the first step to problem solving. Being attentive and showing that you understand are often all that is needed to correct a situation. Sometimes you may need to go further and respond with action, but it's surprising how many times children can solve their own problems with the smallest bit of adult support.

Opening Doors

One day, a new child Noah began to attend our morning preschool program. He was three years and two months old. He had a fraternal twin brother who joined the class also. Right away, Noah began to open and look into every cabinet there was in the room. He would open a door and then watch as the door closed slowly. His mom stayed with the two boys the first day, and after a while she began to apologize for her child's "obsession" with doors. She assured me that the doctor had said there was nothing wrong with her son; he was just quirky. Noah's sibling had stayed close to his mom and listened. He said, "Noah opens doors all the time." I began to watch Noah with great interest. What was it he was watching? What was he thinking about? He showed no interest in what was inside the cabinets; he was interested in opening and closing doors. I knew at that moment that we were going to revel in this child's unique interest and help him explore things with hinges, knobs, and latches. We would set up a curriculum that would encourage this child to explore a variety of things that open and close, and we would attempt to find out something about his thinking process. We did this, and eventually we were able to help the mom explore her child's interest by sharing with her our understanding of the way he was processing his world. Parental stress about how their children are developing impacts children in how they see themselves. It impacts siblings also. What Noah was feeling was unclear, yet why take any chances in devaluing a child's excitement about an everyday thing? Knowledge of the everyday can lead to great discoveries.

–Lynne Doherty Lyle

Listening to verbal cues is possible with children who are old enough to talk, but listening to children starts even before they learn to talk. For infants, young toddlers, children who do not speak English, and some children with special needs, listening requires:

1. picking up on nonverbal cues–cries, affect, gestures, and body language
2. looking at the situation
3. giving back the messages you receive in your own words

Putting language into your interaction with the child is important because it establishes a medium for communication. Most children understand a lot more than they are able to express. **Receptive language** (understanding what is said) develops faster than **expressive language** (being able to speak), so your words are not as useless as they may seem. In fact, during this period where there is a gap between receptive and expressive language, many caregivers have begun teaching young infants to use non-verbal communication, such as sign language, as an additional means of communication prior to verbal communication. Research has shown that this enables pre-verbal children to communicate their wants and needs, thus reducing frustration for infants and toddlers. Also, research consistently shows that a second language is learned best early and that this aids caregivers in understanding developing language better.

receptive language
Language that can be understood, though perhaps not spoken. Receptive language develops earlier than its counterpart, expressive language.

expressive language
Language that is produced to convey ideas, feelings, thoughts, and so on. Expressive language develops later than its counterpart, receptive language.

Listening is the first step to problem solving.
Monkey Business Images/Shutterstock

Research has shown that sign language actually accelerates communication and learning by encouraging early language development. It may result in a higher IQ of children who use it and promotes greater interest in reading as children mature. Further benefits of sign language include the decreased frustration in children who feel more understood so fewer acts of aggression, provides children with more constructive ways to interact with each other, promotes active learning, and motivates caregivers to watch infants closely.

Most importantly, by putting the messages you receive into words, you open up a conversation. When a child perceives that you are listening and trying to understand, he or she may give you feedback to lead you further toward grasping the message. Once you demonstrate you understand the situation, the child is likely to continue the conversation (verbally or not). Thus, communication occurs. The following list gives examples of various responses that indicate holistic listening:

"I wonder what you need. Maybe a burp?"–said to a screaming baby who has just been fed.

"You're really upset about your mommy leaving."–said to a toddler sobbing at the window.

"You want to touch him."–said to a three-year-old who is reaching toward a man in a wheelchair.

"You want me to pick you up."–said to a mentally challenged four-year-old who lifts his hands to a passing adult.

"You don't want her to touch your coat."–said to a preschooler yelling "You're yucky!" at another preschooler who is stroking his new soft jacket.

"She upset you."–said to a five-year-old who is yelling "You can't come to my birthday party."

"You're really mad at me."–said to a seven-year-old screaming "I hate you."

As you read the preceding responses, did you find that you wanted to do more than just listen and put into words what you heard? Of course. Some of these messages called for action.

Listening and Responding to Different Situations

Receiving a Crying Baby's Message

"I wonder what you need. Maybe a burp?"–said to a screaming baby who has just been fed.

Crying is communication, so making it go away should not be your primary goal. Instead, you need to work at understanding the message the baby is sending and then to respond to the need. Once the message is sent and the need is met, the crying *will* subside. If the baby is hungry, feeding him is the right response; if tired, rest is what he needs. It may be easier to distract the baby with soothing devices, but if you're really listening, you will not use distraction. You will do what is needed and address the root cause, not simply the symptom.

Sometimes you can't find out what the child needs. Or it may be you know very well what he needs–mommy–but she's twenty-five miles away and not coming back until after work. In this case, listening and being emphatic helps.

A listening response is an attitude, not a set of actions. Of course, while you're listening you can try to soothe the baby. Maybe just a change of position will help; in the case of a baby with special needs, a change of position can make a crucial difference. Positioning can be very important in providing for the comfort of children whose neurological or muscle systems work differently from those of other babies.

But if the baby is crying because he desperately wants his parent, any soothing will probably only have temporary effects. You cannot meet his need, but you can continue to listen. Listening to a baby who continues to cry after you've done everything you can means you accept the fact that the baby is expressing a feeling, a desire–*something*. Instead of treating him as an annoyance, treat him with respect as a human being with a need to express himself. Let him know that you're there and receiving his signals.

Stay calm yourself, and periodically talk to him as you would anybody in distress. If you came upon a screaming adult trapped in a car wreck, you would not tell him

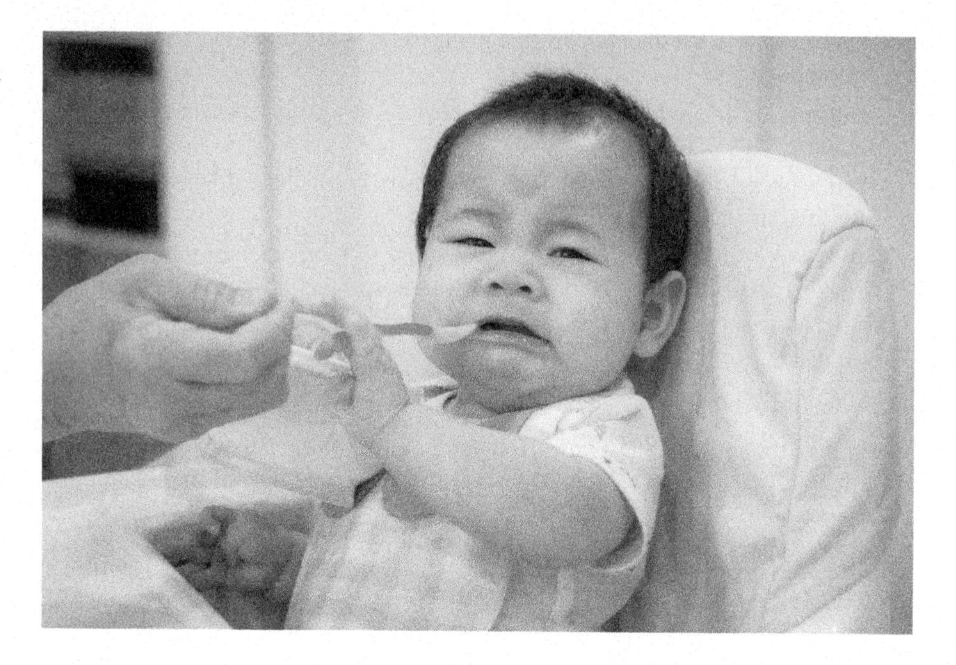

Crying is communication. What do you think this baby is trying to communicate?
Surajet.I/Shutterstock

to be quiet. Think about what you would do. Once you got a grip on yourself, you might ask what he needed from you. You might tell him what you were or were not able to do or suggest what he could do. But if there were nothing either of you could do, you would probably try to reassure or comfort him. If he wanted to talk, you would probably listen. If he continued to scream, you would accept the fact that he has a right to his reaction.

How different is that from the baby in distress? You know the situation is not life threatening, but he does not. Let him know that you accept his feelings and that it is okay for him to express them.

Sometimes the crying baby will "get it out of his system" the way you do by talking to a friend when you are troubled. If the child calms himself on his own, rejoice! Self-calming is one of the most important skills we ever learn–and the earlier the better. Children often learn to calm themselves with a thumb, a fist, or a special object such as a blanket, a soft toy, or an article of a parent's clothing. Imagine the feeling of power to discover you can take yourself out of distress and into a peaceful state with such a small but convenient body part. Now that's empowerment!

Validating the Feelings of a Sobbing Toddler

> "You're really upset about your mommy leaving."–said to a toddler sobbing at the window.

Validating a child's feelings helps her get a sense of reality. She's upset, and you are aware of it. You are not ignoring or distracting her from her feelings; you are accepting them.

It is sometimes hard for an adult to stand by calmly and listen in this kind of situation, especially if it awakens raw feelings in the adult. Who does not have painful memories tucked away somewhere that relate to loss and separation? Children can often trigger an adult's tender areas–the sensitive spots that haven't healed. And it is hard to listen when you feel the child's pain on top of your own. The natural reaction is often to find a way to cut the child off from her pain (usually by distraction) to suppress your own pain. Nevertheless, it is important to recognize when you have unresolved issues that get in the way of being fully available and receptive to children. It is also important to separate the child's feelings from your own.

Once you are aware that you are being influenced by your own feelings, you can then make the decision to switch your attention to the upset toddler sobbing by the window. What the child needs is for you to listen to her and acknowledge that you are aware of her distress. Just being there for her in a supportive way may bring comfort. Being clearheaded and unemotional will help you decide what to do next if just listening is not enough. Perhaps physical affection or a comfort object from home will help. One teacher had children who suffered from separation anxiety dictate letters to their parents, which they then kept in their pockets until they saw them again. The two-year-olds' letters were not sophisticated, but the few sobbed words put on paper seemed to comfort them.

Helping a Screaming Four-Year-Old Find His Own Solutions

> "You don't want her to touch your coat."–said to a preschooler yelling "You're yucky!" at another preschooler who is stroking his new soft jacket.

How would you deal with this scene? Would your response be the same? Try it. It is often quite effective to just state what you think is going on without saying anything else. Such a response often invites the child to say more.

Unfortunately, this kind of scene is usually handled much differently. The more natural responses usually close the communication channels between adult and child. Here are some examples of "conversation stoppers":

- Criticizing: "You always yell about every little thing."
- Giving orders: "Stop that yelling." "Be nice!" "Don't fuss!"
- Discounting: "Come on, that's nothing, she's just touching your coat."

In contrast, if you just say what you think is happening in a matter-of-fact way, without passing judgment as of you are broadcasting, you change the tone of the interaction and encourage the child to say more about what the problem is. Here are some examples of "conversation openers" that encourage children to talk when they are yelling, crying, or complaining about something.

- "You don't want her to push you in the swing."
- "You want to play in the little house by yourself."
- "It hurt when you fell off the tricycle."
- "You don't know where that puzzle piece goes."

Although it's tempting to try to "fix" children's problems, it's more empowering for them to find their own solutions. By doing so, they learn that they are capable of dealing with their frustrations through a variety of problem-solving strategies.

Nevertheless, there will be some situations in which you will have to intervene instead of just letting the children work it out; the box on page 73 illustrates one such situation. The *Tips and Techniques* box on page 74 suggests some prevention strategies.

Helping an Upset Five-Year-Old Put Her Feelings into Words

> "She upset you."–said to a five-year-old who is yelling "You can't come to my birthday party."

One adult tendency is to respond to a situation like this with a lecture or with logic. ("That's not nice to say." or "Your birthday isn't until next year.") What's needed instead is an awareness that the child is using the most powerful non-violent tool she has–rejection. Helping the two children talk it out is usually an effective way to resolve this kind of problem. The adult starts by stating what she perceives the child is feeling:

ADULT: She upset you.

FIRST CHILD: Yes, she took the wagon I was going to use.

ADULT: [*Offering more nonjudgmental feedback*] You didn't like that.

FIRST CHILD: No.

ADULT: [*Getting the two to interact with each other*] Tell her.

FIRST CHILD: I don't want you to have that wagon.

SECOND CHILD: But I had it first.

FIRST CHILD: No you didn't. I did. I just went to get a drink of water.

SECOND CHILD: Well, you can't save it.

FIRST CHILD: I can so.

SECOND CHILD: Who says?

ADULT: [*Realizing the argument is going in circles, offers another nonjudgmental response and a thought-provoking question*] You both want this wagon. I wonder, how can you solve this problem?

FIRST CHILD: She can let me have it.

SECOND CHILD: We can take turns.

FIRST CHILD: Only if I get to pull it first.

SECOND CHILD: And I can ride in it.

Coming up with mutually satisfying solutions is empowering. And children who feel empowered have less need to use threats of rejection to get what they want.

As you can see, a lot of problems can be solved by simply indicating to children you see what is happening and helping them think out loud–or express their feelings. You can do all that just by being a good observer and listener.

An Antibias Response

The following scene calls for more of a response from an adult than simply listening: three boys have made a clubhouse under the climbing structure. Shanti tries to enter, and one boy says loudly, "No girls allowed." The others agree and move to block the entrance.

If they had said, "You can't play with us because you threw sand last time," that would be a different situation. In that case, the adult could listen to the boys nonjudgmentally and help Shanti understand that the problem was her behavior–something she could correct. But Shanti can't change the fact that she's a girl. The boys' behavior reflects out-and-out gender discrimination. In this situation, the adult can't just state what's going on but must intervene. "That isn't fair" would be an appropriate adult response, followed by some discussion about equity.[3]

Let's look at another situation: a four-year-old reacts when another child reaches out to pet the stuffed animal she's carrying in her arms. "You can't touch," she screams. If after careful listening you determine that the child's remark is a response to the other child's race, gender, culture, or ability, you must intervene. If, for example, she says to an African American child, "Your hand is dirty," referring to his skin color, a discussion is in order. Some children equate dark skin color with dirt. If this is the basis for the child's remark, you must clear up her misconception. You must also help her understand how hurtful her remark may have been for the other child.

Not Taking a Seven-Year-Old's Anger Personally

"You're really mad at me."–said to a seven-year-old screaming "I hate you."

"I hate you." Those three little words carry a lot of power. That's why children use them. I've watched adults deal with this situation over the years, and often they respond with a long lecture about brotherly love and the ugliness of hate–a response that only rejects the child's strong feeling.

Why not just acknowledge the child's anger? "You're really mad at me." That's what he's expressing–at least it seems so. By putting his feelings into different words you model a constructive way of expressing anger and, at the same time, validate the child's feelings.

Think for a minute about the strongest way you know to express extreme anger using only words–and none of them obscene. It's hard, isn't it? It may be equally hard for you to name many peers who can express powerful anger using clear, inoffensive words. So why wouldn't a child resort to the one word he knows is strong–hate.

Also, be aware that the child may not be angry at all but is merely trying, for whatever reason, to get a reaction out of you. If that's what you perceive to be happening, say so. You could have a whole conversation about why he needs to stir things up–that is, if you encourage him to talk by using effective listening skills.

TIPS AND TECHNIQUES

Helping Young Children Understand and Appreciate Differences in Skin Color

Early childhood programs have a responsibility to discuss individual differences respectfully and openly. These discussions include the subject of skin color. Racism will live on as long as children continue to grow up thinking that one skin color is better than another. Activities that explore such differences as variations in skin color can help children accept differences. There are a number of ways to help children appreciate different skin colors. Why not make black, brown, and tan play dough once in a while? It's easy to come up with brown; just mix red, blue, yellow, and green until you reach a rich shade of brown. You can also make crayons and paints in various skin tones available to your group. Crayon and paint companies have packaged assortments of colors representing a range of skin tones. As you present these colors, talk about differences and help the children understand that all skin colors are beautiful. Have the children explore their own differences. The child in the picture at right is possibly asking if the picture matches her skin. Having a mirror available helps children to make those comparisons themselves.

To be most effective in helping children appreciate differences and come to view all skin colors as beautiful, teachers must examine their own attitudes and prejudices. One way to do this is to become aware of the way you use language. Do you equate black and dark with evil? Those kinds of images are powerful! They don't belong in early childhood settings. Instead, look for ways to talk about and create positive images for black and other dark colors. This is an area that calls for self-reflection in adults who have trouble being positive about dark colors.

A word of advice about being the target of a child's anger: *don't take it personally!* Often the anger has nothing to do with you. It may be an expression of what's going on inside the child and you're just a convenient target. If you heed this advice, you will be able to deal with an angry child effectively and keep your own feathers from getting ruffled besides.

How to Communicate Clearly

Having been a beginner myself and having watched for many years other beginners operating in all sorts of early childhood classrooms, including family child care homes, I have observed many ways of communicating–some of which cause problems. As adults, each of us has our own natural style of talking to children that says something about who we are and what we've picked up from others, including our own parents. Some of these natural styles of talking produce mixed messages that confuse children and even disconnect them from reality. In this section, we'll explore what to avoid in the name of clear communication.

Ask Real Questions, Not Rhetorical Ones

"Would you like to sit down?" asks the adult at the four-year-old circle time. Such a question is fine if, indeed, the children can choose to sit or stand, stay or leave. But if the adult *means,* "I am going to require you to join the group and sit down," the message may be lost.

"Do you want to wash your hands?" is a fine question if there's no imperative to do so. But if lunch is on the table, the other children are waiting, and the child has no choice, this question would be better phrased as a statement or directive: "It's time to wash hands now; you can play with the blocks again later." An "if, then" statement is another way to say this: "If you wash your hands now, you'll be able to choose what chair you want to sit in for lunch." or, "If you wash your hands now, you can help me set the table." And the ideal is to give a real choice: "Do you want to wash your hands now or play for five more minutes until I call you?" Such token choices allow for children to feel they have some sense of control over what is happening to them and gives them a sense of agency.

Sometimes adults give clear directions but end with a hesitant "okay?": "Wash your hands for lunch, okay?" What the adult wants is to get the child to agree–to comply. However, the child, thinking he has a choice, may say no and then be surprised when the adult becomes irritated.

Validate Feelings and Perceptions Instead of Discounting Them

- "I'm so mad," says the child. "Don't be silly!" says the adult.
- "Ow, that hurts," says the child. "No, it doesn't," says the adult.
- "The soup's hot," says the child. "It's just barely warm," says the adult.
- "I don't like that sandwich," says the child. "Yes, you do" says the adult.

We've all made statements like these. They seem mild enough, but what they really represent is an attempt to control children's sense impressions.[4] **Impression management** keeps children from living in their own reality and creates a false experience of the world. It teaches children to mistrust their senses.

impression management A nonconstructive way of talking to children that discounts their feelings and their sense of reality. For example, a child says, "I don't like that sandwich." In response, the adult says, "Yes, you do." Impression management teaches children to mistrust their senses.

But defining children's reality for them goes beyond just sense perceptions. The child says, "I want to eat outside." The adult responds, "No you don't. It's too cold." A steady diet of such responses can make a child dependent on the adult for understanding his or her own desires. To help children learn to interpret what they want, feel, hear, and perceive, try the following approaches:

- Speak of your own experience and avoid talking in absolutes like "The soup's not hot." If you accept that the child's perceptions may be different from your own, it's better to say, "The soup doesn't feel too hot to me."
- Ask, don't tell: "What don't you like about that sandwich?"
- Validate feelings by repeating what the child tells you or by interpreting his or her meaning: "The soup burned you." "Something about that sandwich bothers you."

"Would you like to sit down for lunch?" is merely a rhetorical question if what you intend to convey is "Lunch is ready and will be served when you are all sitting down at the table."
Ariel Skelley/Blend Images/Getty Images

Living in a sensory world with a full range of perceptual abilities is one of our blessings as human beings. Don't cheat children by trying to convince them their perceptions are wrong. It narrows their experience. Help them learn to talk about what they perceive by accepting their sense impressions instead of imposing your own.

Address Uncomfortable Situations Instead of Ignoring the Obvious

Consider this scene: the guinea pig gets dropped and dies on the spot. The child who dropped it is heartbroken and screams loudly. While one teacher removes the child to another room, the other teacher puts the guinea pig's body in a shoe box, stashes it in a closet, and then puts on a record and starts circle time, carrying on as if nothing had happened.

Here's another example: on a field trip to the post office the children see a woman in a wheelchair, and a little girl asks loudly, "What's wrong with that lady?" The embarrassed teacher ignores the question and the woman in the wheelchair. "Where's the mail slot?" the teacher asks, trying to distract the girl by handing her the letters they brought with them. "Can you find the mail slot and put these letters in it?"

When adults have problems handling awkward and upsetting situations, they often ignore reality. Children then incorporate adults' reactions (such as fear, anger, embarrassment, and silence) into their experience rather than learning more about what makes an issue sensitive. By not being able to ask questions, express their worries, or gather information, children develop misconceptions and learn to question their own perceptions.

Imagine how you would feel if you were at a violin concert and an elephant walked out on the stage. What if no one in the performance or audience acknowledged its presence. Would you wonder what was going on? Would you question your own reality? At the very least you would probably be surprised and confused and wonder if you were crazy. That's how children feel when an event that seems to have some significance is completely ignored.

Adults should clarify what is going on, accept the children's responses, and help them develop a rational understanding of the reality of the situation. Do not ignore the dead guinea pig. Let the children who want to look at it express their sorrow, put words to their questions. If spiritual questions arise, you can say, "You can talk to your parent about . . ." or "Some people believe that . . ."; but you do not have to get into religious matters. Just handling the dead body will help some children grasp the reality of the situation. You might ask the children if they want to create a ritual by burying the guinea pig.[5] Note the children's feelings and accept them by putting them into words: "You're sad about the guinea pig." You can also talk about your own feelings if you want.

A constructive approach to the second example would be to encourage the children to talk to the woman in the wheelchair. Treat her as a fellow human. She may well answer the child's question herself. If she does not, you can point out how she reacted to the question later, explain her feelings to the children, and let them speculate about why she is in the wheelchair. This is also a good opportunity to develop children's competence in relating to people with disabilities. Follow up the chance encounter by inviting a guest to the classroom who has a disability and is comfortable talking about it.[6]

Be Congruent; Avoid Incongruence

When our body language conveys the same message our words convey, that is congruence. When dealing with young children, it is always important to aim for congruence. Yet we have all experienced moments when we felt angry but put on a smiling face–an example of **incongruence.** When adults are incongruent, children get two messages at once and wonder which one to believe. Early childhood educators have to ensure that they are not sending out mixed messages by modeling one thing and saying another. For example, children have been told that they cannot stand on the table, yet the teacher may momentarily forget this rule and stand on the table to put something up on the bulletin board in front of the children. Another example may be a parent who often preaches the importance of honesty to their child, yet when Saturday movie night comes, the parent may say, "Hey kid, pretend you are 10 so you can get the child discount rate."

incongruence A type of mixed message that causes confusion. For example, a person's body language might convey anger while the words contradict the emotion.

Some children are more sensitive to incongruence than others–it depends on whether they pay close attention to both verbal and nonverbal signals. Some ignore the words and pick up mainly the nonverbal messages. They know what the adult is feeling, and those feelings may speak much louder than the words.

Other children take in both the words and the feelings. How confusing for them when the words cloud rather than clarify the meaning! One way some children resolve this dilemma is to ignore the words–a response that may adversely affect both language and social development.

It is important to add a cultural note here: each culture has distinct body language that is not necessarily comprehensible to members of another culture. Learning another culture's nonverbal signals is much like learning a foreign language. For example, a smile has vastly different meanings across cultures (see the *Focus on Diversity* box on the next page). What is considered an appropriate expression of anger also varies among–and even within–cultures, as well as families. Some expressions of anger are so subtle as to be imperceptible to members of other cultures. Some cultures even discourage all forms of expressing anger (see the *Focus on Diversity* box on page 79).

Learning about other cultures' viewpoints and habits is not only interesting, it is also vital for being an effective early childhood educator. In the interest of being a lifelong learner, it is important that you explore, understand, and respect the kaleidoscope of cultural perspectives in our society.

double bind A kind of mixed message that causes confusion. For example, a mother embraces a child and says, "Why don't you go play with the other children?" Her body language says "Stay here with me," but her actual words say the opposite.

distraction A device to keep a child from continuing an action or behavior. Distraction can also be used to take children's minds off of a strong feeling. Distraction works but has side effects as children learn that their energy or feelings are not acceptable to adults who distract them. Distraction is sometimes confused with redirection and may look similar, but distraction is aimed at stopping the energy behind the behavior or the feeling, whereas redirection moves it in a more acceptable direction.

Watch Out for Double-Bind Messages

Like incongruence, a **double bind** is another kind of mixed message. Whenever I hear the term "double bind," I think of something I saw once in an infant program. A mother sat on the floor with her arms wrapped around her child. Her body language said, "Stay here with me." But her actual words were, "Why don't you go play with the other children?" It was not a question, it was a command, and there was no way the child could obey her verbal command without disobeying the nonverbal one that was keeping the child right next to her mother.

Early childhood educators give children another kind of mixed message when they sit them down for story time in a room set up to entice them to play. The teacher says, "Sit still and listen to what's going on." But the physical environment says, "Touch, explore, come try these things out." To take the children out of the double bind, the teacher should put the toys and materials out of sight or move the story time to a less stimulating location. Another example of mixed messages resulting from materials is using materials such as flavored play dough (i.e., bubble gum flavor, chocolate, strawberry, etc.) and asking children not to put the play dough in their mouth nor to taste it.

Use Redirection Instead of Distraction

As mentioned earlier, **distraction** is an ineffective method for responding to children in distress or awkward situations. Yet adults resort to distraction often because it *seems* to work so well.

Consider the following scene: on the "science table" sits a bird's nest with an eggshell in it. It is down low and right by the door so the children can see it when

Focus on Diversity

The Meaning of a Smile

I put a smiling face on a note as a quick way of indicating friendliness. I flash a smile to a passerby for the same reason. I assume that everyone receives the message I intend. But a smile does not mean "friendliness" to everyone.

"When I first came to this country," an Eastern European told me, "I thought Americans were strange because they smiled even when they weren't happy. I couldn't decide if they were all being fake or if they just weren't too intelligent." Likewise Americans who walk along the streets in Eastern Europe sometimes remark on the apparent unfriendliness of the people they encounter. They do not understand that a smile is reserved for expressing joy and that friendliness is displayed in other ways.

Here is what Thanh Binh Duong says about the variety of meanings a smile has for the Vietnamese:

> Almost anyone who has visited Vietnam or come in contact with the Vietnamese has noticed . . . a perpetual and enigmatic smile in all circumstances, unhappy as well as happy. . . . Many foreign teachers in Vietnam have been irritated and frustrated when Vietnamese students smile in what appears to be the wrong time and place. They cannot understand how the students can smile when reprimanded, when not understanding the lessons being explained, and especially when they should have given an answer to the question instead of sitting still and smiling quietly. These teachers often thought the students were not only stupid and disobedient, but insolent as well. One thing they did not understand was that the students often smiled to show their teachers that they did not mind being reprimanded, or that they were indeed stupid for not being able to understand the lesson. Smiling at all times and places is a common characteristic of Vietnamese. There are, however, no guidelines to tell foreigners what meaning each smile represents in each situation . . . the Vietnamese smile may mean almost anything![7]

they come in. A curious toddler goes over to the table and tries to break the shell apart. The teacher removes the child repeatedly and tries to get him interested in something else, but she never explains to him what she is doing or why. That is distraction.

It may seem that distraction and redirection are the same thing, but they are not. Whereas distraction cuts the child off from what he or she is feeling or doing and replaces it with another feeling or activity, redirection acknowledges the child's feelings or energy and helps the child find a related activity that is more acceptable. To redirect the child who wanted to see inside the bird egg, the teacher might have involved the child in peeling a hard-boiled egg. At the very least, the teacher should have acknowledged what the child wanted to do and validated his strong feelings about it.

Here is another example: a child is heartbroken that her mother is getting ready to leave. She cries loudly and runs to cling to her when the teacher intercepts her, whisks her into the air, and says in a loud, jovial voice, "Listen to all the wonderful

Focus on Diversity

Expressing Anger

Many Americans believe that expressing anger verbally and openly is healthy, but not every culture holds that view. Jerome Kagan, a child psychology specialist, says:

> In many cultures–Java, Japan, and China, for example–the importance of maintaining harmonious social relationships, and of adopting a posture of respect for the feelings of elders and of authority, demands that each person not only suppress anger but, in addition, be ready to withhold complete honesty about personal feelings in order to avoid hurting another. This pragmatic view of honesty is regarded as a quality characteristic of the most mature adult and is not given the derogatory labels of insincerity or hypocrisy.[8]

A child who has been taught at home that harmony is more important than saying how one feels still has emotions. Unless you are sensitive to cultural differences, you may miss a particular child's message when she comes to you with what seems to be a very minor complaint–especially if her demeanor appears to be pleasant when she makes her complaint. Even if she is extremely angry or deeply hurt, she may not persist if she is ignored or merely given advice, a quick hug, and sent off to play. "Teacher, he took my doll," says a little girl in a quiet tone, her face not giving away the importance of this event to her. "Well, tell him you want it back or find another one," says a hurried teacher. The child leaves and never mentions again that she is very upset by the loss of the doll. An early childhood educator needs to learn to read each child and to be aware of variations in how children express their feelings.

things we are going to do today." The teacher starts to list the activities, but the child screams louder than she can talk, so she bounces her up and down a few times and walks over to a cupboard. "I have something very special in here for you to play with today," she says dramatically. The teacher sees she has the child's attention, so she continues hamming it up, making funny noises, and whisking her around in the air until they reach the cupboard. The teacher slowly opens the door, reaches in, and finds a feather, which she uses to tickle the child's nose. In the meantime, the mother has slipped out.

It may take a while for the child to remember what she was upset about–or she may not remember at all. But what really happened was that the teacher tricked the child; denied her feelings; and, thus, jeopardized her sense of security, feelings of empowerment, and self-esteem. Distraction not only discounts feelings, it can also set children up for a lifelong pattern of repressing anger, sorrow, fear, and depression. No wonder–they have been taught as children that it is not okay to have feelings.

Here is a more constructive approach to the second example: the child cries loudly and runs to cling to her mother, who is on her way out the door. The teacher stops her by bending down to her level and says in a calm voice, "I know you don't want your mother to leave." The mother blows a kiss, says one more good-bye, and walks out the door. While the child sobs brokenheartedly, the teacher says, "You're really upset." She gently strokes her forehead. The child runs to the window. "Good idea," says the teacher, "Wave to her one more time." The child waves frantically as she sees her mother disappear into the car. Then she throws herself on the floor and begins to kick her feet. The teacher remains close to her without saying anything. The child keeps at it, and the teacher says, "You're really angry right now." Acknowledging her feelings seems to soothe the child. She then gets up off the floor and runs over to investigate the play dough on a table nearby. The single tear running down her cheek is the only sign left of all she has just been through.

If this scene triggered strong feelings in you, reflect on those feelings. In order to effectively support children in emotionally charged situations, it is important that you understand your own feelings and where they come from and that you separate your feelings from the child's. Otherwise the temptation to use distraction will be very great!

Be Sensitive About Questioning Children

Avoid questions that put children on the spot. In some cultures, direct questions are considered rude and intrusive. Even if you come from a culture that tolerates direct questions, you probably understand what it feels like to be interrogated. Like most adults, you may use questions as a device to get to know a child: "Hi, what's your name? How old are you? Where did you get that nice shirt?" The questions themselves are innocent enough, but whether they achieve your goal depends on the child's response. Some children shut up tight when they are questioned. In such cases, it is probably better to find some other way to strike up a conversation.

Direct questions are especially problematic for addressing misbehavior. For example, a three-year-old squeezes a puddle of glue on the table next to his collage project. When the teacher asks, "Why did you do that?" the child is silent. "Don't you know where the glue belongs?" Silence. "What am I going to do with you? Are you going to do that again?" Such questions constitute badgering.

Consider that children often have no explanation for their behavior. Asking, "Why did you do that?" doesn't usually produce a thoughtful reply. Instead, it provokes either a defensive response or utter silence.

Instead of interrogating, the teacher should state the obvious and see how the child responds: "Looks like you enjoyed squeezing that glue." It's the adult's responsibility to correct the situation next time, say, by setting out the glue in something besides a squeeze bottle for a collage project. But perhaps the children are more interested in making glue puddles than pasting down collage bits. If so, the teacher might prepare a thin paste of flour, water, and salt for the children to squirt out paste designs on paper; left to dry, these pasty papers become sparkly pictures. (If you use this suggestion, however, consider that some adults have strong feelings about using food substances—such as flour and salt—as art or play materials. See the *Points of View* box on the next page.)

Let us explore another example: a seven-year-old discovers she can make an interesting effect by spattering paint at the easel. She is having a wonderful time smacking the paint brush and watching the paint fly onto the paper. What she does not notice is that the paint is also going on the floor and even on other children who are passing by her. The teacher arrives on the scene saying, "What in the world do you think you're doing? Do you see the mess you're making? Who do you think is going to have to clean it up? Why are you doing that anyway?"

A better approach would be to redirect instead of firing a bunch of questions. Here are some alternative art projects:

- Instead of using paint and paper, let the child spatter water against a chalkboard.
- Relocate the easel, and let the child continue outside.
- Have the child spatter water onto a piece of paper dusted with tempera paint.
- Using a spray bottle filled with colored water, have the child spray a piece of paper hung from a fence.
- Put leaves or scraps of fabric on a piece of paper and have the child create interesting silhouettes by scraping a paint-filled toothbrush across a screen held or supported above the paper.

Using Observation and Reflection to Improve Communication

Keen observation skills are among the most important tools an early childhood educator can have. Although some people have a talent for seeing what others miss, anyone can improve his or her observation skills through practice. Reflection is a key ingredient of observation. It is not enough to just watch what happens, you must also think about it. Sometimes asking yourself, "what just happened?" will put a whole new light on an incident. You might also ask yourself about cause and effect. What happened that the child reacted that way? A good question to ask when a child's particular behavior is bothering you is, "What is the meaning of this behavior?"

anecdotal record A documentation method that briefly describes an activity, a snatch of conversation, a chant, and so on. Anecdotal records can be based on reflection or written on the spot.

Much of this chapter depends on spur-of-the-moment reactions. Spontaneity is important. You usually don't have time to step back and observe if you are interacting with a child. Thinking has to be lightning quick or the interaction will fall apart. You won't always make just the right decision about the most effective way to communicate, and that's where reflection can help. After something happens, you can rethink it and learn from it. Of course, one of the problems is that you might not remember the incident. That's why **anecdotal records,** written descriptions of an incident or behavior, come in handy. Take a few moments to jot yourself a note

Points of View

What About Using Food for Art?

Is it okay to create a paste out of flour, salt, and water to be used as an art material or perhaps for a sensory experience? What about flour for play dough, rice or seeds for the "sensory table," pudding as finger paint, macaroni for stringing necklaces, or spaghetti for collage work? Here are some of the responses I've collected about using food as an art or play material.

"No. With three-fourths of the world starving, we ought to teach children that food must not be wasted. It's a strong message to the world when American children have so much extra food that they can play with it."

"No. Some of the families in our program don't have enough to eat. It's a shame to send food home glued on pieces of paper."

"No. Waste is a sin."

"Most art and sensory use is okay, but finger painting with pudding confuses children who are being taught not to make messes when they eat."

"I don't see anything wrong with using inexpensive and available materials of all types. With limited budgets, we can't pass up creative opportunities to use a variety of materials. The flour I don't buy for play dough never makes it to starving people anyway."

"I think it's hard to totally avoid using food as a curricular material. Some of our best activities rely on food substances.

People have different opinions about this issue, and each perspective has its own merits. Indeed, you may have your own view on the matter.

when something happens that you want to think more about. You will need paper or pencil or a tape recorder or video. If you are on the floor with children, you will not have time to write much, but afterward, sit down and write out the details of the situation before you forget them. Sometimes it helps to note what happened right before the incident and what happened afterward. That kind of information can be useful in figuring out what exactly happened and how to work with it better next time–if it happens again.

Another use of anecdotes as an observation device is to jot notes about each child. Taking a few minutes out of each day to write something down can prove valuable. Just jot down whatever sticks out in your mind that day. It may be insignificant, or something extraordinary that jumps out at you. All of these unconnected notes can create patterns and answers to questions. Keeping anecdotal records can also tell you when a child is being ignored–for whatever reason: when you sit down to write an anecdote and you cannot think of anything to write about a particular child, that is a clue that you have not been noticing that child. Some children manage to become invisible–perhaps purposely. When that happens, it is important to put special effort into paying attention to and communicating with those children and *making* them visible.

A Story to End With

One day during my second year as a preschool teacher, my four-year-old son visited my classroom for the first time. I was proud of the way I handled some pretty difficult situations using my very best communication skills. I noticed my son watching me several times with a strange expression on his face. Finally the day ended; the cleaning was done and the furniture was stored for the Sunday school class that would use the room on the weekend.

I locked the door behind us, and we started up the stairs to the parking lot when my son stopped. "Mommy," he said, "I want to ask you a question." "Sure," I replied, "Shoot." "How come you were talking so funny to those kids?" I realized at that moment that my teacher self was using my new and improved communication skills, but that I had not tried them out at home yet.

Summary

The relationship of warm, responsive care to healthy brain development links feelings to learning and cognition. That link is one of the important findings of brain research. This chapter is about using communication skills to form relationships with children. To communicate effectively with children, the most important thing you can do is to listen–to both verbal and nonverbal cues–and to let them know you hear their message. Communication begins when you validate children's feelings (rather than deny or judge them) and help them put their problems into words so they can find their own solutions. By being respectful and communicating effectively, the adult can promote children's self-esteem, empower them, and support their psychological well-being in general.

Some adults have a style of talking to children that can harm a child's growing sense of self and his or her perceptions of reality and feelings of security. To communicate clearly with children, avoid rhetorical questions or other questions that put children on the defensive or make them uncomfortable. Do not discount their feelings and perceptions, and do not avoid addressing awkward or uncomfortable situations; avoiding awkward situations can teach children to mistrust their own impressions. Do not send incongruent or mixed messages, which confuse children.

Reflection Questions

1. Think of a time when you were in an emotional state and someone really listened to you. How did that person let you know that you were heard? Can you use your memories of that experience to help you listen to children?
2. How do you feel when children are expressing feelings of separation? Can you remember a time when you had those feelings? When was it? How did you feel? Does reflecting on your own experience help you respond to a child who is having those same feelings?
3. How do you feel about being told to give a nonjudgmental response that merely reflects a child's feelings rather than responding with a lecture or logic in the face of a conflict?

4. Think of a time when someone did something just to get a reaction out of you. How did you respond? If it had been a child, might you have responded differently? What do you think the effects of your response would have been?

5. Can you think of a time when someone discounted or ignored your feelings and/or perceptions or defined your reality for you? How did that feel? How did that affect you? Can you use your own experience to help you work with children?

TERMS TO KNOW

How many of the following words can you use in a sentence? Do you know what they mean?

holistic listening 66
self-esteem 66
receptive language 67
expressive language 67
impression management 75
incongruence 77
double bind 78
distraction 78
anecdotal records 82

FOR FURTHER READING

Brault, L., and Brault, T. (2005). *Children with Challenging Behavior: Strategies for Reflective Thinking.* Phoenix, AZ: CPG Publishing.

Derman-Sparks, L., and Edwards, J. O. (2010). *Antibias Education for Young Children and for Ourselves.* Washington, DC: National Association for the Education of Young Children.

Epstein, A. S. (2009, January/February). Think before you (inter)act: What it means to be an intentional teacher. *Exchange, 31* (1), 46–49.

Espinosa, L. (2015). *Getting It Right for Young Children from Diverse Backgrounds: Applying Research to Improve Practice with a Focus on Dual Language Learners* (2nd ed.). Upper Saddle River, NJ: Pearson.

Friend, M. D., and Bursuck W. D. (2019). *Including Students with Special Needs: A Practical Guide for Classroom Teachers* (8th ed.). Upper Saddle River, NJ: Pearson.

Jacobson, T. (2008). *Don't Get So Upset! Help Young Children Manage Their Feelings by Understanding Your Own.* St. Paul, MN: Redleaf Press.

Jalongo, M. R. (2008). *Learning to Listen, Listening to Learn: Building Essential Skills in Young Children.* Washington, DC: National Association for the Education of Young Children.

Kovach, B., Patrick, S., and Briley, L. (2012). *Being with Infants and Toddlers.* Tulsa, OK: Laura Briley.

Mindess, M., Chen, M., and Brenner, R. (2008, November). Of primary interest: Social-Emotional learning in the primary curriculum. *Young Children, 63* (6), 56–67.

Reynolds, E. (2006). *Guiding Young Children: A Problem Solving Approach* (4th ed.). New York: McGraw-Hill.

Facilitating Young Children's Work and Play

IT Stock Free/Alamy Stock Images

In This Chapter You Will Discover

- why the adult should not be the "star of the show."
- how play relates to cognition.
- how the physical act of playing connects to matters of the mind.
- how pretend play shows the child can create mental symbols.
- why some children need help to learn to play.
- how to make play accessible to all children, including those with varying abilities.
- how work is also a medium for learning.
- how adult attitudes toward work affect children.
- what is child-centered learning.
- why the "project approach" is a way of developing cognition.
- how adults can facilitate children's work and play.
- what the difference is between encouragement and praise.
- how you can use praise appropriately.

NAEYC Program Standards

Program Standards 2 and 3: Curriculum and Teaching

This chapter is about how play enhances development in all areas–physical, social, emotional, and cognitive. Young children learn as they play, though it may not look like the kind of learning that goes on in educational programs for older children. In fact, some adults visit early childhood programs and complain, "All they do here is play. Don't they ever learn anything?" That kind of remark demonstrates a misunderstanding of how young children learn. They learn while they play. They can also learn while they work, but their "work" may not look like traditional academic "school work."

The program looks different, and the adult's role looks different too. Early childhood educators have numerous roles that change according to the children's needs. The adult obviously has a prominent part in the overall scheme of the program, but does that mean he or she is the "star of the show"?

Who Is in the Spotlight—Adult or Children?

Some people who enter the early childhood field hope to emulate the teachers they remember from their elementary school years: they will stand in front of a class and impart knowledge or go out on the playground and lead games such as kickball and freeze tag. Some envision themselves as orchestra leaders, directing the group action to create counterpoint, harmony, texture–all the elements that combine to make beautiful music.

The problem with these two models is that the spotlight is on the teacher. But in early childhood programs, as you will soon discover, the stars of the show are the children.

To some extent, a classroom or family child care home *is* a little like an orchestra, but as the teacher or provider you will spend very little time each day standing behind a podium. Instead, you will be busy in all the roles it takes to create the music–both up front and behind the scenes and right down to setting up the chairs and getting the instruments out of their cases.

When the children arrive and the music starts, you might not even get the chance to orchestrate the group; the group might orchestrate itself, or perhaps a child will take that role. Edward Hall, an anthropologist, wrote a startling description of how a single child took over a whole playground! One of Hall's students made a film of children in a school yard during recess. At regular speed, the film showed an ordinary playground scene with children involved in separate activities throughout the yard. But upon closer examination, at different speeds, it became apparent that:

> one very active little girl seemed to stand out from the rest. She was all over the place. . . . Whenever she was near a cluster of children the members of that group were in sync not only with each other but with her. Many viewings later, he [the student] realized that this girl, with her skipping and dancing and twirling, was actually orchestrating movements of the entire playground! There was something about the pattern of movement which translated into a beat–like a silent movie of people dancing.[1]

I have my own example of such a phenomenon. I was once involved in the production of a videotape of a group of children making tortillas. When I first viewed the tape, the scene appeared rather chaotic. Some of the children were pounding corn with mortars and pestles; one was using an eggbeater to mix up the ground corn in a bowl. Others were kneading dough, and still others were shaping masa balls and putting them on the tortilla presses. It was a noisy scene with the sounds of the activities and the chatter of the children. Two teachers were facilitating what was going on but not obviously directing the action.

The more I watched the tape, however, I realized that the rhythmic pounding of the pestles related to the rhythm of the group. It was not at all clear to me who was orchestrating, if indeed it was only one person; perhaps in subtle ways the teachers were taking that role. I needed more study to come to any definite conclusions; however, what was clear was that the group rhythm influenced every single person there. It felt good to watch this activity; I got caught up in rhythm myself!

The Teacher as Director and Star

Contrast the following scene with the two that have just been described: a student of mine worked in an infant-toddler program set up for teen parents. I observed this student regularly and was there on the morning she was told to "keep the kids on the rug" while the rest of the staff got the van packed for a field trip. Picture this scene: a 9- by 12-foot rug covered with squirming, crawling, and toddling babies in the middle of a playroom surrounded by toy-filled shelves along the walls. From the word go, this young woman's assignment was virtually impossible.

Giving it her best try, she assumed the role of a "real" teacher. She began circle time to keep the children within the confines of the rug; nevertheless, the children began to escape within the first few seconds. With the help of other staff members, the children were carried and enticed back to the rug. This time she tried both herding and entertaining–but to no avail. She soon discovered that keeping a group of children that age together is like trying to keep popcorn inside a lidless popper.

My student continued to need help from the other staff members, who gave it to her grudgingly, complaining all the while that she was delaying them in their preparations for the field trip. It took forever before the group was finally ready to get into the van.

My student learned a good lesson that day: that traditional teacher skills can't keep the attention of that young or varied an age group for very long. Hopefully the other staff members learned a lesson too: to prepare *before* the children arrive.

Perhaps if the scene had been filmed there might have been a rhythm, just not one any adult would try to orchestrate. And although it's hard to say what the children learned during that period, it certainly wasn't what the adults had in mind. Remember: the teaching-learning process is always going on, every minute the children are present.

The Teacher as Responder, Protector, and Facilitator

My student in the preceding scene was trying to be entertainer and herder at the same time. But if she hadn't been under such specific, unrealistic orders, she could have taken on some different and more appropriate roles. Take a look at the scene again as it might have been.

The room is set up for play, and the children are not confined to the rug in the middle of the room. The caregiver [my student] sits on one corner of the rug watching the children actively engaged with the toys and each other. She is alert and ready to intervene if necessary, but at the moment there is no need. A nine-month-old crawls over to her, grabs her blouse, and pulls himself to a standing position. The caregiver puts an arm around him and gives a little hug, being careful not to put him off balance. She smiles and says, "Hunter, look how you're standing up." The child smiles back, makes a sound deep in his throat, and pats her face. Then he reaches for the caregiver's glasses. "Not my glasses," she says, turning her head out of range of the child's grasping fingers.

And so it goes, the caregiver supervising the group of busy youngsters while being appropriately responsive to the children as necessary. She moves to the other side of the room, ready to intervene when a struggle erupts over a toy; however, the struggle is over in an instant with no adult help needed. The scene continues until the field trip preparations are finally ready. Quite a contrast to the first run of this scene!

The first scene wouldn't have been so bad if the children had been older. Four-year-olds can certainly be trained or enticed to sit as a group on a rug, though with toys in sight all around them, it would still be difficult for some to focus their attention on the adult no matter how entertaining he or she was. School-age children are more capable as a group of paying attention to an adult, but even they get restless if asked to sit for very long (some adults have the same problem).

Of course, children of any age will sit longer if they are entertained. The question is: "Should the early childhood educator's main role be to entertain the children? Do children learn best when they are sitting in an audience? Is learning something fun that adults dish out?" The answer to all these questions is no. Learning in the early years is an interactive process–between the child and the environment, the child and another child (or group of children), and the child and an adult. This kind of learning is sometimes referred to as **child-centered (or child-directed) learning.**

child-centered (or child-directed) learning A teaching-learning process in which the child learns from interacting with the environment, other children, and adults. This type of learning contrasts with a classroom in which the educator's main role is to teach specific subject matter or formal lessons.

PLAY, COGNITION, AND LEARNING

Play provides for a varied and rich medium for learning. Infants and toddlers learn best through free-choice, play-based experiences in the context of close relationships. Watch a group of toddlers slipping down a little slide, wrestling and rolling playfully on the rug, and then climbing back up the slide. They are interacting with the environment and with each other. They are gaining physical skills as well as social skills as they engage in rough-and-tumble play and yet not hurt each other.

Observe preschoolers building a huge wall of blocks around a menagerie of plastic farm animals. Listen to them plan what they are going to do next; tune in to their self-talk and their dialogues with their building partners. See the give-and-take as two

Play is not always lighthearted fun. Sometimes deep concentration is required.
sbworld8/123RF

children go for the same block. Imagine what they are learning: physical skills, size and shape discrimination, basic physics, social and language skills, and so on. Watch as these same children climb into the structure they have built and begin to move the animals around in a make-believe fashion, explaining to each other what they are doing. Subsequently, children's play is valued as children's work. Play is children's primary mode to understanding their world. It is their primary means of relating to others, developing concepts, and understanding adult roles through imitation. When we empower children to be responsible for their own learning, we foster independence, competence, and self-esteem. Providing children with the opportunity to be exactly who they need to be at any given moments helps develop a positive sense of self. Further, when children's needs and ideas are validated and given significance through space and time, children develop a strong sense of self-worth and confidence.

The learning processes infants and toddlers use in their early life are similar to processes human beings use to construct knowledge throughout life. School-age children creating a play are also learning as they make costumes, write the script, and assign parts. They are gaining mental, physical, social, and emotional skills. They too are interacting with their environment and with each other.

The advantages of learning through play abound and they need to be recognized for older children as well as younger ones. For many years, the NAEYC has advocated play for children as a **developmentally appropriate practice (DAP)**–that is, a practice that is specifically appropriate for a child's stage of development. According to educator Gaye Gronlund, in her article "Bringing the DAP Messages to Kindergarten and Primary Teachers,"[2] it's important for kindergarten and primary teachers to recognize how DAP is a practice that applies to the education of all children, including those aged five to eight. Developmentally appropriate practice includes

developmentally appropriate practice (DAP) A set of practices that directly relates to a child's stage of development as defined by such theorists as Piaget and Erikson.

THE THEORY BEHIND THE PRACTICE

Piaget on Play

Jean Piaget saw play as vitally related to cognitive development and an important way for children to learn. When children imitate what they see, such as driving a car or cooking dinner on a play stove, they show that they are beginning to understand how such things work. Engaging in what is called symbolic play like using a block for a cell phone helps them construct knowledge and make sense of the world. It is said that Piaget was fond of the expression "construction is superior to instruction." Any program that stresses play as a major way of learning owes a debt to Piaget's work.

Carol Garhart Mooney, *Theories of Childhood* (St. Paul, MN: Redleaf, 2000).

figuring out ways for children to be actively engaged when learning–engaged with objects, ideas, their teacher, and their peers. Children learn by doing, so creating active learning environments should be the focus of all early childhood educators, including kindergarten and primary school teachers.

meaning-making A practice through which children construct knowledge by finding meaning in their experience.

Play is educational for young children, including school-age children. One of the advantages of play is that it helps children practice **meaning-making.** This means that they learn or build knowledge by finding meaning in their experience instead of by passively receiving information from their teachers. Play lends itself to a variety of questions, answers, and conclusions rather than one correct answer, as is often the case when the teacher is the only one who asks the questions. Through play as purposeful engagement, children answer questions that they themselves ask. See *The Theory Behind the Practice* box above for how Piaget's work relates to play.

Using play as an important learning activity also helps children manage themselves, instead of depending on the teacher. Children at play take risks and gain a feeling of power. Play is far less likely to result in failure than many more traditional "school-like" activities, such as structured lessons and worksheets. Thus, children gain confidence and increase their self-esteem through play.

So what do teachers do if they aren't instructing, giving lessons, or teaching in the traditional sense? They set up the environment, work on relationships with the children, and spend some time stepping back to observe what's happening when the children are playing. When appropriate, the adults use what's called *reflective practice,* which includes spending a good deal of time really tuning in to what's going on and even wondering about it. The point is to figure out just when and how to respond in the most appropriate way. It may be that they will discover that a response is not necessary. The benefits of reflective practice are that adults play close attention and are sensitive to what the children need and what might be the next step in furthering their learning. Observing, stepping back, reflecting, and figuring if a response will help or not is all much harder than just instructing, teaching, and showing children how to do things. But the benefits are much greater!

Moving, Talking, and Learning

I keep reading in the newspaper that preschool and kindergarten aren't play-based any more, the way they used to be. They're all about drilling young children on the alphabet and phonics, teaching them to sit at tables and fill out worksheets. But the young children I have known are wigglier than that. They learn best when their bodies are moving and their mouths are talking. What happens to the ones who can't sit still or haven't had time to play with all the possibilities first? It's silly to say they aren't "ready to learn"; they've been learning since they were born.

When I visit early childhood programs—one of my favorite things to do—I'm delighted whenever I discover teachers with creative ideas for teaching literacy. The other day I was observing Thelma and Max.

Thelma keeps a little pad of paper with a pencil in her pocket and pulls it out to write on whenever she sees a play activity she wants to remember. Sometimes she draws a quick picture of a block structure or a child's drawing. "What you doing?" four-year-old Shanda asks her. "I'm writing about children playing," Thelma explains. "Can I do that too?" asks Shanda. "Me too!" says Paulo. Fortunately, there's a pile of little note pads in the cupboard. And soon there are four observers—one tall, three short—writing and drawing notes to be shared at circle time.

Max doesn't like little notepads; he likes clipboards. They're too big to carry around all the time, so he leaves them here and there in the classroom, each with a list of the children in the class and an attached pen. Any adult can grab one to contribute to the growing collection of notes about every child's accomplishments. And by popular demand, there are clipboards and paper available in the writing center too, for children's use.

What happens at circle time? There's singing, with words and movements that give everyone practice in remembering patterns of pitch and rhythm and story and rhyme.

> Bought me a cat,
>
> My cat pleased me,
>
> Fed my cat
>
> Under yonders tree.
>
> Cat went fiddle-I-fee—fiddle-I-fee.

There's a story book, with pictures and words between its covers. Max reads that aloud, roaring at the wild things, and rowing his boat as he sails away for a year and a day. The children roar and row too. Many of them can read the pictures and tell the story to each other.

Then it's Thelma's turn to tell a story. She pulls her notepad out of her pocket and leafs through its pages, looking at what she wrote down during playtime this morning. "Know what I saw?" she asks. "I saw Terryl building a tall, tall block tower. It looked like this." (She holds up a tiny drawing.)

"I can't see!" is heard from all sides.

"Oh, let me draw a bigger picture," she says. Moving to an easel, she copies her drawing on the easel paper with a fat marker as the children watch, absorbed. "He started at the bottom with these blocks—and then he put three more blocks on top, like this. . . ."

"Know what else I saw? I saw Marisol and Ana baking cookies in the playhouse. Were they good? Who ate them? Did you, Stevie? Were they yummy? Know what else happened? Shanda and Paulo and Aaron saw ME writing on my notepad and so THEY wanted notepads to write on and pencils to write with. And all four of us were writing and drawing about what we saw." Sometimes teachers are *observers*—they take notes while children play. Sometimes children are observers too!

—Elizabeth Jones

Elizabeth Jones, who has a talent for reflective practice and is a well-known advocate for play and faculty member at Pacific Oaks College, reports two teachers supporting play and enhancing children's literacy skills at the same time in the *Voices of Experience* story, says the teacher's job is to observe, act as a resource, interact, and intervene and mediate when necessary. While talking to children at play, the teacher helps them define the purpose of the play and encourages individuals to add their own creative ideas and pursue areas of interest. Assessment of the children and what they are learning is also part of the teacher role as he or she observes the children. A talk between the teacher and the children after their play helps both the teacher and the children with assessing what has gone on.

Concept development grows out of rich play as children use their creative and investigative opportunities to examine important issues. What they learn and how they learn depends on the age of the child. In play, children of all ages can increase knowledge and skills by using manipulative materials and by exploring creative activities. Such activities and materials can help school-age children learn the concepts that are part of the curriculum framework.

Through play, children also practice their newly found knowledge and understanding of the world. Part of a teacher's job is to add complexity to the play and challenge children to try their new knowledge in different situations. Not all children will tackle such challenges on their own, so sometimes a teacher must help or encourage them to explore different ways to experience their knowledge.

One aspect of play that is getting more attention than in the past is the growing lack of outdoor play—not on asphalt playgrounds as during recess, but in nature. The problem today, as Richard Louv has named it, is "nature deficit disorder," and he is very clear that it must be addressed! He puts forth many good arguments in his book *The Last Child in the Woods* and he commands huge crowds and lots of attention when he gives talks at various early childhood education conferences. Children now spend much more time indoors and when outdoors they are on cement and asphalt. As they grow into school age, outdoor time is recess—10-minute intervals to run around and burn off energy so they can get back to the serious work of learning. If Richard Louv has his way, all that will change. As Linda Gillespie explains in "It's Never 'Just Play'!," play is the most developmentally appropriate, meaningful method for children to learn and make sense of their world. In doing so, children begin to seed the 6 C skills that support successful and productive cognitive growth in both childhood and adulthood: collaboration, communication, content, creative innovation, critical thinking, and confidence.[3]

Focus on Inclusion: Making Play Available and Appropriate for All Children

Children with disabilities or particular challenges may need more support and assistance than other children who know how to make use of a rich environment. For *all* children activities and materials must be carefully selected to be appropriate for the child's age and/or developmental level. You cannot go by age alone because for some children age and stage do not coincide. You may need to adapt some toys and materials for a particular child who cannot seem to get involved. Or another child who is roaming listlessly around the room may need more challenges than the environment offers. If some activities such as circle time involve only talking or reading aloud to the children, those with language delays or attention deficits will lose interest. If a child with a visual impairment does not explore, he or she may need some simple lessons about how to move around the room. Then, of course, the room arrangement must remain the same, once the child gains knowledge of it. If children with hearing impairments seem unable to engage with toys, materials, or other children, consider that the noise level in the room or play yard may be too high and chaotic for the child to make sense of what is going on. All of these considerations are important if all children are to play and learn in the program.

NAEYC Position Statement

Position on Inclusion

Here are some simple suggestions for modifications in the environment. A tabletop easel can work for a child who does not have the strength or ability to stand. For a child who uses a wheelchair and is too high to be on the level of other children playing on the floor, a beanbag chair may work. The child who has problems with activities that require more than one step, such as collage, can be paired with another child who can encourage through modeling and commenting.

Some children need to be taught how to play. Reflective practice can help the adult know what to do. For example, rather than giving specific instructions, such as do this and do that, it may work better to observe the child and comment on what he is doing. Speculating on what the child might do next can serve as a simple

These children playing "row your boat" are including their friend who has special needs.
Jacobs Stock Photography/Photodisc/Getty Images

suggestion designed to help him go further. "Oh, I see you put the man on the truck. You're looking at the block, are you going to put that on the truck too? What else can you put on the truck?"

You can judge how effective your play areas support inclusion by observing how the children use the materials. Are all children using them appropriately and creatively to the best of their abilities? Are all the children able to choose, access, and use materials as independently as their abilities allow? Do all children become involved and stay with what they are doing for a period of time? Some, of course, will persist much longer than others. The degree of engagement and the length of time it lasts can vary widely from child to child. How do children show they are playing successfully? It takes good observation skills to determine who is feeling successful and who is not. Facial expressions alone won't work because some children have little control over their muscles and a smile may appear to be a grimace. You have to learn the individual ways children show their feelings to decide if they are feeling successful or not.

Although what was said above about play and the teacher role in play applies to preschoolers as well as school-aged children, that does not mean that the two age groups are the same. Somewhere between the age of five and seven, a developmental shift takes place that makes a big difference in how the child thinks and acts. Theorists like Piaget, Freud, and Erikson tell us about this shift, but we do not need theorists to point out these noticable changes! Around the world, cultures that have never heard of those theorists also recognize the shift. For some there is a rite of passage as the child enters a new, more grown-up phase of life. In the United States, the rite of passage could be considered the move from preschool to what some consider "real school."

You can judge how effective a play area is by watching to see how involved the children are and how the play materials are being used.

One difference between children before the five-to-seven shift lies in the ability of a child to comprehend abstractions. Whereas the preschool child may have problems with abstractions, the school-age child is beginning to handle them. Consider the following example, a discussion between a father and his two daughters about the concept of freedom. The eight-year-old asked questions that indicated she had some understanding of the concept, while the preschooler just listened. At the end of the discussion, the father turned to the younger child and asked, "Well, did you understand what I was saying?" The child smiled broadly, held up three fingers, and said with confidence, "Yes. I'm free! And after my birfday I'm going to be four."

Playing to Get Smart

Betty Jones and Renatta Cooper, professors at Pacific Oaks College, used the term "Playing to Get Smart" as the title for their 2006 book. They point out that play gives children practice in choosing, doing, and problem solving. The whole time they are playing, children are thinking, creating, and negotiating. Even though playing make-believe may look as if children are only using their physical or social capabilities, nevertheless, children are developing cognitive skills as they learn to deal intelligently with the world. "Smart" according to Jones and Cooper is not about memorizing facts or learning the skills needed to pass a one-size-fits-all test. In the world of early childhood education, being smart means being skillful in curiosity and critical thinking by, for example, inventing classification systems and solving problems through interactions with others. Children at play are learning to like new things as they encounter the unexpected and acquiring the dispositions and skills to live in an ever-changing world. Who knows what it will be like when they grow up? Play is a good way to prepare them for the unknown.

Is Play Always Fun?

Play can be a lot of fun, but it is not always. I have seen children work up quite a sweat on the jungle gym, bite a lip as they intensely try to balance on the edge of a sandbox, or show signs of stress as they glue together a scrap-wood sculpture that just will not remain upright. These children are playing and learning, and although they may derive pleasure or satisfaction from their efforts, they are not smiling and laughing while doing it.

I remember watching a video of one of my sons digging a hole in the play yard of his preschool. He was sweating and straining. I could not remember ever seeing him put that much effort into anything before that. When I remarked to him later about how hard he was working on that hole, he looked up at me beaming and said, "That's the best thing I ever did in preschool." I did not take that as a criticism of the school but as an indication of how intense involvement in play can give children deep satisfaction. Playing is the child's single most beneficial activity in the early childhood program.

How Does Play Differ from Work?

Was my son really playing when he was straining to dig that hole? It looked like work, but what made it play? Five characteristics distinguish play from work:

Active engagement

Intrinsic motivation

Attention to the means rather than the ends

Nonliteral behavior

Freedom from external rules[4]

For my son, digging the hole was play because it entailed all of the five preceding characteristics. First, my son was obviously actively engaged; he could not have produced the hole without actively putting shovel to dirt. Second, he was intrinsically motivated; no one made him dig the hole, nor did anyone pay him to do it. Third, the act of digging, not the resulting size of the hole, was what my son was focused on; if the hole had had to meet certain specifications, then the digging would have constituted work rather than play. My son, who is very process oriented, made it obvious when he was a young child that the fun was in the doing, not in producing something notable. He filled every hole he ever dug and knocked over every sand castle he ever built, even the ones I suggested he save for a while. In fact, one day he came home from second grade with his papier-mìaché science-project volcano and asked me to help "erupt" it. We erupted it all right. I let him touch a match to it, and the volcano–with the blue ribbon still attached–went up in a cloud of smoke. Where the proud volcano had once stood was now only a handful of ashes. My son proved to me that for him process was a lot more important than product, even at an age by which many children have become product oriented in some areas of their lives.

Finally, it is difficult to say for sure how the fourth and fifth characteristics–nonliteral behavior and freedom from external rules–applied without being able to tune in to what my son was thinking while he was digging his hole. But, knowing him, he probably was a pirate digging for treasure. Imagination takes what could be a purely physical activity and turns it into a playful one.

Stages of Play–Infants and Toddlers

The stages of play in young infants and toddlers emerge quite organically. First, early practice play with objects begins at around four months of age and peaks around nine to thirty-six months of age. This type of self-initiated play is the primary form of play during this period of life.

Also, at around two months of age, you begin to see play around social routines. Around this time, the same time that their social smile first develops, infants begin to respond to turn-taking, and by seven to nine months of age, infants invite play.

Finally, infants begin to engage in various types of early pretend play or symbolic play. This type of play, which will be discussed in more depth shortly, is a kind of play in which objects, actions, and social roles are transformed from literal to nonliteral states. Simple pretend play schemes are present in most children by eighteen months of age. Young infants need realistic objects, but as children develop, they are able to initiate pretend play with more abstract or imaginary objects.

The next section will delve into more depth on the different types of play that emerge during these early stages of life.

Types of Play—Cognitive and Social

Try an exercise. Think back to your own childhood to a time when you were playing. What were you doing? How were you feeling? What were you thinking about? Take a few minutes to relive that time. Try to be yourself as a young child playing. Come as close to *reexperiencing* this event as possible. The chances are the type of playing you remembered constituted either **sensorimotor play,** which focuses more on the body, but with direct connections to the mind, or **symbolic play,** which involves imagination and thought processes. Both these types of play are considered cognitive play.

sensorimotor play A form of play that involves exploring, manipulating, using movement, and experiencing the senses. It is sometimes called "practice play" or "functional play." In sensorimotor play, the child interacts with his or her environment using both objects and other people.

Sensorimotor Play. According to Monighan-Nourot, sensorimotor play starts with the infant interacting with his or her environment using both objects and humans. Sensorimotor play involves making things happen and imitating, for example, holding a seashell to one's ear to hear the sounds of waves. For older children, examples of sensorimotor play include fiddling with things, doodling with pencils, roughhousing, and playing pointless chase games.[5]

"Fiddling" is a term that describes sensorimotor play, but it is not the best term because it suggests that this type of play is aimless activity. It may seem that children engaged in sensorimotor play are doing nothing, but in reality the sensory discoveries they are making contribute to language and cognitive development in important ways. Although sensorimotor play seems to occur only on a physical level, one never knows what is going on inside the child's imagination. Examples of sensorimotor play are:

symbolic play A form of play that uses one thing to stand for another and shows the person's ability to create mental images. Three types of symbolic play are dramatic play, constructive play, and playing games with rules.

An infant fingering a string of large plastic beads

A toddler playing in the sink while washing his hands

A preschooler swinging on a tire swing

A kindergartner playing chase

A school-age child doodling on a piece of paper

Symbolic Play. Another category of play, symbolic play, involves make-believe. There are three kinds of symbolic play: dramatic play, constructive play, and games with rules.

Think again about your childhood play experience. Were you pretending something? Were you building or creating something? Or were you playing a game?

In symbolic play, children use objects to pretend, or they take on roles themselves. Dramatic play is a common type of symbolic play. You can visit most any preschool

Sensorimotor play contributes to language and cognitive development in important ways.

and see children in the dramatic-play corner acting out various parts: "I'll be the mommy. You be the baby." Of course, the dramatic-play area does not have to be a "house." According to Vivian Paley, notable author of many books on children, children can easily play rocket ship in what may be designated the "doll corner."[6]

Symbolic play has many important functions in a child's life. For example, empathy may be too abstract a concept for young children; "How do you think he feels when you hit him?" doesn't always work for a two-year-old. It takes many years of living plus some dramatic-play experiences for some children to put themselves into someone else's shoes. Playing make-believe is how they practice.

Constructive play, such as building structures with blocks, Lincoln logs, toothpicks, and so on, involves symbolic thinking and even fantasy, as children create an image in their minds of what they are building. Building with blocks is a slightly different form of constructive play than using them as props for fantasy play, such as using a block as a microphone.

Play dough and paint are other materials children often use in a form of constructive play that involves symbolic representation. When a child sits and pounds a wad of play dough for the enjoyment of the body movement and the feel of her hand hitting dough, that is sensorimotor play. When she makes a play dough birthday cake for her teachers, she is engaged in constructive play.

Games with rules require an understanding of predetermined structures and an ability to use strategy. Group or circle games like ring-around-the-rosy, a la vibora de la mar, and duck-duck-goose are appropriate for children who do not yet use abstract thinking but are able to mimic what others are doing. Older children are able to play a variety of games with complex rules and strategies, including card games, board games, and sports.

Games may be competitive or cooperative, depending on the teacher and his or her culture and approach to children's games. However, most early childhood educators frown on competition of any sort with young children. They worry that focusing on winning creates losers. Young children are still in the process of discovering who they are and developing a sense of their own self-worth; competition works against rather than enhances this process for most young children.

If you have never seen a game played noncompetitively, you may be surprised how easy it is. For example, children playing bingo or lotto do not have to know that there is a way to win. There's no need to stress who finishes first. If the object is to finish (rather than to win), all the children can finish. It does not matter if some finish sooner than others. Most young children are not born competitors–they have to be taught. And there's plenty of time for them to learn (if indeed they ever need to).

Social Play: Solitary, Parallel, Associative, and Cooperative Play. Think again about your own childhood play experience. Who was there with you? Perhaps you were playing all by yourself, an activity called **solitary play.** Children of all ages engage in solitary play, though it is often thought of as more characteristic of infants and toddlers. But even for older children, solitary play has many benefits: some children need to be by themselves to explore their own thoughts and feelings; some need to get away from excess stimulation to focus; and some just like their own company.

solitary play A form of play in which a child plays alone even though other children may be present.

Have you ever seen a child playing alone and at the same time watching other children? Although you may worry that the child has been left out, assess the situation carefully before urging the child to participate. Each child has his or her own timetable for entering group play. Some children are content to spend time observing other children, and, indeed, they learn a good deal by watching. When they are ready, they will join in. Observation is an important skill to learn. Encourage it!

If in your childhood play experience you recalled there being one or several children playing nearby, it's possible that you were engaged in **parallel play.** Parallel play is common among toddlers but occurs at all ages. To illustrate parallel play, imagine two-year-olds playing in a sandbox. One is pouring sand and carrying on a monologue about making cakes. The other is running a dump truck over a hill, making engine noises. If the children are engaged in parallel play, the first child's monologue might incorporate a truck theme now and then, yet neither child will directly acknowledge the other child's play. In fact, if they intrude on each other, the play ceases altogether.

parallel play A form of play in which two or several children are playing by themselves but within close proximity of each other. Each child's play may be influenced by what another child is doing or saying, but there is no direct interaction or acknowledgment of the other child.

Children of all ages engage in parallel play. Take, for example, two older children sitting side-by-side at computers. During parallel play, children are close to each other, influenced by what each other is doing, but not directly interacting. If you watch slightly older children in parallel play, you may discover that they coordinate with each other, even to the extent of imitating each other's gestures.

Playing next to but not *with* another person is some children's way of getting involved with another child or working their way into a group. The child moves close to the group she wants to join, plays alongside for a while, and eventually finds herself incorporated in their play. This is a less risky way to approach children in play than walking up and asking, "Can I play?" because the other children might say no. Researchers once viewed parallel play as less sophisticated than other types of play, but they now see it as a device that children with advanced social skills use to enter into play with others. Parallel play allows children to warm up to each other slowly; it allows them to prove themselves as compatible players.

associative play A form of play in which children use the same materials, interact with each other, and carry on conversations. It is not as organized as cooperative play, in which children take on differentiated roles.

cooperative play A form of play that involves a significant degree of organization. Interactive role-playing and creating a joint sculpture are two examples of cooperative play.

NAEYC Position Statement
Learning to Read and Write

Perhaps in your memory of a childhood play experience you were interacting with other children. There are two types of interactive play. **Associative play** involves children interacting in a loosely organized fashion, such as working on a craft project side-by-side. The children associate with each other in a give-and-take way; they carry on conversations and pay attention to each other. In contrast, **cooperative play** involves a significant degree of organization: "Let's play house. You be the mommy, and I'll be the daddy. Now, who will be the baby? Oh, I know, let's ask Julie." School-age children are often very organized when they play. Creating a joint sculpture out of cardboard boxes or putting on a circus are two examples of sophisticated cooperative play. See *The Theory Behind the Practice* box on page 103 to understand how play space can be set up to accommodate the different types of social play.

Benefits of Play

Play offers so many benefits that it's hard to list them all. One major benefit of play is that it increases children's ability to deal with the world on a symbolic level—a skill that falls mainly in the realm of intellectual development and is the foundation for all subsequent intellectual development. Only by using symbols can children learn to talk, read, write, and understand mathematic and scientific concepts. Through the use of symbols, children eventually become proficient users of logic and reasoning as well. There is a link between symbolic play and later reading, writing, and other academic achievements. When parents understand this link, they sometimes become more appreciative of the time their children spend playing pretend.

Of course, playing does not enhance intellectual development alone. Through play children also gain a variety of social and emotional skills. Remember, these skills are also linked to intellect. They learn to get along and cooperate with each other. They work out fears, anger, and emotional conflicts through pounding play dough, through hugging and then scolding dolls, and through climbing to the top of the jungle gym. The practice of these social and emotional skills in turn creates neural pathways in the brain that directly affect intellectual development. During play, children also gain physical skills and practice them over and over until they master them: balancing precariously on the edge of the sandbox and then walking it like a tightrope or tipping and falling on a two-wheeler until, finally, learning how to keep the bicycle upright and moving forward.

Go back to your own childhood play experience once again. How did you benefit from this experience? Try listing the benefits by category: cognitive (intellectual), psychomotor (physical), and affective (social-emotional). In other words, explore how your play experience benefited the "whole you"—your mind, body, and feelings.

As an early childhood educator, it is important that you understand the benefits of play. Eventually, you will have to answer the perennial parent question, "Do children learn something in this program or do they just play?" It is worth taking some time now to formulate your answer for when the occasion arises.

Play Spaces

Some classic research at Pacific Oaks College has helped a generation of teachers organize their play spaces so that there is enough room for everybody and the right number of choices of things to do. With careful adult planning children can find spaces to play alone, with one other person, and in small and larger groups. The guides developed by Kritchevsky, Prescott, and Walling aid teachers in creating environments that work to encourage active play. The environment also helps children group appropriately instead of staying in a large group that needs to be "herded." Movement between play areas is important too, if children are to have choices. Clear and visible pathways lead and even entice children to change play areas when ready so they gain benefits from being in a rich environment.

From L. Kritchevsky, E. Prescott, and L. Walling, *Planning Environments for Young Children: Physical Space* (Washington, DC: National Association for the Education of Young Children, 1977).

Through play, children explore the physical world and gain a foundation in math and physics.

Work: A Way of Learning

When children are involved in dramatic play, they are almost always imitating adult work. Children know that work is important. They are anxious to become workers themselves, or they would not play at it so much. Adults encourage a healthy attitude toward work when they involve children in chores. Sweeping, wiping off tables, setting up for circle time and meals are routine chores that not only satisfy children but also contribute to their intellectual, social, physical, and emotional development. Helping out is an important part of the early childhood curriculum.

Adult Attitudes Toward Work and Their Effect on Children

"Play is children's work" is a saying that once guided the early childhood field. The saying is still true, but some question whether it denies children responsibility. When children are focused on play, adults often hesitate to ask them to pitch in to help. Nevertheless, young children are often glad–indeed, consider it a privilege–to help. At the Pikler Institute, the residential nursery founded by Dr. Emmi Pikler in Budapest, Hungary (see Chapter 1), helping starts early in life as babies learn in one-on-one situations to cooperate with their caregivers during caregiving activities. That is the beginning. Later, once they can walk, they learn to take responsibility for others by being the designated helper of the day. Anna Tardos, Pikler's daughter and present director of the institute, says that such responsibility is not a privilege, but a right. One of the advantages of group care is that children can learn what it means to take responsibility. Some children also learn that at home, but not all.

Attitudes toward work are learned early. Adults who regard work as something disagreeable can easily pass that attitude on to children without even saying anything, through facial expressions and body language. Therefore, it is vital for adults to model positive work attitudes. Cleanup time is a case in point. Toddlers, for example, love cleaning up as much as getting things out, especially if no one puts heavy pressure on them. Putting toys away is similar to working a puzzle and can

Giving children responsibility and involving them in chores encourages them to be participants in their own lives as well as builds a healthy attitude toward work.
Oksana Kuzmina/Shutterstock

be just as much fun. Building a pick-up routine into an early childhood program is one way to develop a positive attitude toward work.

One program used the concept of cycles to explain picking up. The cycle of playing with something is to (1) get it out, (2) play with it, and (3) put it away. If a child abandoned a toy, the teacher explained that he had not finished the cycle and was not done yet. Children who appreciate closure learn this pattern fast.

Other kinds of work include having children take responsibility for their own things, hanging up their jackets, putting their toothbrushes away, clearing their own place at the table, and folding their napping mats. Doing things for the group is another kind of work, such as setting the table, sweeping, arranging chairs for circle time, and helping get the tricycles out of the shed.

It is true that many people see work and play as opposites. Just remember, young children don't need to be taught that.

Children's Observations of Adults at Work

One way to introduce children to positive work attitudes is to let them observe adults at work–at all kinds of jobs. A plugged toilet or broken pipe, for example, might be a headache for the teacher or provider, but it can be the highlight of the day for the children if they are allowed to watch the plumber at work. Children also enjoy watching daily tasks, such as an adult preparing and cooking a meal. (One advantage of family child care is that it provides many opportunities for children to see adults at work; in center-based programs, adult work is often hidden from children, such as meals being prepared in an isolated kitchen or even off-site.) Likewise, construction work lures curious children–not just because of the equipment but because of the workers in action.

In one family child care program, the children watched through a garage window while the provider's teenage daughter worked on her car. At a center-based program, a teacher brought his motorcycle to work and tinkered with it in the play yard while four-year-old "mechanics" tinkered with their tricycles around him.[7] And in yet another program, the director, who enjoyed woodworking, set up a workshop for school-age children. While he worked on various pieces of equipment for the center, the children either helped him or worked alongside him on their own projects.

The Project-Based Approach to Learning

The **project-based approach** to learning lies somewhere between the realm of pure work and pure play. Conducted in a child-centered environment, a project is not the same as free play and exploration. The project-based approach allows for in-depth studies of concepts, ideas, and interests that organically arise within the context of the group. A project starts with an idea that emerges from either adults and/or children and is carried out by a small group of children over a period of time. This approach emphasizes depth over breadth.

project-based approach An in-depth teaching–learning process that emerges from an idea—thought up by either a child or an adult—and is carried out over days or weeks. Unlike free play, project work emphasizes product as well as process. Documentation of the process (during and upon completion) is an important element of the project-based approach.

The project approach came out of the University of Chicago in the 1930s, yet it is gaining attention today from its current application in the Reggio Emilia schools in Italy, the early childhood programs mentioned in Chapter 1. In Reggio, every

POINTS OF VIEW

Two Views on Child-Centered Learning

The preceding examples involved adults doing adult work in child-centered environments, yet some early childhood educators feel strongly that children's space should be *just for children*. Because they believe purely in child-centered learning, they disapprove of adult work in early childhood education settings. They see learning as interactive and believe that children learn best as they play and explore freely in a rich and responsive child-centered environment, not when they just stand around and watch adults or imitate them.

Jayanthi Mistry, a child-development researcher, holds another point of view and questions the necessity of child-centered learning:

> In some cultural communities children learn by simply being present as adults go about their jobs and household activities. Adults do not create learning situations to teach their children. Rather, children have the responsibility to learn culturally valued behaviors and practices by observing and being around adults during the course of the day. Children are naturally included in adult jobs and activities. For example, a toddler may be cared for by a mother who runs the family store. In this setting, the child is assured a role in the action, at least as a close observer. The child is responsible for learning through active observation and gradual participation.[8]

Plenty of children today are cared for in family-run businesses, such as stores and restaurants. Whether this is an appropriate environment for their early years depends on one's point of view. The difficulty arises when people with these opposing views are responsible for the care of the same child. As has been said before, there is room in early childhood education for multiple perspectives–as long as early childhood providers remain flexible and willing to negotiate.

school has a room dedicated to project work–a combination workshop-laboratory. In this workshop, the children have room to spread out and work undisturbed. (Unfortunately, space is a rare luxury for most early childhood programs in this country. The dilemma is often finding a place to store the projects in progress when the tables must be cleared for meals and snacks every few hours and the whole room must be cleared for nap time once a day.)

Projects provide an element of continuity that is sometimes missed in programs that focus mainly on free play or isolated, unrelated curriculum activities. Projects continue for days or weeks or even years and build from one thing to another in a connected way. Having children bridge and build on concepts develops an important cognitive skill. This is because throughout a project, teachers and children make decisions about the direction of the study, the ways in which the group will research the topic, the representational medium that will demonstrate and showcase the topic, and the selection of materials needed to represent the work. These threads of work eventually encompass opportunities to support learning through several domains of development.

Projects are designed to have outcomes, which makes them slightly different from free play, in which often the process is all and the product nonexistent. Of course, process is also important in projects. In fact, process is so important that the teachers in the Reggio schools and elsewhere document the process as a project unfolds. Through the documentation process, adults recall what happened and suggest ideas for planning. Documentation can take many forms: writing, audiotapes, photos, videos, drawings. Furthermore, a written record can be re-told to children to help them remember what happened and to demonstrate to them strategies that people use for remembering. Through documentation, children can revisit their own learning process; they sometimes even learn that the best way to go forward is to go back and review. Revisiting work can also stimulate a debriefing process: What happened? Let's remember it, look at it, and talk about it. Documentation is covered in more detail in Chapter 12.

THE ADULT'S ROLES IN CHILDREN'S WORK AND PLAY

Although the adult roles have been discussed previously in this chapter, this section puts them into categories and focuses on project work.

The Adult as Observer

In project work, the role of observer is a very important one. By observing, you learn what direction the project is taking and how the children as individuals and as a group think about what they are doing. Nevertheless, whether working on projects or not, the early childhood educator must always be a careful observer to learn what is going on and to step in when needed. Observing children individually and in groups points out their needs and is an important means of assessment. A way to focus on observation and record what you see is to keep anecdotal records or what are called running records. A **running record observation** is a detailed description of what is happening while it is happening. Such an observation can be made with or without adult interpretation and speculation about the meaning of the behavior observed; however, when recording both objective data and subjective comments, keep the two separated. One way to do this is to fold a piece of notebook paper in half. On one half, write down simply what you see in objective terms; on the other half, jot down your ideas about the meanings of the behavior you have recorded. Keep time while you are writing a running record observation, and note it in one-, two-, or five-minute increments. It may be important later to know whether the behavior was fleeting or lasted for a while. A series of accurate and complete running record observations can help you gain insights into a child's play and what the child is learning from it.

running record observation A method of documenting that gives a detailed, objective, written description of what is happening while it is happening. A running record can include adult interpretations about the meaning of the observed behaviors, but it must separate objective data from subjective comments.

Of course, when working with children, you cannot always step back and write notes for a running record observation. The main use of the observer role is to really pay attention to what is going on. When in the observer role, you must observe yourself as well as the children to get information that can guide you in your supervision and interactions. Are you interfering too much? Are unfortunate things

One way to interact effectively with children at work or play is to be interested in what they are doing and offer nonjudgmental commentary.
Jupiterimages/Stockbyte/Getty Images

happening because you are not close enough to the action? Are you facilitating but not taking over?

descriptive feedback A form of nonjudgmental commentary. Adults use descriptive feedback to put children's actions and feelings to words to convey recognition, acceptance, and support: "You're putting a lot of work into that drawing" or "Looks like you don't like him to touch your painting." Descriptive feedback should be used to facilitate rather than interrupt children's exploration.

One way to interact effectively with children at work or play is to offer nonjudgmental commentary, or **descriptive feedback,** often also referred to as broadcasting. Teachers put words to actions and feelings for children when they say things such as:

> "I see how you are pouring that sand into your shoe."
>
> "Looks like you don't like him to touch your painting."
>
> "You're putting one block on another."
>
> "You've been picking up walnuts for a long time now."
>
> "You're putting a lot of work into that drawing."
>
> "You're really scrubbing that table."

Just talking about what a child is doing or what you perceive he or she is feeling conveys recognition, acceptance, and, thus, validation. Although this way of communicating may seem awkward to you as a beginner, once you learn how to give descriptive feedback, you can use it effectively in the role of facilitator. Kathleen Grey, an infant-toddler teacher, calls such a way of talking to children "active listening." Read what she has to say about the subject in the *Tips and Techniques* box on page 109.

Once you learn how to give descriptive feedback, you may offer a lot of it. Just be careful to watch for comments that disrupt. The idea of giving descriptive feedback is to facilitate, not to interrupt and draw unnecessary attention to your presence or add to the ambient sound.

TIPS AND TECHNIQUES

A Respectful Way of Communicating

I longed for a way to "just be" with children. I wanted to be a positive force in their lives, but I wanted to give up the exhausting and useless push to mold them in the images in my own mind. I wanted to communicate expectations for them that they simply be the best of whoever they were capable of being, not the best of whomever I thought they should be. But how in the world could I communicate that without going all the way into total permissiveness where anything goes and everything is okay?

Then I learned about active listening, an accepting, reflective way of communicating respect for other people . . . and the world opened up for me. I discovered that reflecting back to children what they were doing, and what it looked like they were feeling, reinforced their sense of themselves in such a way that they felt strengthened and validated as potent, competent, worthwhile human beings. I could trust them to want to grow. The active listening style of communicating felt so clear, uncomplicated, honest, and real that I just sank into it with a sense of great relief.

It was like dropping a pebble into a still pond. From that time, the ripples have traveled outward in wider and wider circles. I began to have frequent experiences of connecting with the children, of watching their dawning understanding and their evident pleasure in that understanding. Even during many of the times when I had to set limits, I experienced the companionship that comes with genuine connection and with shared meaning.

Adapted from Kathleen Grey, "Not in Praise of Praise," *Exchange,* 104 (July-August 1995), 56-59.

The Adult as Stage Manager

It is surprising what a creative mind can invent when setting up an environment for children. The setup can be elaborate or simple. It can also result from your creativity alone or incorporate some of the children's ideas. Try to encourage children's inventiveness. When a child suggests a new use for, say, a piece of equipment, ask yourself "Why not?" before automatically saying "That's not what that is used for" (see the *Tips and Techniques* box on page 110).

You can boost children's creativity by letting them change the environment from time to time. Remember, though, that if anyone who uses the environment has visual impairments, the rearrangement may create hazards for that person. Some environments invite rearranging more than others because they have fewer built-in features, and some materials invite rearranging because they consist of many loose parts (such as giant building blocks that children can use to create large structures they can actually enter). Some programs have boards, boxes, ladders, and other components that fit together so children can design their own play equipment. It is a little like letting children play at a construction site—but much safer.

Besides setting up the environment—deciding what materials, activities, equipment, or toys will be provided and how they will be arranged—you must constantly

TIPS AND TECHNIQUES

Why Not?

I was stationed in the dramatic-play area, which was set up as a play store, when the children started hauling empty Cheerios boxes and empty milk cartons into the block corner. As a new parent in a co-op preschool, I wasn't sure what the rules were, so I made some up. "The store stuff has to stay over here," I announced cheerily in what I hoped was also a firm voice. The children ignored me. I repeated myself. One little boy said, "But we want to make our store over there." I took a big breath and was about to make a fuss when the teacher arrived at my side. I told her that they wanted to move the store, and all she said was, "Why not?" I was stumped. I couldn't think of a good reason. Later, when we discussed the issue at our end-of-the-morning meeting, the teacher explained that she always asked herself "Why not?" when children wanted to do something she disapproved of. If she couldn't think of a good reason, she put her disapproval on hold. It made good sense, and I've been guided by her rule-of-thumb ever since.

Adapted from Janet Gonzalez-Mena, "From a Parent's Perspective," *Napa Valley Register,* February 10, 1995.

be aware of how the stage is affecting the play. Is something missing that would facilitate the activity, project, or routine? Is a particular area getting too chaotic? Do I need to put the setup back in order? Is this particular area being ignored? Do I need to add something of interest or even stand here myself in order to draw the children's attention?

The Adult as Teacher

NAEYC Program Standards
Program Standard 3: Teaching

An important goal of a teacher should be to help all children become good at playing. Many of us grew up acquiring such skills naturally and have never thought about what it takes to be good at playing. Most children following a typical pattern of development learn to play on their own. Nobody has to teach them, and play is second nature to them. But other children benefit from a teacher's or provider's help in learning how to play. In fact, some children are desperate for such help and will not learn to play without it. Some of the children who need help are those who have special needs or particular challenges, but not all. Some children just have not had the chance to learn play skills. Read the next two paragraphs with both those groups of children in mind.

In teaching a child to play, the basic goals are to get a child to begin to play with one or more children and to give him the skills to keep on playing. There are several steps that a child may go through in learning play skills. Playing with the teacher or provider is a preliminary step for many children. Playing in pairs is another early step because it is easier to play with one another than to try to enter a group that is already playing. Playing in parallel—when two or more children play side by side–is a step toward interactive play. Learning to imitate another child can be another step toward becoming a skilled player. Imitating can be a forerunner of learning to follow the lead of another. Infants and toddlers are very skilled at imitation as well as delayed imitation (accommodation), yet they start elaborating on imitated actions after repeating them several times (assimilation). What this means for caregivers is that they need to provide frequent, warm, and prolonged opportunities for children to interact with them and remember to model behavior that they want to endorse.

Being able to take the lead periodically once part of a play group is another skill that children need to learn. Knowing when to lead and when to follow in a give-and-take way is vital to interactive play. Being able to pretend is still another important play skill. All these skills can be taught to most children if they are developmentally ready for them. Of course, basic social skills are important too, skills such as sharing, turn taking, empathy, and the ability to share one's own feelings and thoughts.

To promote the skills that help children play, start by observing. Determine who needs what skills. Guide children and model for them. If they need words for entering into a group, give them those words–cue them and prompt them. For some children, words are not enough; they need a lot more. In that case, use physical guidance. Some children need the early childhood educator to actually move their hand to get that first piece of the puzzle or roll the ball across the floor. If the adult holds the child's hand and performs the movements for her, the child will eventually take over and do it herself. For example, say three children are sitting on the floor with a guinea pig and two are holding and petting the little animal gently. If the third child seems to want to touch the guinea pig but has no idea what to do, demonstrating or explaining may be all that is needed to help the child interact. If the child still does not respond, the adult can physically help the child to hold and pet the animal.

A sense of security is another important factor in play. Children who lack a sense of security may not demonstrate their play skills. There is the emotional sense of security that is discussed at length in Chapters 3 and 9. But there is also a need for a physical sense of security. Sometimes this need is satisfied simply by proper handling and positioning of a child who lacks varying degrees of control over his or her body, as in the case of a child with cerebral palsy or another type of physical disability (see the *Tips and Techniques* on the next page). A child who feels uncomfortable, out of balance, or afraid of falling over may lack the concentration needed to play (or do anything else).

Observation and communication are important. It may be that the teacher does not know how to provide the physical security the child needs and that the child is not able to express himself well enough to explain or show how he feels. Parents are the best sources for this kind of information. Physical therapists may have some information as well. Some children who are beyond infancy cannot move out of the position they have been put in. Be sure they are securely supported. To help them stay comfortable, change their position periodically. Try to find positions that leave the child's hands free to manipulate objects. If the child has a hard time handling things, modify toys and materials so that the smallest movement will have a big

TIPS AND TECHNIQUES

Helping Children with Varying Abilities Play with Other Children

Play is a way that children learn skills, and play itself is a skill that sometimes has to be taught. Some children may not have had much experience playing, and others may lack skills or capabilities. A child who has a hearing or a language challenge may have problems communicating with peers, thus hindering play. Adults can teach children ways of communicating with a hearing- or speech-impaired child and in turn can help that child communicate with other children as well as adults. Some children lack control over their behavior and other children do not want to play with them. Here again, teacher help is needed. The information on guidance alternatives to punishment in the next chapter is useful for changing children with out-of-control behavior into desired playmates. Even adaptive equipment, such as a wheelchair or a walker, can be a problem and can hinder play when it becomes a barrier to social interaction. Information in Chapter 8 on the physical environment can be of some help here. Whether adaptive equipment makes a child more or less popular with peers is not necessarily predictable. For example, I heard one story about a four-year-old who asked Santa if he could have a walker for Christmas like his friend in preschool. He did not want a bike, he wanted a walker!

effect. Do not underestimate children's ability to learn and improve their skills in the right environment and with the proper help.

The Adult as Encourager

Adults facilitate both work and play (whether differentiated or not) through their encouragement.

There is compelling evidence that praising children for success hampers them as compared to children who are acknowledged for their continuing effort. When adults focus on success by saying things such as "Good job!" and "You are so smart," children eventually begin to avoid trying things they do not think they can do well. Adults instead can say things such as, "You did it!" Children should be encouraged to do activities for their own pleasure and not for adult approval. When adults give attention to children for trying hard and keeping at it, those children become more effective learners. They are much more likely to take on complex problems and enjoy the process of trying to solve them. Peter H. Johnston wrote a book entirely about how teachers can help children reach their full potential by thinking carefully about the words they use. Called *Opening Minds: Using Language to Change Lives,* the book cites a good deal of research including that of Carol Dweck (2006) about how children develop theories of intelligence and what it means to be smart. Dweck points out that intelligence is, in reality, dynamic–it grows and changes. Dweck concluded that children praised for their intelligence or talents alone tend to develop a "fixed mindset," or a belief that their intelligence or talents cannot grow through hard work, better strategies, or help from others. They are more likely to avoid or give up in the fact of difficult tasks. In contrast, children who are praised for their process or effort tend to develop into motivated learners with a "growth mindset" and are more likely to persist and even thrive during challenging circumstances. Dweck's research inspired Angela Duckworth's award-winning work on grit, or the perseverance to put forth concentrated effort on difficult or unfamiliar tasks to achieve long-term goals, which requires a growth mindset. Setbacks are inevitable; both grit and a growth mindset are required for both children and adults alike to overcome challenges in life.[9] Adults have the power to start cultivating these characteristics in young children through the words they use to encourage them to take on and work to solve problems. The following scene shows how encouragement works during a free-play situation in an infant center.

Ashley crawls over to the toy shelf, where she spots a little wooden wagon containing blocks. The string on the wagon is tangled around the wheel of a truck, and when Ashley pulls the wagon, the truck comes too. Frustrated, Ashley jerks the wagon hard, but that does not solve the problem. She bangs it. Still no good. She takes the blocks out one by one, but the string is still tangled. She jerks the wagon again by the string and lets out a frustrated yell, which brings an adult over. The adult puts Ashley's feelings into words, "You want that wagon loose."

Ashley hands her the string, but the adult hands it back. When Ashley pulls on the string again, the adult, using nonverbal encouragement, puts one finger on the taut string and traces it to the source of the problem. Then she adds words to describe the problem: "The string is tangled."

Ashley throws the string in frustration and, by doing so, loosens it around the wheel. "Look," encourages the adult. "I bet you can untangle it." Ashley grabs the string again, and it comes free. Off she toddles with the wagon following her. The adult remains silent, figuring that Ashley's success is reward enough.

The adult did not solve the problem for Ashley; she encouraged her to solve it herself. She gave her small bits of help and just the right amount of verbal encouragement, a process called "scaffolding."

Encouraging children is an important adult role. It takes some determination to stick with encouragement rather than solve a child's problem and thereby rescue him or her from frustration. For example, it would have been easy for the teacher to untangle the string for Ashley. But that might have given Ashley the message that she is too little to be an effective problem solver. Children who are constantly rescued often become reluctant to even *try* to fix things themselves. They look to an adult to take care of everything, and they fail to learn how to deal with the frustration that often accompanies problem solving. As such, it is also important for adults to validate children's emotions and not uses phrases such as, "It's OK," "You're OK," and "Don't cry". Rather, adults should say, "I know that must have been really scary when you fell off the hammock." Try not to discourage crying but help children identify, label, and validate their emotions.

Encouragement Versus Praise. Verbal encouragement plays a role in empowering children; however, many adults use praise instead of encouragement. To illustrate the distinction, see what might have happened if the adult had praised Ashley instead of encouraging her.

Ashley pulls the wagon off the bottom shelf, but because the string attached to the wagon is tangled around the wheel of a truck, the truck comes too.

The adult says, "Good job, Ashley, you got the wagon down," without acknowledging the problem. Ashley looks frustrated. What effect will the praise have on her when she has not yet accomplished what she wanted. She jerks and bangs the wagon to no avail. When she lets out a frustrated yell, the adult has a hard time finding something to praise her for. At this point, Ashley just gives up. The praise alone just did not work.

Many adults do not understand that inappropriate praise can be as addictive and debilitating a response as rescuing a child. When children are hooked on praise, they lose their ability to judge for themselves, they become dependent on outside opinions, and they lose intrinsic motivation. Compare these scenes.

In the first scene, Trevor is working hard to build a very tall tower, and he announces his intention to the student teacher assigned to the block area. He carefully stacks a block on the teetering structure. It falls off, and the whole thing nearly collapses. He looks quickly to the adult, who registers disappointment in her face. He tries again. Success! Again he looks at the teacher, who says, "You made it! Good job." It is only after hearing from her that he sits back and glories in his success.

In the second scene, Brian has made a tall stack of blocks and is trying to knock the top block off with a homemade pendulum consisting of a string hanging from the ceiling with a tennis ball attached to one end. He is oblivious to everyone else around him. Brian swings the pendulum and misses; swings again and misses again. He then adds another block to the stack. The next swing does the job. The block tumbles. Brian barely pauses before he begins experiment two: knocking a plastic pear off the stack.

intrinsic motivation Inner rewards that drive a child to accomplish something. Intrinsic motivation contrasts with extrinsic motivation, in which rewards are given to the child in the form of praise, tokens, stickers, stars, privileges, and so forth. in response to accomplishing a task or behaving in a particular manner.

Which of the two children is more motivated by praise and which finds rewards in what he is doing without receiving recognition? Brian has what is called **intrinsic motivation;** that is, he feels good *without* someone praising him. He's also more focused on the process than the outcome. He needs neither encouragement nor praise.

Intrinsic motivation is what learning should be based on—not an external reward system. Yet many adults in the early childhood field are caught up in dishing out rewards for success in the form of praise, stickers, privileges, and other devices that are designed to motivate children from the outside.

It is easy to detect children who are hooked on recognition. They do not enjoy their successes unless shared with someone. In one infant program, I noticed that some children searched the room for admiring eyes or clapping hands every time they accomplished some little feat. Their faces showed disappointment when no one cheered or clapped.

Praise has a lot of power—and also some potentially disastrous side effects, as illustrated in the following experiment conducted at a university lab school. The experiment occurred at the time felt pens were first introduced to the market; they were very expensive, and most children had never seen them before. The experimenters brought a bunch of felt pens into a preschool and left them on a table next to some sheets of paper. Try to imagine how different those felt pens were from crayons, chalk, or pencils—the only drawing materials the children had ever used.

The children were allowed to play freely with the felt pens for about a week or so; after that, the pens were removed from the classroom. In the meantime, the children were randomly divided into groups, and a reward plan was set up. Systematically, during the period when the felt pens were not available in the classroom, the children were taken to another room in small groups and given the felt pens to play with. Some groups were rewarded for playing with the pens, and other groups were not. After a month or so, the felt pens were reintroduced to the classroom for the children to play with freely once again. What the researchers found was that the children who had not been rewarded flocked again to the table and played with the pens to their hearts' content, but the children who had been rewarded now held back when it was clear that there was nothing in it for them besides just the pleasure of using the pens.

I know from my own experience that activities that are rewarding for their own sake lose their fun when adults add external rewards to the intrinsic ones. So when a toddler is playing in the sand and an adult interrupts to say, "Oh, I like that big pile you made," the toddler may be inclined the next time to make a pile for the adult rather than for himself. Or, much later, when a child who is eager to learn to read gets stickers or tokens for each step, the reading process loses some of its intrinsic value for the child. It is almost as if the adult were saying, "I know there's nothing in this for you, so I'll make it worth your while."

When children are motivated in their play by adults' praise, they are not engaged in true play. Remember that one of the five characteristics of play is intrinsic motivation. Children who are driven by praise are producing for rewards rather than for the pleasure of the activity.

Indeed, we all need some praise, attention, strokes, recognition, and acknowledgment. But it is important to know how to use praise judiciously–to build relationships, not tear them down. To use praise cautiously and wisely, follow these guidelines:

- Avoid using praise when a child is obviously intrinsically motivated in her activity.
- Help children tune in to their own good feelings about accomplishing something. Say "You must feel good about that" when it seems to apply.
- Do not praise just successes but also attempts and risk taking, even if they are unsuccessful.
- Avoid broad value judgments like "Good girl!" or "Smart boy!" It is important that children always think of themselves as good and smart–not just when they please adults.
- Distinguish between encouragement and praise. Try to use more of the former and less of the latter. Take note when a child needs a little boost, and learn to use encouragement effectively.
- Give children recognition regularly, not just when they have earned it by accomplishing something.
- Giving positive attention in the form of a pat, a smile, or conversation conveys your appreciation of the child himself. Some people call this kind of attention "unconditional love."

- Be generous in your attention every single day. Be attentive when you interact with children and during diapering, feeding, and mealtimes. During play time, however, allow children to go about their own business at times; children who are well fed on attention will not hunger for it while playing. Play is richer when it is undisturbed by intruding needs.
- Remember, praise is addictive. If you suspect a child of getting hooked on praise, help him learn to experience his own inner rewards and satisfaction.

Praise is often used to build self-esteem, but, ironically, it tends to have the opposite effect. Self-esteem grows when children can realistically appraise positives and negatives and decide there are more of the former than the latter. Overabundant praise only clouds reality and can give children a false sense of themselves. One final note about such communication: it is important to note when communicating with children that how you say something is just as important as what you say. Your tone and language should always show respect when speaking to children, and also always keep body language in mind. For instance, it is important to remember that when speaking to children we do not want to be towering over them, we want to get down and speak to them at their eye-level. As this section has demonstrated, we have to be mindful of the language we use with children because it can be instrumental in helping shape children's growing sense of themselves.

A Story to End With

A little girl once brought me a truly remarkable painting, stuck it under my nose, and waited for my reaction. I gave her one. "That's beautiful," I gushed spontaneously. Then, remembering I should not be judging children's art, I gave my honest personal reaction. "I really like your painting!" The child beamed and departed, painting in hand. A few minutes later she brought me another painting. This one was not nearly so remarkable. She obviously wanted more praise. I felt trapped. I lied a little and said, "I like that one too." She left again, only to return a minute later. There she was, holding out a splash of paint across a crumpled piece of newsprint. I finally had to tell the truth, so I said I thought she was making pictures now just so I would praise them. She was crushed and stopped producing pictures for me. I felt terrible. It took her weeks before she started painting again, *for her own satisfaction.*

I could have responded differently to the first picture by talking about it in objective terms instead of praising it. I thought it was all right to react emotionally as long as it was an honest reaction, but all I offered was a value judgment ("That's beautiful"). And although sharing my feelings ("I really like your painting!") was an improvement on my first reaction, I should have used a different response.

The most effective reaction of all, I have learned, is to discuss the process with the child, comment about the painting in objective terms, and encourage the child to express *her* feelings about her accomplishment: "I see how much time and effort you put into that picture. I bet you enjoyed doing it. I see you have red in this corner and yellow across the bottom. Look where they ran together." Honesty from the beginning would have given this story a different ending.

Summary

In early childhood education, the spotlight belongs on the children. Though it may be tempting for the teacher to put himself or herself in a central position, it is important to understand the benefits of child-centered learning. Play provides a rich medium for children's mental, social, physical, and emotional development. Some categories of play are sensorimotor, symbolic, solitary, parallel, and interactive play. Work also offers a medium for children to learn about the environment and taking responsibility for themselves. Adult attitudes toward work have a significant impact on children's attitudes. Children love to watch adults at work, but experts disagree on the appropriateness of exposing children to adult work in early childhood programs. The project approach to learning combines both work and play and involves the joint efforts of adults and children. Three of the major roles adults assume to facilitate children's work and play are the observer role, the stage manager role, and the encourager role. Descriptive feedback and encouragement (as opposed to praise) are two effective devices early childhood educators use in fulfilling these three roles.

Reflection Questions

1. What roles did play and work take in your own learning? Can you think of a time when you learned something through either play or work? Can you relate your personal experience to that of children in early childhood programs, including family child care?
2. Do you believe that it is necessary to distinguish between work and play (and make them opposites)? How much do you think your answer to this question relates to your culture? If it does, how?
3. Think of a time when you experienced a sense of rhythm in your life. Describe it carefully. Do you see any benefits in this experience? Do you see how your experience might apply to early childhood education?
4. What part has competition played in your life? Have you found it beneficial? What is your reaction to the text's strong opposition to introducing or allowing competition in early childhood programs?
5. What part have praise and intrinsic motivation played in your life? Have you found one to be more beneficial than the other? What is your reaction to the text's strong warnings about the possible negative effects of praise?

Terms to Know

How many of the following words can you use in a sentence? Do you know what they mean?

child-centered (or child-directed) learning 90
developmentally appropriate practice (DAP) 91
meaning-making 92
sensorimotor play 99
symbolic play 99
solitary play 101
parallel play 101
associative play 102
cooperative play 102
project-based approach 105
running record observation 107
descriptive feedback 108
intrinsic motivation 114

For Further Reading

Gillespie, L. (July 2016). It's never 'just play! *Young Children, 71* (3), 92–94.

Johnston, P. H. (2012). *Opening Minds: Using Language to Change Lives.* Portland, ME: Stenhouse.

Jones, E., and Reynolds, G. (2011). *The Play's the Thing: Teachers' Roles in Children's Play.* New York: Teachers College Press.

Leong, D. J., and Bodrova, E. (2012). Assessing and scaffolding: Make-believe play. *Young Children, 67* (1), 28–35.

Louv, R. (2008). *Last Child in the Woods: Saving our Children from Nature-Deficit Disorder.* Chapel Hill, NC: Algonquin Books of Chapel Hill.

Luckenbill, J. (2012). Getting the picture: Using the digital camera as a tool to support reflective practice and responsive care. *Young Children, 67* (2), 28–36.

McDermont, L. B. (September 2011). Play school: Where children and families learn and grow together. *Young Children, 66* (5), 81–86.

Riojas-Cortez, M. (September 2011). Culture, play, and family: Supporting young children on the autism spectrum. *Young Children, 66* (5), 94–99.

Stroll, J., Hamilton, A., Oxley, E., Mitroff Eastman, A., and Brent, R. (March 2012). Young thinkers in motion: Problem solving and physics in preschool. *Young Children, 67* (2), 20–26.

Weatherson, D., Weigand, R. F., and Weigand, B. (2010). Reflective supervision: Supporting reflection as a cornerstone for competency. *Zero to Three, 31* (2), 22–40.

Wirth, S., and Rosenow, N. (2012). Supporting whole-child learning in nature-filled outdoor classrooms. *Young Children, 67* (1), 42–48.

Design credits: Tips and Techniques: ©Ingram Publishing; Focus on Diversity: ©Pixelic/Getty Images

5 Guiding Young Children's Behavior

In This Chapter You Will Discover

- what to expect of children at each developmental level.
- why spanking and other forms of punishment don't work.
- what you can do instead of punishing.
- why time-out doesn't always work.
- how children learn from experiencing consequences.
- why children need limits.
- what "testing the limits" means.
- how to use redirection and avoid power struggles.
- why it is important to accept children's feelings.
- how important it is for adults to set good examples for children.
- how to read behavior as communication.
- how to prevent misbehavior.
- how to deal with particularly challenging behaviors.
- why relationships are important to effective guidance strategies.

NAEYC Program Standards

Program Standard 2 and 3 Curriculum and Teaching

To guide the behavior of young children in positive ways, the early childhood educator must have knowledge and skills. To start out, the beginner must understand that the goal of guidance is to teach children how to control themselves and to act in socially acceptable, respectful ways. In this chapter, you will learn about several guidance tools and how skillful guidance promotes self-esteem and enhances the adult-child relationship. You will also discover how skillful guidance presents children opportunities to learn, so it is not just a chapter about social development, but also about cognitive development.

Appropriate Behavioral Expectations

A prerequisite for developing guidance skills is an understanding of behavioral expectations for each stage of development. Before you can ever determine what behaviors need to be "guided," you need to learn the behavioral norms for each stage and how to interpret them. For example, the cry of a six-week-old is communication, not a manipulative device, and the defiance of a two-year-old is a step toward autonomy, not the sign of a mean spirit.

Knowing what is appropriate behavior for each age group is vital for anyone who works with children. Former preschool teacher and director Marion Cowee, now a community college teacher educator, shares a story in the *Voices of Experience* box on page 124 about a child whose behavior didn't fit that of other four-year-olds in the class. Following is a brief summary of realistic behavioral expectations for infants and young children.

- Six-week-old babies cry not to manipulate adults but to communicate their needs. Crying is their way of getting adults to care for them in a timely manner. When babies are able to communicate with sensitive and responsive caregivers, they develop a sense of basic trust, which is the major developmental task of the first year of life, according to Erik Erikson.[1]
- By nature, two-year-olds are defiant. According to Erikson, the major developmental task of children at this age is to experience their own autonomy.[2] By behaving defiantly, they aren't trying to be "bad," but rather, they are learning to assert themselves in the world. Although some of their behavior may call for gentle guidance and direction, it is important that the adult understand the behavior and its purpose.
- It is common for three-year-olds to tell untruths. But rather than characterize these untruths as "lies," the adult needs to realize that understanding the difference between fantasy and reality is a cognitive task for the early years. According to Jean Piaget, a three-year-old is still learning to discern one from another and may engage in wishful thinking, hoping that fantasy is, in fact, reality.[3] (More about this subject appears in the *Tips and Techniques* box on page 125.)
- A four-year-old who "steals" something is not exhibiting signs of a criminal nature; instead she may be demonstrating that she does not yet

understand right from wrong. For her, trying out different behaviors and seeing which ones bring negative consequences helps her learn this distinction. According to Lawrence Kohlberg and William Damon, two theorists who have researched moral development, she is in an early stage of moral development.[4]

- Arguing in school-age children does not signal a lack of socialization; rather, arguing is their way of sorting out what they think and asserting themselves socially. Through childhood squabbles, young children learn about the give-and-take of social relationships. Occasionally, guidance is necessary to help them learn to argue more constructively.

This chapter is designed to help you see that **guidance** is much more than just responding to children's behaviors. In fact, this entire book is about guidance in early childhood education. Indeed, everything influences children's behavior:

guidance Nonpunishing methods of leading children's behavior in positive directions so that children learn to control themselves, develop a healthy conscience, and preserve their self-esteem.

- The way adults initiate interactions
- The way adults talk to children
- The way the early childhood environment is set up
- The early childhood curriculum
- The choice of equipment, materials, activities, and projects
- The way adults deal with unsocialized behaviors

Some examples of behaviors that require guidance include biting, kicking, hitting, grabbing, name calling, and destroying materials. Such behaviors also require some consideration about how to prevent them. Prevention is an important part of any early childhood program's guidance system.

This book uses the term "guidance" instead of the more common word "discipline" because the latter term is often associated with punishment. Guidance methods focus on preventing misbehavior, not controlling behavior, by teaching children positive alternatives so they can learn self-control. Thus, guidance is a major goal of the early childhood program and its curriculum.

Because many equate "guidance" with "discipline" and "discipline" with "punishment," we will first discuss why punishment, especially spanking, has no place in early childhood guidance systems.

Punishment, Including Spanking, Is a No No

Beginners in the field of early childhood education sometimes feel at a loss when faced with difficult behavior. They recognize that they need to act quickly, but they do not know what to do. Some of these adults grew up in settings where punishment—hurting a child in some way—was the major form of discipline. Not surprisingly, these adults are unequipped to handle situations in which no one is allowed to inflict physical or emotional distress.

How you approach guidance and discipline depends on your perception of basic human nature. Here are two contrasting points of view on what children need to

Trying Hard *Not* to *Be* a Baby

Four-year-old Gabriel joined our class mid-year. His family were friends of another family that had a child enrolled in the same classroom, so we had expectations that Gabriel would just fit right in. He exhibited no signs of separation anxiety and seemed to dive right into the class routines and curriculum. He was a bright, inquisitive, and highly verbal child. But within a month, Gabriel wasn't fitting in. He was causing pandemonium wherever he played, showing aggressive and destructive behavior toward other children and play materials. Soon he had the teachers wondering about his "fitness" for a play-based school. Gabriel's mom said that this behavior was atypical, but we knew what we were seeing. Fortunately, she raised the possibility that Gabriel was trying too hard to be the big brother (he had a six-month-old sibling) and, because of his powerful feelings of trying not to be like a baby, was not able to live up to the expectations of self-regulation. She was right. His mom arranged to spend an hour a day at the school, allowing Gabriel to enjoy sharing his new school with her for about a week. Her willingness to give him the time to develop his emotional sturdiness helped Gabriel to relax into authentically experienced emotions, instead of struggling with the emotions that he felt were expected of him.

–Marion Cowee

become caring adults who can get along with others. One view is that children are not born bad or good, but they are born "uncivilized." This does not mean they are antisocial, indeed just the opposite–they are born with a need to relate to others, or what theorists call "attachment." Babies need a relationship, but they do not know how to behave according to cultural rules to survive, and they start to learn and adapt their behavior according to the social and cultural rules in their environment for their first day of life. Misbehavior gives adults a chance to teach them to be upstanding members of society in general and of their culture in particular. The most developmentally appropriate form of discipline is guidance and teaching.

In contrast, some still believe that children are born of "original sin," or that children are born bad and therefore misbehave. The teaching then is not through guidance strategies, but through punishment. A person with a punitive mindset may rationalize punishment with how he or she was raised or how society normally functions. For instance, those who break the laws that uphold society are punished accordingly. From this point of view, punishment is a form of discipline that eliminates misbehavior. I remember a bumper stick once that read: "Help Stamp Out Violence." Some people do not see the incongruity of that statement.

What is Wrong with Punishment?

Have you ever said or thought, "I was spanked as a kid, and I still turned out OK"? If so, you might feel that punishment in general and spanking in particular are

TIPS AND TECHNIQUES

Why Young Children Tell "Lies"

Children sometimes say things that are not true. We all do. We tell a joke. We tell tall tales. We make excuses. Yet when we say make false statements for negative reasons or to avoid blame, we are lying.

Young children gradually learn about the power of words and their ability to manipulate reality. They sometimes engage in wishful thinking, which means that they believe *saying* something is true *makes* it true. So when a child tells a lie, he may be more interested in changing reality than deceiveing an adult. When adults understand this characteristic of the young mind, they can manage such situations constructively.

Young children gradually learn to distinguish what is more real from what is not from their interactions and experiences with the people in their environment. We hamper this learning when we do not separate fantasy from reality. When we tell children that the tooth fairy puts the money under the pillow, why would we not also expect children to state fantasy as truth? Fantasy is not bad for children. However, when children are trying to sort things out, they may become frustrated when an adult perpetuates a fantasy under questioning. In addition, children may imitate that same approach to explain why things happen: "I didn't drop the glass. The wind knocked it out of my hand."

Sometimes a lie is really a difference in perception. When two children are arguing, each may firmly believe his or her own version of the story. Instead of intervening to decide which version is truth, educators can use this opportunity to teach them how to distinguish reality from truth by guiding the children to sort it out themselves.

The adult can encourage children in conflict to give each other feedback and to explain their own perceptions to each other. This kind of situation provides excellent practice for problem solving and conflict-resolution skills. Let children learn these skills early on. Do not decide for them what happened and who was right and who was wrong. You will not have to deal with lies if you let them sort out their disagreements on their own.

Tips for Dealing with a Child Who "Lies"

- Understand that for young children there is a fuzzy line between fantasy and reality. Children do not perceive the world in the same way as adults. Gently help them sort out the truth.
- Be truthful yourself. Honesty is taught best through modeling. If you say there is no more dessert when the freezer is full of ice cream, you are teaching lying. If you model that kind of behavior, you must expect children to engage in it as well.
- Do not back children into a corner when you know they have made a mistake. Do not ask, "Who did this?" if you already know. Most children, unknowingly for younger children, will try to save face or escape the consequences by going off into fantasyland if provoked.

A sensitive, understanding approach is more effective than heavy-handed confrontation when dealing with children who depart from the truth. And who knows–maybe their reality is more valid than ours anyway!

FOCUS ON DIVERSITY

Views of Authority

The guidance methods discussed in this chapter are based on a particular view of authority. The author of this text and many other early childhood professionals believe that children should be encouraged to think for themselves rather than blindly obey any adult. Such a belief is based on a cultural view that corresponds with the rebellious and independent spirit of the founding of this country.[5]

Yet some cultures do not share the same values or views. Instead, they believe that children should never question adults. They value conformity instead of independent thinking. They believe that proper behavior is clearly defined and that an adult's job is to teach children accordingly. They do not see life as a series of problems to be solved, but as a matter of learning how to act in each situation. They stress basic skills, patterns, ritual, and tradition over innovation, problem solving, and self-expression.

When people who hold opposing views of authority and discipline are charged with guiding the behavior of young children, clashes are bound to happen. In such circumstances, it is important that adults respect each other's views, discuss their differences, and reach consensus so the children in their care do not suffer.

effective means to control children's behavior. Nevertheless, spanking is *never* an effective guidance tool.

Children are most open to learning right after misbehaving or making a mistake. It is during this "teachable moment" that children should remain calm and receptive to understand the consequences of their misbehavior or to accept feedback on their mistake. Yet spanking and other forms of punishment destroy that calm, receptive state; they throw children into emotional turmoil and distract them from the teachable moment.

The primary lesson learned from spanking and other forms of punishment is that using force is okay. Children learn that using force is an acceptable means to get what they want, and the last thing any early childhood program wants is a child who uses force to get his or her way.

Spanking and punishment also convey the message "You had better obey or else." But consider the goal of obedience. Children who learn to be obedient are in danger of always conforming and never questioning authority. Teaching obedience hinders children from developing intellectual and critical-thinking skills. (See the *Focus on Diversity* box for a different view of authority.)

Another drawback to teaching blind obedience to authority arises as children enter adolescence. Although in the early years children see adults as the unquestioned authority, eventually many children begin to see their peer group as the authority. Children regularly taught to conform to authority may later become hesitant to speak up and defy their peers during questionable circumstances. Young children who are not encouraged to think for themselves may have difficulty making decisions about their own behavior in later years.

Side Effects of Punishment

Like strong medicine, spanking and other forms of punishment have both obvious and unexpected side effects. For instance:

- Abusive spanking and other forms of harsh punishment are humiliating and hurt self-esteem. It is important to note that the research conducted by Lindner Gunnoe suggest that some spanking is associated with authoritative parenting: "Claims that authoritative parents do not spank constitute an explicit misrepresentation of Baumrind's exposition of authoritative parenting."[6] One of the best predictors of academic success is the degree of self-esteem a child has at the elementary school age.
- Abusive spanking and other forms of punishment and humiliation leave a child hurt and angry, often inciting strong urges to retaliate against the punishing adult. Moreover, some children lacking healthy coping mechanisms or support may suppress such urges and anger only to lash out unexpectedly at others–adults, children, and objects alike–at a later time. Other children, however, may turn their anger inward and take it out on themselves.
- Children imitate the adults in their lives. When they spend time around punishing adults, they are more likely to use punishment to control other children. Early childhood professionals see this phenomenon all the time–how children are reared at home has an impact on their behavior at school. On the Baumrind spectrum of parenting styles, the most aggressive children are often raised by authoritarian parents or those with strict, inflexible views who administer severe punishments as discipline. However, children with few boundaries and under little supervision at home, from families that practice permissive or uninvolved parenting, may also become aggressive. See Table 5.1 for a summary of parenting styles according to Diana Baumrind.
- Guidance works best when the adult and child have a good relationship. Adults should avoid punishment because it erodes the relationship and diminishes the respect the child has for the adult.
- Frequent punishment can create brutal unrelenting cycle of increasing degrees of force because some children can become immune to punishment. As the punishments escalate, the adult-child relationship drastically deteriorates, eroding the bridges of communication. Child abuse is often an unfortunate outcome of this situation. Moreover, children sometimes resort to breaking the law as a way to punish the adult.

Current social norms and cultural rules enforce acceptable views on punishment that are often incongruent with the conclusions drawn from science on its efficacy. Our prison systems reflect this contradiction, we try to "rehabilitate" prisoners, yet our methods are often cruel, humiliating and ineffective. Furthermore, though adult corporal punishment is no longer tolerated in the military, prisons, and many other institutions, society has yet to apply the same restrictions to protect children. Corporal punishment is legal for parents in all states as long as they do not leave

TABLE 5.1 PARENTING STYLES ACCORDING TO THE RESEARCH OF DIANA BAUMRIND

Parenting Style	Authoritarian	Authoritative	Permissive	Uninvolved
Demand/ control level	High	High	Low	Low
Responsiveness level	Low	High	High	Low
Characteristics	Parents are strict and inflexible with expectations of unquestioning obedience. Reciprocal communication is discouraged.	Parents maintain high expectations with affection, understanding, and reason. Guidance strategies are used to redirect behavior. Communication is welcome.	Parents provide warmth and affection with little demand to obey any rules. Boundaries are few or inconsistent. There is weak parental control.	Parents are emotionally detached or self-absorbed, providing little to no attention, affection, supervision, or discipline. Boundaries are nonexistent or inconsistent.

Sources: D. Baumrind, "Child Care Practices Anteceding Three Patterns of Preschool Behavior," *Genetic Psychology Monographs* 75.1 (1967): 43–88; D. Baumrind, (1966). "Effects of Authoritative Parental Control on Child Behavior," *Child Development* 37.4 (1966): 887; Donna Hancock Hoskins, "Consequences of Parenting on Adolescent Outcomes," *Societies* 4.3 (2014): 506–531, https://doi.org/10.3390/soc4030506.

physical marks, and many states still permit corporal punishment in public schools.[7] Ultimately, the psychological wounds persist long after any physical pain and can adversely impact the mental development of young children.[8]

GUIDANCE ALTERNATIVES TO PUNISHMENT

As an early childhood student, you must learn many alternatives to physical and mental punishment. Since the goal of guidance is to teach children the skills for self-control while helping them develop a healthy conscience, the first thing you must learn is to be sensitive to the effects your actions have on children's behavior. Ask yourself: What teaching and discipline methods prevent or mitigate inadvertent negative development and learning outcomes? What guidance tools promote positive self-esteem and leave self-respect intact? What methods work without harming my relationship with the child?

It is also important to remember that there is no one approach that can always be applied to every child. You must approach each situation with a fresh mind and identify the appropriate response for the particular situation and for the particular child at his or her particular stage of development. Keeping that in mind, consider the following six alternatives to punishment: time-out, consequences, setting and enforcing limits, redirection, teaching expression of feelings, and modeling prosocial behaviors.

POINTS OF VIEW Opposing Perspectives on Time-Out

Time-out makes sense to people who come from a culture in which individual needs are the focus. When time and space alone are seen as a basic need, removing a child who is out of control is not punishment but can be seen as a form of help. A child eventually comes to understand that when overwhelmed, being alone can restore balance and calmness. Children may seek refuge in privacy when they find that it works for them. Privacy becomes an important need, but not everybody comes from a culture that stresses individual needs. Some people live in a culture that prioritizes interdependence over individual needs. For these people, time-out can be an extreme punishment because isolation from the group is seen as the worst thing that can happen to a person. In these cultures, shunning is a severe punishment and is reserved for serious offenses. When the subject of time-out came up in a diversity workshop, the group was divided sharply on its use. Both sides were emotional about their perspective. Several of the participants worked in a child care program where time-out was outlawed as a guidance measure. They saw it as too harsh. Other participants spoke strongly in defense of time-out as an effective and humane way to help children improve their behavior.

Time-Out

Removing a child from a scene in which she has misbehaved is called **time-out** by some. A time-out is a nonviolent alternative to spanking and other forms of punishment. Nevertheless, removing a child from the group is one of the least useful alternatives because it is effective only under very specific circumstances.

time-out a nonviolent alternative to punishment that removes a child from a situation in which he or she is behaving in an unacceptable way. Time-out is an effective guidance measure when the child is truly out of control and needs to be removed to settle down. Used as a punishing device by controlling adults, however, it has side effects—as does any punishment—including undermining self-esteem.

Time-out is effective only when a child is truly unable to control herself and the adult approaches the situation as a helper instead of a punisher. The child may actually *appreciate* being taken out of the situation and placed in time-out as a chance to regroup, but it is important to give the child as much time as needed to regroup by letting him or her decide when to leave. Many children can easily decide for themselves when they are ready to rejoin the group or go back to whatever they were doing before they "went out of control." Removing a child in these circumstances benefits both the adult and the child.

Unfortunately, time-out is seldom used this way. Many adults who grew up with punishment tend to systematize time-out and use it as their exclusive method to address misbehavior. You can almost envision a dunce cap on the child's head as the adult's attitude conveys "This is a bad, dumb, or unworthy child who must sit in the time-out chair for all to see." When used this way, a time-out chair is no better than the stocks used in public squares long ago. Consider the negative effect of such practices on self-respect and self-esteem!

Carefully used, time-out provides an alternative to punishment, but there are other, more effective alternatives to consider. (See the *Points of View* box for a look at different perspectives on the use of time-out.)

Learning from Consequences

Children learn from experiencing the consequences of their actions. For example, the family child care provider says, "Don't fill your milk glass too full. It will spill over." The child does it anyway and the milk drips when he picks up his glass to drink. The provider, who is tempted to say, "I told you so!" hands the child a cloth without a word. Thus, she lets him experience the consequence and allows him to correct the problem himself.

When Sarah starts pouring the water out of the water table onto the floor, she is asked to clean it up. When she refuses and keeps on pouring water out, she is told to find another activity because the water needs to stay inside the water table for safety reasons.

Kyle, who sits at the table idly playing with his food, is told, "I see you aren't hungry anymore, so it's time to put your dishes away and clean up your place now. You can either get a book when you are finished or go play quietly by your mat until nap time."

In each of these cases, the consequences of the child's actions may cause disappointment. Sarah may be distressed that she continued to pour water onto the floor and now cannot play at the water table. Kyle may regret that he did not keep eating and now his plate is in the dishwasher. The hardest part of using consequences to teach is wanting to protect the child from suffering. Nobody likes to see a child make a choice that he will regret, but that is one of life's best lessons. (See the *Focus on Diversity* box for a different view on using consequences as a guidance tool.)

Of course, adults must protect children from choices that harm them or cause injury. You would not allow a two-year-old climb over the play yard fence and think to yourself, "If you insist on playing in the parking lot you'll find out what happens to you." Yet it would be appropriate to say to a four-year-old, "It's cold outside. If you choose to go out without your jacket, you might not like it." Of course, you might hesitate giving the child this choice because you have a responsibility to the child and his parents, but this approach certainly avoids a lot of arguments. Children quickly learn how uncomfortable it feels to be outside inadequately dressed.

Setting Limits

limits Boundaries placed on children's behavior. They can be physical boundaries or verbal boundaries.

Young children need limits more than rules. **Limits** are different from rules–they are restrictions, not regulations. Some people think of them as boundaries. Limits fall into two main categories: physical limits and adult limits.

Physical limits consist of structural and security measures used to keep children safe. They are the gates at the top of stairs, the locks on medicine cabinets, or the fences around play yards. There may be rules to back up these limits ("Children may not leave the play yard without an adult"), but you can rest assured that children too young to understand rules are safe because they do not have a choice of whether or not to break the rule. There is a physical limit.

FOCUS ON DIVERSITY

Differing Perspectives on Discipline: A Personal Story

"Your children aren't like ours," an African American mother once told me in anguish when I was trying to discuss discipline with her. "Your ways don't work with our children." It has taken me a long time and some research to understand this mother's views on discipline.

"Just reprimand them," the mother would say when I was using a consequences approach. I saw her way as negative, and she saw mine as cold and unfeeling. This mother wanted to shelter the children from disappointment, but her approach to misbehavior was to shake a stern finger, issue a warning, and stop the children from acting inappropriately. She did it all in the name of love and let them know it. My approach was to offer the children choices and let them experience the consequences, even when I knew they would not like them. I also acted in the name of love, but I did not talk about it that way.

Reading an article by Cindy Ballenger—a researcher who has worked with Haitians—helped me understand some basic cultural differences in guidance goals and techniques.[9] Ballenger contrasts "mainstream" early childhood educators' ways of managing behavior with accepted Haitian methods: "The North American teachers are concerned with making a connection with the individual child, with articulating his or her feelings and problems." The North Americans use consequences to explain why not to do something. No behavior is intrinsically good or bad, it merely has consequences, which the child must learn in each situation. Consequences are the issue, not shared morals and values.

Haitian teachers do not refer to feelings or consequences; rather they "emphasize the group in their control talk, articulating the values and responsibilities of group membership." They do not differentiate specific misbehaviors but lump everything into "bad behavior." Haitian adults are clear about good and bad and so are the children. They know why they need to be good: so they do not bring shame on their families. It is a system of shared values. According to Ballenger, Haitian teachers view a reprimand not as a negative response but as one that defines and strengthens relationships.

This observation really hit home with me. In my early childhood training, I learned not to scold or reprimand. I learned how to approach discipline matters in a positive or at least impartial way. I avoided words like "good" and "bad." I never used love as a reason for doing something. And I could get through a whole day in preschool without ever saying no. I am proud of my skills, but what I have come to realize is that my approach could be misinterpreted by some children who are used to a more stringent, controlling, and therefore, to them, more loving approach.

One example of a physical limit in society is modern freeway design. For example, many off-ramps are designed so that there is no way to drive on them in the wrong direction. There's just no access. The decision is not yours; it was built into the design to protect all drivers.

NAEYC Program Standards

Program Standard 9: Physical Environment

As soon as you introduce restrictions to children, you will encounter the phenomenon called "testing." Adults understand why it is dangerous to enter a highway from an off-ramp; they do not need to try it out. However, young children are less knowledgeable than adults and need to test the physical limits they encounter. They jiggle the gate across the top of the stairs, and they bang on the locked medicine cabinet. If these safety devices are properly installed, they will hold.

When physical limits hold, young children usually do not continue to test them. Seeing and feeling the limits is all the feedback they need. Children may express frustration when they bump up against these limits, but once they discover that the boundaries are solid, they eventually go on about their business.

One big advantage of good, firm physical limits is that they provide a sense of safety. Just knowing they are there permits children to operate freely within the environment, like a horse in a fenced pasture. Without the fence, the horse must be tied or penned so it cannot wander off. With good, firm boundaries, the horse is free to move around safely within the confines of the fence.

I think of limits when I drive across the Golden Gate Bridge, near where I live. I appreciate the protection rails along the sides. Although I have never needed them to actually keep me on the bridge–that is, I have never bumped into them–I would refuse to drive across the bridge if they were not there! I need those limits to make me feel safe.

Physical limits can be seen and touched, yet adult limits are a different story. Adult limits are like an invisible fence; they are the restrictions adults impose on children for their own good, for the good of others, and for the good of the environment. For example, when children walk out to the car in the driveway, there are no physical barriers to keep them out of the street, but the adult may establish the limit by saying, "I don't want you to go near the street. It's dangerous. Hold my hand." If the children let go and wander off, the adult will call them back or go get them.

Using physical limits keeps children safe without imposing rules.
Ronstik/Shutterstock

Children only come to understand these limits through our words and our behavior. Challenging and testing adults are the ways in which they learn the strength and boundaries of these verbal limits. They continually test our limits to determine the shape, size, and strength of the invisible fence.

Testing Intangible Limits. Children spend more time testing adult limits than physical limits. The result is a good deal of misbehavior. Let us look at an example of a child testing adult limits.

A child throws a plastic toy across the room. The adult explains gently but firmly that it is not okay to throw toys. The child decides to test the limit to see if it really holds. She says to herself, "Does she really mean that?" She retrieves the toy and throws it again. The adult goes over and picks up the toy, which the child runs to grab. Holding the toy out of her reach, the adult repeats the limit.

The child stands frowning in front of the adult with her hand stretched out. "If I give it back, I won't let you throw it!" states the adult firmly. "Okay," says the child. She takes the toy and the adult remains close, watching to see if she must physically stop the child from throwing the toy. When the adult is satisfied that the throwing is over, she relaxes.

Then the child asks herself, "If I drop the toy straight down on the floor, will she stop me?" She tries it, but there is no response. She decides that the limit has to do with throwing not dropping.

She is still not through testing, though. Now she asks herself, "Can I throw some toys but not others?" She picks up a rubber ball and throws it. The adult comes to her side immediately. "You can go outside if you want to throw the rubber ball, or you can throw this foam ball inside." The child considers her choices, then takes the foam ball from the adult's hand and begins throwing it against the wall. The adult smiles at her and goes to another part of the room.

This process of testing may not occur all at once nor is it always conscious. It is, however, very real. This child is not being "bad." She's just testing. Her sense of security depends to some extent on discovering that there are steadfast limits.

Testing stops when children discover just how far they can go. They do not need to keep checking once they discover where the invisible fence lies, and they gain freedom once they know where the boundaries are.

As with physical limits, it is important for children to know that the invisible fence doesn't move when they bump up against it. Adults must be consistent about setting limits. When they are not, children cannot prove anything by testing and, therefore, must keep on trying. If they never get a clear set of results, they test excessively and run the risk of being labeled problem children.

Children never stop testing. As they grow and develop, their world expands and adult limits change to correspond with their new abilities. They continually face new boundaries to explore (that is until they reach adulthood and have to set their own limits, as well as deal with the limits set by society).

At times, power struggles occur when children are unexpectedly blocked by adult limits. While most children would surrender to a locked door or a stone wall, many will also argue with, tease, whine at, fuss at, and sometimes directly attack an adult until they prove to themselves that the verbal boundary is firm. You may be tempted to back down when faced with an upset child. Nevertheless, remember that children will continue to test you if they think they can get you to change your mind. It is important to be clear about how reasonable the limits are before you set them, not afterwards. Only set limits that you can follow through on.

Finally, as important as it is to set boundaries and keep limits, it is also important to be wise about them. Remember, when faced with enforcing limits, you are not necessarily giving in when you see things from the child's point of view–you are showing respect. Do not change the limit just because you take a different perspective; however, do consider whether you need to reevaluate your stance. If necessary, alter the limit when the situation arises at a later time. Think of adult limits as movable barriers rather than permanent stone walls.

Guidelines for Thinking About Limits. There are three questions to consider when setting limits.

Has the Child Outgrown the Limit? Things change, and what was once an appropriate restriction when the child was two may no longer apply. Obviously, a two-year-old should not be allowed to use a chair to reach something on a high shelf, but the same child at eight years old is capable of climbing a stepstool to clean out the top of a closet. Remember that many limits deserve periodic review.

Is There a Valid Reason for the Limit? Sometimes adults establish limits and hold to them no matter what, even when there is no legitimate justification for the restriction. In order to avoid that problem, when faced with unconventional behavior, ask yourself, "Why not?" If you cannot think of a good reason, lift the restriction.

Can I Rearrange the Environment to Eliminate the Restriction? One infant center had a problem with children climbing up a bookcase that held their toys. Even though unconventional behavior was respected in this program, this climbing had to be prohibited because it was dangerous. However, the caregivers could not stop the children from climbing that bookcase; they were constantly removing children off the shelves. Then the head teacher came up with a creative solution: she emptied out the bookcase, put it on its back on the floor, and let the children crawl all over it–safely.

Redirection

Redirection is an alternate approach to enforcing limits that tend to create power struggles. When you can deflect a child's challenging energy in positive directions, firmly upholding limits is unnecessary. For instance, you can redirect a child splashing water in the bathroom on a warm day to the water table. Exuberant children confined indoors on a rainy day can be organized into a rhythm band in a lively, noisy parade. The toddler who seizes a torn book page can be given a piece of scrap paper to rip to his heart's content.

Redirection differs from distraction. distraction is a sharp change of focus designed to get a child's mind off whatever he was doing or feeling. In contrast, redirection respects the child's energy and feelings but shifts them in a direction or to an activity that is more acceptable.

Teaching Children to Express Their Feelings

Children often get into trouble when they do not know how to express angry feelings. The four-year-old cannot pull apart the Lego blocks, so she throws them on the floor in frustration. The two-year-old hits another child who has just snatched a book from his hands. The five-year-old who is reminded to pick up his jacket and hang it on the hook screams, "I hate you!" at the teacher and runs outside.

What would you do in each of those situations? To begin with, you should accept the child's feelings and put them into words. "You're really upset that you can't get those blocks apart." "You're angry because you wanted that book." "You're mad at me."

After acknowledging the child's feelings, you can deal with the inappropriate behavior. You will be more effective if you take a problem-solving attitude instead of reacting with your own display of emotion. (It is interesting how one person's anger can so easily trigger that of another person, even when the second person is not a target of the anger.) Here are some problem-solving responses:

> "I wonder how you could get those blocks apart without throwing them. I worry that you might hurt someone if you throw them–or maybe break the plastic."
>
> "Tell him how you feel. It's okay to let him know you are upset that he grabbed the book away, but it's not okay to hit him."
>
> "I see you are really angry with me. Let's sit down and talk about it. Is it just the jacket that's the problem, or is there something else going on?"

In the preceding examples, the adult is not expressing her personal feelings. But what if she is feeling emotional? What if she is not able to step aside from her own

Teaching children how to express their feelings through guidance or modeling in moments of frustration helps them process negative feelings appropriately, without harming oneself, others, or the environment.

feelings? If she too is upset and angry, it is better to be honest than to pretend. Here are some appropriate ways for the upset adult to express her feelings:

"I don't like it when you throw the blocks."

"I get angry when I see you hit him like that."

"It upsets me when someone says 'I hate you.'"

Do not expect that simply expressing your feelings will change the immediate situation. Just because you are angry does not mean the child will change his or her behavior. Also, do not express your feelings to manipulate the child's behavior; there are other ways to guide misbehavior. Express your feelings just because it is honest communication and because you are modeling appropriate ways to let the child know how you feel.

Managing angry feelings is a challenge for both children and adults alike. When is it appropriate to act on those feelings? What are some ways to express them that won't hurt anyone? When faced with a child's anger, it is important

- to accept and acknowledge it.
- to model appropriate expression.
- to teach children the difference between emotions and behavior.

Let us examine those points. When a child is angry, it is important to help the child understand that it is always all right to *feel* something. By acknowledging and accepting the child's feelings, you give him or her permission to accept them as well. Have you ever heard anyone say to a child, "You shouldn't feel that way"? Such a statement conveys a message that the child's feelings are not appropriate and that the child should bury those feelings.

The feelings themselves do not create difficulties; it is the behavior that sometimes accompanies them. How the child chooses to express or act on his or her feelings that may not be appropriate. This is when it is important for you to help the child understand the difference beween feeling something and acting on that feeling. You can teach the child that there are a variety of means to express emotion other than simply using words. Some alternatives to saying how one feels include expression through body language, vigorous movement, and art and music. Guidance and adult modeling are important to teach the child how to accept his or her feelings and to express those feelings appropriately–without harming oneself, others, or the environment. (See the *Points of View* box for a discussion of two cultural views of approaches to handling tantrums–one way some children express feelings.)

Modeling Prosocial Behaviors

To prevent misbehavior, you must model appropriate behavior yourself. Before any child ever had a chance to pound on the family dog, one family child care provider gave lessons on how to pet him. While demonstrating, she said, "See, this is the way he likes his fur rubbed. You have to do it gently."

Unfortunately, some inappropriate behavior cannot be prevented. In such cases it is important that you model a constructive response. For all early childhood

Expressing Feelings: Two Views of Tantrums

The current practice in early childhood education is to accept all feelings as valid. Some practitioners even see the importance of children completely exploring a feeling such as rage. They encourage children to "work it through" and regard the process as something that should not be interrupted until it is finished. The theory is that children's unexpressed feelings may remain unfelt and go underground, only to pop up again and again when triggered by some minor incident. Tantrums are acceptable to these early childhood practitioners because they are a means for children to "get it out of their system." The adult's job, as they see it, is to keep the child safe during the tantrum but not to distract or otherwise stop the process once it starts.

These educators do, however, also believe in preventing tantrums. One prevention method is to examine the child's frustration level and see if some of the stress can be removed. Early childhood educators also prevent tantrums by not making a big fuss over them. When a child learns that anger brings attention, he may learn to use a tantrum as a means to put a spotlight on himself.

Some tantrums are attempts to manipulate adults. When an adult rewards a tantrum by "giving in," the child will try the same means to get something else she wants. Early childhood educators who believe in allowing tantrums know how to avoid being manipulated by a small screaming child.

On the Other Hand . . .

Not everyone believes in the benefits of tantrums, even those who know how to avoid being controlled. Some people believe that the individual expression of feelings (such as a tantrum) is not as important as respecting authority or maintaining group harmony.

Jerome Kagan, in *The Nature of the Child,* looked at the subject from a cross-cultural perspective. He said:

> Americans place greater value on sincerity and personal honesty than on social harmony. But in many cultures–Java, Japan, and China, for example–the importance of maintaining harmonious social relationships, and of adopting a posture of respect for the feelings of elders and of authority, demands that each person not only suppress anger but, in addition, be ready to withhold complete honesty about personal feelings in order to avoid hurting another. This pragmatic view of honesty is regarded as a quality characteristic of the most mature adult and is not given the derogatory labels of insincerity or hypocrisy.[10]

educators, modeling is a valuable guidance tool. Although modeling can be used as a teaching technique, it is also an important guidance alternative to punishment, as shown in the example of modeling in the following scene.

Two little girls are sitting on the family-room floor playing happily when suddenly and for no apparent reason two-year-old Shelby reaches out and shoves Amanda, her baby sister. The caregiver hears a thud as Amanda's head hits the carpet. Startled, Amanda starts crying loudly.

The caregiver's instinct is to grab Shelby angrily and correct her firmly. This is a no-nonsense situation. But she knows that responding to aggression with aggression just creates more aggression. She reminds herself that if she responds in anger, the lesson in nonaggression will be lost. The caregiver sets her feelings aside for the moment so she can approach both girls calmly. She is not faking calmness; she actually feels it. After a lot of practice, she is able now to remain emotionally detached once she remembers what her goals are–to teach nonaggression.

The caregiver gets down to Shelby's eye level and begins talking quietly while stroking Shelby lightly. She lets touch and tone add meaning to the words. "Gently, Shelby, gently. You hurt Amanda when you push her."

She turns to Amanda and says soothingly, "It hurt when you bumped." She touches her head where it hit the floor. Then, just to get the facts into words, the caregiver says, "Shelby pushed you over."

She turns back to Shelby and says softly, "See how Amanda's crying. You hurt her." She touches Shelby on the head the way she touched Amanda. Neither her tone nor her touch are accusing. She is stating the facts and modeling gentleness.

The caregiver turns her attention back to Amanda but does not say anything. She sits quietly by Amanda and allows her calming presence to help the baby get herself back together again.

In a surprisingly short time, Amanda calms herself, rolls over, and crawls away. She goes straight for a drum on the floor nearby. She is happily banging on it when Shelby starts for her, saying more to herself than to the caregiver, "Careful, not hurt Amanda!" Her tone is the same gentle tone she just heard the caregiver use. She bends over her little sister and gives her a big kiss on the head.

This caregiver responded to aggression with gentleness, not because it was a natural response but because she had learned that gentleness breeds gentleness.

While you are modeling prosocial behaviors, pay special attention to the way you talk to children. The *Tips and Techniques* box gives some ideas about using words as guidance strategies.

This chapter has set forth six guidance tools to use as alternatives to punishment. To summarize, the tools are

1. Removing the child (time-out)
2. Allowing children to experience the consequences of their actions
3. Setting limits and enforcing them
4. Redirecting inappropriate behavior
5. Teaching appropriate expression of feelings
6. Modeling prosocial behaviors

Nevertheless, these tools are useless if the early childhood educator does not learn how to read behavior to understand its meaning and to know how to appropriately respond. (See page 139 for more tips on handling children with challenging behaviors.) We move now to a section designed to help you identify the causes behind some inappropriate behavior.

TIPS AND TECHNIQUES

Words as Guidance Strategies

These suggestions were inspired by Tom Udell and Gary Glasenapp in their 2005 article Managing Challenging Behaviors: Adult Communication as a Prevention and Teaching Tool.

- When a particular behavior needs correcting, be specific, clear, and positive. "No, that's dangerous!" is one response to watching a child climbing where she should not. "Keep your feet on the floor" or "Feet belong on the floor" are positive ways to guide the child rather than "Don't climb on the table."
- If you are giving a direction, do not ask a question. "Do you want to eat now?" is not the same as saying, "It's time to sit down for lunch." Of course, if the child has a choice about eating, that is different, but often adults sound as if they are giving a choice when they are not.
- Avoid comparison among children and stay away from competitions. "Who can be first to put the books away?" invites comparisons and competitions.
- Notice when a child is behaving appropriately and mention it. Though it may be hard to remember that, Udall and Glasenapp suggest that you should give four positive or encouraging feedback comments to every comment about inappropriate behavior. It takes real effort to manage to meet that goal, but it's worth it!

Udell, T., and Glasenapp, G. (2005). Managing challenging behaviors: Adult communication as a prevention and teaching tool. In B. Neugebaurer (Ed.), *Behavior* (pp. 26–29). Redmond WA: Exchange Press.

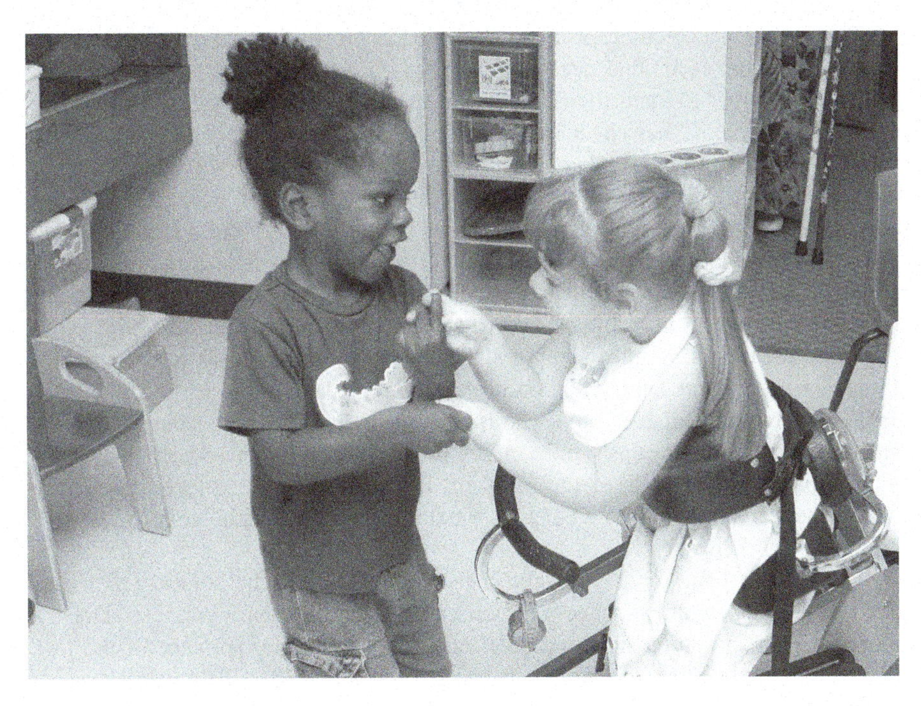

A lesson in gentleness and prosocial behavior.
Realistic Reflections

Focus on Inclusion: Children with Special Needs

NAEYC Program Standards
Position on Inclusion

Children Who Are a Handful! Some children arrive in early childhood programs with an array of challenging behaviors. Some of these children are identified as children with special needs. They may have already been labeled as "emotionally disturbed," or be said to have "attention deficit disorder" (ADD) or "attention deficit with hyperactivity disorder" (ADHD). Examples of challenging behaviors include impulsiveness, aggression, excessive movement, and lack of self-regulation. Some children lack the ability to concentrate. Others lack the ability to follow rules or keep within the limits. Nevertheless, ECE educators should try to evaluate these behaviors through the lens of the child to better understand its origins. These difficult behaviors will be most effective if the teacher tries to understand and respond to what the child is trying to communicate. What might the child be conveying by hitting or grabbing, for example? His lack of social skills? His inability to wait? His frustration? That there is too much stimulation for him? Not enough stimulation? That he needs attention? The task of the teacher or provider is to read the behavior, understand its meaning, and respond with what the child needs. If there is too much stimulation, reduce it. If there is too little, find ways to enhance the environment with activities and materials that will engage the child. Is there little regard for rules? Perhaps the limits are too tight or perhaps too loose. Is attention needed? Give it to him! If the child needs attention and he has been getting it in all the wrong ways, then change the situation. But do not just withdraw attention for misbehavior and not replace it. The need for attention does not subside when attention is removed. It is still there. The teacher or care provider must help the child substitute the difficult behaviors with acceptable alternative behaviors.

Reflective practice comes in here. Observation provides an important means of determining what each child needs and what to do about it. Next, a warm relationship between the child and the adult is vital to successful work with the child. If a child feels bad about herself, it is hard to work with her. Also, pick one behavior or issue at a time; do not tackle everything at once. Note the times when the challenging behavior is likely to occur and figure out ways to stop it before it starts. For example, do not keep a high-energy child who has trouble with self-control standing in line. If a child faced with choices goes out of control, limit choices and restructure the environment. During free play periods, stay close, if possible, so you can get the child involved in play and so you can help out during interactions with other children. Assure the child and others that you will maintain control if needed. Protection is the issue here. Children need to be protected, which is the adult's job. Take notice when it might be you who is causing the problem. For example, getting angry at a child who is upset only makes things worse. To stay objective, in your mind separate the child from the behavior. Every child needs to feel accepted by her teacher no matter how she behaves. Finally, do not underestimate a child's ability to grow and change!

Challenging behavior has always been a problem, but the solution to it has become suspension and expulsion. According to data collected by the U.S. Department of Education on public preschools, 2.6 million children experienced 1 or more suspensions and over 110,000 were expelled during the 2013–2014 school year.[11]

Research from the Yale University Child Study Center further suggests that expulsion rates in private preschools may be twice as high as those of public preschools. What is happening? One answer is that preschool teachers don't have access to mental health consultants to work with them on problem behavior. Another answer is that teachers don't have the skills to guide into positive pathways the children who challenge them. Recent research from the Yale University Child Study Center suggests that rates of expulsion and suspensions may be driven by implicit bias. Preschool teachers of all races tend to observe African American children more closely, particularly boys, when expecting challenging behaviors.[12] The study correlates with the uneven rates of discipline experience by African American boys, which in 2013–2014 represented only 19 percent of the male public preschool enrollment but made up 45 percent of the male suspensions.

The lack of developmentally appropriate interventions or responses can adversely effect a child's development. Suspensions and expulsions can not only squander opportunities for a child to learn positive behaviors, but may also exacerbate the existing maladaptive behaviors. Parents may even erringly be coerced into medicating their child or placing their child in special needs classes,[13] inflating the rates of learning disorders such as attention deficit with hyperactivity disorder and conduct disorder. Barbara Kaiser and Judy Sklar Rasminsky see the importance of including children with challenging behaviors in child care communities, and they offer some approaches to use to keep children in the programs instead of rejecting them.[14] They focus on strengths and recognizing appropriate behavior. They also suggest looking at the environment for possible modifications. Teaching social-emotional skills and problem solving are other approaches they suggest.

The Division for Early Childhood (DEC) of the Council for Exceptional Children published a position statement in 2017 on Interventions for Challenging Behavior, which not only acknowledges that challenging behaviors can interfere with the development of social-emotional competencies, but also that children who lack such skills can develop challenging behaviors to communicate their needs.[15] The DEC also adamantly acknowledges that punitive practices such as corporal punishment, suspensions, and expulsions hinder a child's ability to learn important social and communicative behaviors and disproportionately affect young children of color. Therefore, they emphasize positive and preventive interventions and strategies that support the development of all children, including:

- preventing challenging behavior through designing the environment and activities.
- using effective behavioral interventions that are positive and address both form and function of the behavior.
- modifying the curriculum and using accommodation strategies to help young children learn appropriate behaviors.
- seeking external consultation and technical assistance or additional staff support.
- training all staff in skills necessary for effective prevention and intervention programs.

- practicing activities that support children's home culture to help them develop healthy identities.
  - -assessing children from infancy through early childhood to screen for social-emotional competencies, mental health, and challenging behavior.
  - -forging strong relationships with families.

Families are part of any intervention team when it comes to designing and carrying out effective interventions for challenging behavior. Success comes from a coordinated effort between family members and professionals so that approaches address both child and family needs and strengths.

Interpreting Children's Behavior

Behavior is communication; it tells us what the child needs. Adults must correctly interpret those messages, but getting the right interpretation takes practice. To help you examine what may be behind inappropriate behavior, consider the following six questions.

Are the Child's Basic Needs Met? Examples: A four-year-old throws a tantrum every day at about 10:00; on the day after Halloween, the children are fussy and tired; on a rainy day, the children are restless from being cooped up indoors.

In each of these examples, you must consider if the children's needs are being met. For example, does the four-year-old get hungry before snack time? Instead of using guidance tools to get her back on track, try giving her something to eat at, say, 9:30. Chances are, the tantrums will vanish. Likewise, try creating a soothing environment that helps the post-Halloween children calm down and relax; and, in the case of rainy-day restlessness, create a gross-motor area to relieve the tension and reduce the need for constant redirection and other guidance measures. Although, you cannot meet all the basic needs of all the children all the time, keep in mind that unmet needs play a part in misbehavior.

Does the Environment Fit the Child? Example: A two-year-old in a preschool classroom runs into limits at every turn. He has to be stopped from dumping the collage materials, puddling the glue and finger painting in it, and scattering puzzle pieces. The teacher spends the day reminding him of the limits and redirecting him. The problem is that the materials and activities are inappropriate for his age. The solution would be to find an area where he can play freely and explore appropriate materials on his own.

Example: A five-year-old has been in child care since birth, and the preschool environment holds fewer challenges for her than she needs. To overcome her boredom, the child creates her own diversions only to get in trouble with the staff. The solution to this problem is to adapt the environment to add novelty and meet the child's developmental needs, as well as those of the other children. Such a solution can go a long way to eliminate misbehavior and the need for guidance.

Is the Child's Behavior a Cry for Attention? Example: A child has been acting up all morning but is behaving appropriately at the moment. The teacher asks, "Jorgito, how would you like to help me plant these seeds?"

In this genuine distress or a bid for attention?
Radius Images/Alamy Stock Photo

Do not discount the need for attention; respond to it. Be aware of how children use misbehavior to get adults to respond. Make a clear plan to lavish attention on children who need it when they are *not* misbehaving; use the opportunity to give them a "time-in"–the opposite of a time-out. Set aside periods when you focus on a single child or a small group of children who need more intense attention than they can get in larger groups. Stanley Greenspan calls this approach "floor time" because the adult gets right down on the floor with the child or children, not to direct them but to remain fully available to respond to them.[16]

Is the Child's Behavior a Response to Feeling Powerless? Example: A toddler bites whenever she gets frustrated or wants something.

Children who feel powerless need to get in touch with their own power. In this case, the child uses the most powerful muscles in her body in the most powerful way she knows. (See more on biting and what to do about it in the *Tips and Techniques* box on page 144.) There are several ways to help such children, such as teaching them skills, giving them responsibility, encouraging them to express themselves, and giving them the language to do it.

Did the Child Learn This Behavior by Being Rewarded for It in the Past? Example: When asked to come inside for lunch, Taylor screams no and runs in the other direction. The teacher pays a lot of attention to Taylor, talking, scolding, and threatening until, finally, she manages to get him inside by promising that he can sit next to her at lunch.

When such a situation becomes a pattern, children learn that defiant behavior often gets them special attention. The way to change the pattern is to remove the reward. It is important to understand the principles of learning theory. The best approach to changing unwanted behaviors is not to get them started in the first place. If the teacher in the example had been aware of the effects of her response to Taylor's defiance, she might have considered alternative ways of getting him to lunch. But it is too late; now she has to retrain Taylor by not paying so much attention to his defiance and by rewarding him instead when he cooperates. Such a system of retraining is called **behavior modification.**

behavior modification
A form of systematic training that attempts to change unacceptable behavior patterns. It involves reinforcing acceptable behavior rather than paying attention to and thus rewarding unacceptable behavior.

TIPS AND TECHNIQUES

Biting: A Universal Toddler Problem

Screams filled the hallway of the child care program I was visiting. "Sounds like the gators are snapping," remarked the director, who was showing me around. "That's the toddler room," she explained in answer to my questioning look. "I just hope Jaws isn't after the Princess again," she added cryptically.

She went on to explain the alligator reference: "We call the toddlers 'the gators' because they are always biting each other." She did not have to tell me that toddlers bite. I knew that from experience.

Toddlers bite because they cannot yet express their thoughts and feelings very well, so they use their mouths in more direct ways to gain power. Biting is a behavior that is easier to understand than to control. Yet control it you must–not after the fact but before it happens. *Prevention* is the key word. You would not ignore a toddler with a loaded gun in his hand, and a mouth full of teeth in some toddlers is a lot like a loaded gun. Since you cannot take the teeth away from him, you *must* keep him from using them on other people. Here are four tips to prevent biting.

Be vigilant. Whenever two toddlers are together and one of them is a known biter, supervise closely. You must prevent biting until the children learn other ways to express their desires and affection, touch and explore each other, get their needs met, ask for attention, and feel powerful.

Help toddlers feel powerful by giving choices and by offering challenges that require *strength and skill.* Feed a child's need to feel powerful rather than deny it. When children are made to see themselves as small, weak, and inadequate, their power needs grow, sometimes to a monstrous size. The more helpless they feel, the more likely they are to use the strongest muscles in their bodies–the jaws!

Provide props for the children to act out their aggressions symbolically. Toy alligators can be useful for this purpose. Or try hand puppets that can "bite" without hurting.

Teach toddlers how to defend themselves–to keep themselves from being bitten. By this I do not mean to teach children to bite first or slug it out. The best way to explain this tip is to offer my own version of the "Jaws and Princess" story.

Princess is innocently playing with a yellow ball when along comes Jaws, the child who has bitten her many times in the past. Jaws wants the yellow ball–or perhaps he just wants another taste of Princess's tender arm. He approaches her with his mouth open wide. But today is different. Princess is now armed with some nonviolent self-defense skills.

First, she puts up one hand and says in her firmest toddler voice, "Stop!" At the same time, with the other hand, she takes a plastic teething ring out of her pocket. Stretching out her arm, she gently places the teething ring into the oncoming mouth. Surprised, Jaws bites down on it instead of her arm.

Does This Child Clearly Understand Why Her Behavior Is Inappropriate? Example: Mike is toasting a play waffle. He pops it out, and Stephanie comes over and grabs it out of his hand. He protests angrily, but Stephanie ignores him and holds the waffle out of his reach. The teacher says to Stephanie, "Mike doesn't like it when you grab things away. He gets angry. Just look at his face! Mike, tell Stephanie how you are feeling."

Children have to learn about how their actions affect other people. The early childhood educator's job is to help them gain this understanding by facilitating communication–on both sides of the conflict.

But what if the teacher had gotten angry at Stephanie's grabbiness and intervened in a more demanding way to try to get Stephanie to give back the waffle. When Stephanie refuses, the teacher finds herself in a win-lose situation. She now has a stake in the outcome. Will she get the waffle and give it to Mike, or will Stephanie win and end up keeping it? The teacher may remember moments of unfairness from her own childhood and want to settle an old score by getting that waffle and punishing Stephanie.

As you can see, it is important to rely on self-reflection and awareness to move away from imposing adult judgments and punishment to using guidance tools. When faced with a misdeed that triggers strong feelings, pause and ask yourself two important questions: Do I really want to change the behavior, or am I more interested in seeking revenge for what the child did? Am I more interested in winning this conflict than in changing the behavior?

Revenge is a powerful motivator. Sometimes adults feel strongly about "giving the child a taste of her own medicine" or "giving her what she deserves." But taking out angry feelings on a child is an ineffective way to change behavior.

Some adults see winning power struggles as the best way to gain children's respect. Just remember, when there is a winner, there is also a loser. Losing can damage a child's self-respect and self-esteem, as well as destroy any relationship you have built thus far. In a confrontation with a child, it's important to take a problem-solving attitude instead of a power stance. The goal of the problem solving should be for both the adult and child to come out feeling okay about the solution. Win-win solutions are the very best solutions in any conflict.

If you really want to change inappropriate behavior rather than dispense "justice" or assert your power over the child, consider some of the alternatives to punishment that we have explored in this chapter.

Remember that you are the child's ally when it comes to guidance. Your goal is to help children grow into socialized beings by preventing or transforming problem behavior and aiding them in developing their inner controls to regulate their own behavior.

A Story to End With

As a teacher, I learned early that giving children choices prevents many head-on confrontations. Instead of saying, "No, you can't do that" I learned to say, "If you want to play with the guinea pig, you have to be very gentle. If you want to roughhouse, go play with the stuffed animals." I was an expert at giving choices. "Do you want to take your medicine yourself, or should I hold the spoon?" "Are you ready for a diaper change, or do you want to wait until you finish climbing on the slide?" "Do you want to climb up on the diaper table yourself, or should I put you up?" Having a choice gives a child a feeling of power–a way out in the face of an insisting adult.

It is not that I never said no, but I saved the no's for important occasions so they were dramatic enough to have some real meaning. And because I also knew that children imitate adults, saying no to every little thing would have come back to me in the form of children saying no to me.

Knowing about modeling, then, I should not have been surprised the day I picked up my four-year-old son from his preschool. He climbed into his car seat, waited for me to buckle him in, and then looked me straight in the eye and announced in a voice that sounded a lot like my own: "You have two choices, Mom. You can either take me to play at a friend's house or take me to the store and buy me a toy." Going home was not one of my choices. But luckily, one of the alternatives he named was acceptable to me, so I arranged a visit to a friend's house.

I have thought about that occasion since and worried about the manipulating aspects of my son's approach. It felt different when I was on the other end of the choices. I don't like to be manipulated! But then I decided I could see the episode in a different light: I could regard my son's offering me two choices as a means of opening up a dialogue. If I had not liked either of his options, I could have responded with two more of my own, and we could have negotiated back and forth until we reached a mutual agreement.

SUMMARY

Guiding young children's behavior in positive directions starts with an understanding of developmentally appropriate behavior expectations. It's tempting to respond to misbehavior with punishment, but spanking and other forms of punishment–both mental and physical–have side effects such as anger, humiliation, revenge, loss of self-esteem, and damage to the adult-child relationship. The goal of guidance in the early childhood program is to help children learn to control themselves and, at the same time, develop a healthy conscience. There are six guidance tools that the early childhood educator should use in lieu of punishment: time-out, consequences, setting and enforcing limits, redirection, teaching expression of feelings, and modeling prosocial behaviors. Children's behavior is their way of telling adults what they need. There are six questions to consider when trying to understand any misbehavior's underlying message: (1) Are the child's basic needs met? (2) Does the environment fit the child? (3) Is the child's behavior a cry for attention? (4) Is the child's behavior a response to feeling powerless? (5) Did the child learn this behavior by being rewarded for it in the past? (6) Does the child clearly understand why her behavior is inappropriate?

REFLECTION QUESTIONS

1. How do you feel when a child tells you something that is not true? Do you think of the child as lying? How do you react? Do you ever tell lies? Do you think that there are times when lies do not hurt–or actually do good? What is your view on honesty as the highest value? Do you think your ideas, feelings, and views relate to your culture?
2. Were you spanked as a child? How do you feel about spanking? What is your reaction to the oppositional stance that the text takes on the subject?
3. What are your views of and feelings about authority in the lives of young children? What do you mean when you use the word *authority?* Do you regard your views, feelings, and meanings as a cultural issue?
4. What is your experience with using time-outs? Were you ever placed in time-out as a child? Have you used this approach as an adult? How do you feel about it? Do you agree with the view of the text?

5. Did you have any reactions to the *Focus on Diversity* box on page 131 (differing perspectives on discipline)? What were they?
6. What are some examples of limits in your own life? Have you ever tested these limits? Why or why not?

Terms to Know

How many of the following words can you use in a sentence? Do you know what they mean?

guidance 123
time-out 129
limits 130
behavior modification 143

For Further Reading

Brault, L., and Brault, T. (2005). *Children with Challenging Behavior: Strategies for Reflective Thinking*. Phoenix, AZ: CPG Publishing.

Carter, D. R., Norman, R., and Tredwell, C. (2011). Program-wide positive behavior support in preschool: Lessons for getting started. *Early Childhood Education Journal, 38* (5), 349–55.

Elliot, E., and Gonzalez-Mena, J. (2011). Babies' self-regulation: Taking a broad perspective. *Young Children, 66* (1), 28–33.

Florez, I. R. (2011). Developing young children's self-regulation through everyday experiences. *Young Children, 66* (4), 46–51.

Gonzalez-Mena, J., and Shareef, I. (2005). Discussing diverse perspectives on guidance. *Young Children, 60* (6), 34–38.

Hancock, C., and Carter, D. (2016). Preschool: Building environments that encourage positive behavior: The preschool behavior support self-assessment. *YC Young Children, 71* (1), 66–73.

Hirschland, D. (2008). *Collaborative Intervention in Early Childhood: Consulting with Parents and Teachers of 3–7 Year Olds*. New York: Oxford University Press.

Jacobson, T. (2008). *Don't Get So Upset! Help Young Children Manage Their Feelings by Understanding Your Own*. St. Paul, MN: Redleaf Press.

Kersey, K. C., and Masterson, M. L. (2011). Learn to say yes when you want to say no to create cooperation instead of resistance. *Young Children, 66* (4), 40–44.

Luckenbill, J. (2012). Getting the picture: Using the digital camera as a tool to support reflective practice and responsive care. *Young Children, 67* (2), 28–36.

Marion, M. (2018). *Guidance of Young Children,* 10th ed. Upper Saddle River:, NJ: Pearson.

Nelsen, J., Foster, S., and Raphael, A. (2011). *Positive Discipline for Children with Special Needs: Raising and Teaching all Children to Become Resilient, Responsible and Respectful*. New York: Three Rivers.

Price, C., and Steed, E. (2016). Culturally responsive strategies to support young children with challenging behavior. *YC Young Children, 71* (5), 36–43.

Reynolds, E. (2006). *Guiding Young Children: A Problem Solving Approach*. New York: McGraw-Hill.

Udell, T., and Glasenapp, G. (2005). Managing challenging behaviors: Adult communication as a prevention and teaching tool. In B. Neugebaurer (Ed.), *Behavior* (pp. 26–29). Redmond, WA: Exchange Press.

Weatherson, D., Weigand, R. F., and Weigand, B. (2010). Reflective supervision: Supporting reflection as a cornerstone for competency. *Zero to Three, 31* (2), 22–40.

6 The Teacher as Model

Rob Van Petten/Photodisc/Getty Images

Modeling Nonviolent Problem Solving

Seeking Information

Recognizing Alternatives

Considering Consequences

The Many Roots of Violence

Modeling Self-Esteem

Modeling Virtue

Modeling Power

Modeling Significance

Modeling Competence

Modeling Equity

Modeling Learning, Development, and Cognition

Reflective Practice and the Importance of Observation

Creating an Emergent Curriculum

A Story to End With

In This Chapter You Will Discover

- how teenage violence has its roots in the preschool years.
- what adults can do to help young children find alternatives to aggression during a conflict.
- how to teach young children to consider the consequences of lashing out aggressively.
- how adult self-esteem affects children's self-esteem.
- four aspects of self-esteem.
- what "antibias focus" means.
- examples of adult behavior that give children gender-biased messages.
- why a teacher also needs to be a learner.
- why observation is an important skill for the early childhood educator.
- what an "emergent curriculum" is.

Modeling as a Teaching Strategy

Modeling seems like such a simple approach; does it really work? Yes! Especially when you don't want it to. Modeling is about imitating and is also known in scientific circles as observational learning. Children watch and listen to the people around them and then exhibit the same behaviors. In the famous Bobo Doll study, Albert Bandura and his colleagues found that young children would mimic both aggressive and considerate behaviors of adults when given similar situations. Bandura's research demonstrated that observation is a powerful learning tool for children and can contribute to negative learning outcomes if adults fail to set boundaries on their own behaviors. Fortunately, the reverse is also true; children also learn positive behaviors by observing assertive adults who use positive strategies to navigate their relationships. Nonaggressive behavior is modeled too. Albert Bandura named what he was studying social learning theory which he later renamed to social-cognitive theory because children do not simply imitate. They take what they see, hear, and remember and make inferences about the norms and rules of their social environment. They also do not model after just anyone but are more likely to imitate people they perceive as powerful and whom they see as being rewarded for their behavior.

Albert Bandura, "Social Cognitive Theory." In *Annals of Child Development*, Vol. 6, ed. R. Vasta. Greenwich, CT: JAI Press, 1989, pp. 1–60.

modeling A teaching device and guidance tool in which an adult's attitude or behavior becomes an example the child consciously or unconsciously imitates.

"Actions speak louder than words" and this is true for all people, and particularly for young children. Imitation is one of the many ways children learn, and children copy adult behavior. Adults model for children all the time, whether they know it or not. Marion Cowee says that if you want to see how you look as a teacher, watch your children play school. See the *Voices of Experience* box on page 153. Consciously using **modeling** as a teaching method is one of the most powerful ways to influence children's behavior. The problem is that the modeling effect cannot be turned off and on at will. Whether or not we are aware, our actions teach all the time, even when we are setting bad examples.

As an early childhood student, you must realize how important it is to be aware of your own behavior. If you yell at a group of children to be quiet, they will probably just get louder. If you give a lecture on sharing and then do not share your own grapes, your lecture will have less impact on the children than your own behavior.[1]

What you *do* carries a stronger message than what you *say*, which is why preaching to children doesn't work very well. It is important to support your words with your actions–by setting an example. See *The Theory Behind the Practice* box above for a peek at the research that supports this idea.

The thought of so many eyes watching your actions all the time can bring out the perfectionist tendencies in anyone. Perfectionism, or a fear of making mistakes, can create such stress that it clouds one's decision making. Lilian Katz, an early

TIPS AND TECHNIQUES

Children Are Fragile

A little boy ended up with a dislocated elbow when his teacher tried to keep him from running out in the street. It was not anybody's fault. The teacher was trying to protect the child, but when she grabbed his arm and jerked him back, the tendons gave. She had no idea that would happen, and she felt terrible.

Children are fragile, and they can be hurt when we treat them roughly. Shaking tots, even a little, can cause brain damage. The muscles are not strong enough to control the weight of the head when it snaps back and forth.

It is hard to grasp the concept of being firm but gentle. Sometimes we have to be hard and unyielding, but we *never* have to be rough. Some adults who are rough on children are also rough on themselves. They have not yet learned that being gentle on themselves is the first step to being gentle with children. And it is very important to be gentle with children. After all, we want them to learn to be gentle, but how can they if they have models who are rough?

childhood education researcher, calls this condition "analysis paralysis."[2] While we would all like to be on our very best behavior when working with children, we are humans, not saints. We want to model our humanness and children also need to learn how to manage both the highs and lows of life.

This chapter explores the many ways in which adults influence children through modeling, including the development of self-esteem in the areas of virtue, power, significance, and competence. We will examine how adults convey messages about respecting and valuing diversity through setting an example. Finally, this chapter explores how adults model being learners themselves. To begin with, however, let us look at some methods for modeling nonviolent problem solving.

Modeling Nonviolent Problem Solving

The violence in America today is a concern for all, and it is important that we take a close look at the role of the early childhood practitioner in preventing violence. Violent tendencies start early: aggressive preschoolers risk becoming violent teens when no one teaches them constructive, peaceful ways of responding to conflict. In order to teach children the difference between being assertive and being aggressive, child care providers and teachers must model nonviolent approaches to problem solving. When children are being aggressive, many adults tend to intervene in ways that are as aggressive as–if not more aggressive than–the children's behavior. Instead of exhibiting gentle calmness, they get angry and lash out verbally, and sometimes even physically. (The subject of modeling gentleness appears in the *Tips and Techniques* box above.)

This teacher is attempting to show her students a problem-solving approach that is better than the strong-arm tactics they were using.

Although we do not label violence as such during early childhood, the roots of harmful aggression that grow into teen violence often lie in preschool experiences. When young children do not learn problem-solving skills, they resort to strong-arm tactics. If they want a toy, they grab it from someone else. If they are bumped accidentally, they shove back the offender.

Although these behaviors may eventually transform into violence, grabbing, shoving, and even hitting are normal for the very young. Children who exhibit such aggression are not bad, they just need guidance from adults to learn other ways to solve problems and express their feelings. (You can find some suggestions in Chapter 5, under *Guidance Alternatives to Punishment*).

Three patterns of thought that start in the early years have been identified in violent teens. During conflicts they (1) neglect to seek out information, (2) possess a narrow vision of how to respond, and (3) are unable to consider the consequences of their actions. All these thought patterns are present in young children as well. Whether these patterns become ingrained depends on early childhood experiences. Let us review some ways adults can model behaviors to replace these maladaptive modes thinking.

Seeking Information

Violent teenagers neglect to seek information about what is really happening in a conflict; they take an act-first-ask-later approach. They seldom give anyone the benefit of the doubt but rather see everyone as a potential adversary. This habit of thought starts early. The preschooler interprets an accidental bump as being intentional, for example. Trying to convince him otherwise usually will not work, which is why modeling information seeking is so important.

Children as Mirrors of Their Teacher

Do you wonder how children experience you as teacher? I discovered the best way to find out was to pay attention to children when they play "school." Watch who wants to be the teacher and listen to how they play act. They mimic the sound and the body language of the teacher, sometimes clearly enjoying the power of a teacher's role. I've had children who actually thought the way to read a book was to hold it open facing outwards so all the children could see the book like at story time! Just as we learn about children's conceptions of domestic life in the housekeeping corner, we learn about their conceptions of the classroom when they play school. Sometimes I have initiated playing school with a child in order to learn about the way in which that child perceived the world of school and the power relations within it. Lillian Katz once said that children are natural-born anthropologists. In the sociodrama of playing school, we use their natural mimicry to learn how children experience us as teachers and to get a sense of the behavior we are modeling.

–Marion Cowee

In order to teach children to avoid hasty conclusions, you must avoid the same habit yourself. By seeking information before making up your mind, you model an important conflict-resolution approach for children.

Consider this example of a teacher who is not an information seeker and does not know how to help children resolve their conflicts: a child screams, "He shoved me!" The teacher arrives on the scene promptly. She believes the accuser and takes his side, though she did not see what happened. To the "offender," the teacher says sharply, "Don't push kids! It isn't nice. They don't like it! If you push him again, you'll have to go into time-out."

This approach did not address what really happened, how either child felt about the situation, what was behind the feelings, or what happened before the scream. The children received no support to help them prevent the same incident from happening again.

Now let us look at a contrasting example: The teacher sees the shove and immediately comes over as an information seeker. She does not interrogate the children but puts into words what she sees.

TEACHER: Looks like you shoved Jerad, Nick.

NICK: Yeah, I shoved him. Look what he did to me.

Nick holds out his arm with a red mark on it. The teacher turns to Jerad to hear what he has to say. She just waits.

JERAD: Come on, I didn't mean to do that. You were in my way and I didn't see you.

NICK: Well, why didn't you look?

Jerad is silent. He doesn't have an explanation. The teacher helps keep the conversation going to draw out more information about what happened.

TEACHER: So you bumped Nick?

JERAD: [*Speaking quietly*] Well, I kind of ran into him with a block.

TEACHER: With a block . . .

JERAD: [*Demonstrating*] I was carrying it like this.

TEACHER: And you hit Nick . . .

JERAD: But I didn't mean to.

NICK: Yeah, I bet!

TEACHER: [*To Nick*] And it made you mad when the block hit your arm.

NICK: Yeah.

TEACHER: So you shoved Jerad.

NICK: Yeah.

TEACHER: I wonder how else you could have let him know how you felt about the block hitting your arm.

JERAD: He could have just told me he was mad.

NICK: But he hurt me.

TEACHER: You got hurt . . .

JERAD: I didn't mean to hurt you, Nick.

NICK: Well, be more careful next time!

It is tempting to make a quick judgment and settle a problem in your own way, but that approach is almost always less effective than talking it through to discover each child's perspective on the situation. By getting the two parties to talk it out, you model an information-seeking, problem-solving approach that the children will eventually adopt themselves to replace the habit of jumping to hasty conclusions.

Helping children learn to clarify situations by seeking more information must be done on the spot, when difficulties arise. It is more effective if the "talking through" occurs before a fight begins. It takes diligent supervision to be on the spot early enough to prevent aggression, but it is worth it. But even if you cannot get there fast enough, and the fight gets physical, it is still important to sort things out once you have stopped the angry parties from hurting each other.

Recognizing Alternatives

During a conflict, many children have a narrow vision of how to respond to the problem. They may only see one way out–physical aggression.

When adults show physical aggression in the face of a conflict–such as grabbing a toy back from a child who grabbed it from another child–they model the same aggression they are trying to prevent. To model a problem-solving approach, you must consider the range of alternatives to responding aggressively. Once you have a view of the alternatives, you can help by stepping in with a remark like, "I wonder what you could do if he has the toy and you want it." Let us look at a toy-grabbing scene as it plays out with the adult encouraging the children to think up alternatives to aggression.

Blake grabs a small truck from Haley, who is playing with it in the sandbox. Haley jumps up and tries to grab it back. Blake tosses it over the fence. Haley screams and pounces on Blake. The teacher arrives and gently but firmly separates the two, placing himself between them. He holds Blake's hand when Blake tries to leave the scene.

The teacher starts by explaining what he sees–two angry children. He states the facts, without interpretation or judgment. Blake and Haley both start talking at once, to explain their perspectives to the teacher.

TEACHER: Wait a minute. I can't understand you. [*He turns to Haley*] What's going on here?

HALEY: [*Whining*] He took my truck.

TEACHER: Tell Blake how you feel.

HALEY: [*Screaming*] I'm going to beat you up for taking my truck.

TEACHER: [*Speaking in a calm, understanding voice*] You're very angry at Blake for taking your truck.

BLAKE: It wasn't yours!

HALEY: [*Screaming*] Teacher . . .

TEACHER: Talk to him.

HALEY: [*Speaking in a threatening tone*] Give me that truck!

BLAKE: I can't.

TEACHER: You both want the truck.

The teacher is squatted down, gently holding both children's hands. He has moved from his position between them and has left them facing each other. "I wonder how you can solve this problem," he says and then waits.

The scene goes on, with the children redefining what the problem is–neither has the truck now. They discuss their feelings and what they can do about the situation. The teacher does far more listening than talking and remains calm and nonaggressive throughout, modeling for the children equanimity in the face of conflict.

Eventually, the two come to a tentative understanding of the other's feelings and begin to work on a solution. The first issue is how to get the truck back into the play yard, which, of course, involves the teacher, since neither child can leave the play yard. The teacher says he will help them get the truck back but not until they settle the problem that started the conflict.

The two children talk it out, with the teacher helping them see the possible solutions. The three eventually decide that Haley gets the truck since she was the one playing with it, but Blake is not altogether happy with this decision. He wants to play with it too. So they do some more brainstorming. Will another truck do?

Will taking turns work? The teacher keeps asking, "How will you solve this problem?" Finally, they agree to taking turns. The teacher makes a phone call to the neighbor, who agrees to hand the truck back over the fence.

It may seem that this problem solving took more time than it warranted. Indeed, problem solving with children does take a lot of time at first, but it is time well invested. The payoff comes bit by bit as children get better at solving their own problems until they no longer need the teacher to help resolve every squabble.[3]

To increase children's problem-solving skills, do not provide alternative solutions; rather, encourage children to solve it for themselves. If the children cannot think of a constructive solution, keep asking, "What are some things that you can try to solve this problem?" At first, the children may think you are quizzing them and that you have some right answer in mind; they may respond to what they perceive as an "adult game" with silence. However, if you keep encouraging them to come up with their own approaches, the children will eventually figure something out. Remember, the children will learn more deeply if you remain quiet and refrain from offering suggestions immediately. The goal is to help children learn how to solve their own problems, which requires thinking and sometimes creativity.[4]

Some teacher trainers advocate never offering solutions, but in my experience it doesn't hurt to nudge a little by making small suggestions (scaffolding). Nevertheless, I suggest doing this only following periods of silence, to give children a chance to respond on their own; silence usually creates a vacuum that needs to be filled.

It also helps to rehearse problem solving during group time. Take either made-up conflicts or real ones and discuss the alternatives to aggression and the possible consequences of each alternative. Discuss each solution in terms of its acceptability to you and to the children involved; you need to be open to children's creative ideas, but you also need to point out solutions that are inappropriate or nonconstructive. Rehearsals in a nonemotional setting can prepare children for responding to real conflicts in nonaggressive ways.

Your tone of voice in a conflict is extremely important. It should be *firm* so the children know that you mean it when you say, "I won't let you hurt anyone or hurt yourself." Your tone of voice should also convey *empathy:* "I know how much you want that truck. You don't like for him to grab the truck you're playing with." You must assume a *problem-solving attitude:* "What can you do instead of grabbing or hitting?" And, finally, you must be *persistent.* Wait it out; let the children come up with solutions, and insist they go back to the bargaining table when the solutions don't work out. Ask, "I wonder what else you could try" often to continue to nudge the children to think of a solution. Ultimately, children increase their self-esteem and confidence when they gain the ability to do things on their own.

Considering Consequences

Violence-prone children do not consider the consequences when they lash out. Through calm dialogue, you can help children understand the effects of their actions. In the previous example with Blake and Haley, the teacher connects the consequences to the behavior with a direct and succinct statement: "When you take the truck from Haley, you can't play together because she gets mad. You threw the truck over the fence, and now you don't have it."

It is not easy to refrain from lecturing; just remember, a sentence or two is all that is necessary. And, of course, *never* say "I told you so." No child (indeed, no adult) likes to be told that someone else knows more than he or she does. This statement belittles children, and belittled people are apt to lose confidence in their abilities to solve problems.

Two approaches to considering consequences work well. One is to help children understand beforehand what consequences might result from aggression. This approach has been called **feedforward.** Of course, feedforward only works if all the adults in a program have a zero-tolerance attitude toward aggression. If supervision is lax and inconsistent, children learn that as long as their aggression is hidden from adults, they can get what they want, and their aggression will likely increase.

feedforward A guidance tool that helps children understand beforehand what consequences might result from certain behavior (often unacceptable behavior). It is only feedforward if it is presented in a neutral tone and is neither judgmental nor threatening.

In order to talk things out *before* aggression can occur, adults must be vigilant, anticipate problems, and get to the scene of a brewing conflict immediately. Once there, they can help children sort out the effects of their actions on others. Giving children the idea that foresight prevents problems is important!

A second approach is to let children experience firsthand the result of their actions, when it is possible to do so without promoting further aggression. Blake learned one lesson when he threw the truck over the fence: he did not get to play with it. He might have learned another lesson if the neighbor had not been home or had refused to give it back.

Did you notice that the teacher helping Blake and Haley problem-solve their conflict was male? You will find male teachers throughout this book; perhaps they appear at a greater rate than in real-life early childhood settings. The lack of men working with young children is a serious problem and one that needs to be addressed by the field and the society. Think about the modeling effect. Children who grow up in single parent families without men may lack real-life male models. If you watch the stereotypes of men portrayed in the media–television programs, movies, cartoons, and commercial advertisements–you find very few examples of men being nurturing, gentle, or caring. How can boys grow up to be all they can be if their models of masculinity are overly tough, aggressive, and violent? How can girls learn what to expect in a man if the men she is most acquainted with come from the media? How can both boys and girls broaden their view of gender roles if their role models–real or media generated–only fit the narrow definitions of masculine and feminine? These are social, gender, and equality concerns that remain prevalent in our society today.

The Many Roots of Violence

Adult modeling and skillful intervention go a long way toward teaching children nonviolent approaches to conflict. Unfortunately, other factors counteract such lessons, such as when children see violence at home, on the streets, online, or through other forms of media. Likewise, children who are victims of abuse experience a stronger modeling effect than that exhibited in the classroom and subsequently risk growing up to become perpetrators of child abuse themselves.

There is no one solution to violence. If we are to live in a peaceful society, we must take a multipronged approach. Effectively modeling nonviolent problem-solving techniques in a high-quality early childhood program is the first place to start.

In this section, we have closely examined three patterns of thought that are at the root of violence: (1) a failure to seek information during a conflict, (2) a narrow vision on how to respond to conflict, and (3) an inability to consider the consequences of aggression. Once we recognize that violent behavior starts as a way of approaching problems in the early years, we can see the importance of our work with young children in creating a future nonviolent society. Unless we model for children and help families learn the alternative means of solving problems, they will continue to use the aggressive ways that come so naturally to them in their childhood.

Some early childhood educators report seeing more instances of violence and aggression during outdoor play. This may be due to several reasons. In some programs, the outdoor area is shared with other groups and the numbers of children can be overstimulating and unmanageable. Another problem arises when outside time is thought of as recess and few resources are provided. Also, the least trained staff members or even volunteers may be supervising, rather than more experienced teachers who know how to facilitate problem solving.

Some people think the solution is to eliminate outdoor time altogether, which then becomes a health issue as children do not get the fresh air and exercise they need when cooped up indoors all day. The Pikler Institute, mentioned in previous chapters, sees fresh air as vital. From infancy on, children spend a good part of their day outdoors–summer and winter. They eat outdoors in the summer and nap outdoors year round. Although it is time consuming to dress children in winter clothes to go outside every day, the time spent dressing for the outdoors is not considered wasted but is part of the way children and adults establish close bonds. It is a one-on-one time with the caregiver that is rich in personal attention, language, touching, and all the other behaviors that build relationships, which is an effective approach to violence prevention. Fresh air and exercise also helps avert violence. The effects of fresh air on behavior are noticeable to even the most casual observer. The children get along with each other remarkably well and incidents of aggression are far lower in the Pikler Institute than in the average infant-toddler program. The children are extraordinarily healthy and their appetites amazing!

Children also need more than just fresh air and exercise from their outdoor play environment. When children are sent out to run around on a fenced-in asphalt playground and climb on bright colored plastic equipment, their experience is very different from when the outdoor area is a natural setting. When the outdoors is considered a learning environment and exploring nature is part of the curriculum, children's experience is enhanced, and children develop into healthier, more well-rounded people–people who are less prone to violent behavior. At the end of this chapter are suggestions for reading more about this subject.

Another factor in violence is low self-esteem. Here again, the modeling effect is powerful. The next section looks carefully at four aspects of self-esteem: virtue, power, significance, and competence.

MODELING SELF-ESTEEM

Although there are an increasing number of books and articles on how to raise children's self-esteem, some neglect to say how much the degree of adult self-esteem influences that of the child. Adults with low self-esteem provide poor models for children and set poor examples.

What is a poor example? Does being a good early childhood educator mean *never* making a mistake? Of course not! We do not want children to have models they'd never be able to live up to. As humans, we all have areas of weakness and we all make mistakes, but it is how you accept your human frailty that provides a living example for children. Do you accept yourself, or are you overly critical of yourself? Do you hide your mistakes from the children? Do you forgive yourself for mistakes and then figure out how to resolve them? The answers to those questions provide clues regarding your level of self-esteem and what kind of model you provide for children.

Adults with high self-esteem who function well in their lives model attitudes and behaviors that can contribute to children's success and their own degree of self-esteem. They say that you cannot give children what you do not have, and this is certainly true with self-esteem. However, if after reading this section you decide you are a person with low self-esteem, do not give up. Most of us go through periods of self-doubt. The fact that you are reading this book means that you are working to improve yourself. Self-improvement is a self-esteem booster and there is no shortage of resources if you need help. Check out your local bookstore or library.

The following sections explore four aspects of self-esteem and how you can model them. These four elements come from the work of Stanley Coopersmith, one of the pioneer researchers of self-esteem. He wrote a classic book–the first of its kind–in 1967 called *The Antecedents of Self-Esteem.*[5] Coopersmith's theory, is discussed in *The Theory Behind the Practice* box on the next page. The next sections apply his theory to practice by examining its four elements: virtue, power, significance, and competence.

Modeling Virtue

Virtue is an old-fashioned word not often found in educational textbooks. It is a word that evokes other concepts, such as integrity, morality, honesty, character, decency, respectability. These concepts that relate to "being good," but what does it mean to be good?

Consider this situation: the teacher has some bright red paper that she is saving for a Valentine's Day project. It is tucked away in the top of a cupboard. A child who is cutting scraps of colored paper asks for some red paper, which is not one

The Dimensions of Self-Esteem

In his pioneering book *The Antecedents of Self-Esteem,* Stanley Coopersmith broke down self-esteem into four elements. What follows is a list of the elements and a short description of each.

- Virtue: Doing what is right, whether by a set of standards or moral code, or by one's own standards and ideas of morality.
- Power: A feeling that one has the ability to live one's own life, get needs met, and be oneself. Also, a feeling that one is able to influence others.
- Significance: A feeling of being loved and cared about by those who are important to one's life.
- Competence: The ability to be successful, especially in the areas that one cares about.

of the colors set out on the table. The teacher feels bad about denying her wish but doesn't want to get out the special paper. She is afraid the other children will want some and there will not be any left for the project she has planned. Instead of telling the truth and dealing with the possible problems, she pretends there is not any red paper. This, of course, is not a big lie, but it can backfire. If she gets caught, think of what this teacher is modeling.

What message does this response give to the child who happens to know that there is more red paper?

Here is another situation that involves doing what is right according to the law. A teacher and a group of five-year-old children are going on a walk. They want to cross the street to take a closer look at a tree with beautifully colored leaves. The teacher looks up and down the street. No cars are coming. However, half a block away is a crosswalk at a signal. If the teacher chooses against walking down to the crosswalk, what message does he give the children, who are just beginning to learn that laws are made for their own safety?

Although it is sometimes tempting to take the convenient or easy way out of a predicament, it is important that you *always* opt to do the right thing. Self-esteem depends to some extent on seeing yourself as virtuous. When you care about your integrity, you provide a good model for children, which relates to their integrity and, therefore, to their self-esteem.

Modeling Power

A second aspect of self-esteem is power. "Power" is a word I never heard as a young teacher-in-training, yet we all experience power in many areas of our lives. Let me start out by explaining two views of power. The kind of power most people think of we will call "dominating power," which gives a person power over something or someone, or what is known as control. Dictators, for example, exploit this kind of power. If this were the only meaning of the word "power," weaker, gentler people would by definition be powerless.

However, there is another kind of power, one that comes from within that enables us to live our own lives, get our needs met, and be who we are. This type of power we will call "personal power"; it has nothing to do with dominating or controlling others. The English word "power" actually comes from the Latin word *poder,* meaning "to be able."

Dominating power differs from personal power as **aggressiveness** differs from **assertiveness.** An aggressive person just pushes forward, with little regard for other people. In contrast, an assertive person stands up for what he or she needs and expresses those needs and desires in a way that recognizes and respects the needs and wants of others.

aggressiveness The quality of dominating power that results in pushing forward (sometimes in hostile, harmful, attacks) without regard for the welfare of the other person or persons.

assertiveness The quality of standing up for one's own needs and wants in ways that recognize and respect what other people need and want.

You want to be clear about which kind of power the children learn from you. If you have serious control issues and keep children under your thumb, you exhibit dominating power. Even if you dominate them in kind ways by continually manipulating them, you are still modeling a dominant manner of relating to people.

This discussion of manipulation brings us back to the issue of using praise to manipulate children. Consider the following scene: the early childhood educator is trying to conduct group time, but the children are noisy and restless in spite of her efforts to get them to settle down and be quiet. She is getting frustrated and nervous when she finally resorts to a manipulative device. In a saccharine voice she says, "I like the way Jesse is sitting so quietly and listening. What a good listener you are, Jesse!"

The teacher's primary goal here was not to commend Jesse's behavior but to manipulate the other children to settle down and listen. The teacher's attempt was dishonest and disrespectful. Just remember, if you are respectful of the children and their needs and, at the same time, respect yourself, you are exhibiting personal power.

Adults who develop a good sense of their personal power and are assertive in their interactions with others are good models for children. They show constructive ways of effecting change. For example, during a field trip to a fire station, a firefighter is going on and on about "response times." Sensing that the children are losing interest, the teacher intervenes and says, "Unfortunately, we are running out of time, and before we leave the children were hoping to get a closer look at your hat." In this example, the teacher is responding in a constructive way to restore the children's interest while still respecting the firefighter's feelings.

We have explored ways early childhood educators use scaffolding to encourage children to solve their problems and, thus, develop personal power. Children constantly experience their personal power when they are supported in their explorations of the world, each other, and themselves. Rather than overpower children, adults help children clarify their needs, wants, and inclinations. By supporting them in finding solutions when there are obstacles, adults teach children to respect others as well as assert themselves when appropriate. The more personal power is used, the more it grows.

By using the pedestrian crossing and not jay walking, teachers model respect for the law.
Image Source

Modeling Significance

Coopersmith identified the third aspect of self-esteem as "significance," or the extent to which we feel loved or cared for by others.[6] The core value is acceptance, and it is tied to personal power, because the more we care about ourselves in a healthy way, the more likely others will care about us.

Coopersmith's research suggests that children's self-esteem is initially shaped by their parents and home environment. Because the home is where children experience their first relationships, their initial perceptions of worthiness for love and acceptance is, in part, determined by how they are treated by their parents.[7] However, because children are increasingly coming from dual-income families and are spending more time in early childhood children centers, early childhood education (ECE) educators have also become significant models of positive self-esteem.

The emotional rewards of working with young children are many.
Rawpixel.com/Shutterstock

Adults provide good models for children when they show that they feel significant themselves. Taking pride in your work as an early childhood educator and showing respect and care for other workers in your profession are two of the best ways to model significance to young children.[8]

A child comes to feel significant through his day-to-day interactions with others. A teacher can increase his feeling of significance by treating him with respect, caring for him appropriately, and helping him create healthy attachments with other children. But how the child chooses to accept what is given him is his own decision.

A cautionary note: sometimes the notion of self-respect gets in the way of gaining significance in the eyes of others. We must recognize that some people are drawn to the early childhood field because it increases their feelings of significance to work with children. There is nothing wrong with enjoying the emotional rewards of working with young children; it is one of the best parts of being an early childhood professional. It feels great when a child arrives in the morning with a big hug for the teacher. It can also feel rewarding when a child does not want to leave with her parents at the end of the day. But the teacher who personalizes this show of affection may not recognize the possible reasons for the child's not wanting to depart: she may have a hard time with transitions, she may be angry with her parents for leaving her all day, or she may be engaged in a power play.

Regardless, it is important for early childhood educators to examine how much significance they derive from their relationship with the children. For example, a vacationing teacher who returns to the class and is disappointed that things went smoothly in her absence may rely on the job too much as a source of her self-esteem. Teachers who use children to meet their own needs for feeling loved do a disservice to the children. It is important that teachers have a social and emotional life outside the program that is in balance with the loving relationships they have with the children in their care.

Modeling Competence

The fourth element of self-esteem is competence. Many questions arise when discussing competence. What if I do not see myself as a very competent person? What does it mean to be competent anyway? What will happen to the children's competence if they are cared for by someone like me who lacks competence?

Remember, you have responsibility for your own self-esteem, and you can make some choices, including how you feel about your own level of competence and what you can do about it. On the surface, the aspect of competence seems to be the most unfair of all. Some people seem to be born competent; they seem to have more brains, skills, and talent than others. The inequity, however, is the result of a value system that places more importance on certain skills and talents over others.

The child who teaches himself to read at four is the marvel of the preschool program, and his teacher who is enrolled in a Ph.D. program may be the envy of her coworkers. The seven-year-old math wiz may excite her teacher and care provider more than the child who has superior social skills. The budding artist may get some attention for his creativity, but probably not as much as the academically gifted child. The original thinker, who sees potentials far beyond what the teacher can envision, may receive an entirely different response–resentment and scolding–rather than being valued for her abilities.

Research now shows that competence is not just a matter of talent, though talent can help people pick up new skills more easily. Some people have to struggle over everything they learn, but that does not mean they can never meet or exceed the skill level of the talented person. With grit and a growth mindset, a person dedicated to hard work can achieve great things, even with no extraordinary talents.[9]

The element of competence as it influences self-esteem is not really a matter of how skilled or talented you are but how you approach challenges. The best models for children are adults who approach learning as a welcome challenge instead of something to be avoided or a dreaded chore. In fact, if you are one of those people for whom figuring things out does not come easily yet you try hard and become more skilled at a particular task, you are an even better model than the person who finds learning easy.

The modeling effect is most productive when children see adults learning. For example, a family child care provider who swallows her fears and bravely sits down at the brand new computer teaches the children a valuable lesson in patient, persistent learning, as does the teacher who tries to unplug the toilet before calling a plumber. The adults in these examples model competence through perseverance and problem solving. Sometimes the roles are reversed, and children model competence for adults (see the *Tips and Techniques* box on page 166).

MODELING EQUITY

As children watch how adults approach (or run away from) learning situations, they often discover unspoken gender roles. A screw comes loose from a wooden toy and the teacher puts it away until her husband can fix it. The children get the message that only a man can use tools– which conveys a strong message about gender roles. Likewise, if a teacher says that bread making has to wait until his co-teacher, a woman, has time to do it, or if he refuses to change diapers, he may be sending the children messages about "women's work."[10]

When you demonstrate broad concepts of gender roles and capabilities, you help children see themselves as being capable in a wide variety of ways. To model equity, you should avoid perpetuating gender and cultural stereotypes and model learning new activities without fear of failure. Also, try to avoid such statements as "I need a strong boy to help me carry this heavy board." Ask instead for a "strong child."

One program had a problem with some of the oldest boys who spent their time showing everyone how mean and tough they were. The director decided they needed to be exposed to a different idea of manliness. The boys were assigned to help out in the infant room on a regular basis. The director's goal–which succeeded–was to give the boys a chance to drop their tough, masculine personas by expanding their views of what males can do and enjoy.

Louise Derman-Sparks and Julie Olsen Edwards (2010) have written about other aspects of diversity in addition to gender.[11] Race, culture, language, and ability are other areas where an **antibias focus** becomes important. An antibias focus is an activist approach to valuing diversity and promoting equity. The goal is to cultivate tolerance and help children accept, respect, and celebrate diversity. They must come to understand that bias is not fair and learn how to respond to it.

antibias focus An activist approach to valuing diversity and promoting equity by teaching children to accept, respect, and celebrate diversity as it relates to gender, race, culture, language, ability, and so on.

The first important step of an antibias approach is to model a positive attitude toward diversity yourself. Louise Derman-Sparks laid out a number of areas of bias to help adults uncover their own biases in such areas as race, gender, culture, and abilities. Early childhood educators need to understand what an antibias attitude entails and they need to work on it. To do this, become aware of your own unconscious biases; bring out hidden attitudes and examine them. Sometimes, working with another adult or getting outside training can help uncover such biases.

For example, one teacher willingly wiped runny noses when necessary. But when the runny nose belonged to an immigrant child, she said, "Go get a tissue and wipe your nose." When her differential treatment was pointed out to her, she was surprised. She said she was not aware that she was treating one group differently from the other. Upon questioning, she dug deeper and admitted that she was afraid of the diseases she thought the immigrant children might carry. She had not thought about how the two groups of children might view her actions. When it was pointed out that the major difference between the two groups was skin color, she was shocked. She never intended to pass a racist attitude on to the children. In another instance, a teacher-director who felt uncomfortable around people with disabilities refused to enroll a five-year-old who had developmental delays. Her explanation was that the child was not toilet trained, so the program would not be able to take the child. When it was pointed out

TIPS AND TECHNIQUES

Modeling Effect Can Work Two Ways

Sometimes the child is a model for the adult. My son Adam taught me about confronting the unknown with courage and curiosity. The lesson occurred one day long ago in a mall. Adam (who was about eight) and I stopped to see why a crowd was gathered inside a computer store. All we could see were heads, but somewhere in front a screen flickered. Curious Adam wiggled his way up to the front of the crowd. A man at a computer motioned for Adam to take his place. My son, who had never touched a computer in his life, stepped forward without hesitation. By then I had worked my way up close enough to see what was happening.

Adam read a question on the screen: "What's your name?" He carefully searched out the letters on the keyboard and using one finger punched in the letters J-E-R-K.

"Hi, Jerk," the words on the screen responded cheerily. The crowd snickered. The screen continued to scroll. "Are you ready to make something delicious to eat? Press 'Y' for Yes or 'N' for No." Adam pressed Y. A list of recipes appeared. He picked one at random.

"Okay, Jerk, I'm going to tell you how to make chicken cacciatore," said the words on the screen. The crowd laughed again.

"How many people are you going to feed?" the screen asked Adam. Looking over the numbers at the top of the keyboard carefully, Adam picked a one and then punched in six zeros after it.

The computer never batted an eye. "Okay, Jerk, take 250,000 chickens, wash them, and cut them up." The crowd around the computer roared with laughter.

The computer continued, oblivious of the ridiculous portions. "Then take 500,000 cloves of garlic and dice them fine." The computer continued, calling for thousands of gallons of olive oil and millions of pinches of spices. The crowd around the computer was thoroughly amused. My son had found his niche–and it was not cooking for a multitude.

I still marvel at the way Adam walked straight up and started pressing keys on that computer. I would never have had the confidence.

Today Adam is an engineer, and he is still teaching me. I see him take on projects he knows very little about. I watch him teach himself what he needs to know. He is a good model for his mom, who often feels fearful in the face of the unknown.

Adapted from: Janet Gonzalez-Mena, "In Praise of Children," *Child Care Express* (March–April 1995), 2.

to her that the program had facilities for changing diapers and indeed had a number of younger children in diapers, the teacher became aware of her own biases.

Much biased behavior is often unconscious, and a good deal of inequity is not even the result of individual behavior but of institutional bias. If a program has only white middle-class applicants when there is a job opening, the bias may exist at an institutional level; the bias may exist within the program itself, or it may be imposed by cultural or societal trends. Bias also applies to the person in a wheelchair who is a qualified early childhood educator but cannot get in the door for a job interview.

When you become aware of your own hidden attitudes, you take the first step toward becoming a good model for children. The next step is to do something about those attitudes. Tolerance.org, a nonprofit organization created by the Southern Poverty Law Center, is a good place to start. It provides abundant resources for educators to recognize and remedy issues of hidden bias, including learning plans and a hidden bias test (Implicit Association Test).[12]

MODELING LEARNING, DEVELOPMENT, AND COGNITION

It may seem strange to think of the early childhood educator as a learner. Of course, you have to learn to become an early childhood educator–a teacher. So your major role should be to teach or at least facilitate learning, right? Yes, but in order to do that you must also be constantly learning. This chapter has already examined several areas of adult learning, such as increasing your competence and discovering your own attitudes about diversity. The rest of the chapter will look at how to plan an emergent curriculum–that is, a curriculum that grows out of the interests of the children *and* the adults. You'll see how you as an early childhood educator must not only be a willing and able learner, but also a master learner, because you will be setting the example for the children.

What does it mean to be a master learner? What do you have to learn anyway once you've completed your early childhood training? Simply put, you have to learn each child–inside out, a highly cognitive task. What does Trevor need at this stage of his development? What in particular does he need this week? How about today? What does he need right now in this situation? How about Nicole? How similar are her needs to Trevor's? How will she react if I approach her a certain way? What are her fears? What are her strengths?

Is this child playing teacher?

Teachers must "learn" each child through close observation and careful listening.
IT Stock Free/Alamy Stock Photo

Reflective Practice and the Importance of Observation

Reflective practice helps you develop keen observation skills, which will allow you to know each child as an individual and the group as a whole. When you pay close attention to children, you promote attachment, increase respect, and open new vistas. Children who are around observant adults may pick up that way of learning themselves. The *Tips and Techniques* box on page 169 tells what some students learned just from observing an apple closely.

One of the goals of observation is to discover what each child's passions are. We all learn best if what we are learning relates to something we deeply care about. Beyond just the obvious interest level, a child's passion may constitute the foundation of future learning. An example from Seymour Papert, a researcher in artificial intelligence who studied with Jean Piaget, shows how his interest in gears related to his studies in math.

> Before I was two years old, I had developed an intense involvement with automobiles. The names of the car parts made up a very substantial portion of my vocabulary: I was particularly proud of knowing about the parts of the transmission system, the gearbox, and most especially the differential. It was, of course, many years later before I understood how gears work; but once I did, playing with gears was my favorite pastime. I loved rotating circular objects against one another in gearlike motions, and, naturally, my first "erector set" project was a crude gear system. I became adept at turning wheels in my head and at making chains of cause and effect: This one turns this way so that must turn that way so. . . .[13]

TIPS AND TECHNIQUES

Observation Skills: The Apple Exercise

To illustrate observation skills in my college classes, I hand out an apple to each student with the instruction, "Get to know your apple." This exercise serves as an icebreaker and, at the same time, teaches students to focus carefully. When the students are thoroughly acquainted with their own apple, I ask them to "introduce" their apple to two other people in the room, describing its unique characteristics. Finally, I have the students put their apples into a basket. At the end of class, I pass the basket around and ask the students to find their apple.

Even when there are 40 other apples in the basket, the students can always identify their own apple. What started as any old apple became a special apple once the students did some close observation. Furthermore, the students get "attached" to their apple after just a few minutes of observation. Observation is a powerful tool!

Creating an Emergent Curriculum

Children are encouraged to feel passionately about something when their teachers and providers demonstrate their own interests. For example, a friend of mine, Joan, is known as the cat lady at the preschool where she teaches because she talks about her cats to the children. In another program, a teacher shared her passion for and knowledge of rocks; the children learned to identify different types of rocks, and some even became rock hounds themselves. When we share our interests with children, either verbally or physically, we show them that it is okay to bring our own individuality to child care. We show them that the early childhood program is a place where they can bring their passions and maybe even follow them further.

Finding out what children are interested in as individuals and as a group is the first step in developing an emergent curriculum. As mentioned earlier, an emergent curriculum is a plan for learning that comes from the children's interests and needs rather than from a book or from the adult's head alone. An emergent curriculum depends heavily on the adult being a learner.[14] How else can you know what the children's interests and needs are? You have to *learn* about them–through observation.

The best way to explain the concept of an emergent curriculum is to show it. The following example takes place in a family child care home, but an emergent curriculum is equally possible in a center-based program.

Julie is busy helping two-year-old Brianna get used to being in her family child care home for longer hours. Although she has been coming part-time for a year, today is her first full-day experience. Her mother is in the hospital following complications from having a baby, and Brianna is not taking the changes in her life well.

Brianna clings to Julie constantly and cries for long periods, making it hard for Julie to pay as much attention as usual to the other children. Luckily, Julie has plenty for them to do, so they manage quite well while Julie tends to Brianna's needs. The two preschool-age boys spend a long time with crayons, paper, and scissors, Julie notices. At one point in the morning, she watches them trying to fold paper airplanes. They do not really know much about aerodynamics, but they have fun making random folds and trying to fly their crumpled pieces of paper.

On the following day, several things are different about the environment when the children arrive. One area of the playroom is set up as a newborn nursery, complete with several baby dolls and accessories, such as a crib, bottles, pacifiers, and blankets. Brianna is intrigued by the new setup and starts out far less clingy and weepy. Julie notices that Brianna plays the baby instead of playing *with* a baby. She tosses the baby dolls out of the little crib, climbs in herself, pulls up the baby blanket, and lies there sucking her thumb. She obviously is not ready for the mother role yet; she needs to explore what it means to be a baby for a while. But Julie will be ready to encourage her to change roles when the time is right.

When the two boys breeze in a little later, they pick up one of the dolls Brianna tossed aside earlier. She jumps up shouting, "Mine!" and climbs out of the crib to pursue them. Julie observes but does not do anything yet because the boys are headed for the area she has set up especially for them. She waits to see what will happen next. When the boys see the "airplane activity," they toss the doll back in the direction of Brianna and begin to explore the materials and books Julie has assembled for them from the library and her personal collection of books.

Brianna, hugging the doll, is also interested. Together, the three children try out the plastic airplanes sitting on the table, flip through a picture book with airplanes, and then settle down with a book on folding paper airplanes. They each try a fold or two with the paper Julie has set out on the table. Soon, however, their interest lags because the folding is too hard. Julie makes a suggestion or two, but they lose interest and migrate, with the plastic airplanes in hand, to the block area, where all three begin to build an airport.

Julie does not see the two activities as a failure, only as a first step. Later that day, after nap time, the school-age children arrive and find the two activities invitingly arranged. "Oh, boy, airplanes!" says one enthusiastic eight-year-old, racing to the table with the paper, books, and plastic airplanes. She immediately grabs a piece of paper and starts folding a plane. The two preschoolers migrate to the table to watch her. Soon, they are both folding too—one copying her every move. The other preschooler makes more random folds and ends up with a little wad of paper. "Start like this," says the master folder, demonstrating. When he misses a fold, she reaches over and does it for him. (See *The Theory Behind the Practice* box on page 172 to learn how the master folder is showing an example of the theory of Lev Vygotsky.)

In the meantime, two other school-age children head for the dolls. Brianna grasps the baby doll tightly to convey her "mine" message. They find two other baby dolls and a supply of bottles, and the three of them contentedly sit and feed their babies for a long time, talking all the while about the difficulties of being parents.

But neither activity ends there. Throughout the week, Julie keeps the doll corner supplied with new items of interest. On one day, there are jars of baby food and bibs, which several children try out. After feeding the dolls, they try feeding each other, with Julie standing by to ensure the activity remains sanitary. On another day, Julie borrows strollers, and they take each other for rides.

Julie also follows up on the airplane activity. The airplane table is set up both morning and afternoon all week long. When she has a chance, Julie hangs around the table and listens to the children's conversations. At times, she writes down what the children say to get clues about what they already know and the areas where she can help move them forward to learn more. The children understand what she is doing and why. She is showing herself to be a learner–a researcher even. Julie is also being a mediator of the children's learning as she assists them in making meaning out of their environment and experiences.

By the following week, everyone in her program is able to fold one type of plane, and several have learned a number of designs. The children have also seen Julie working on some complicated folding patterns. Several of the children have even learned to distinguish between gliders and airplanes and have made drawings showing their concepts of what makes them fly.

The children have also built an airport out of blocks and painted a mural of their airport. Julie gets her digital camera out to take a photo to hang up next to the mural.

They all take a field trip to the park to fly their paper airplanes from the top of the slide. Setting up an experimental atmosphere, Julie challenges the children to observe carefully by asking questions. "I wonder, does size make a difference?" "Do the heavier ones fly faster than the lighter ones?" "Does the color make a difference?" It is not clear whether Julie knows the answers to these questions herself, but she is obviously interested in finding out. She does not push it, however, because the children are more interested in the process of flying the planes than in analyzing their flight. They delight in climbing up the steps of the slide, tossing their airplanes in the air, then whooshing down the slide and chasing off after them.

They are not doing science experiments, but Julie is taking note of what happens. She wants to know which ones fly farther and faster. On the way home, she learns from the children's conversations that they were paying attention and making some comparisons of their own. One girl takes Julie's hand and asks her, "Does color make a difference?" Julie says she does not think so, but then the two go into a long discussion about the effect of an airplane's weight on flight, which leads to a whole new area of inquiry. Once home, Julie gets out a set of scales, and the two begin to experiment with design and weight. Julie is as interested as the girl. They take notes on their findings and share them with the other children.

Julie also takes pictures of their airplanes and their experiments. She writes down their comments and encourages them to write (or, in the case of the younger children, to dictate) their thoughts on the airplane drawings that they are continuing to produce. Julie's interest in what they are doing gives the children the message that these are important activities. Furthermore, she does not send the drawings home for the parents' refrigerators. What the children are producing is research data, not artwork. They need to review their data periodically to examine earlier hypotheses, throw out what doesn't work, and refine their thinking. Julie sees the collection of drawings as a way of making the children's mental processes visible–not only for her benefit, but also for their own.

In the meantime, another area of inquiry arises: One of the preschool boys who got this project going in the first place notices that the seed he picked up on his way back from the park also "flies." "Look a helicopter!" he says, dropping it and watching it spin its way down to the floor.

NAEYC Program Standards

Program Standard 2: Curriculum

So now Julie has other avenues to pursue as the group studies flight. They go together to the library to find books about helicopters and about seeds. They plan

Zone of Proximal Development

The scene of one child helping another fold paper airplanes illustrates an important concept identified by Russian theorist Lev Vygotsky, which is called the **zone of proximal development** (ZPD). Vygotsky described how older children can help younger ones perform in ways they would not be able to on their own, which is what happened in the prior example with Julie. When the older child actually helped one of the boys fold his airplane, she demonstrated another of Vygotsky's concepts–**assisted performance.** The assisted-performance principle suggests that children cannot perform as well on their own in some cases as they can with a bit of help from a more skilled person. Although some early childhood educators, especially those who follow Jean Piaget's work, frown on helping children do some things that they aren't able to do on their own, others see nothing wrong with assisted performance. See the *Points of View* box on page 173 for more on this controversy.

zone of proximal development (ZPD) Older children help younger ones perform in ways they wouldn't be able to on their own.

assisted performance This principle suggests that children cannot perform as well on their own in some cases as they can with a bit of help from a more skilled person.

walks to find seeds and set up experiments to see which ones fly best. They start talking about a trip to the airport.

Then one day, before the children arrive, Brianna's grandmother telephones Julie. It turns out that Brianna's mother, who came home from the hospital a week ago, had to be taken back to the hospital by an ambulance in the middle of the night. The grandmother worries about Brianna's reaction to her mother being hauled off in an ambulance with sirens screaming. When Brianna arrives at Julie's house the following day, she finds an ambulance, a police car, and some other cars in the sand table. She will not talk about her feelings or her mother, but she spends a long time making siren noises and crashing the cars together. Finally, Brianna buries all the cars in the sand and goes to look for the baby doll.

And so it goes in the home of Julie, family child care provider, teacher, and learner. As you can see, her emergent curriculum fed off the interests and needs of the children. It not only addressed the children's development and learning in the areas of cognition and social-emotional development, but it expanded Julie's knowledge as well.

Julie exhibits the qualities of a master learner and provides an excellent model for the children. Can you sense the richness of learning that the children–and Julie–experienced in this example of an emergent curriculum? Think of how different Julie's long-term, ongoing flight project differs from a weeklong unit on transportation. In the former, each project connected with and "emerged" from the children's interests and their response to the activities Julie provided; an example of the latter is a "canned" curriculum in which children sing songs about planes, trains, and buses, color printed pictures, and glue precut art projects. The purpose of exploring Julie's emergent curriculum is to demonstrate how adults can model positive learning attitudes by actively involving themselves in learning *from and along with* the children.

Points of View

How Much Should You Help Young Children?

For many years, early childhood training programs tended to teach a hands-off approach to helping children. It did not matter so much if children were successful or not in their endeavors; the goal was for them to do things by themselves. The process was more important than the outcome anyway. The only exception to this approach was when safety became an issue.

Then along came Reggio Emilia, the famous early childhood program in northern Italy. The teachers there are not trained to keep their "hands off." Part of their approach is due to their culture; Italians tend to be a hands-on kind of people. They are also more collectivist in their social orientation and less individualistic. But the theories of Lev Vygotsky–which are gaining renewed attention–also enter into the picture. Vygotsky believed that helping children accomplish something was more important than letting them struggle and possibly fail.

Art is an area where the comparison of the two opposing viewpoints raises important questions. Should children be shown how to draw realistically? Should children be given formal art lessons and work with models? Or should they be left alone to paint, draw, or sculpt however they want? If art is self-expression, how much does adult teaching add to or detract from the child's own style? Is it "pushing" children to help them?

The prevailing early childhood viewpoint in this country is that children should not be shown how to draw realistically through the use of models. Instead, they should be left alone to paint, draw, or sculpt however they want. Art is considered self-expression, and adult teaching detracts from the child's own style. Nevertheless, as previously mentioned, differing viewpoints predominate elsewhere. Children in the Reggio schools create amazing drawings and sculptures with the help of their teachers. In China, adults give children formal art lessons and teach them specific painting techniques rather than leave them alone to explore the materials and express themselves freely.

A Story to End With

"This is a strawlegged mosquito," my daughter told me. "Hmmmm," I said. "Very interesting . . ." I took a brief look at the flattened creature in her hand, wrapped it in a scrap of paper, packed my daughter and the bug in the car, and took off for the library. It was science lesson time.

"We're going to learn more about insects," I announced, nudging her through the doors of the library. As I looked for the section on insects, she headed for the picture books. While I researched mosquitos, she escaped outside to play on the grass.

I searched and searched but could find no "strawlegged" category of mosquitos. Finally, I found a name likely to be that of the creature she had shown me. Calling her back in, I gave her what I assumed was the correct name. "Hmmmm," she said and wandered off to look at the display cases.

I checked out three books, hoping she would look at them at home when there were not so many other distractions. My daughter never touched the books. Finally, when the books were due, I brought up the subject again and suggested she learn more about mosquitos and other insects, including their proper names.

"Oh," she answered, "I don't care what they are really named—I like to make up my own names. I named that one the strawlegged mosquito because of the stripes on his legs. Remember how it looked?" I did not. I had not paid attention to either the legs or the name she had invented once I found out it was "incorrect."

I found the wrapped specimen stuck in the pages of one of the books. Indeed, it did have stripes on its legs, just like the old-fashioned drinking straws.

"Oh, now I see," I said, "you're a good observer," I added—a little too late.

My heart was in the right place, but my attention was elsewhere when it had counted. Instead of being sensitive to what my daughter was trying to tell me in the first place, I became overzealous about teaching a lesson that she wasn't interested in. She did not care about someone else's classification scheme of insects; she was more interested in inventing her own. She was being scientific in her observation skills, and I should have picked up on that fact. As it turns out, I was less of a skilled observer of her than she was of the mosquito.[15]

Summary

It is important for adults to set an example for children in their care. Modeling is a powerful way of teaching children. In early childhood programs, adults should model (1) gentleness and nonviolent problem solving; (2) self-esteem in the areas of virtue, power, significance, and competence; (3) an acceptance of and respect for diversity; and (4) a positive attitude toward learning. The early childhood educator must fill the role of the teacher as well as the role of the master learner. To be a master learner, the early childhood educator must observe and learn from and along with the children to create an emergent curriculum that facilitates physical, social, emotional, and cognitive development.

Reflection Questions

1. It has been said that all we ever teach is ourselves. If that statement is true, what is it that you are or will be teaching young children? Please be specific.
2. Have you ever said to yourself, "I sound just like my mother (or father)"? Are you aware of how modeling works in your own life? We take after people even in areas where we'd rather not. Reflect on how modeling has influenced you.
3. If you believe in teaching children to share, how do you model sharing for them?
4. Are you a perfectionist? Do you wish to model perfectionism for children?
5. Recall a time in your life when you were treated more roughly than was good for you. Did you learn something from that experience that can help you in your work with young children?
6. How well are you able to have your own needs met today? What might your answer have to do with working with young children?
7. How much do you feel loved by the people in your life? Would your answer affect your work with young children?
8. What part has bias played in your life and how has it influenced you? How do you feel about taking on the role of antibias educator as an early childhood professional?
9. Are you a learner? How can you model being a learner to young children?

TERMS TO KNOW

How many of the following words can you use in a sentence? Do you know what they mean?

modeling 150
feedforward 157
aggressiveness 161
assertiveness 161
antibias focus 165
zone of proximal development (ZPD) 172
assisted performance 172

FOR FURTHER READING

Adams, E. J. (2011). Teaching children to name their feelings. *Young Children, 66* (3).

Carter, D. R., Norman, R., and Tredwell, C. (2011). Program-wide positive behavior support in preschool: Lessons for getting started. *Early Childhood Education Journal, 38* (5), 349-55.

Derman-Sparks, L. (2011). Anti-bias education. *Exchange, 33* (4), 55-58.

Derman-Sparks, L., and Edwards, J. O. (2010). *Anti-Bias Education for Young Children and Ourselves*. Washington, D.C.: National Association for the Education of Young Children.

Derman-Sparks, L., LeeKeenan, E., & Nimmo, J. (2015). *Leading Anti-Bias Early Childhood Programs: A Guide for Change.* New York: Teachers College, Columbia University.

Dombro, A. L., Jablon, J., and Stetson, C. (2012). *Powerful Interactions: How to Connect with Children to Extend Their Learning.* Washington, D.C.: National Association for the Education of Young Children.

Galinsky, E. (2010). *Mind in the Making: The Seven Essential Life Skills Every Child Needs*. New York: HarperCollins.

Gartell, D. (2011). Aggression, the prequel: Preventing the need. *Young Children, 66* (6), 62-64.

Jones, E., and Nimmo, J. (1994). *Emergent Curriculum.* Washington, D.C.: National Association for the Education of Young Children.

Pelo, A. (Ed.). (2008). *Rethinking Early Childhood Education*. Milwaukee, WI: Rethinking Schools.

Weatherson, D., Weigand, R. F., and Weigand, B. (2010). Reflective supervision: Supporting reflection as a cornerstone for competency. *Zero to Three, 31* (2), 22-40.

Wien, C. A. (Ed.). (2008). *Emergent Curriculum in the Primary Classroom: Interpreting the Reggio Emilia Approach in Schools.* New York: Teacher's College Press.

Wien, C. A. (2014). *The Power of Emergent Curriculum: Stories from Early Childhood Settings*. Washington, D.C.: National Association for the Education of Young Children.

7 Modeling Adult Relationships in Early Childhood Settings

SW Productions/Photodisc/Getty Images

Working with Each Other: Relationships with Other Early Childhood Educators

Being Sensitive to Cultural Diversity

Recognizing Some Differences in the Way Adults Approach Problems

The Importance of Being Authentic

Handling Adult Disagreements Through Dialoguing

Teachers Dialoguing: An Example

Working with Families: Professionals' Relationships with Families

Making Families Feel Part of the Program

Honoring Diversity

Focus on Inclusion: A Special Kind of Partnership

Recognizing That Parents' and Providers' Roles Are Different

Handling Conflicts with Parents

Facilitating Communication with Families

Supporting Families

A Story to End With

In This Chapter You Will Discover

- how adult relationships model ways for children to interact with others.
- that adult modeling teaches children how to deal with conflict and appreciate diversity.
- why adults should be authentic.
- how to distinguish an argument from a dialogue.
- how coworkers can use dialogue to work out differences.
- why early childhood educators have to focus on families and not just children.
- some ways to make families feel included in the early childhood program.
- why early childhood educators need to be sensitive to parents who complain.
- how to handle conflicts with parents.
- what kinds of cultural and language issues can arise between providers and parents.
- how to use the RERUN process for resolving conflicts.
- the four possible outcomes to parent-provider conflicts.
- how early childhood educators facilitate communication with families.
- the purpose of parent support groups.

NAEYC Program Standards
Program Standards 1 and 7: Relationships and Families

Early childhood education is about relationships, which reflects NAEYC's first standard. In a nutshell, NAEYC wants programs to promote positive relationships among all children and adults. Although this particular chapter focuses on relationships between adults, children are still at the heart of the matter. Children learn about interactions from watching adults model healthy interactions between each other, whether coworkers, support staff, outside specialists, or parents. Therefore, this chapter is about relationships among professionals, and relationships between early childhood professionals and families. This chapter is in the first half of the book, not at the end because the subject is so important. In keeping with the nontraditional approach this book takes, parent education is not a separate chapter but is part of this chapter on adults developing and maintaining relationships with each other. Although the focus of early childhood programs is the children, adult relationships are of paramount importance to the care and education of those children.

The importance of adult relationships is not always recognized in all early childhood programs. In some programs adults are discouraged from talking to each other and, thus, children cannot see how adult relationships form and sustain. This tradition of placing all the focus on the children dates back to when half-day programs greatly outnumbered full-day programs; at that time, children spent most of their waking hours at home with their families, where they could observe adults relating to each other. Today, however, many children spend a good deal more time in child care; in fact, many are spending their childhoods in child care. It is imperative, then, that these children see adults relating to each other in the early childhood setting–indeed, it may be some children's only opportunity to do so.

Relationships among adults in a child care situation influence the environment even though young children may not be aware of those relationships.
mangostock/Shutterstock

Tips and Techniques

Soft Eyes

Experienced child care teachers and caregivers develop powers of observation that enable them to see the whole scene and, at the same time, focus on one detail. Called *soft eyes* by George Leonard, an educational philosopher, this ability could also be called dual-focus and is used in dual-focus supervision. The term "soft eyes" refers to the skill that allows an early childhood educator to talk to another staff member or a parent while still being aware of what is going on in the rest of the room or play yard. This ability to pay attention to detail while still taking in the larger picture is often a distinguishing characteristic between beginning teachers and veterans. Beginners need to practice the soft-eyes technique to remain alert to everything, even during a conversation.

Soft eyes, as Leonard defines it, is more than just seeing; it also includes hearing and sensing what else is going on. Soft eyes is a valuable technique–basketball players, football players, and Aikido experts have it. Early childhood educators need it too.

Furthermore, children begin to understand maturity by watching adults. If they only see adults interacting with children all day every day, they miss out on learning how adults relate to each other. Moreover, if their primary models of adult relationships come from TV, imagine what they learn about adult behavior.

This chapter is divided into two parts: relationships among staff members and staff members' relationships with families.

Working with Each Other: Relationships with Other Early Childhood Educators

Unless child care providers work alone, adult relationships are part of the early childhood environment. Children watch adults to see how they build relationships with each other, express affection, communicate anger, listen empathetically, define problems, figure out solutions, and respond to challenges and triumphs. You may wonder, How do staff members relate to each other while they are supervising the children? The *Tips and Techniques* box explains how a dual-focus supervision method allows teachers to model proper adult relationships while simultaneously ensuring the safety, care, and learning of the children in their classroom.

The best work environment, of course, is one in which staff members work together as a team and the relationships are healthy and functional. This does not mean that the staff members all must think and act alike or come from the same background; diversity is a benefit. Nevertheless, even adults with healthy relationships periodically have disagreements, and healthy resolutions to conflict

These girls may be reflecting the healthy, equitable relationships they see their teachers modeling with each other.
Spass/Shutterstock

provide valuable life lessons through the modeling of appropriate behaviors for the children in the classroom. In contrast, intolerance and suppression of conflict models unhealthy behaviors and teaches the wrong lessons to children. Unhealthy models can promote the development of maladaptive behaviors that often hamper a child's ability to develop stable relationships and healthy coping mechanisms to life's challenges.

In our diverse society, it is important for children to learn not only tolerance but also how to appreciate diversity. The lessons are more effective when children have adult models who show them how to deal with conflict and appreciate differences. Even when staff members and families come from similar backgrounds, it is still possible to explore and appreciate differences because no two people are exactly alike. You do not have to bring out Old World costumes or exotic recipes to demonstrate differences. Feelings, attitudes, tastes, inclinations, values, goals, ways of thinking, and ways of approaching problems all show individual variations.

The basic question for each adult in an early childhood program is, "How do I take care of myself so that I'm available emotionally and physically for the children in my care?" The second question is, "How do I get along with other adults who have to meet their own needs as well as relate to me and provide for the children?" Those questions must be asked continually–not only on a long-term basis, but also on a day-to-day basis, and sometimes even on a minute-to-minute basis. The *Focus on Diversity* box on page 181 shows how one family child care provider balanced the needs of the program with the needs of the families.

Family Child Care: Meeting the Needs of the Family

When discussing adult relationships, family child care providers have a different set of issues from center-based staff members. Think about the family of the family child care provider. What must be taken into account to ensure that the rights, needs, and desires of the members of the household are not neglected for the sake of the child care program?

One provider decided that she had to have the full consent of every family member before she went into business. She realized that her business would have an impact on each of them because it would take place in their home. To begin, she drew up a contract outlining the rights and responsibilities of each family member. She presented it to them, and they sat down and discussed and revised the contract until they were all willing to sign it. The negotiation process considered the following questions: How much do we have to share our space (both communal and private rooms) and our possessions (such as toys, games, and equipment)? How much will we have to help out? Furthermore, as a parent of young children, the provider needed to ask herself two questions: How will I differentiate between my roles as "mom" and "provider"? How will I handle issues of "fairness"?

Thinking about these issues before they became bones of contention made sense. As it turned out, this family took their rights and responsibilities seriously. This child care business did not disrupt the family nearly as much as it does in other households where issues are not thought through ahead of time.

This exercise was not only valuable for family relations, it also gave the child care provider practice in drawing up contracts for the parents to sign. When she opened her business, each parent received a written contract with everything laid out clearly. She still uses the same contract, but it has been modified over time as needs arise.

Being Sensitive to Cultural Diversity

Early childhood educators differ from one another in their perspectives on the nature of childhood, what children need, how they should be cared for, and even what "education" means. For example, in English, the word "education" is associated with academic matters; in Spanish, however, its cognate "educacíon" refers to a person's upbringing. For English-speaking people, a well-educated person is someone who has attained a relatively high level of academic achievement; for Spanish-speaking people, a person who is "bien educado" is someone who has many social graces and is sensitive to and respectful of other people.

Imagine, then, how differently two adults–one from an English-speaking background, the other from a Spanish-speaking background–might approach the "education" of a child. Imagine also how they might approach the care and education of a group of young children. To work well together, both teachers will have to actively explore their cultural differences and use constructive communication and conflict-management skills. If successful, both teachers will provide excellent models for the children on how to be culturally sensitive.

Recognizing Some Differences in the Way Adults Approach Problems

A teacher who works in a child care center has a problem with another teacher. Let us say she thinks her co-teacher, Dolores, coddles children too much. Dolores wants to do everything for the children rather than encourage them to do things for themselves. How might the first teacher solve this problem? Following are six approaches that she might use. She might:

- boldly confront Dolores and say she wants to talk about the problem.
- talk to Dolores but just hint at the problem without telling her exactly what is on her mind.
- talk to the director about Dolores and leave it in her hands.
- ignore the problem, put on a happy face, and not talk to anyone.
- talk to a third party about the problem and ask that person to act as a mediator.
- talk to everyone but Dolores.

Notice how some of these approaches are similar to the way children solve problems: They (1) take the person in question head on, (2) withdraw and pretend the problem does not exist, or (3) run to the adult in charge. There are some interesting parallels here!

How you feel about each of the preceding approaches is probably determined by two things: (1) how you feel problems *should be* approached and (2) how you are used to dealing with problems. Your culture and values in general will also dictate how you view each of these approaches.

For example, you might label the teacher a "gossip" if she talks to everyone about the problem except Dolores. Or, you might believe it is important for the teacher to get a variety of opinions and feedback before making any decisions about the problem with Dolores.

Power can be an issue when dealing with problems. If, for example, you fear the consequences of direct confrontation, you would probably avoid such an approach. If, however, you believe that straightforward communication is important, you may be critical of people who skirt issues rather than take them head-on. You might even say, "Please get to the point," without realizing that the other person thinks it is up to you to figure out what the point is.

Cross-cultural miscommunication can occur in numerous ways. For example, speaking indirectly about an issue or problem is an accepted way to communicate in some cultures. It is insulting to lay everything out in the open and thus deny the other person the opportunity to make the point. It is like doing a puzzle and leaving the last piece for the other person; some believe it is rude to put that piece in if you are the person presenting the problem. In this case, person A thinks person B is "beating around the bush," and person B thinks he is taking the normal path to problem solving.

Here is another example of cross-cultural miscommunication: Person A believes that if you have a problem, you need to state it clearly; she becomes annoyed when other people expect her to "read their minds." "Put it into words!" is her motto. When she meets up with person B, who uses nonverbal and indirect ways to communicate a problem, person A ignores the message completely.

Person B is used to people who are good at "reading her mind"; she expects everyone to do it. When she meets up with person A, however, she is surprised when she gets no response to her nonverbal messages. After repeated attempts at indirect communication, she eventually puts her problem into words, but her words are not strong or forceful enough. To person A, the complaint sounds minor. Once again, person A ignores the complaint, thinking it is unimportant.

These kinds of problems present opportunities for adults to explore cross-cultural communication and for children to observe them. Of course, the children will not always be present to observe; some discussions will occur at meetings, breaks, or after the children have gone home. Yet it does not hurt for some problem solving to proceed while the children are present.

Naturally, adults want to be sensitive that their exchanges do not trouble the children. At the same time, however, remember that if conflict is palpable but it is not discussed, the children would feel it anyway, and what they imagine may be worse than the real thing. They may even think that the problem has something to do with them when it is really between two adults.

Many people believe children should never be exposed to adult arguments, but if they are not, how will they learn healthy conflict and problem resolution skills? Furthermore, such protections can create misconceptions that adults always get along. Unhealthy and inauthentic relationships often rob children of opportunities to learn how to solve problems constructively, as limited by the poor models available in their environment. Teach children constructive problem-solving methods, you need to use them yourself, with other adults.

The Importance of Being Authentic

What does "authenticity" mean in the early childhood setting? It means being yourself around the children as well as other adults. Be authentic and *compassionate* at the same time. To do that, check out Marshall Rosenberg's view of what he calls nonviolent communication in the *Tips and Techniques* box. Rosenberg adds compassion to communication, resulting in more harmonic relationships with other adults. His approach works for children too! Children need to be around "authentic" and compassionate adults to experience genuine feelings and human interactions, not around adults who exhibit contrived feelings, such as the adult who always speaks in a honeyed voice and has a smile on her face, even when she is annoyed or unhappy. Children cannot learn how to create and navigate genuine relationships if they are continually exposed to sacchrine shells that conceal the adult's authentic self.

TIPS AND TECHNIQUES

Communication with Compassion

In his book *Living Nonviolent Communication,* Marshall Rosenberg divides communication into four components: observation, feeling, needs, and request. He defines observation as viewing the action of another whom you have feelings about, and he connects the observation to the underlying needs and values driving these feelings. The last component, request, puts into words a request for an action that you need. These components make up half of the process. The other half involves listening to the other person empathically and to his or her response to what you have observed—your feelings, needs, and request. This is not a step-by-step guide, but rather a change of consciousness. It could be thought of as communication with compassion for the other. The point is to move from communication that attacks and provokes a defensive response to communication that moves the conversation toward mutual understanding.

Rosenberg, M. (2012). *Living Nonviolent Communication: Practical Tools to Connect and Communicate Skillfully in Every Situation.* Boulder, CO: Sounds True.

Handling Adult Disagreements Through Dialoguing

dialoguing An approach to conflict whose goal is to reach agreement and solve problems.

What happens when two adults strongly disagree? How can they handle their conflict and still provide a good model for the children? One of the best alternatives to arguing is **dialoguing,** an approach to conflict with the goal of reaching agreement and solving problems. The idea behind dialoguing is not to win the conflict, but to understand the other person's perspective and find the best solution for all concerned. Here is a summary of the differences between arguing and dialoguing:

- The object of an argument is to win; the object of a dialogue is to gather information.
- The arguer tells; the dialoguer asks.
- The arguer tries to persuade and convince; the dialoguer seeks to learn.
- The arguer considers her point of view the better one; the dialoguer is willing to understand multiple viewpoints.

Try watching two people in a disagreement; without even hearing what they are saying, you can guess from their body language whether they are having a dialogue or an argument (see the *Tips and Techniques* box).

Most people approach a conflict by trying to set the other person straight instead of exploring the problem or hearing the perspective of the other. Whenever you think someone is wrong, stop and ask yourself, "What might the reason be for this

TIPS AND TECHNIQUES

The Body Language of Conflict

After repeatedly viewing a videotape of role-played arguments, I began to identify certain types of body language the two people used to convince each other that their viewpoint was the correct one. They tended to stand firm and tough when listening–assuming a defensive posture. During their turn to talk, they leaned forward and made cutting or pushing hand gestures. It was obvious, from their body language alone, that the two were embroiled in a heated argument in which neither side was willing to lose.

I also viewed segments of role-played dialoguing and noticed how the body language differed from that displayed in the argument segment. While dialoguing, the people were highly emotional and equally firm in their own stance, but they used their bodies and voices differently. The goal of dialoguing is to open avenues of communication to share and understand different perspectives; indeed, the role-players' gestures reflected their attitude–their hands especially. Instead of waving fists or making strong pushing or cutting movements, their hands tended to be open; their hands reflected their minds–or maybe it was the reverse.

So how does one switch from an argument to a dialogue in the heat of the moment? Start by noticing your body language. Some times, all you have to do is adjust your body language and the rest will follow.

Then it is a matter of simply listening to the other person. To truly listen, suspend judgments and focus on what is being said rather than just gathering ammunition for the next attack. Really hearing someone is extremely simple–but it's not easy.

behavior or point of view?" Consider that you might have a gap in your knowledge rather than assume that the other person is wrong. Do not rush to conclusions; use dialogue to sort things through.

Let us go back to Dolores, who was perceived to be "over-coddling" the children. Imagine being the other teacher who disapproved of her methods. Would you understand why she did so much for children rather than encourage them to do things for themselves? Maybe her approach was culturally motivated. Maybe she viewed children as precious beings that need to be catered to. Perhaps she has a different idea about independence and is more concerned with keeping children closely connected and attached, or maybe it is not cultural at all but rather something from her personal history. Perhaps she had a child who died at a young age, or maybe she wasn't cared for as a child. It is surprising what you might uncover through dialoguing.

The point of dialoguing is to see a perspective different from your own. Realize, however, that gaining perspective is not the same as giving in. After you both *understand* each other, you can work with the other to find a mutually beneficial solution.

Often what starts out as an either-or situation ends up looking quite different at the end of a dialogue. Although it may have seemed at the start that there were only two options, by talking things out, people in conflict often discover a third or even a fourth and fifth solution that transcends the dichotomy of "If I'm right, you have to be wrong" or "If we don't do it my way, then we have to do it your way."

Teachers Dialoguing: An Example

Let us look at how dialoguing works between two child care teachers who differ in their approach to art activities. Dawn and Amy have been working in separate classrooms until now; they are about to begin team-teaching in the same classroom, and they need to plan the art projects together.

Dawn is an advocate of free-expression art. For her, the purpose of art is to let children explore the materials in any creative way they like. In her old classroom, she always put out easels, play dough (sometimes clay), paper, scissors, felt pens, tape, glue, stickers, and hole punches–all for the children to use whenever they wanted. Dawn did not tell them what to do; instead, she stood back to observe how the children used the materials. She feels strongly that children should use their imagination and not be inhibited by directions or ideas of right and wrong.

Amy disapproved of the kind of "art" produced in Dawn's old classroom. Rarely did she find the pieces interesting or attractive; many consisted of bits of randomly ripped and cut paper that had little aesthetic or expressive value. Amy's goal is for the children to create something they feel good about which is why Amy prefers thematic art activities that are representational rather than abstract. In her old classroom, she spent a lot of time cutting out shapes and had the children glue them on paper or use them in collages. She liked the idea of sending something home that parents would hang on the refrigerator. She made a point of telling the children that they had made something their parents could be proud of. Amy's theory is that when parents compliment their children, the children feel rewarded and motivated to try even harder next time.

Amy and Dawn sit down together to talk about their differing approach to art. Here is what happens when they argue instead of using dialogue.

AMY: Dawn, your kids just mess around at the art table. They don't learn anything when they sit and snip and scribble. You don't provide any structure, and you don't have any activities. I propose we use this book to plan our art curriculum.

DAWN: What, so my kids do those silly little crafty projects that end up all looking alike? That's not art! That's a waste of time.

AMY: [*Getting angry*] Talk about a waste of time. Your kids are the ones wasting time.

Amy gets up and walks out of the room. Her feathers are ruffled. Planning time is over for the moment.

Now look at a planning scene that incorporates dialoguing.

Amy: I've got a book with art ideas. Why don't you borrow it and see what you think. It's a different approach from what you're using.

Dawn: Are they craft ideas? I'm always looking for good ideas, but I like free-expression art, not craft projects.

Amy: Well, these are the kinds of projects I use. I guess you don't exactly approve of them, do you?

Dawn: I think they stifle the children's creativity. You don't agree with me, do you?

Amy: How can children be creative if they don't develop any skills?

Dawn: So you think all those directions and the models you use help children learn how to do things?

Amy: Yes. Otherwise they just fiddle around with the glue and paint. But I guess that's OK with you. You're more into process than product, right?

Dawn: Right. I think it's the doing that counts, not the outcome. But it seems you want them to have something to take home that they feel proud of.

Amy: Not only the children but the parents. Most parents like to see a product and have something they can exclaim over. It makes them think we're doing a good job here. They feel good about their children, and they feel good about us.

Dawn: So are feeling good and being proud your main objectives? That seems superficial to me compared to the satisfaction that comes from exploring, experimenting, and taking initiative. I don't think children get that kind of satisfaction from directions and copying a model.

Amy: I never heard you put it quite that way before. I don't agree with you, but I see where you are coming from.

These two have just begun to manage their conflict. It will probably require more discussions like this to decide how they will plan their art curriculum, but the dialogue helped them discuss their differences, begin to understand each other's point of view, and build on their relationship. The argument did not have any of these benefits. In fact, it got in the way of the relationship.

Amy and Dawn acknowledged each other's feelings and perspectives by listening to each other. Listening is an important skill for the early childhood educator to develop, not only for the purpose of hearing what children have to say but also for hearing what peers are thinking. The dialogue approach for understanding diverse perspectives and building healthy relationships works well among professionals and equally well between professionals and families, the subject of the next section.

WORKING WITH FAMILIES: PROFESSIONALS' RELATIONSHIPS WITH FAMILIES

Beginners in the field of early childhood education are often unaware of the family's role in child care programs. They tend to focus on the child without realizing who the client really is–the family. A program cannot educate or care for the child without taking the family into consideration. In the *Voices of Experience* box on this page 189, Ethel Seiderman talks about how much it meant to an immigrant family to find a "family-friendly" center for their child. Seiderman is founder of Parent Services Project, which helps child care programs to become family centered.

Two common approaches to focusing on families are called "parent education" and "parent involvement." These two approaches may have considerable overlap, as shown in the descriptions that follow.

Parent education can be as formal as traditional classes, or as informal as occasional workshops or evening sessions on particular subjects, such as discipline. Parent education can also come about as parents work in the classroom as teacher aides. Most parents have something to gain from parent education; however, the curriculum must be culturally sensitive and responsive to parental goals. The best parent-education programs continually assess parent needs and interests instead of imposing a predetermined curriculum on the group.

Parent involvement may or may not also include parent education. Programs taking the parent involvement approach may require parents to help out in the classroom and/or help with maintenance of the program's equipment, building, or yard. Signs of a parent-involvement approach are parents fixing tricycles, taking aprons home to wash, or spending a morning a week in the classroom. Parent education, if it is included, may come about as parents work in the classroom with the children. It may also include separate classes, workshops, or informal parent meetings. Parent involvement in some or all of these activities may be a requirement or a choice, depending on the program.

A third approach is called a "parents-as-partners" approach or a "families-as-partners" approach. Though parent education and involvement may both be part of this approach, the focus is mainly on the partnership rather than the education or involvement. Education is the result of the partnership, not the other way around. Involvement that naturally arises out of partnerships has a sense of equality and means a lot more than requiring parents to do such tasks as fix tricycles, wash aprons, or spend a morning a week in the program. When programs see families as partners, they involve them in decision making. Collaboration is a key part of this approach. Further, these programs include entire families as well. As a result, the families may still carry out tasks like fixing tricycles, but the decision to do so grows out of the partnership rather than originating in a requirement set by the program. The challenge for many programs is to figure out ways to create the partnership. For information on partnerships with families of children with special needs, see the "Focus on Inclusion" section on pages 192–3.

NAEYC Program Standards

Program Standard 7: Families

Feeling at Home in Child Care

Lejana and Davko came to the United States in 1995. While in Bosnia, Lejana was a chemical engineer and Davko was a truck driver. Needless to say, their lives are much different here in America. Lejana now works in an office and goes to school at night. Davko works for a cheese company.

Lejana and Davko were both in child care when they were young, as was usual in their country. Both their parents worked, yet still spent time with them and their friends. In America, they miss the closeness of family and the lifelong friendship of neighbors. Here they live in an apartment complex with many children, but the children don't play outside. Here there are playgrounds, but no one seems to play in them. Neighbors don't know each other. People don't stay in one place long enough to nurture lasting friendships. Most kids change homes three or four times before they are 10 years old.

What made a big difference in their lives was finding the right child care center for their son, Gianni. They describe the center as "family friendly," "very European." It reminds them of the child care centers in their country because it "has heart." They appreciate the fact that Gianni is in a warm place and has a warm meal every day while he is away from his parents at school. They enjoy the family picnics, camping, and parent meetings. They will always remember the first time they made gingerbread houses with all the children and their families at the center.

The family first chose this center because it was affordable and close to Davko's work place. The center is so special that they continue with it even though they have changed jobs and now work farther away. They stay because it "feels like home." They hope their new child who is due in December will find an opening when it is time. Their advice to parents looking for child care is to find a center that has heart. It will become a place where your child and family can grow and be happy together.

–Ethel Seiderman

Making Families Feel Part of the Program

If families are to be partners in a program, they must feel a part of it. Here are some ways early childhood educators can make family members feel included:

- Make the intake interview a two-way *exchange* of information. Avoid jargon that families do not understand. If the interview cannot be conducted in the family's language, use translators.
- Create a welcoming environment and provide some adult furniture and gathering places. The environment should be set up for children who are not enrolled in the program so they too feel comfortable and welcome when accompanying parents or other family members.

- Find ways to reflect the family's home language and culture in the program.
- Offer a variety of parenting-related information on a counter or bulletin board near the entrance. Be sure that there is something for every language group in the program.
- Consider family needs in all matters.
- Include family representatives in decisions that affect their child or the program. Use translators when necessary.
- Create a parent-education and support program based on what the families need and want. Use translators when necessary.
- Provide links to pertinent community resources when needed.

The intake interview is a good place to start to make the family feel welcome. The goal of the interview is to learn about the child and the family and to teach them about the program. It is also an opportunity to start building the relationship that is so important to teacher-parent communication and future problem solving.

As mentioned earlier, the intake interview should be a two-way *exchange* of information. At some point, you should expose the family to the program philosophy and explore with them how well it matches their ideas of appropriate care and early education.

An intake interview is a good place to start making the whole family feel welcome.
Digital Vision/Getty Images

Running into Your Own Prejudices

Have you ever found yourself in this situation: you think you are open and respectful of everyone's differences until something unexpected comes into your life? This happened to me.

I met Thomas's mom, Jane. Jane was very tall and had broad shoulders for a woman. Her voice was deep and low. She was friendly, articulate, and clearly loved her son. As I listened, Jane explained to me that she had been Thomas's dad, Joseph. Jane said she had always felt she was female inside. As Jane grew older, she decided to become on the outside what she had always known she was on the inside. Jane explained the long process she experienced to become female.

I began to feel uncomfortable as I listened. Jane was bringing up questions I hadn't asked before. Questions like, what does it mean to be a woman? How are we defined by our gender? Then I thought of Thomas. What was his experience? Was he shunned by his friends? And *then* I began to wonder about the relationship Jane had with her spouse.

At the end of our time together, I thanked Jane but walked to my car full of questions, confusion, and sadness. Now my challenge is to put aside my questions, confusion, and sad feelings so I can relate as respectfully to Jane as I do to any other parent.

–Holly Elissa Bruno

Honoring Diversity

You should enter the intake interview with a broad view of what constitutes a family. In the *Voices of Experience* box above, Holly Elissa Bruno, an ECE consultant and an internationally known presenter, shares a story about struggling to accept a family that did not resemble the "traditional" nuclear family, one composed of a mother, a father, and a child or children. Nowadays, there are single-parent families, families in which the grandparents raise the grandchildren, same-sex families, extended families, and kinship networks. Keep an open mind, and do not hold one model above the others. Avoid using terms such as "a broken family." Also, think about common practices such as making mother's-day or father's-day gifts. Will some children feel left out? What can you do to make them see that their family is recognized and respected?

The provider, teacher, or director needs to take a broad view of family structures along with the authority structure of each particular family. If the mother is the main contact with the program, but she has no decision-making power, it is a good idea to find out who does. Sometimes, it's the father, or it might be a family elder. It is important to have this information when scheduling conferences.

Finally, it is important to recognize that you may disapprove of the child-rearing methods, ethics, or lifestyles of some families. You may even feel the urge to rescue a child from such parents. Be assured that this is a natural stage of teacher development, however, you must move beyond it. Saving a child from his or her parents is not your job. Keep in mind that to teach the whole child, you cannot separate the child from the context of his or her family, culture, or background. Your primary role as an ECE educator is to work *with* families to provide the very best care and education for their children.

FOCUS ON INCLUSION: A SPECIAL KIND OF PARTNERSHIP

NAEYC Position Statement on Inclusion

When children with identified special needs come into a program, a particular process is required by the federal Individuals with Disabilities Education Act (**IDEA**) in order to create a plan that responds to the child's needs. The plan generated for infants and toddlers and their families is called an "individualized family service plan" (**IFSP**). Preschoolers and school-age children need what is called an "individualized education program" (**IEP**). The IFSP or the IEP is a legally binding document protected by the Child Find mandate under IDEA, which requires all schools–public and private–to find and help children from birth to age 21 with special needs. To create these documents, a team of multidisciplinary specialists come together to work in collaboration with the family to create appropriate and meaningful goals and objectives for the family and child with special needs. The goals and objectives depend on the individual needs, challenges, and expectations. For example, a goal for a child with emotional issues might be to establish and maintain a friendship with a child in the program. A goal for a child with physical challenges might be to be toilet trained. A goal for a child with language delays might be to increase vocabulary. Objectives relate to the goals and describe exactly what will be accomplished, to what extent, and how to evaluate whether the objective has been met. The family has a say in what these goals are and can give input on how to meet them. The decisions relate to education but, depending on the degree of disability, may not look like ordinary educational goals or objectives.

The aim of this process is to let families use the expertise of the specialists while the family itself remains the primary decision maker. In some ways, the outcome of the plan or program is less important than the process itself. When the process works, parents come into early childhood programs with some experience at partnering with professionals. Partnering with professionals, when it works well, empowers parents to take charge of their own child. When parents or families have to take a backseat to professional decisions, they may feel helpless and even disconnected from their own child. When unempowered families disagree with professionals about what the child needs, they may give up their advocacy role and leave the child in the hands of people who don't know their child, family, or culture the way the family does.

Early childhood programs play an important role in providing early intervention services to children during a period when their brain development is most receptive to change. While the brain remains plastic for the course of the lifespan, brain development is most prolific during early childhood. This means that effective implementation of the services outlined in an IFSP or IEP created in early childhood can mitigate, or even eliminate, the complications that may arise from conditions such as language delay or autism spectrum disorder that may otherwise fully develop if interventions were not in place.[1] Research suggests that the benefits may be more pronounced for children and families from the lower end of the socioeconomic spectrum.[2]

Parents of children with special needs are not all the same, of course, but one thing they have in common is the need for increased understanding and reassurance from early childhood staff members or providers. Many of them have been through tremendous emotional upheavals. Imagine the many feelings that parents of children with special needs must work through: sadness, anger, fear, disappointment, and frustration. Few parents are prepared ahead of time for the birth of a child with special needs or disabilities, and most struggle to understand and deal with the unique set of challenges they have been presented. When they reach the early childhood program, they may have already endured a number of traumas, spoken with endless professionals, and possibly met with rejections from other programs that were unable to meet their child's particular needs. They may feel anxious, stressed out, and perhaps isolated. They may be worried about juggling the program's requirements and schedule with the child's illness and early-intervention services that may include medical or therapy appointments.

Some parents cope well with their feelings; some even express them openly. Others who are not so capable of coping and who may hide their feelings (even from themselves) may express their emotions in unexpected ways, sometimes focusing on something that has nothing to do with what's really bothering them. It is important that you keep their particular circumstances in mind when working with parents of children with special needs. They can use all the understanding and support you can give.

If the parents of a child with special needs complain and criticize, it is important to understand what is behind their dissatisfaction. Of course, not all parents of children who have special needs complain and criticize; they are as varied as other parents. However, one thing they have in common is the need for support from early childhood staff members or providers.

Recognizing That Parents' and Providers' Roles Are Different

Families and child care providers fill two distinct roles, but both roles call for an attachment to the child. The family has a long-term, close attachment to the child. The family also provides the links to the child's past and the vision for his or her future. In contrast, the provider's role is to maintain an optimum distance that enables a short-term but not too intense attachment to the child.[3] Unlike the family, the provider has no links to the child's past or future (aside from the lasting effects of his or her care and education). The focus of the relationship between the provider and the child is in the *present*. The relationship is temporary; the child often moves up to the next class as he or she gets older, and sometimes the association ends abruptly with the family moving away or the provider quitting.

Therefore, it is important to recognize that you should avoid too close an attachment to a child. That is not to say that you should avoid attachment altogether; attachment is a feature of child care. Children need a relationship with someone they care about–someone who cares about them.

The most important adults in a child's life–those at home and those in child care–should work together.[4] When both sides understand that each other has the child's best interests at heart, they can weather most disagreements because of that trust.

Handling Conflicts with Parents

This chapter began with a look at how staff members can approach and resolve disagreements with each other. Conflicts, however, occur not only between staff members; they sometimes involve parents as well. Provider-parent conflicts are one of the most trying aspects of the early childhood educator's job.

When families arrive in a program, the early childhood educator begins the process of establishing a warm and friendly relationship with them. Unfortunately, it is not unusual for a family to complain about and criticize the program. You may have difficulty maintaining a working relationship with a family that is critical of your program. However, be aware that the complaints are often the way the family expresses its distress. Guilt may be at the core of their criticism; maybe they wish the child could be at home full-time or perhaps the parents are unconsciously jealous because they think the child is getting better care away from home. Regardless, be careful not to feed on distress, guilt, or jealousy. If you become competitive with the parent, the child will be caught in the middle and may feel agony over split loyalties. (Handling conflicts with families who have children with special needs requires careful consideration on the part of providers. The previous section discusses the particular challenges these families face and how the early childhood educator should remain sensitive to their needs and feelings.)

Some families present no problems at all but simply arrive and trustingly turn their children over to the child care provider or staff member without a question. But other parents arrive with their own set of ideas about what they want for their child. They have their own beliefs about child rearing, early education, and discipline. Conflicts often arise between parents and early childhood professionals when there is a breakdown in understanding the other person's perspectives on early childhood issues. The *Points of View* box on page 195 lists some common complaints expressed by both parents and providers. Use this box as an exercise to test your ability to move from one perspective to another. If you have difficulty shifting perspectives, interview some parents and providers and let them explain their viewpoints to you.

Diversity: Being Sensitive to Cultural and Language Differences. Communication between parents and early childhood providers is not always easy, but it is vital. When the provider and family come from different cultures, communication may present a challenge, whether they speak different languages or even the same language.

Taking Another Perspective

Below are some common complaints that providers and teachers have about parents. See if you can take the parent's perspective in answering each of the following questions. Can you imagine a dialogue in which the parent was able to get the provider or teacher to see his or her point of view?

Why do some parents

- refuse to leave promptly after saying good-bye?
- sneak away without saying good-bye?
- always seem to be in a rush?
- always criticize the program?
- seem so lenient with their children?
- seem so strict with their children?
- get upset when their children get dirty?
- push for academics?

Parents also have issues with teachers and providers. Try to take the provider's perspective in responding to each of the following questions. Imagine a dialogue in which the provider or teacher was able to get the parent to see his or her point of view.

Why does my child's teacher/provider not

- make it easier for me to leave in the morning?
- stop insisting that I say good-bye when I leave?
- keep my child cleaner?
- teach my child to read?
- demand that children be more respectful of adults?
- make the children sit down more?
- make the children be quiet?
- make the program more like "school"?

In the case of the family and provider speaking different languages, it is important to find ways to communicate. A translator may be necessary. Although a child may be able to handle some minor communication gaps, it is not developmentally appropriate to do so. Relying on a child to translate for his parents puts the child in the position of being the capable one. Such a role reversal can upset family relations. Also, if the subject of the conversation is the child, it puts the child in an awkward position.

Translation is a delicate job. It takes a special set of skills. There is always pressure on the translator to convey an accurate message–the type of pressure that should not be placed on children. Moreover, children do not have the maturity to discuss adult concepts in two languages. They can easily make mistakes that interfere drastically with the communication between adults.

A single word mistranslated can present an enormous problem. Take, for example, the story of a child who needed surgery. Her parents did not speak English, so a translator was called in. The translator explained the situation and asked the parents to sign the consent forms. They refused. After hours of talking, they still would not budge. It was not until the situation got desperate and a second translator was called that they finally agreed to the operation. As it turns out, the first translator mistakenly used the word "butcher" instead of "operate" in his translation. No wonder the parents refused. Imagine what they must have been thinking! An alternative to managing language discrepancies between families and providers is to ensure that the program staffs teachers who can speak the home language of the children under their care. Knowing your community, getting to know the families through hosting open houses or conducting home visits and using intake forms, and observing the children in your program all help inform staffing needs.

Communication can be a problem even between two people who speak the same language. Let us look at an example of a parent and a provider who come from different cultures and hold dissimilar views on toilet training.

A parent enrolls her child in an infant program, having started toilet training before her baby could even sit up. She is not crazy, she just has a different way of looking at toileting. The mother's method of toilet training is to anticipate her infant's need to urinate and hold him over the potty. She has been successful so far and has eliminated the need for diapers in the daytime hours.

From the provider's perspective, the mother is the one who is trained. The caregiver's view of toilet training is very different. She characterizes the process as "toilet learning" and does not believe it should start until toddlerhood. The caregiver's approach is to wait for the child to show signs of readiness, like being able to hold his urine for longer and longer periods. She thinks toilet learning takes a certain maturity and regards it as an important step toward independence.

One set of values lies behind the parent's approach and another behind the caregiver's approach. The parent values *interdependence*, the provider *independence*. The term "interdependence" is also sometimes called "mutual dependence." Many educators–especially those having been raised to place a high value on independence–wouldn't think of setting dependence as a goal of their early childhood curriculum.

Obviously, this parent and the provider are far apart in their ideas. They need to talk with each other and begin a dialogue. They will have an easier time communicating if they have already established the foundation of a relationship, which is why it is so important for early childhood educators to work on building a relationship with each family from day one.

The concepts of independence and interdependence also tend to clash in the area of eating. Imagine a provider and a parent who differ in their views of self-help skills. The parent values a clean, neat, orderly mealtime, which she accomplishes by spoon-feeding her young child. For her, feeding times are special moments that allow mother and daughter to connect with each other. She feels good about feeding her child and is not about to give it up.

The provider, however, is shocked. Her goal is to get children, even babies, to feed themselves in the name of independence. Independence is so important to her that she is willing to suspend her concern for neatness and order while they are learning. She is critical of spoonfeeding except in the beginning and considers it a practice that hampers a child's development.

"Right" and "wrong" methods are tied to culture and values, so there is no way for the parents or the providers to "win" either the toilet-training or the

Sensitivity to language, culture, and age differences is always important.
Lizardflms/Shutterstock

spoonfeeding argument. The only answer is communication. Dialogue is essential for the adults to understand each other well enough to share the care of the child. Again, if they have already developed a relationship before they find themselves in conflict, communication should come easier.

Resolving Conflicts. In order to sort out these issues, the parties in question need to practice dialoguing. The primary elements of dialoguing are talking and listening and, in a problem-solving situation, negotiating. The elements needed to resolve a conflict can be broken down further into

Reflect

Explain

Reason

Understand

Negotiate

RERUN An acronym that lists all the elements needed to resolve a conflict through dialoguing: reflect, explain, reason, understand, and negotiate. RERUN is a holistic process, and, as such, it is not a series of steps that must always occur in the same order; but as the acronym suggests, the process can be repeated as often as necessary.

Notice that the first letters of each element spell **RERUN.** The acronym is easy to remember and suggests that the sequence can be repeated as often as necessary. It is important to point out, however, that RERUN is a holistic process, not a series of steps that always occur in the same order. The elements are listed sequentially only to make them easier to remember. Let us now examine each element individually.

- *Reflect.* During a conflict, be sure you let the other person know that his or her feelings are received and accepted. Reflect them back with such words as, "I see how upset you are about this situation." Also reflect back thoughts and ideas: "I guess what you mean is . . ." Reflection opens up communication and prompts dialogue. Continue to reflect until you understand the other person's perspective. Reflect also stands for self-reflection. Notice your own reactions. What are you thinking? What are you feeling? What is this discussion bringing up for you? Hold off on expressing self-reflections to the other person until you have finally grasped what her perspective is–what it is like to walk in her shoes, as the saying goes. Remember that communication involves listening–not just talking. Remember also that the word *listen* has the same letters as the word *silent*.

- *Explain.* At some point during the conflict, put your own thoughts into words and/or feelings: "Here's what I think (what I feel) . . ."
- *Reason.* Provide your reasons for what you think or feel: "And this is why . . ."
- *Understand.* Try to see the conflict from both points of view. You do not have to say anything; just aim for clarity. Listen closely to really understand the other person. Also work to understand yourself: if you are ambivalent, you are more likely to become defensive and hamper the RERUN process.
- *Negotiate.* Begin to look for solutions when both parties are clear about the issues and their differing perspectives. "What can be done about this situation?" is a good opening line for this step of the RERUN process.

If negotiations break down and communication gets blocked, go back and start the RERUN sequence again. Reflection is always in order and helps to open things back up again.

The RERUN process leads to solutions. Arguing leads to discontent. Let us examine the two approaches to conflict in the following example.

The scene is a child care center at around 5:00 p.m. A parent new to the program arrives to pick up her daughter and finds her clothes spotted with food stains. She complains to the provider.

PARENT: Why does my daughter have food all over her clothes? I don't like her to look like this at the end of the day!

PROVIDER: I'm sorry. We had spaghetti for lunch and grape juice at snack time. She refused to wear a bib.

PARENT: [*Looking puzzled*] I still don't understand why food got on her clothes.

PROVIDER: [*Mystified*] Young children aren't neat eaters, you know.

PARENT: Of course, I know that. That's why you have to feed them! If you're careful when you feed them, their clothes should stay clean with or without a bib.

PROVIDER: Feed a three-year-old? We don't feed children here, only infants. Preschoolers can feed themselves.

PARENT: Feed themselves? That's ridiculous. They just make a mess with their food.

These two are just warming up for a good argument. Here is what might happen if they decided to use the RERUN process instead:

PROVIDER: I guess we have different ideas about young children and what they need. Tell me more about your approach at home.

PARENT: Well, I feed my children until they are able to handle tableware and not dirty themselves.

PROVIDER: Doesn't it bother you to have to do that at every meal for so long?

PARENT: Of course not. I love doing things for my children. It makes me feel good.

PROVIDER: You feel good feeding your children.

PARENT: Yes, don't you?

PROVIDER: I don't think so. I never fed a child as old as your daughter. I don't mind feeding babies though. It's appropriate because they need my help.

PARENT: So you don't think it's appropriate to help a three-year-old?

PROVIDER: Yes, helping them handle buttons, tie their shoes, roll up their bedding after nap. That feels OK.

PARENT: OK but not good?

PROVIDER: I don't feel good about children being dependent on me. I feel good when they are independent.

PARENT: Well, we do have different points of view. I love babying my children, and I love to be babied myself. It feels great.

PROVIDER: That's a different way to look at it. So what are we going to do about our different approaches to feeding?

PARENT: And about the soiled clothes?

Four Outcomes to Conflicts. The two adults in the RERUN example are listening to each other. Their differences may be resolved in one of four ways.[5] Any one of these resolutions is satisfactory, as long as *both* parties agree. The goal is a win-win situation.

- *Resolution through parent education.* In this outcome, the parent changes her opinion by gaining knowledge or understanding of the provider's views. (Many early childhood programs offer parent-education courses that are aimed at informing parents of the early childhood theories and practices that are the basis of the program's philosophy and curriculum.) If the parent understands the provider's view and agrees that it makes sense, she may be willing to change her own practice. If she is happy with the change, the two have arrived at a win-win solution. However, they would not have achieved a win-win solution if (1) the parent had to compromise a deeply held value or (2) the parent felt disconnected from the rest of the family or other members of her culture.

- *Resolution through provider education.* In this outcome, the provider changes her approach to the issue by gaining knowledge or understanding of the parent's views. The parent might convince the provider that, say, her method of spoonfeeding is important to her daughter, to her sense of her own culture, and to her value system. The conflict might be resolved if the provider can see that independence and interdependence are not mutually exclusive values or if she can suspend her own value in the name of cultural responsiveness. The two parties cannot achieve a win-win solution, however, if (1) the provider abandons her own values and feels uncomfortable or (2) agrees to spoon-feed the child but feels resentful.
- *Resolution through mutual education.* In this outcome, both the provider and the parent change their approach to the issue by gaining knowledge or understanding each other's views. If both can see the other's point of view, they may come to some agreement that respects both perspectives. Maybe they will agree that the provider can continue teaching self-help skills at school and have the child dress in old clothes at mealtime and that the parent can continue with spoonfeeding at home. Maybe the parent will come in at noon and feed her child. These are win-win solutions if both parties feel good about them. If one or both parties feel they have to give in or compromise, they may have uncomfortable feelings about the outcome of the conflict. However, they can also use what is called **third space,** which is an approach named by Isaura Barrera, professor of special education at the University of New Mexico, Albuquerque. Third space goes beyond compromise. The parties in question reach third space when they move from dualistic thinking—"my way or your way"—to holistic thinking that then becomes "our way" and feels satisfying to everybody!

third space The process or ability to move beyond a dualistic or exclusive mindset ("me versus you") to an inclusive perspective that focuses on the complementary aspects of opposing values, behaviors, and beliefs ("sum is greater than its parts"). Without making concessions, both parties try to reduce emphasis on the differences while validating each others' language, values, and ideas. By doing so both parties can identify and integrate the strengths of each others' views to come to a satisfactory agreement.

- *No resolution.* In this outcome, neither the provider nor the parent changes her approach to the issue. Such an outcome is a no-win situation when neither adult respects the other's viewpoint and the conflict continues or escalates. In some circumstances, a no-resolution outcome *can* be a win-win situation. Both parties can be sensitive to and respectful of each other and, because of their differing values, agree to disagree. In this situation, *conflict-management* skills are vital. Both parties learn to cope with their differences above board. As long as they both trust that the other has the child's best interests at heart, this isn't a bad outcome. In fact, when cultures bump up against each other, this may be the best outcome of all. Neither party has to give in. Neither side has to abandon her cultural values.

Many forces work to push our society toward homogeneity. By valuing cultural diversity, we resist those forces. It is important to recognize that the goal of our pluralistic society is not separatism. A society is a group of people who live together, not split apart. The goal of pluralism is to create a unified (not uniform) society in which diversity is acknowledged, accepted, and celebrated.

Early childhood education programs are miniature societies where providers, staff members, families, and children can practice being part of society at large. The early childhood society takes a healthy view of diversity by respecting and even encouraging differences.

Facilitating Communication with Families

The way to establish relationships and provide a supportive atmosphere for families is through communication. This means that families and providers should communicate on a regular basis, yet considering the nature of the early childhood educator's job, that is not always easy. Below are some ideas about how to facilitate and promote communication with parents.

- *Be available.* Quick exchanges at arrival and departure time can add up to significant minutes of communication time. To facilitate adult communication in the morning, for example, try having an interesting activity set up for the children when they arrive with their parents.
- *Be informative.* Parents appreciate knowing what went on during the day. Keep notes so you can be specific: the time the baby last ate, the chant

Teachers and families should talk to each other on a regular basis, but this can be difficult when the only time to talk is during arrivals and departures.
Michael Hall Photography Pty Ltd/Corbis

the toddler made up while swinging, the four-year-old's excitement at watching the butterfly come out of its cocoon, the seven-year-old's thrill at making his first basket ever in the tall hoop. It is important to talk about problems too, but do not focus solely on problems.

- *Be receptive.* Communication is a two-way process. Provide information and be open to receiving it. Make parents feel comfortable about exchanging, through your affect, body language, and vocal tone. Make sure to welcome and respond to e-mails, calls, and/or texts in a timely manner. Also, ensure that you know the names of all the parents of the children you serve so you can be more personable and directly address them by name.
- *Develop listening skills.* Try listening beyond parents' words to uncover unspoken messages. If the parent expresses a feeling, pick up on it and feed it back to the parent to open up communication channels: "You sound worried." This approach works just as well with parents as it does with children.
- *Document.* Documentation is the magnifying glass through which the staff observe children's development and growth while they are at school. Documentation can vary in frequency and detail, depending on the type of information conveyed. All forms of documentation are a means of communication, a way for teachers to highlight the children's learning to parents, and to create dialogue among the readers. An example of documentation is sharing daily anecdotes in writing with parents. This type of daily communication, where each day teachers write a brief anecdote detailing some aspect or event of a child's day, is intended to help connect the children's school and home life by providing a starting point for conversation between child and parent. Other forms of documentation include teachers sharing information regarding the specifics of children's needs and providing a detailed, written charting system to convey information about the details of children's routine care (i.e., meals, sleep schedule, diapering/toileting).
- *Figure out problems together.* Although you will be the expert on child development, parents are the experts on their own children. When problems arise, initiate a dialogue rather than decide on a solution and present it to the parent. Working together allows for the best possible outcome.

Supporting Families

Family is the first and foremost teacher in a child's life. Therefore, communication, cooperation with families, and meeting children's individual needs within the group setting are of the utmost importance to high-quality caregivers. Parent involvement begins in the classroom through established relationships between caregivers and parents. In this chapter, we have explored some specific behaviors that make parents feel supported. One of the most important ways you can support parents is to *respect differences* through listening. Alicia Lieberman, a researcher with years of experience working with families and children, says, "Be aware that parental practices are related to their values. If you try to change the practice, you may well be treading on the family's values."[6] Take great caution before you "educate" parents about your ideas of what's "right" for their children. Understand their value system and respect it.

Recognize also that it is your responsibility to *create a relationship* and keep it going. Take steps toward developing a relationship from day one, and continue to build on it, no matter how many obstacles you encounter. Parents should always be seen as welcome visitors in the child care center and be encouraged to participate and share their family traditions. Moreover, parents should have the opportunity to give feedback through written annual evaluations or communication with the staff.

Although you might think it would be so much easier to just take care of the child and forget about the parents, you can never do that. According to Lieberman, you must *have concern for the parent's welfare* because it relates directly to the child's welfare. If, for example, a parent continually has trouble saying good-bye when she leaves her child, you might begin to think, "If she'd just leave faster, her child wouldn't get so upset. She's the one that is causing the problem!" Adults who suffer from leaving their children often have unresolved separation issues. Those issues affect the child beyond just the morning agony over saying good-bye. Therefore, it is important that you not discount but rather acknowledge the parent's distress and find ways to support her so she can work through the issue.

Likewise, when a child has a problem and a conference is called for, focus on the parents, not the child. Without interrogating or giving advice–two methods that put parents on the defensive–open up communication by asking, "How do you feel about what is happening?" or "Is this problem affecting you?"

Another way to be supportive is to *help parents further appreciate and enjoy their children*. Sharpen your own observation skills so you can help parents see all the wonderful things that are going on with their child. This is an important function of relating to parents. Even when there are problems with the child, stress the positives and resist the temptation to focus on the negatives.[7]

A perceptive teacher or provider can help parents increase their understanding of their child and their anticipation of his or her needs, and a perceptive parent can help the teacher or provider do the same thing. It is a matter of sharpening your observation skills and learning to read behavioral clues. What does it mean when the baby pushes away the bottle? Has she had enough? Is she distracted? Does she need a burp? Or is it something else? Why does the toddler resist lying down on his cot? Is he not sleepy? Is he missing a special blanket or stuffed animal? Does sleeping make him feel vulnerable? Why does the four-year-old hate the play yard and avoid going until the inside is off-limits during mopping time? What is it about her experience out there that turns her off? Why does the seven-year-old fight doing homework every single day? Is this a new power struggle or is there something else going on?

Many of these questions cannot be answered by the early childhood educator alone; he or she needs parent input and insight. Because parents may see a whole side of a child that the provider does not see, and vice versa, it's important to share information.

One more component of parent and family support is to *help parents support each other*. The early childhood educator cannot do it all. The ratio of parents to staff is often almost double that of the child-to-staff ratio. So how can the staff relate to the children and at the same time give each parent the support he or she needs?

The answer is to get the parents to support each other. You may need to facilitate their coming together by giving out names and phone numbers of other parents (with their permission, of course). By referring them to each other, parents have the opportunity to come together with other parents who have dealt with or are dealing with similar issues–both big and small. Just talking with other parents can provide the emotional support necessary to deal with even overwhelming issues.

A Story to End With

When I started in the early childhood field as an aide in a half-day preschool, I learned right away that the classroom and play yard were set off as special–almost sacred–spaces for the children. The focus was on the children's needs, their education, and their feelings. The sounds filling the classroom and play yard were to be sounds of children, not of adults. It was understood that adults were there for the children, not for themselves and definitely not to talk to each other. Adult time came at meetings, sometimes at breaks–if, indeed, there was a space away from the children. Adults who found themselves in the kitchen together were able to talk to each other, but if children were around, adult conversation was frowned on.

This experience as an aide was different from my habits as a young mother. I used to spend a lot of time sitting around talking to my friends while our children played around us. In my new role, I had to get used to the idea that I was surrounded by interesting adults but that I was to ignore them unless some concern about a child might bring us together.

I remember how the play yard and classroom were designed. There was no adult furniture, either indoors or outside. No benches invited us to sit around the play yard. The idea was to stand and move to where the action was. Adults were placed in scattered "stations" to ensure that the whole space was supervised.

But adults are like magnets; they gravitate toward each other. Every so often two would find themselves close enough to talk. Every time I indulged myself in such behavior I felt guilty because I knew it was inappropriate. And, of course, if a teacher was watching, I felt unspoken criticism and quickly moved away from the other adult.

So imagine my surprise to discover that some all-day child care programs were different from my half-day preschool experience. In some all-day programs, adults supervise children and talk to each other at the same time. In most programs, however, such behavior is not sanctioned, but it tends to occur.

Adults in child care programs exhibit varying degrees of discomfort about adult talk in the children's space. I've reflected much on this subject and have come to the conclusion that one must relate to other adults as well as to children. Of course, there has to be a balance: adults must not neglect children because they are too busy chatting with each other, but for adults to relate only to children all day every day is not natural or beneficial for adults or children. I hesitate to make such a statement in a book designed to train early childhood teachers and providers. My own training strongly instilled in me the mandate to deal with all adult-related matters outside working hours, but I now question this mandate.

SUMMARY

Adult relationships in the early childhood program are extremely important. Children learn ways of interacting, showing feelings, and resolving conflicts through watching adults. Adults model mature relationships and problem solving for children. Adults approach problem solving in many ways. When adults work on accepting each other's differences, they also model a respect for diversity. Finally, adults who act naturally and behave in a genuine way model authenticity.

Adults in a working relationship with other adults must learn to distinguish between arguing and dialoguing in a conflict. The point of an argument is to win; the point of a dialogue is to gain information and understand the other's point of view. Dialogues are more likely to lead to satisfactory conflict management or resolution than arguments. When adults teach children dialoguing and demonstrate it in their own relationships, the lesson is twice as strong.

Early childhood educators create relationships with each other and also with the families in the program. It is important for early childhood educators to see the family as the client and to include them in the program. To create relationships with families, it is important to recognize the ways in which the provider and the parent roles differ.

Occasionally conflicts arise between parents and caregivers, and it is important for the early childhood educator to recognize that many conflicts are due to emotional, cultural, and language differences. A process for resolving conflicts is called RERUN, which stands for reflect, explain, reason, understand, and negotiate. Conflicts addressed by the RERUN process can result in four possible outcomes: (1) resolution through parent education, (2) resolution through provider education, (3) resolution through mutual education, and (4) no resolution.

It's the early childhood educators' job to support families by (1) respecting differences, (2) creating and maintaining a relationship, (3) showing concern for the parents' welfare, (4) helping parents further appreciate and enjoy their children, and (5) creating parent support groups.

REFLECTION QUESTIONS

1. Are you a person who seeks relationships and is good at creating and maintaining them? If not, how do you feel about the statement that early childhood education is about relationships? What can you do to upgrade your interest and skills in relationship building and maintaining?
2. Are you collaborative or competitive? How will your inclinations one way or the other possibly affect your career as an early childhood educator?
3. Think of a time when you had a problem with someone. How did you approach the problem? Did you use one of the approaches listed on pages 199–200?
4. Did you attend an early childhood program as a child? If yes, what do you know about what part your parent(s) played in the program?
5. Did you ever play translator for your parents? If yes, how did that make you feel? What advice do you have for early childhood professionals who need to communicate with parents who do not speak English?
6. How would you feel about an adult spoon-feeding a three-year-old? Explain your view of the situation. Do you think your feelings and your perspective relate to your culture?

Terms to Know

Can you use this word and these acronyms in a sentence? Do you know what they mean?

dialoguing 184
IDEA 192
IFSP 192
IEP 192
RERUN 197
third space 200

For Further Reading

Barrera, I., Kramer, L., and Macpherson, T. D. (2012). *Skilled Dialogue: Strategies for Responding to Cultural Diversity in Early Childhood* (2nd ed.). Baltimore, MD: Brookes.

Feeney, S. (2012). *Professionalism in Early Childhood Education: Doing Our Best for Young Children*. Upper Saddle River, NJ: Pearson.

Grefsrud, S. (2011, March–April). Room at the table: Parent engagement in head start. *Exchange, 33* (2, Serial No. 198), 57–59.

Hillman, C. B. (2011, November–December). Home visits: Building relationships by revisiting home visits. *Exchange 33* (6, Serial No. 202), 80–85.

Lally, J.R. and Mangione, P. (2017). Caring Relationships: The Heart of Early Brain Development. *Young Children, 72* (2). Retrieved from: https://www.naeyc.org/resources/pubs/yc/may2017/caring-relationships-heart-early-brain-development

Manning, M. (2010). Family involvement: Challenges to consider, strengths to build on. *Young Children, 65* (2), 82–88.

McWilliams, S. M., Maldonado-Mancebo, T., Szczpaniak, P. S., and Jones, J. (2011, November). Supporting native Indian preschoolers and their families: Family-school-community partnerships. *Young Children, 66* (6), 34–41.

Patterson, K., Grenny, J. McMillan, R., and Switzler, A. (2012). *Crucial Conversations: Tools for Talking when Stakes are High* (2nd ed.) New York: McGraw-Hill.

Rosenberg, M. (2012). *Living Nonviolent Communication: Practical Tools to Connect and Communicate Skillfully in Every Situation.* Boulder, CO: Sounds True.

Souto-Manning, M. (2010). Family involvement: Challenges to consider, strengths to build on. *Young Children, 65* (2), 82–88.

U.S. Department of Education. (n.d.). Sec. 300.323, When IEPs must be in effect. Retrieved from: https://sites.ed.gov/idea/regs/b/d/300.323.

Vesely C. K. and Ginsberg, M. R. (2011). Strategies and practices for working with immigrant families in early education programs. *Young Children 66* (1), 84–89.

PART

2

Foundations in Supporting Development and Learning

The chapters in Part 2 focus on plans for development and learning in both traditional and nontraditional ways. Traditionally environments play an important role in planning for development and learning in early childhood education, so this section starts with a chapter on the setup of the physical environment. The next chapter goes beyond the physical and deals with the social, emotional, and cultural conditions that also make up the environment. To dedicate an entire chapter to the social-emotional environment is not very traditional. You will see the reason for it once you discover that these conditions such as nurturing, respect, protection, and responsiveness are foundational to development and learning. The latest brain development research validates this chapter. You will not necessarily come by these qualities through rearranging the furniture; you have to know how to plan for them. The chapter deals with the details of the how-to's. Some of these details relate to the special issues of children in inclusion programs.

You may have noticed by now that cultural diversity is woven into every chapter of this book and is also in the *Focus on Diversity* boxes. In the chapter on the social-emotional environment, culture occupies a whole section. More than one person suggested to me that this long section on culture ought to be in its own chapter and placed at the end of the book. I think it ought to be at the beginning, but I compromised by folding it into a middle chapter. Culture is the major force in all our lives and, of course, goes with us to work in early care and education programs. The issue is that many of us do not notice because it is so much a part of us. Similarly, it has been said that fish can tell you little about water. I have similarly found that people can tell you little about their physical environment. I hope that in this next part of the text you will begin to see the physical world through new eyes. If I had encapsulated the subject of culture and placed it at the end of the book, I would have been giving a stamp of universality on everything that came before. The message would have been, "Oh, and by the way, there are cultural differences." That's like writing P.S. on a letter–something forgotten until it is almost too late. So culture is spread throughout and is also a prominent feature in the chapter on the social-emotional environment.

A whole chapter on caregiving routines is unusual when the subject is about planning for development and learning. Those who work with infants and toddlers may appreciate it, but others have complained that the chapter is not pertinent to older children. They simply see no relevance to them if they do not work with children under three years of age. I maintain that routines and physical care are important subjects for anyone who works with young children. For one thing, the age at which physical care by adults is no longer important varies greatly depending on the individual family and the culture. Further, with the movement toward full inclusion in high gear, children with special needs who are with their typically developing peers in preschool, kindergarten, or the primary grades may still need the kind of physical care described in this chapter. All early childhood professionals must know how to provide physical care in ways that promote relationships, learning, and development.

The information on developmental ages and stages is in Chapter 11. The subject matter is traditional, but the placement of the information is not. Although the ages and stages information is fundamental, it does not come until Chapter 11. Why? The reason for this is because I want students to look closely at children for a long time before learning about developmental stages. Otherwise they get locked into stage theory and have trouble thinking beyond it. My teacher and mentor, Magda Gerber, used to tell students and professionals alike, "Come to the child with a blank mind and let the child teach you." If you come with all the information about ages and stages, you see what you expect to see and miss other important aspects. Besides, if you have been observing for awhile, you have probably already noticed some patterns that will be familiar as you read Chapter 11. You have been building your own theory as you observed. You are becoming a stage theorist. Creating theory is even better than memorizing it, according to another one of my teachers, Betty Jones of Pacific Oaks College.

By Chapter 12, you are ready to learn something about assessment–a huge issue right now in the early childhood field. This chapter focuses mostly on the kind of ongoing assessment that helps you plan and report on the kinds of experiences that relate to both individual needs and interests as well as those of the group. The assessment chapter is the culmination of Part 2.

8 Setting Up the Physical Environment

Hero Images/Getty Images

Setting Up Activity Areas

Focus on Inclusion: Modifying the Environment for Special Needs

Physical-Care Centers

Infant Play Areas

Interest Centers

Gross-Motor Learning Spaces

Other Considerations for Early Childhood Environments

"Dimensions"

Space

How Much Should There Be to Do?

Circulation Patterns

Balance

A Safe and Healthy Environment

Ensuring Developmental Appropriateness

Providing Protection

Focus on Inclusion: Safe Environments for All

Assessing the Environment for Safety

Sanitation and Cleanliness

The Environment as a Reflection of Program Goals and Values

Individuality

Independence and Interdependence

Cooperation

Authenticity

The Outdoors and Nature

Exploration

Aesthetics

Environments for Various Types of Programs

Full-Day Child Care Center

Half-Day Parent Co-op

Half-Day Head Start Preschool

School-Age Child Care

Family Child Care Home

Kindergarten and Primary Programs

A Story to End With

In This Chapter You Will Discover

- what an "activity area" is.
- what to consider when setting up an early childhood environment.
- what a "play space" is.
- how to plan for circulation patterns.
- how to create balance in the environment.
- how to create a safe, healthy environment that is developmentally appropriate.
- how to adjust the environment to facilitate supervision.
- how to assess an environment for safety.
- how the environment reflects a program's goals and values.

THE THEORY BEHIND THE PRACTICE

"The environment is the most visible aspect of the work done in the schools by all the protagonists. It conveys the message that this is a place where adults have thought about the quality and the instructive power of space."

–Lella Gandini (2002), author & Reggio Children liaison

The Importance of the Environment

Some of the great scholars of early childhood education have written about the importance of the environment to young children's development and learning. For example, Loris Malaguzzi, the visionary and guiding light of the early education programs in Reggio Emilia in northern Italy, regarded the environment as a teacher. So did Maria Montessori, first woman physician in Italy who started a movement with a preschool in Rome called Children's House in 1907. Specialized Montessori materials are still found in preschools around the world, as are the little chairs and tables she designed to fit the size of the children.

Emmi Pikler, who established the residential nursery now known as the Pikler Institute in Budapest, was very particular about the environment and what was in it. One unusual feature is that furniture and equipment for eating and napping are duplicated outside so caregiving activities can easily take place both inside and out of doors. Inside, Pikler gave careful attention to the kinds of floors and coverings because she wanted the babies to be as free to move as possible. The wooden floors give a solid surface. The youngest babies have a thin cloth covering–but no rugs or pads to hinder their movement. The toys at the Pikler Institute are simple–designed for exploration and learning rather than for entertainment.

At higher levels of education (grade school and up), teachers spend a lot of time creating lessons. In early childhood programs–whether family child care, center-based, half-day, or full-day programs–educators put more thought and energy into creating environments, environments that enhance the teaching-learning process. The "lessons" come from the young child's interactions with other people and things in the environment.[1] This chapter focuses on the physical aspects of the environment. Two of NAEYC's standards relate to this chapter:

NAEYC Program Standards

Program Standard 5: Health

Program Standard 9: Physical Environment

- Program Standard 5 on Health requires that programs protect children from illness and injury, which is a function of a well-thought-out and maintained environment.
- Program Standard 9 overlaps with Standard 5 and requires that the program provide appropriate and well-maintained indoor and outdoor physical environments, including facilities, equipment, and materials, to facilitate child and staff learning and development. To this end, a program structures a safe and healthful environment.

In other words, the environment is a teacher, so that means it is not enough to simply clear a space for children to play and dump a box of plastic toys in the middle of it. Thought must go into planning what goes into the environment and matching it to the needs, developmental levels, and interests of the children. You need to plan not only *what* to put into the environment but also *how* to arrange it. See *The Theory*

Many of the "lessons" in an early childhood environment come from the young child's interactions with objects and materials placed there by adults.

Behind the Practice box for examples of theorists who emphasized the importance of a carefully planned and arranged environment for young children.

The first step in planning an environment is knowing what the licensing standard is for your state; these regulations will dictate many of your planning decisions. It is also important to understand the implications of the environment as they relate to liability issues. Licensing regulations differ for each state. In California, for instance, there must be at least 35 square feet of indoor play space and 75 square feet of outdoor play space per child based on the total licensed capacity.[2]

Thus, when setting up the physical environment, program planners must envision the type of atmosphere they want to create. Whether the child care setting feels like a warm and cozy home, an impersonal institution, or a place of chaos depends on a number of factors. Both center-based and family child care environments often resemble "schools" or "institutions" rather than "homes." Some program planners believe that an institutional setting is the way to promote learning. This text, however, suggests that a homelike environment is more suited to young children (see the *Points of View* box on page 214 for some revealing scientific evidence).

The trend seems to be leaning more toward the school model–look at the history of kindergarten. From its name, which means "children's garden," one can imagine that the early kindergartens were more natural and beautiful than a square room with an asphalt play yard outside the door. Some kindergartens today are more homelike than others, but most public school kindergartens consist of a basic, cold schoolroom.

Unfortunately, many preschool and child care facilities are modeled after schools. Some programs even operate in public school classrooms, and the challenge the providers face is to soften and warm up the environment. Other programs are based in more homelike environments, but many program planners try to give them a school-like atmosphere–perhaps because they never stopped to consider any other model.

Points of View

Child Care Programs: Should They Be More Like Home or School?

A classic study conducted by researchers at the Pacific Oaks College in Pasadena looked at the advantages of early childhood programs operated out of homes and small homelike centers.[3] Compared to large school-like programs, children spent far less of their child care day waiting–waiting in lines to use the toilet, to wash their hands, to go outside. In the homelike setting, the waiting took up only 3 percent of the children's time; in the school-like setting, 25 percent–two out of eight hours–was spent waiting.

Decision making was another area of advantage of the homelike program: here, children made 80 percent of the decisions about what they wanted to do; in the school-like programs, children made only 42 percent of the decisions and adults made the rest. Furthermore, it was adults who started and ended the activities in the school-like programs, while in the homelike programs children decided when to initiate an activity, how long to stick with it, and when to go on to something else. In other words, the children were allowed to follow their own interests and pace themselves in the homelike setting. Some of the important tasks of childhood are (1) discovering who you are, what you like, and how to make choices and (2) developing a sense of time. These tasks are accomplished better in a homelike setting, where children can practice structuring their own time.

The Pacific Oaks research also showed that in homelike programs children had five times greater one-on-one or shared contact (along with two or three other children) with an adult. In the school-like programs, children were much more apt to be in groups of 10 to 12 children for most of the day. It is not hard to see the implications of children having little privacy, few one-on-one interactions, and very little personal access to adult attention.

Adults were more available in the homelike centers, and they were also more likely to facilitate learning than demand compliance. Children's behavior also differed markedly in each setting. In the school-like programs, the children tended to interact with adults in a one-way fashion; they spent much more time resisting or responding to adult expectations. In contrast, children in the homelike programs exhibited a greater variety of healthy behaviors: They initiated contact with as well as responded to adults and other children, and they were more likely to be both physically and socially engaged as they gave orders, chose activities, playfully and aggressively intruded on each other, asked for help, and expressed their opinions.

Setting Up Activity Areas

Clearly defined spaces is a key element to the effective design of any early child care program. This conveys the purpose of spaces, sets limits on behavior, indicates how many children can comfortably use an area, establishes boundaries, invites possible combinations of play, and encourages quiet or active involvement. Family child care providers and some centers have to plan their space to serve several purposes over the course of a day, so their environments will differ from those of programs designed and built for group care. For example, in some centers and family child care homes, the tables used for toys, games, manipulative materials, and art activities may have to double for mealtime tables. The play space might also fill up with cots at nap time (although such a plan will not work with infants, who sleep according to their own schedules; they need to have a separate napping space that does not depend on taking over play or eating areas). A few centers now are designed with separate eating and napping facilities; some even have separate workshops and art areas.[4]

In an early childhood setting, the materials, space, and equipment for each of the areas should be appropriate for the specific ages of the children served. For example, an infant program has different environmental requirements than a toddler-program environment which will differ from a school-age-program environment. The size of the furniture should match the size of the children. It will not work to put toddlers in preschool chairs, with their feet dangling off the floor, and wait for them to grow into the chairs. Comfortable, appropriately sized furniture should be provided for children of every size and age group. Likewise, when planning the environment, consider the children's specific needs and abilities. Climbing structures must be scaled way down for infants and toddlers as compared to those provided for older children. Materials for preschoolers and school-age children can include numerous small pieces and parts, but infants and toddlers should not have access to items small enough to choke them. In infant programs, attention should be given to the floor, where infants lie or crawl. How safe is it? What does a baby see when he or she is lying on the floor in a safe place space. The way to find out is to lie down and try it. Is there an overhead light glaring in the baby's eyes? What about a skylight? Ceilings become important for immobile babies who lie on their backs. This is also true for immobile children of any age. A mistake sometimes made in infant centers is the use of overhead lighting in areas where infants lie on their backs and look up at the ceiling. It is important to take the view of the youngest children. Again, try taking the perspective of each child who will use the space, especially those who can't move to reach the objects and materials meant for them to play with. What can they reach? Availability and accessibility of toys and materials is also important.

FOCUS ON INCLUSION: MODIFYING THE ENVIRONMENT FOR SPECIAL NEEDS

For some children, taking ages and stages into account is not enough. Children of differing abilities need different kinds of adaptations. For example, a child may be large enough and old enough for a preschool environment but unable or unwilling to keep small things out of her mouth that she can choke on, or a child with severe physical challenges may be at the stage when other children are mobile, but he is unable to move. Provisions should be made in the environment so that he gets around in it, whether by being wheeled, carried, or put on some kind of scooter

NAEYC Position Statement on Inclusion

that he can push himself in. Careful positioning is an important requirement for children who cannot sit securely. Nonslip pads on chairs or cushions to fill in gaps can help. A rolled towel can do wonders for a child who needs support at the back or sides. A footstool can give added security to a child if there is no way her feet will reach the floor or ground.

As you can see, there are many details to consider when planning the early childhood environment. To begin, let us look at the program broken down into specific activities–activities having to do with physical care and meeting needs, and activities having to do with the children's work and play.

Physical-Care Centers

physical-care centers Areas of the early childhood environment that are designated and equipped for cooking, eating, cleaning up, hand washing, diapering and toileting, and napping.

The full-day program is brimming with activities: cooking, eating, cleaning up, hand washing, toileting, changing, sleeping. Each of these activities requires space (which I will call **physical-care centers**) and equipment. The half-day program entails a modified version of this range of activities since children go home for naps and, perhaps, for meals. Still, attention to the children's physical needs, must be a focus of any early childhood program.

In early childhood education, learning and teaching are not restricted to "academic" matters; the learning and teaching process is *always* at work. Learning and teaching occur in bathrooms, kitchens, sleeping areas, diaper-changing areas, and places you never even thought of. Perhaps you never considered that physical-care centers can be outdoors as well as inside. At the Pikler Institute in Budapest where they have been using an internationally known approach for caring for children under three years of age since 1946, the outdoors is as important as the indoors–not only for learning and development, but also for physical care. For example, in the summer, eating is done outdoors–not just sometimes, like a picnic or a special occasion–but all the time. Children nap outdoors year round–they have two sets of beds: those outdoors and the ones inside. Even newborns are exposed to fresh air on a screened sleeping porch. The health benefits are remarkable.

Infants and toddlers in many programs in the United States are the ones who spend the least time outdoors. The complaint is that especially in cold weather it is just too much trouble to get them dressed to go out and then undressed when they come back in. At the Pikler Institute, the dressing and undressing are an important part of the program where several goals are met. One goal is attachment to the caregiver; the time spent together is considered valuable and carefully carried out so the interactions promote the relationship. Another goal is that children learn to cooperate with the caregiver from birth on, so as they grow they begin to take on more and more of the dressing themselves–not because they are pushed to do so but because they want to help and they are also interested in learning to do things by themselves.

Children learn every minute. What they learn is determined to some extent by how the environment is set up. Deb Curtis, an internationally known trainer, speaker, and consultant as well as the writer of many books, discovered how the placement

A Place for Diaper Parties

I have always envisioned the diaper-changing table of a toddler room located against a wall, protected for health, safety, and privacy, providing a quiet place for intimacy between child and caregiver. My ideas changed dramatically the year I spent as a toddler teacher in a room where the diaper table was positioned near the middle of the room.

It was a gorgeous, wooden table with stairs for the children to walk up. Because it couldn't fit against any of our walls, the licensor directed us to attach a four-foot tall, clear plastic wall around three sides of the table. Initially I found this arrangement disheartening, but then I began to watch the children's relationship with it. They loved to climb the stairs by themselves, and when they got to the top they relished the view from that high place. I think it felt powerful and exhilarating to them; like reaching the top of a mountain, standing on the edge and seeing the sweeping expanse below.

Needless to say, it was a popular place and with every diaper change, many other children gathered around the table, waiting for their turn to climb up high (whether they needed diapering or not!). The children's interest in exploring this bird's eye view encouraged me to linger and look with them. It became my time to slow down, be present with the children, and delight in their point of view. Eventually I began to help two or three of them spend time at the top sharing the view before getting on to diapering.

When I did encourage one of them to lie down for a diaper change, and guided the others to the bottom, a new game began. The children waiting their turn would interact playfully, watching, talking, and laughing with me, the child I was diapering, and each other through the plastic wall. They would stand up on their toes to see, until one of them figured out to use the stool from the sink. In time, I found a short bench and another stool, to provide a permanent platform around the changing station for the children to stand on.

This small environmental change, based on observing the children's interests and joys, led to many wonderful "diaper parties" throughout the year. This experience helped me learn so much about how to see the environment. Although I was initially daunted by this diapering arrangement, which didn't fit my idea of the diaper-changing process, the children taught me the joy of sharing this regulated, often thought of as mundane or distasteful, diapering task. I've come to realize that the environment cannot be a static place. It must evolve with the constraints and possibilities of a space and most importantly, to reflect the ways children and adults live together in it. When we see it in this way our days with children can be calm, rich, and rewarding!

–Deb Curtis

of the diapering table made changing toddlers a social event instead of an intimate ritual. See the *Voices of Experience* story.

If the environment is chaotic, with adults racing around in a disorganized fashion, the children will learn lessons that differ from those learned by children exposed to a calm, ordered environment. Notice the difference in the following two scenes.

In the first scene, the caregiver is about to change a diaper. She approaches a six-month-old who is on a rug exploring a soft textured ball. The caregiver explains that it is time to be diapered, and holds out her arms to the child. She waits for a response from the child, then slowly picks him up and carries him across the room.

Once the child is on the diapering counter, the teacher can concentrate on the child because she has already prepared the diapering area. Everything she needs is right at her fingertips. So the diapering proceeds at a relaxed pace, with the teacher involving the child in what is going on. The baby, even from a young age, is a participant rather than just a recipient. The two are partners in the process, and the environment supports them. The child learns more about his own body and its processes. He also learns something about teamwork.

As demonstrated in this scene, it is important for caregivers to take the time for one-on-one interaction during diapering. It is also important that the rest of the children remain safe and secure. An environment set up for play helps adults by occupying children who are not being diapered with interesting things to do. When more than one caregiver is present, the person diapering can pay full attention to the child she's working on. That is an advantage of staff teamwork. When the caregiver must work alone, she must take advantage of the dual-focus technique, which allows her to focus on the baby while still being conscious of what else is going on in the room.

When everything is not organized it is easy for both the child and teacher to be distracted, and it is quite likely the teacher will just hurry to finish up the job and put the child down. This child did not learn about teamwork. He did not learn about concentrated focus. Perhaps what he did learn was to consider diapering an annoying interruption rather than an opportunity to learn more about his body and have time alone with his teacher.

What these two examples demonstrate is that the environment can facilitate the teaching-learning process during caregiving times.

Infant Play Areas

After the newborn period when infants are still settling in, they do not need much space, and their main activities relate mostly to the routines involved in their physical needs. They need quiet places where they can sleep and wake securely without undue stimulation. As they mature, they need larger spaces where they can lie on their backs to stretch, wiggle, and move around to the extent they are able. These can be thought of as play areas and should be protected from mobile infants and toddlers. The surface on which the infants lie should be firm, and the space large enough to hold more than one infant so they can be aware of their peers around them. They will begin to interact with other babies if they have access to them. They will also interact with any adult who is down on the floor in the play area. The adult's role at this point is to be responsive to what the babies initiate and to protect babies who need it from peers who want to explore ears and eyes.

A few simple objects should be in the play space close enough to the babies so they can be grasped. A cotton scarf about the size of a large dinner napkin makes a perfect first toy for the infant to grab, hold, and manipulate. Although the trend is to put infants upright and strap them into various infant holders, there is huge advantage to allowing them to be on their backs unhindered so they free to move. Research from the Pikler Institute in Budapest shows evidence of the developmental advantages of freedom of movement for infants.

The play space should expand as the infants' capabilities expand. Careful observations of freely moving infants show that they become somewhat mobile long before turning over or beginning to crawl. Observations at the Pikler Institute show how young infants change location by using snakelike and/or inch-worm movements. They also use their feet and legs to move in a circle keeping around the head. This ability to change location can happen long before the infant begins to turn on her side and from that turn over. Babies who are propped up or strapped into car seats, infant carriers, swings, high chairs, and other devices do not have the advantage of developing the sets of muscles that freely moving babies do.

When babies begin to crawl, they need larger play spaces and objects to crawl over, into, and around. When they get up and walk, they need even larger spaces and can eventually take advantage of the interest centers that are the focus of the next section.

Interest Centers

In early childhood programs, people traditionally think of the **interest centers** as the primary teaching-learning sites. They include the floor space, the equipment, and the materials for exploring, such as soft blocks for younger children and for older ones wooden unit blocks and props; manipulative toys, including puzzles; sensory materials; writing materials and projects; reading materials; housekeeping/dramatic-play equipment, furniture, and props; science and nature activities; toy animals; art materials; musical instruments; movement props; and math and cooking activities. All of these activities should fit the ages and developmental stage(s) of the children who are in the group. There should also be an area with materials for ongoing projects that don't fit into the above categories. Classroom pets can also be included in interest centers. (See the *Points of View* box for two views on having animals as classroom pets in early childhood programs.) A family child care provider would not have all of the preceding materials and equipment out every day, but over a period of time most categories on the list should be available to the children.

interest centers They include the floor space, equipment and materials for play, interaction, and exploration.

In temperate climates, some or all of the interest activities can be set up outside. Even physical-care centers can be set up outside.[5] Depending on your values, you may believe that fresh air and sunshine are vital to one's health. You may want to arrange for children to eat and sleep outside or on open-air porches. To some extent, that notion is a thing of the past on this continent, but other countries still have sleeping porches or outdoor napping areas.[6] See the *Tips and Techniques* on page 221 for more about children's inclinations toward being outside.

POINTS OF VIEW

Two Views on Animals in the Classroom

Some people believe that children need to learn to relate to animals and that the early childhood classroom is a good place for this relationship to develop. Animals provide wonderful, responsive, sensory experiences. Children learn about responsibility when they help feed and care for animals. Some children also benefit emotionally through relationships with animals. Moreover, families have pets, and having animals in the classroom makes it seem more like home.

All kinds of pets are appropriate, from rats, hamsters, or guinea pigs in cages to pet bunnies whose cage doors stay open so they can hop around and use the cage as a home base. Most center-based programs don't have cats or dogs, but some family child care programs do. Some programs even keep farm animals when they have room.

The opposing view is that some people believe that keeping pets, particularly caged pets, is cruel. In this view, keeping pets is limiting and demeaning for the animal and sends the wrong messages to children about humans' place in nature. According to this viewpoint, by keeping pets, humans use animals for their own pleasure without regard for the animals' feelings. Classroom animals are dragged around, held awkwardly, and petted when they are trying to sleep. Animals should be free to be what they are, *not* caged, tied, leashed, penned, and made into people's toys.

Gross-Motor Learning Spaces

Gross-motor activities for infants and toddlers are much simpler than for older children and they were discussed briefly in the section on Play Areas for Infants. Gross-motor activities for older toddlers and preschoolers include running, stretching, climbing, jumping, rolling, swinging, ball throwing, and (in the case of older children) game playing–in other words, activities that use the large muscles of the arms, legs, and trunk. The category of gross-motor learning includes vigorous exercise as well as skill building.

Weather and climate dictate whether gross-motor learning occurs outdoors, indoors, or both. Some programs can make year-round provisions for gross-motor activities both indoors and outdoors. Other programs have to limit gross-motor activities to the outside because there is no indoor space for active movement. In areas with prolonged insufferable weather, such as extreme heat, cold, or precipitation, programs adjust by incorporating indoor areas specially designed for gross-motor skill building and/or vigorous play involving large-muscle activities. Some have indoor/outdoor activity rooms similar to four-season porches. Others set up climbing equipment as needed or provide other indoor gross-motor activities on an occasional basis only.

gross-motor spaces Indoor and outdoor areas specifically designed for gross-motor skill building and/or vigorous play involving large-muscle activities, such as running, stretching, climbing, jumping, rolling, swinging, ball throwing, and (in the case of older children) game playing.

No matter what the setup or climate, all early childhood programs must arrange for **gross-motor spaces,** where children can use their bodies in many different ways. The older the children served, the larger the space must be. As already pointed out,

TIPS AND TECHNIQUES

What Was Your Favorite Environment as a Child?

Think of your favorite childhood place. Was it a secluded place outside, away from the scrutiny of adult supervision? Think about what made that particular place so attractive to you, and then use your own experience to guide you in creating special environments for young children.

Most of the people I have asked report that their favorite childhood spots were outdoor hiding places: out behind the garage, out in a back lot, or up in a tree house. Designing hiding places for young children can be problematic. Where they like to be is not necessarily compatible with the logistic and safety requirements of group child care. Nevertheless, It's good to remember that children like to be able to get away from the crowd, and they like to be outdoors.

crawling infants and young toddlers need only a small amount of space to practice their gross-motor skills. They can even ride small scooters and trikes indoors. School-age children, however, need a good-sized playground–and a gym in areas with extreme summer or winter weather.

Other Considerations for Early Childhood Environments

How much space is enough? Is there an appropriate amount of equipment and materials for the children to choose from? What about circulation patterns? How do I keep the children from walking through each other's play space? What about balance? Are the goals and values of the program reflected in the environment? All of these questions are pertinent to planning the environment for the care and education of young children.

"Dimensions"

Elizabeth Jones, an early childhood researcher, discusses the following five dimensions of a learning environment in her book *Dimensions of Teaching/Learning Environments.*[7]

- *The soft/hard dimension.* Children need a balance of soft and hard surfaces and objects in their environments. While hard surfaces and objects are easier to clean, be mindful that young children need plenty of thick rugs, soft blankets, stuffed animals, cozy furniture, mattresses, pads, cushions, and laps. Outside, softness comes in the form of grass, sand, water, soft balls, pads, and, once again, laps. A soft environment is *responsive.*

Areas of seclusion should be available to children who need or want to be alone.

Children also need hard objects (such as wooden toys) to experience a different feel. Hardness belongs in early childhood settings, along with softness. There should be a balance. Keep in mind the younger the child, the farther the balance must tip toward softness.

- *The intrusion/seclusion dimension.* The environment should provide for both optimal intrusion and seclusion. To illustrate this dimension, let us begin with the term "intrusion." Intrusion is anything from the outside environment–physical, visible, or audible–that comes inside, providing interest and novelty. For example, a low window is a valuable form of intrusion; it allows children to see what is happening outside and, if open, to hear, say, workers repairing the street. Another desirable form of intrusion are outsiders who come into the children's environment–the telephone repair person, parents picking up their children, visitors. Realizing that too much intrusion can be disruptive to the child care environment and upsetting to some children, you should understand the children's need for stimulation and strive to maintain an optimal level of intrusion.

  Seclusion, the second element of this dimension, should also be provided so that children who need to be alone–either by themselves or with another child–can find spaces to get away from the larger group. Of course, supervision must always be a concern, but there are ways to make private spaces that you can still see into. One simple way is to move a couch out from a wall. Platforms, lofts, hidy-holes, and nests also create secluded environments.

For some children with special needs, especially those who are easily overstimulated, having a place to escape to may be imperative. Be aware of all the children's needs for seclusion and optimal intrusion, and provide for those needs.

- *The mobility dimension.* High mobility and low mobility should balance out in early childhood programs. Children should be able to move around freely without having to wait for outdoor time to engage in vigorous movement. Likewise, children need some down time to relax, listen to a story, or play quietly.
- *The open/closed dimension.* This dimension has to do with choices. An example of openness would be low, open shelves containing toys for the children to select from if they wish. In contrast, the environment should have some closed storage–usually up high–that is used to regulate choices, get rid of clutter, or lock up poisonous or hazardous items.

  Openness also has to do with the arrangement of furniture and dividers. A good arrangement is to have openness from your waist up to facilitate supervision and some closed space from the waist down to prevent young children from being overwhelmed by large expanses.

  The open/closed dimension also has to do with whether a toy or material has one right solution or use (like a puzzle or graduated stacking rings) or whether it encourages many kinds of exploration (such as stuffed animals, play dough, and water play). Children under two need many more open materials and toys than closed ones. By three, children enjoy and benefit from both open and closed materials.
- *The simple/complex dimension.* The more complex a material or toy (or combination of materials and toys), the more ways children can use it. Sand, water, and utensils combined stimulate more ideas and uses than any one of the three by itself. Caregivers have found that when children are engaged in complex activities, their attention spans increase.

  Nature has built-in complexity, and play yards should be as natural as possible. Rusty Keeler, author of the book *Natural Playscapes,* is an advocate for bringing nature into play yards. He likes to see play yards where children can run, climb, dig, play make-believe, and perhaps best of all, children can use their "outside voices" at will. Activities that usually occur indoors, like art, music, and story time can happen outdoors as well. A clever idea that Keeler suggests is planting a "tree circle" that changes with the seasons. It can become a gathering place for groups, a place for pretend play, even a place for teachers and parents to meet.[8]

Space

The environment's space should match the size of the group and take into account the ages of the children who use it. Infants need very little space compared to school-age children, who need enough space to run around and play organized games. The recommended space requirement is a minimum of 35 square feet per child.

Group size is another important consideration. It is important here to distinguish between adult-child ratios and group size. A program can observe recommended adult-child ratios and still have groups that are too big. For example, five rooms with 2 adults and 12 children are very different from one room with 10 adults and 60 children, yet the adult-child ratio is the same in both.

Groups need their own discrete spaces that are separate from other groups. The environment should be set up with barriers to cut down on the visual and audible distractions of the larger group and help each child focus on the children and adults in his or her own group.

Louis Torelli and Charles Durrett, specialists in designing child care environments, point out that with a small group–say, six–the mandated 35 square feet per child is too little. They recommend more–up to 50 square feet per child–so the environment feels more spacious.[9]

As you can see, the space in proportion to the size of the group is important. Too much space has its own problems. Put 12 school-age children in a gym and you get chaos unless you limit the use of space. Twelve preschoolers in a huge play yard are also hard to supervise. Ideally, the amount of space should enable a freedom of play and exploration that does not compromise the supervision, health, and safety of the children. Of course, because there are regulations to follow, sometimes there is no choice about how much or how little space to provide.

How Much Should There Be to Do?

When setting up an environment for play, it is important to provide children with choices. The question is, How much choice? To some extent, that may be a cultural question. A room or play yard stocked with too many toys or activities can overstimulate and distract young children; likewise, a sparsely equipped environment can create problems when children have to resort to inventing ways to keep busy that aren't always positive.

One way to evaluate the environment is to count "play spaces." For example, there are six play spaces at the play dough table with six chairs and a wad of play dough big enough to be divided six ways. The easel provides two play spaces–one on each side (or four if children share paper and paint). The dramatic play corner, which holds four children comfortably, counts as four play spaces. And outside, the swing set provides four play spaces, and the tricycles count for six more. When planning the environment, use the following rule of thumb: When a child changes activities, there should be two to three open play spaces to choose from; if there are fewer, choice is too limited and the children will have to wait too long for a turn at something they want to do.[10]

Circulation Patterns

Good house design dictates that people should not have to cross through the middle of one room to get to another. The same principle can be applied to setting up family child care and center-based environments.

In any sort of child care setting, the circulation patterns should not run through activity areas or interest centers if at all possible. For infants and toddlers, one recommended layout is to place activity areas along the walls, with the center space left open for traffic.[11] For preschoolers, arrange "paths" that clearly draw the children from one activity to another.[12] And for children of all ages, it's important to set up areas that are enticing to children. The arrangement and availability of equipment, materials, and supplies are important. (Figures 8.1, 8.2, and 8.3 illustrate sample floor plans for infant, toddler, and preschool child care classrooms.)

Balance

Think about the environment in terms of balance: there should be some quiet areas and some noisy areas, as well as areas for large-group, small-group, and solitary activities. Consider, too, how to balance the activities according to the children's physical, cognitive, and social-emotional development.[13]

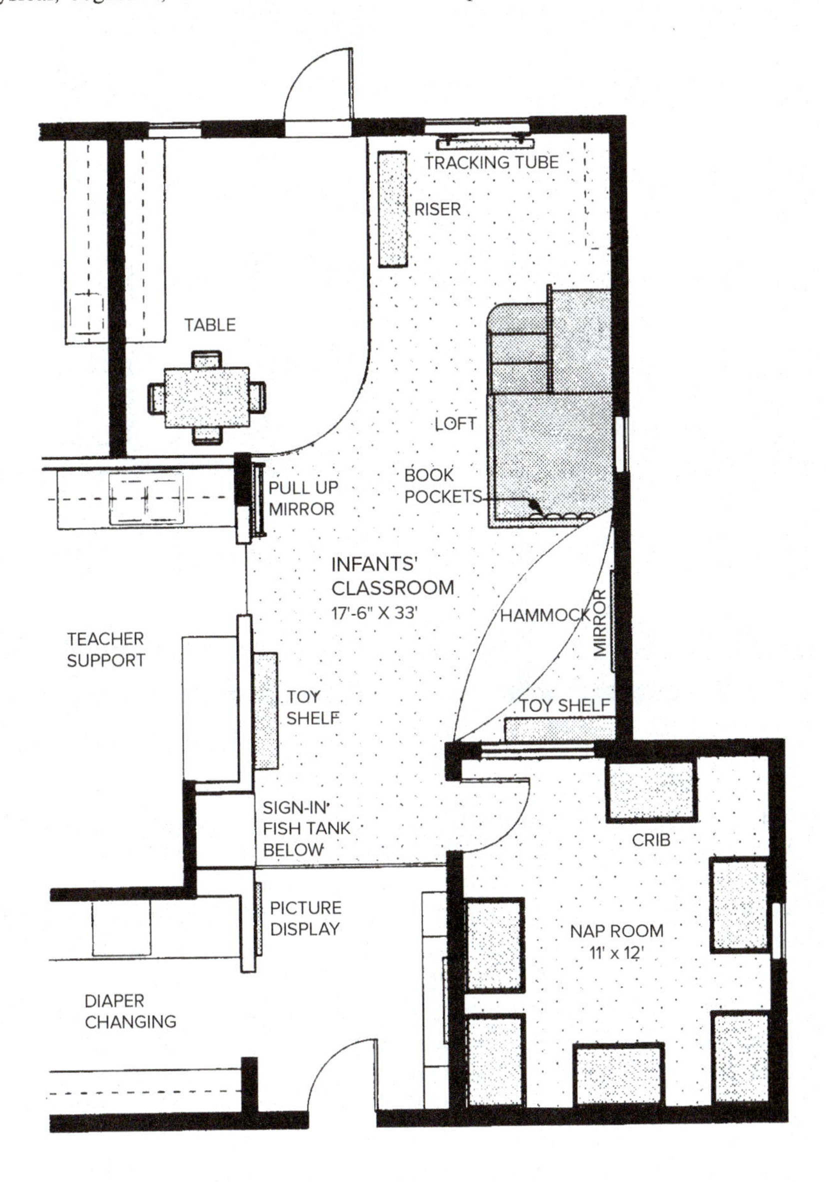

FIGURE 8.1 Infant classroom. It is important to highlight that in this design, the infant classroom is well equipped with a separate nap room, so infants—who require much sleep during the day—can have uninterrupted and quiet sleep space. Also having a riser to help define a space as well as serve as a surface for infants to use to pull to stand or cruise on is helpful in scaffolding gross-motor development for mobile infants. The loft with wide steps and and a not very steep slide is also conducive for providing just the right amount of scaffolding for the infant without overwhelming more mobile infants.

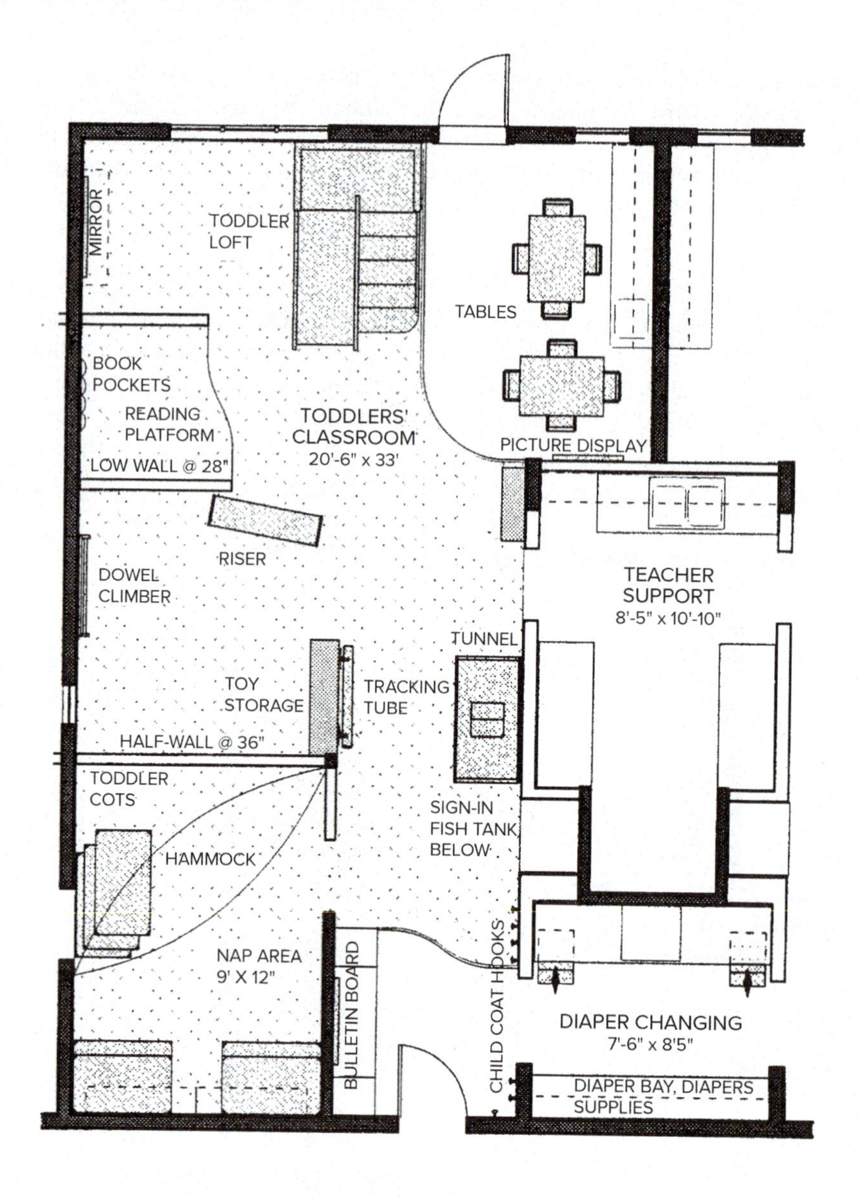

FIGURE 8.2 Toddler classroom. This space again incorporates a well-defined and separate sleeping, eating, and diapering area. The loft is different from the infant loft in that the slide is steeper and the steps narrower. Making it more appropriate for scaffolding the gross-motor development of toddlers. The dowel climber is perfect for helping early walkers pull to stand and more advanced walkers to climb and their peers to join.

Balance is a key consideration when evaluating the environment. For example, the environment should encourage both quiet and noisy activities, but it should be arranged so the quiet spaces are separated from the noisy spaces; don't put the book area next to the music area. Also try to balance small- and large-muscle activities. Provide tools, toys, materials, and equipment that encourage both gross- and fine-motor skills. Again, separate the two kinds of activities; puzzles do not belong in the sand under the climbing structure.

Provide spaces for one child, several children, or, if appropriate, a large group of children. Not every program will have ample space to allow large groups to gather

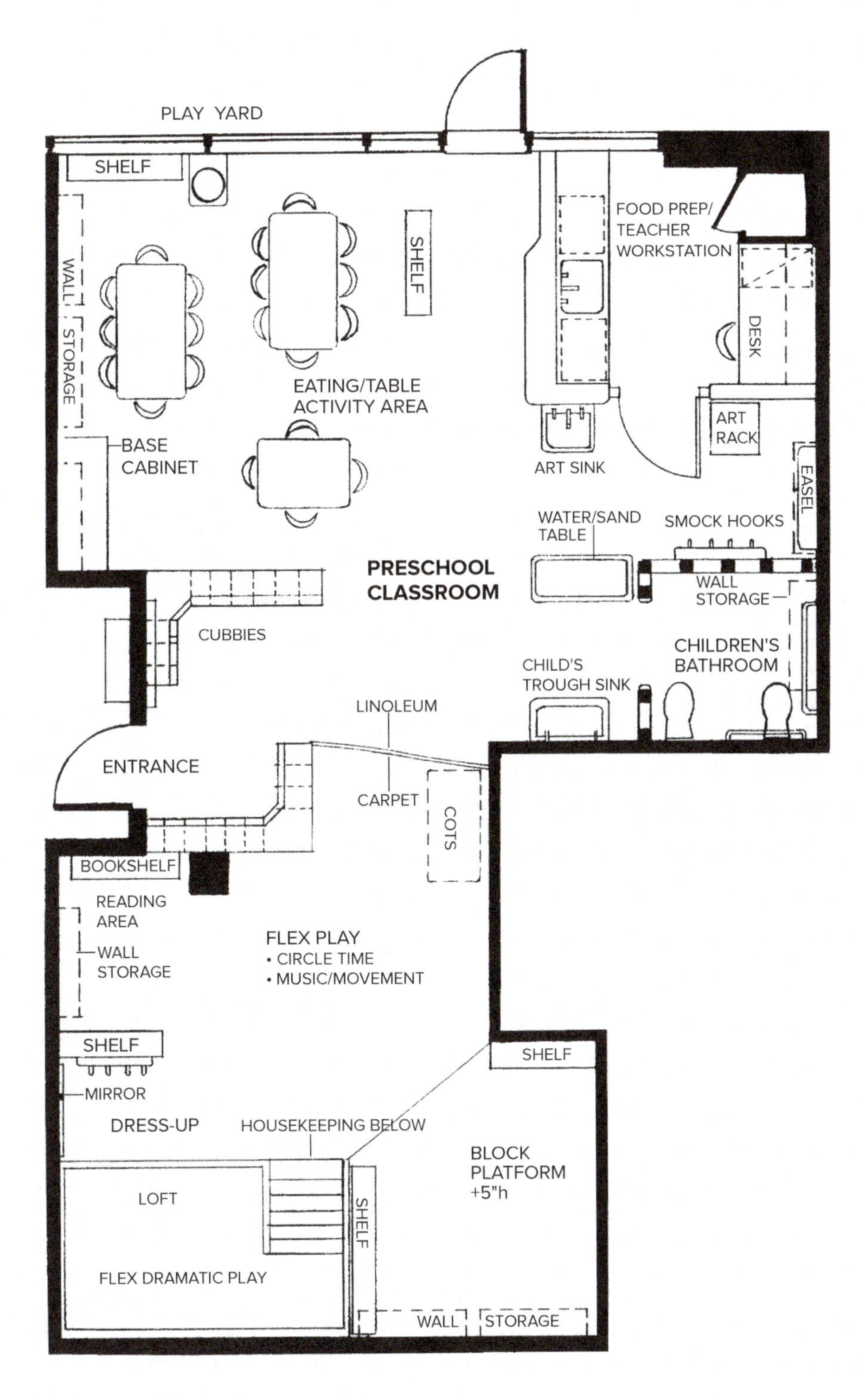

FIGURE 8.3
Preschool classroom. This floor plan conveys a clear purpose of the various spaces. The clearly organized play space and clear paths can result in fewer disruptions and more goal-directed behavior. The area allows children to participate in small-group activities without being disturbed and is designed in a manner to keep loud and quiet activities apart, while the large spaces allow for active, large-group activities that are more boisterous and noisy.

indoors; in this case, consider going outdoors or rearranging the furniture to accommodate larger numbers.

Consider all the children's needs when setting up the environment. The concept of the "whole child" should guide your environmental planning. Child care is not about physical care alone. Early childhood enrichment programs are not just about intellectual development. Likewise although some parents send their children to early childhood programs primarily for social experiences, no program is purely social either.

You cannot separate the "whole child" into parts; each area of development links together to form the individual child. Because the individual child comes from a larger social context, the only sensible approach to meeting all the children's needs is to take a holistic and balanced approach to environment planning.

A Safe and Healthy Environment

NAEYC Program Standards

Program Standards 5 and 9: Health and Physical Environment

Sound health and safety measures are a primary requirement for an early childhood program, regardless of the setting. In this section, we will take a close look at the specific elements of a safe and healthy environment.

What measures are necessary to make an environment safe depend on the age and capabilities of the children. Each environment must be assessed for its developmental appropriateness. Even within a safe environment, adult supervision and protection are needed. Let us explore each of these factors and revisit some of the information from Chapter 2. For information on adapting an environment for children with special needs, see the *Focus on Diversity* box.

Ensuring Developmental Appropriateness

One of the first requirements of a safe environment is that it should be developmentally appropriate. Toddlers who play in a yard set up for school-age children risk injury when they try to climb to the top of high slides, for example. Infants crawling around a room set up for preschoolers are apt to find any number of dangerous objects to pop into their mouths.

When children are separated by age groups, childproofing the environment is simply a matter of determining the proper scale, size, and type of toys, equipment, and materials for the particular age and stage. However, it is quite possible to care for children safely in mixed-age groups. People do it in homes all the time–when they raise families or when they provide family child care. It takes a little thought and attention, but it is not hard to set up a safe environment for a range of ages.

Sometimes childproofing involves protecting younger children from equipment that is too big for them: toddlers cannot get to the loft bed where school-age children go to read because the first step of the ladder is too high for them. Infants cannot chew on cardboard puzzle pieces for older children because they are stored high up and used on the kitchen table instead of the low play tables. The sculpture the big kids are creating is inside a playpen so crawlers cannot reach it. A fence across the opening between the family room and the kitchen gives older children access to the oven while protecting younger ones from getting burned. As you can see, organization

Adapting an Environment for Children with Special Needs

It is important to recognize that the way an environment is set up influences how well children learn and interact. If a program includes children with special needs, certain modifications can be made to the physical environment so that the environment has the same positive influence on children with disabilities that it does on their peers.

In adapting an environment for children with special needs, one of the first things educators and care providers should consider is the amount of space the program provides the group. Some programs require bigger rooms or more space to accommodate extra adults who may be needed to help the children with special needs, or they may need extra space for such devices as wheelchairs, walkers, and ramps.

Another aspect of room design that should be taken into consideration is where materials are placed. All children should be able to get at materials in the room, and all children should feel encouraged to get materials for themselves. If children are not ambulatory, they may have to crawl or wriggle toward the materials. If they are ambulatory, they may rely on furniture to help them get to the materials. Sturdy furniture is always important, but it is a necessity for children who need to hang onto shelves and tables in order to stand or balance. Nonslip floor coverings are also essential for these children.

The level of disorder in a room is another important consideration. For many children, chaos and disorder suit their style. However, not all children thrive in such an environment. For example, children with moderate to severe visual impairments need a clutter-free environment. They also need the furniture and room arrangements to stay exactly the same so that they can feel secure about moving around without bumping into things. Large groups and a high noise level can also distract and upset some children. Those with visual impairments have trouble reading sound clues when there is a lot of noise. They do better when groups are kept small and when the noise level is controlled.

Finally, some programs include observation as an ongoing activity. The inclusion of an observation room with darkened glass is one of the most useful ways to keep observers from disrupting the class.

of space and the use of barriers are important factors in keeping the environment safe. It is important to underscore the point that while it is important to ensure safety, accessibility is also an important element in designing spaces. Young children's spaces will enhance children's self-esteem when it is designed with their needs and development in mind. As mentioned earlier, these spaces promote competence by allowing children to function independently. Children's burgeoning sense of independence is fostered when they can competently access and use materials in their environment. Subsequently, materials should be arranged in a manner so young children can see what is available and make autonomous choices. Research shows that children are more productively involved in activities when the purpose of classroom spaces is clearly defined and when materials are developmentally appropriate.

Furthermore, keep in mind that children with physical challenges may require special safety considerations. Can a child in a wheelchair or walker get around safely to the various areas of the building and yard? Are wheelchair ramps at the proper slope?

This room demonstrates the good organization of space and the proper use of barriers that keep children safe.
Mogddl/Panther Media GmbH/Alamy Stock Photo

Providing Protection

In the name of health and safety, provide environmental protection for children who need it. For example, infants who are not yet mobile should be kept out of the way of toddlers. Imagine how scary it would be to lie unprotected while larger people trooped around and over you. Fence off a corner of the room, or use furniture as barriers, but do not isolate babies in cribs; cribs should be reserved for sleeping. Infants often learn to distinguish sleeping time from waking time from environmental cues, and keeping infants in cribs to protect them may throw off their environmental cues about sleeping.

If there is only one nonmobile infant in a program, using a playpen is a good idea. The purpose is not to confine but to protect the infant. Once he or she is old enough to crawl, however, the playpen will no longer be appropriate because it is too limiting. The crawling baby needs to be able to get around. Some programs create built-in "playpens" large enough for several babies and an adult to sit inside. This kind of playpen has a very different effect from the small portable ones that restrict babies' movements.

One challenge is to provide for safe, private, tucked-away spaces and yet keep every area of the room or outdoor play area visible to the adult's watchful eye. One way an infant center met that challenge was to build wooden boxes with small entry holes. Each box held one or two children. They felt alone in there, but they were not unsupervised. The tops of the boxes were left off so the adults could look in and see what was happening. A card table with an old sheet draped over it serves the same purpose if you leave the back undraped, pull it out from the wall, and hang an unbreakable mirror by it.

Climbing structures, slides, and swings, whether indoors or outside need safety surfacing to soften any falls. It is important to know if the surface under consideration has passed safety tests. Not all mats, cushioning material, and rugs are able to absorb the force of a fall and prevent injury. Be sure what is selected meets safety standards for crash or landing surfaces.

Focus on Inclusion: Safe Environments for All

NAEYC Position Statement on Inclusion

Both the indoor and outdoor environment must be analyzed so they can be adapted for all children to safely participate as fully as possible. That means children who are physically challenged need to feel secure in the face of rough-and-tumble play common among preschoolers and school-age children. Close adult supervision is called for here. Children need to learn to be careful of classmates who are not as sturdy or agile as they are. Adults must teach them.

It goes without saying that children in wheelchairs need ramps and wide spaces in which to maneuver. They also need hard surfaces to roll on. Thick shag rugs and sandy paths will not work. Soft surfaces are challenging for children with braces, crutches, and walkers as well.

Children who are visually impaired need protected, consistent space. If the early childhood educator constantly changes the furniture around, children with visual impairments get easily confused or lost. It is better to leave things alone or, if necessary, just change one piece of furniture and help the child get used to that change before switching anything else around. Other suggestions for safe outdoor play for a child with visual impairments is to take the child to a playground while talking about where things are located and where friends are. Give the child a chance to orient to the location–the sandbox is near the path, the picnic table is beside the tree. There probably is a specialist working with the child and family who can give you other ideas about how to support the child's independence in new settings. Help the child to the area of choice or have another child help. Think of toys that make noise, such as a ball that makes a beeper noise so it can be easily traced. It is best to have more than one such ball so other children can also play with it. In the sandbox, tell the child what is available and who is playing. Stay around to support the child's entering the play and remaining interactive with peers.

Children with hearing impairments require a particularly interesting visual environment. Also, attention should be paid to the noise level in the room. It is a challenge for children with normal hearing to distinguish one noise from another in a chaotic setting, but for children who are hearing impaired, it may be impossible. They should have a quiet environment that allows them to distinguish voices and understand conversations, if that lies within their capability range. Create a quieter atmosphere by dampening background sound with soft coverings on walls, floors, and furniture surfaces. *However, be aware that whatever soundproofing material you decide on must meet fire-code standards.*

Assessing the Environment for Safety

Ensuring a safe environment requires getting down on a child's level to see what things look like from their vantage point. How would the world look if your eyes were on your ankles or shins (in the case of infants) or on your knees? Crawl around the room or play yard, and see how things look after you have set them up, or move around the environment with a video camera held low. You may find safety problems you never saw before from your lofty view. You will probably also find aesthetic problems as you begin to notice the backs and undersides of furniture and equipment that aren't visible from above.

Sanitation and Cleanliness

Chapter 2 examined some important sanitation procedures. Here, we will focus on six environmental considerations.

1. *Hand washing.* Many diseases are spread by dirty hands, from the common cold to other serious illnesses. The environment should be set up to provide for easy hand washing at any time. Having to go off to another room to wash may leave staff short-handed, so the temptation may be great to skip some hand washings. Hand washing, however, should *never* be skipped. The National Association for the Education of Young Children (NAEYC) puts forth guidelines for hand-washing in child care centers. It recommends that children and adults wash their hands:

   - on arrival for the day
   - after diapering or using the toilet
   - after handling bodily fluids (e.g., blowing or wiping a nose, coughing on a hand, or touching any mucus, blood, or vomit)
   - before meals and snacks and before preparing or serving food
   - after playing in water that is shared by two or more people
   - after handling pets and other animals or any materials such as sand, dirt, or surfaces that might be contaminated by contact with animals

   Adults should also wash their hands:

   - before and after feeding a child
   - before and after administering medication
   - after assisting a child with toileting
   - after handling garbage or cleaning

   Always wash your hands with liquid soap (bars spread germs) and warm water. Proper hand-washing procedures are followed by adults and children and include:

   - using liquid soap and running water
   - rubbing hands vigorously for at least 10 seconds, including backs of hands, wrists, between fingers, under and around any jewelry, and under fingernails; rinsing well; drying hands with a paper towel, a single-use towel, or a dryer; avoiding touching the faucet with just-washed hands (e.g., by using a paper towel to turn off water).

A Place for Families, Too!

By the time Miss Dee Dee's program did the self-study for NAEYC accreditation, her room and teaching reflected most aspects of our professional definitions of "best practices." But, as I began to show her photos of other classrooms I was visiting around the country, Miss Dee Dee and I started viewing her room with new eyes. Yes, it met all the quality standards, but what about it was unique to her and her group of children? We agreed that it wasn't a good thing for all early childhood classrooms to look alike. I suggested she think about how chefs in restaurants create "signature dishes," their own special blend of quality ingredients put together in a unique way. Together with her coworkers we began to explore the particular values they wanted to reflect in their rooms. We took field trips to galleries and shops that had nothing to do with early childhood, but which had interesting features, displays, and arrangements for us to analyze. We asked ourselves, why do you think they put that there? What do the people who created this have as a set of values? If you spent your days in this environment, how would you feel about yourself?

Miss Dee Dee decided that she wanted a family-centered classroom and this meant rethinking how her room was furnished and arranged. If parents and aunties, grandpas, cousins, younger and older siblings were to feel comfortable in her space, what would need to change? Miss Dee Dee realized that first she had to get them through the door and relaxed enough to want to hang out and talk. She got adult-sized, stuffed chairs donated for her room and gave them a fresh look with slip covers. Slowly she moved them farther away from the door and into the room, adding a foot stool, nearby plants, a lamp, and book rack. She started requesting framed family photos and with each arrival made a fanfare of finding a special spot for it to be seen. Of course, this generated interest from the children and other families, and not only more pictures, but family stories and artifacts also became part of the environment. Her clean-up song was even changed to the Pointer Sisters' recording, "We Are Family." In subsequent years Miss Dee Dee moved her sign-in book farther into the room and created routines where children invited family members to stop and play, look at a documentation story of their activities, or ask the adults to "tell us about your day." Miss Dee Dee's signature value was clearly evident—this was a place that nourished the whole family.

—Margie Carter

Be sure to use running water, not communal buckets. If there is no running water (e.g., outside), use pitchers to pour water over hands. Anyone who observes a child care program should see hand washing going on continually throughout the day.

2. *Diapering and toileting.* Establish and use a sanitary diapering procedure in an area designed specifically for diapering. Disease is spread through feces, and the way the diapering area is set up can make a difference in protecting everyone from coming in contact with fecal matter. The diapering area should be located next to a sink (it must not double as a food-preparation sink). There should also be ample storage so that all the materials necessary for diapering (diapers, wipes or washcloths, liner paper, paper towels, lidded waste container, and so on) are located within easy reach.

For older children, a toilet should be located nearby so that it can be used according to the children's needs. Ideally, there should be two sinks—one for adults and one set low enough for children.

Staff should also be trained in how to handle blood and other disease-spreading body secretions. AIDS and hepatitis B are examples of diseases that can be spread when precautions are ignored. The environment should be set up to minimize health risks; for example, latex gloves should be accessible to staff at all times. They should not be stored in a separate room or building. In one program, each head teacher had a fanny pack containing plastic gloves that was hung by the door and put on when going outside.

3. *Cleaning and sanitizing.* Cleaning is removing dirt and soil with soap and water and includes wiping dust or other visible dirt from objects with a clean rag that has been dampened with a mixture of water and soap. Sanitizing is removing dirt and certain bacteria so that the number of germs is reduced to such a level that the spread of disease is unlikely. Sanitizing is achieved by using a spray bottle of bleach solution made fresh daily. Every early childhood program should have an established cleaning and sanitizing procedure. All toys for infants and toddlers, who tend to put things in their mouths, must be either wiped daily with a fresh bleach solution or run through a dishwasher set up with water hot enough to sanitize. Store all cleaning items in a conveniently located, locked cabinet or closet.
4. *Personal possessions.* Children need to be taught to keep their personal possessions for themselves. Such items as combs, toothbrushes, clothing, bottles, washcloths, towels, and bedding should not be shared. Label all personal possessions and store them in the children's cubbies, in the bathroom (such as toothbrushes), or in separate shoe boxes or bags in a closet (such as extra clothing). Refrigerate children's bottles.
5. *Food preparation and storage.* Proper food preparation and storage are a must. Set up the food-preparation area away from the bathroom and diapering areas. This area should have a sink that is used for food preparation and cleanup only; ideally, there should be two sinks–one for adults and one low enough for children to use. Sanitize bottles and eating utensils. Sanitize tables before meals and snacks. The food-preparation area should be equipped with a refrigerator set at 40 degrees or colder.

   Important to note is that young children are highly susceptible to choking. The following foods can cause choking and should be avoided: whole raw carrots (use thinly sliced or shredded), raw celery, taco shells, foods with seeds, whole grapes (use halved or quartered pieces), whole olives, popcorn, and whole hot dogs (lengthwise cut).[14]

   All food and beverages brought from home need to be labeled with the child's full name and the date (date stickers should be available in each classroom). All food is served together on an individual plate. Once it is served, it must be discarded within an hour due to foods' high probability of becoming contaminated during serving. Bacterial multiplication proceeds rapidly, as much as doubling the numbers of bacteria every 15–20 minutes.[15]

Also, for children with special nutrition needs, the staff should work with the child's family to create an individualized care plan. In order to protect the children with food allergies from contact with the problem allergen, families may be asked to give their consent to post information about that child's food allergy in the food preparation area and in the areas of the facility the child uses to serve as a visual reminder to all those who interact with the child during the program day.

6. *Napping and resting areas.* Napping and resting areas require special attention. Cots, mats, or cribs should be spaced according to regulations. A three-foot minimum is standard. Children should have their own sheets and blankets, which should be washed often. Cots, mats, and cribs should be wiped down with a bleach solution at least every week, or more often if needed. Each child's bedding should be stored separately so it does not touch the bedding of any other child to prevent the spread of germs or parasites (like lice or pinworms).

The Environment as a Reflection of Program Goals and Values

The environment should reflect the goals and values of the program. Look at the *Voices of Experience* entry for a story on that very subject by Margie Carter. Carter is an early childhood instructor, internationally known consultant, and presenter, as well as author of numerous early childhood books and videos. If you were to observe a program, what environmental cues would reflect its values? For example, how would you know if a particular program valued independence and individuality? What about cooperation, authenticity, or exploration?

Individuality

Many programs have individual cubbies where children keep their personal possessions rather than in a large common storage area. In addition to being efficient, cubbies also promote individuality. Likewise, asking parents to sew name tags in their children's clothes is another way to promote individuality. Certainly, name tags make it easier for staff members to keep track of stray items, but they also reflect the idea that each child is an individual with his or her own possessions. Helping build a space for individuality within a collective setting is important because it ensures that each individual voice will be nurtured and fostered and not lost within the group setting.

Another way to promote individuality is to display a child's name and a picture on his or her cubby or in another prominent location. Yes, these devices are conveniences for teachers and help children gain in symbolic development, but, perhaps most importantly, they also send the message to each child that he or she is a unique individual.

Independence and Interdependence

An early childhood environment that is arranged so children can help themselves reflects a program that values independence. For example, when changing tables have stairs, even the youngest child is encouraged to be independent. Some programs have art areas where a variety of materials are always available for children to help themselves. Provisions are made for children to serve their own food and pour their own drinks using scaled-down serving dishes and pitchers.

An environment set up to promote interdependence would look different from the one just described. Instead of being displayed on low, open shelves, materials, supplies, and equipment might be stored up high. (The *Focus on Diversity* box contrasts the environmental aspects of an independence-promoting program versus an interdependence-promoting program.) Children would have to ask for help. In such a program, children might be carried around or held more than they would in a program that focused more on independence. The adult–child ratios might have to be higher because children would need many things done for them. One way to take care of that problem might be to mix the age groups so that older children could take responsibility for younger ones, which is the way it works in many large families that focus on interdependence.

Cooperation

Cooperation is a widely promoted quality that transcends most cultural boundaries. Most independent-minded, individualistic people deeply value cooperation, though they may frown on fostering interdependence in young children.

An example of an environment that promotes cooperation is one in which there are many small areas where two or three children can work and play together. A large open classroom or play yard is less likely to encourage cooperation partly because it invites more fleeting interactions.

There are many inventive ways adults can promote cooperation between children, such as selecting equipment that takes two to operate. Old-fashioned glider swings and seesaws are examples of equipment that demand cooperative effort; you could not seesaw by yourself no matter how hard you tried. Although both glider swings and seesaws are now considered unsafe, some programs have figured out ways to encourage cooperation through the use of other types of equipment.[16] For example, large wooden blocks take two to carry and thus promote cooperation. Tricycles built for two and tricycles with passenger seats are other examples of equipment that encourages cooperation. Likewise, stretchers in the dramatic area require a child at each end; to get a kite in the air takes two children; sleds can accommodate two or three children; and at least two children are needed for a game of catch.

Cooperative art projects include painting murals (for older children) and drawing with crayons on a large piece of butcher paper across a tabletop (for younger children). Children can also fingerpaint directly on a tabletop rather than on individual sheets of paper. A collage can become a cooperative art activity by placing a length of contact paper on a wall or on a table sticky side up. Shared glue bottles also promote cooperation. Clay and play dough can also be used as cooperative materials when process is emphasized over product. Instead of children taking home what they have made, the clay or play dough is put back into the common pot to be used again cooperatively.

Focus on Diversity

How the Environment Reflects Program Goals

Independence and individuality are two fundamental values that support *Developmentally Appropriate Practice.* However, mandates that cultural context is considered when making decisions in the early childhood program. As discussed, some cultures encourage interdependence in the young child. The following lists illustrate how the environment can be set up to promote independence and interdependence, respectively.

Independence-Promoting Environmental Characteristics

- Shelves are low, open, and invite children to help themselves to supplies and materials.
- Coat hooks are placed low, within easy reach of the children.
- Steps up to the diapering counter allow children to climb up themselves.
- Small pitchers are provided for children to pour their own drinks.
- Serving dishes and child-sized serving spoons encourage children to dish up their own food.
- Wipe-up cloths are available for children to clean their respective places at the table following a meal or snack.

Interdependence-Promoting Environmental Characteristics

- Supplies and materials are stored on high shelves and in closed cupboards.
- Coat hooks are placed at adult height.
- There are no steps up to the diapering counter.
- Pitchers, serving dishes, and utensils are adult-sized and reserved for adult use only.
- Cleaning duties and supplies are reserved for adults only.

A person who places a high value on independence and individuality would see the second list as extremely negative because the program characteristics promote dependence on adults. A person from a culture that values interdependence, however, would view the classroom where adults do things for children as a constructive environment: in this view, the children watch adults model cooperative behavior, and the children mimic them by doing things for each other. Perhaps they catch the spirit of helping and even assist adults without being asked to. Wiping tables–not just their own space but the whole table–becomes something they may *want* to do. They may not serve food to themselves, but you can bet some of the children are in the doll corner serving the dolls. In a mixed-age group, the older children may be more inclined to help younger children than in a classroom where independence and individuality are encouraged.

As an early childhood educator, be sensitive to parents whose child-rearing approaches differ from the ones your program promotes. If a philosophical clash arises, realize that a dialogue is in order. The goal is not to "educate" the parents in the hopes that they will abandon their values; instead, your goal is to follow the NAEYC guidelines and see if there is a way to intersect the program's values with those held by the parents. At the very least, you should try to figure out how to make the program fit the child instead of expecting the child to fit the program.

This environment is furnished with accessible shelving that doubles as dividers to separate activity areas, which promotes independence and allows children to easily flow from one activity to another.

Authenticity

Is authenticity a value? If not, perhaps it should be. Have you ever thought about what a child "learns" from man-made replicas of real things? Should children not learn firsthand about their environment and culture through contact with real objects? Should they not have real things they can touch–like a bird's nest instead of just pictures?

If authenticity is an issue for you, set up your environment with real things, such as tools instead of plastic facsimiles. It is far more satisfying to pound real nails with a real hammer than plastic pegs with a plastic, hollow hammer. Most play kitchens are stocked with child-sized utensils and appliances, however, introduce a child to a real kitchen and their excitement and curiosity will mushroom. One program keeps a "kitchen drawer" full of real objects for the children's use in addition to the toy versions in the playhouse area.

At a living-history farm, one of the houses has an attic designated as a play area whose sole purpose is for children's exploration. Another early childhood program brings in broken appliances, such as a dryer or a radio, and lets children "fix" them by dismantling them with real tools.

Aside from exposing children to real objects, a program that promotes authenticity might provide "real" experiences. Outside of child care, children manage to create a variety of fun experiences for themselves that child care teachers and caregivers might never consider. As a child, one of my favorite things to do was to rummage around in drawers and, when asked to, organize and clean them out. My dream as a child was to have an attic full of old-fashioned stuff to play with.[17]

Authenticity also includes reflecting in the environment the interests of the people who use it–children and adults. Reggio Emilia schools provide for this kind of authenticity; their projects are "authentic" and are not made-up activities designed primarily to promote skill building. For example, children can authentically be involved in helping prepare and bake bread daily for their snack and be fully engaged in each part of the process instead of only pretending to cook in the dramatic play area.

Authenticity is also reflected in how the environment is decorated. Consider the difference between decorating the classroom with cute cartoon characters versus displaying art posters, photos, children's work, and natural objects.

The Outdoors and Nature

It used to be that children played freely in nature. Way back when most children grew up in the country, nature was part of their lives. When families moved into cities, nature was not so accessible, but they still found ways to be in nature. Now things are different. Finding nature in cities, among the buildings, asphalt, and cement is increasingly more difficult. Yet safety is also a major reason that many children are confined indoors and lack experiences in nature or even just outdoors in fresh air. If they do go outside, it is only under close supervision by adults.

Everything seems dangerous nowadays; the sun, air pollution, disease-carrying insects like mosquitoes and ticks are among the factors keeping children inside. Instead of exploring nature, many families either keep them at home indoors or drive them to structured classes, lessons, and sport activities. They don't get to explore freely or become acquainted with nature in an outdoor environment.

Experiences of children today make them unaware of and disinterested in environmental concerns. Many of us have become so disconnected from our natural origins that we no longer recognize our basic dependence on nature as a condition of growth and development. Just watching *National Geographic* and the Nature channel is not enough. The message children take from those programs is that nature is exotic and far away–not part of their own lives. When children grow up with no appreciation of nature around them and the benefits they can derive from it, they grow up to be adults who are less likely to advocate for preserving nature. The results are that destruction and exploitation of nature will continue to increase. When children experience nature and come to understand the personal benefits they can derive from it, they have a much greater stake in preserving it.[18]

The outdoors and nature are highly valued at two model early care and education programs: the Pikler Institute in Budapest, Hungary, and the Pacific Oaks Children School in Pasadena. The first is a residential nursery mainly for children under age three, and the other is a care and education program serving families with children from six months of age through kindergarten. Children in both these settings get a good start toward appreciating the outdoors and experiencing a more natural environment. They spend a good part of each day outside where they not only exercise but also explore and learn. Outdoor spaces are considered at least as important an educational environment as the indoor spaces. At the Pikler Institute they have found over the years that children are not only healthier when they spend time outdoors, but their appetites greatly increase. It is a real pleasure to watch young ones relish every bite of their simple and highly nutritious meals rather than pick and complain. Both of these programs are models for how to use the outdoors as a classroom, not just a place to run around at recess.

Exploration

Exploring on one's own is a value related to being in nature and also closely related to authenticity. A program that encourages children to explore should have an environment that calls out "Explore me!" It should be arranged safely and provide plenty of enticing materials and equipment. Free exploration takes time and space to allow children to experience what's there and freely try new ways of doing things. An environment that is not interesting or varied enough takes more adult direction and input and cuts down on free exploration.

If, on the other hand, a program's goal is for children to pay more attention to the adults in the environment, removing interesting things from the space will promote that goal. Consider how you might set up an environment that emphasizes human interaction and deemphasizes interactions with physical objects.[19]

As mentioned under authenticity, the environment should reflect the people in the program–both individually and as a group. A good environment tells the visitor something about both the children and the adults, about who uses it, and maybe even something about the parents; for example, Reggio environments reflect the fact that aesthetics are important to teachers and parents too (see the following section).[20]

Aesthetics

In many centers or play areas of family child care homes, you have to look hard to find even one thing of beauty. Child care and aesthetics do not necessarily go together. Perhaps the lack of aesthetics is a sign of the times–a sign that other considerations take priority over beauty.

Often adults decorate environments with objects that they think reflect children's tastes; they hang commercial cartoon drawings on the wall to make the room appear child-friendly. However, they sometimes end up making the room overstimulating, and they miss the opportunity to help children learn to appreciate aesthetics. Unfortunately, some children never have an opportunity to explore, understand, or appreciate the concept of "beauty" in its most basic form.

But aesthetics is not a low priority everywhere. In the heyday of the British Infant School (a particular approach to early childhood education observed around the United Kingdom in the 1960s), some programs set aside an area just for aesthetics; a special visual display would be created for no other reason than to have something nice to look at. Aesthetics is alive and well across North America as well. I have observed many environments that teachers and providers have arranged in ways that are pleasing to the eye. On the other hand, it is amazing how many children see nothing in their environments but brightly colored plastic objects and equipment that have no aesthetic appeal.

Programs need to create an aesthetically pleasing environment. They should present materials in an uncluttered, aesthetically pleasing manner–be attractive and inviting. For example, in some programs, art supplies are arranged attractively. The art table itself looks like a work of art. Of course, it gets disorganized with use, but then adults enlist the help of a child or two to create order again out of the chaos. Sometimes the children rearrange the materials on their own initiative. Toys should be displayed in a way that makes it appealing to children to use them; choose containers or displays that emphasize the materials.

Today, Reggio Emilia represents the ultimate in aesthetic child care environments. Visual appeal–to both children and adults–is a primary consideration in space arrangements.[21]

ENVIRONMENTS FOR VARIOUS TYPES OF PROGRAMS

As discussed in Chapter 1, early childhood programs vary according to size, location, duration, and so on, and the type of program affects the way the environment is set up. An all-day center-based child care program that runs from the wee hours of the morning until the evening will look different from a half-day preschool. In this section, we'll compare environmental considerations for six types of programs.

Full-Day Child Care Center

In the full-day child care center, some children stay longer hours than any single staff member. In one such program, the beginning and the end of the day are called **surround care.** The predawn group is small and comprises children of different ages. They meet in a special room set up specifically for mixed-age groups. It's more like a family room than a school and has soft furniture, rugs, and a quiet atmosphere. Children who stay late also come to this room, where they can settle down and relax after a busy, noisy day. They can also help with chores for the following day, as there are a washer and a dryer just off from the main section. It's not unusual to walk in and find children folding clothes with adults. Brothers and sisters migrate to each other and can be found snuggling together in the corners of the deep couches.

surround care Child care that extends beyond the regular daily program. It may be offered in a child care center for infants, toddlers, and preschoolers during the early morning and evening, when there are fewer children present. Surround care is also offered in some programs for school-age children who attend before and after school.

During "business hours," children are in age-segregated classrooms and play yards set up specifically for their developmental stage, with small furniture and appropriate materials, equipment, and supplies to promote learning, creativity, imagination, physical development, and social exchange. They eat meals and snacks at the same tables where they finger paint. They sleep during nap time in the same room where at other times they build with blocks, work puzzles, and play house.

Half-Day Parent Co-op

Like the full-day child care center, the typical half-day parent cooperative is divided into age-segregated classrooms and play yards. The children arrive after breakfast and leave before lunch, so the morning is mostly devoted to exciting activities, some of which are very messy. Snack time may be a formal affair in some programs, but in others it is casual and often held outside, so there is little preparation or cleanup. Without needing to constantly clear tables for meals and snacks and rearrange furniture for naps, staff members in half-day programs have a different kind of job. There may be fewer transitions and less emphasis on a strict time schedule because there is less pressure to constantly rearrange the environment.

In co-ops, children sometimes help with the cleanup at the end of the session; then they go home to eat, sleep, and take a bath. The teacher and assistant finish scrubbing up and putting the environment back in order again. When they are through, they have a quiet, peaceful time to prepare for the next day's activities.

Half-day programs such as Head Start often share space with some other kind of program, for example a Sunday School.
Jill Braaten/McGraw-Hill Education

Half-Day Head Start Preschool

Although many Head Start programs now operate out of their own space, there are a substantial number of programs (as there are some half-day co-op programs) that must share their space with another program or institution. Being located in a church's Sunday School room is an example of a shared space; often, the same room has to be emptied on weekdays for evening church meetings, and Fridays are a flurry of activity as staff members rush around to lock equipment and materials away from curious Sunday Schoolers.

Like the full-day center and the half-day co-op, the typical Head Start program is divided into age-specific classes; there is often no surround care. Meals are served—both breakfast and lunch.

School-Age Child Care

Before- and after-school child care environments come in a variety of forms, but they commonly operate out of elementary schools. Children meet before school begins in their own special classroom or perhaps in the multipurpose room. When school starts, the children go off to class. In school systems with half-day kindergarten programs, the children attend before- or after-school child care accordingly. (Children may attend the same school the child care program is housed in or have to be transported to another site to attend their kindergarten program.) At noon, the afternoon kindergartners go off to their classrooms, and the morning kindergartners arrive. When school is out, the rest of the children arrive, and the group grows drastically. Some school-age children are also cared for in family child care homes. Each environment responds to the needs of school-age children in a slightly different way.

Family Child Care Home

The family child care home serves many purposes; it houses the family who lives there and provides a suitable environment for the children who spend their weekdays there. Like the program housed in the church, decisions have to be made about shared space and equipment. Provisions have to be made for storing family possessions, and since the home is commonly turned back into a place for the family at night and on weekends, child care equipment, materials, and supplies also must be stored away (except in the room that is the primary child care environment–often the family room). Because of the size of some equipment, however, it seldom all disappears; the master bedroom may house a crib; the dining room, high chairs; and the bathroom, a stepping stool, a potty chair, and other trappings of young children.

Like many center-based programs, the environment serves more than one purpose and has to be continually rearranged during the day and after hours. Since there is probably no gap between the time children leave and the family comes home, the provider has little time to plan for the next day and get the environment ready.

Kindergarten and Primary Programs

Kindergarten and primary programs come in all shapes and sizes. Traditionally, the teacher worked alone in a self-contained classroom with a group of children all the same age. Today, however, some schools offer programs with mixed-age groups. Such a program might have a combination classroom set up for two grades, like kindergartners and first graders in the same room, or it could be an ungraded primary program that accommodates a group of children of an even wider age span, say five to eight. In these programs, the teacher may still be alone, or there might be an aide, parent volunteers, or specialists in the classroom to help out.

If these classrooms are in elementary schools, the rooms still look much the same no matter how the children are sorted out. Rooms are of similar size, with the bathroom down the hall, unless they were built as kindergarten rooms, in which case bathrooms are in or adjacent to the classroom. Kindergarten classrooms are more likely to have sinks and areas for messy play than rooms built for grades 1-3. In cold climates, there may be indoor play space, where children can move around freely, as well as outdoor playgrounds. In moderate climates, vigorous play tends to happen outdoors.

A Story to End With

I started my preschool teaching career as a volunteer in a Head Start program that shared space with the Sunday School of a Congregational church. Every Monday morning, we worked furiously to take toys and equipment out of the cupboards and replace the pictures of Jesus surrounded by lambs with multicultural pictures reflecting the children and families in the program. It was like the act of a quick-change artist: at the beginning of the day the environment had a distinctly Christian flavor, but before the children arrived, it was transformed into a preschool classroom reflecting the diverse interests of young children.

As children walked in, they passed by a hands-on science display next to a greeting table, where a teacher welcomed them and pinned on name tags. After a conversation and a feel of the science objects, the children, properly labeled with name tags, headed off to one of the many activity areas. A quick glance around the room showed them all: the dress-up center with clothes hung enticingly on hooks; the housekeeping corner with sink, stove, table, doll beds, and plenty of dolls; the easels, standing ready with brushes sticking out of cups containing bright-colored paint the consistency of heavy cream; and the block center, brimming with stacks and stacks of blocks organized neatly on open shelves. On the top shelf of the block cabinet, within easy reach, were block accessories such as little people, cars and trucks, and miniature furniture. A cozy book corner invited children to plop down on a cushion and choose from a variety of books attractively displayed and close at hand. Those inspired to try their hand at writing could get up and go to a nearby writing area fully equipped with all kinds of writing tools and paper.

It was a lot of work to arrange and rearrange that environment every day, but it was well worth it. The looks and sounds of children playing used to make me wish periodically that I were once again a beginning learner in such an environment.

Summary

To know how to set up the environment, the early childhood educator must consider the children's physical, cognitive, and social-emotional needs—all of which overlap. The environment is an important teacher, so it requires careful planning. The layout of the early childhood program should be broken down into physical-care centers, interest centers, and gross-motor centers. Other factors of environmental planning include space, activity choices, circulation patterns, balance, and health and safety.

The environment reflects the values of the program. If the program values independence, the environment will be set up so that children can do things for themselves. If it values cooperation, the room arrangement and choice of toys will encourage children to work and play together. A program that has an antibias focus will have a slightly different look from one that takes a multicultural approach without emphasizing activism. Other values that might be reflected by the early childhood program include authenticity, exploration, human interaction, and aesthetics.

Early childhood environments vary according to the program's enrollment, length of day, location, and whether the children are segregated by age. Although there may be some similarities among the six programs discussed, each type of early childhood program must plan and arrange its environment to fit the specific needs of the children it serves.

Reflection Questions

1. How aware are you of the environments where you spend your time? How much do they affect you? What is an example of an environment that has a negative effect on you? A positive effect?
2. What was your favorite place as a child? How could you re-create that favorite spot in an early childhood program?
3. What are your personal reactions to an environment that promotes independence and individuality? Where do you think these reactions come from? Do you think your culture influences your reactions? What are your personal reactions to an environment that promotes dependence? Where do you think these reactions come from? Do you think that your culture influences your reactions?

Terms to Know

How many of the following words can you use in a sentence? Do you know what they mean?

physical-care center 216
interest center 219
gross-motor space 220
surround care 241

For Further Reading

Duncan, S. (2011). Breaking the code–Changing our thinking about children's environments. *Exchange 33* (4), 13-17.

Liewra, C., Reeble, T., and Rosenow, N. (2011). *Growing with Nature: Supporting Whole-Child Learning in Outdoor Classrooms.* Lincoln, NE: Arbor Day Foundation.

McDermont, L. B. (2011, September). Play school: Where children and families lean and grow together. *Young Children 66* (5), 81-86.

Pikler, E. (2006). *Unfolding of Infants' Natural Gross Motor Development.* Los Angeles, CA: Resources for Infant Educators.

Ranck, E. R., and Anderson, C. (2010). Blocks: A versatile learning tool for yesterday, today, and tomorrow. *Young Children 65* (2), 54-56.

Schein, D., and Rivkin, M. (2014). *The Great Outdoors: Advocating for Natural Spaces for young Children* (Rev. Ed.) Washington, DC: National Association for the Education of Young Children.

Rosenow, N. (2011). Planning intentionally for children's outdoor environments. *Exchange 33* (4), 46-40.

Stacey, S. (2011). *The Unscripted Classroom: Emergent Curriculum in Action.* St. Paul, MN: Redleaf.

Stoltz, D., Conner, M., and Bradberry, J. (2014). *The Power of Play: Designing Early Learning Spaces.* Chicago: ALA Editions.

Tardos, A. (2007). *Bringing Up and Providing Care for Infants and Toddlers in an Institution.* Budapest, Hungary: Pikler-Loczy Tarsasag.

Willis, C. (2009). *Creating Inclusive Learning Environments for Young Children.* Thousand Oaks, CA: Corwin.

Wirth, S., and Rosenow, N. (2012). Supporting whole-child learning in nature-filled outdoor classrooms. *Young Children 67* (1), 42-48.

Design credits: Tips and Techniques: ©Ingram Publishing; Focus on Diversity: ©Pixelic/Getty Images

9 Creating a Social-Emotional Environment

Rob Hainer/Shutterstock

In This Chapter You Will Discover

- what factors go into creating a healthy, safe social-emotional environment.
- how adults treat children respectfully and what it looks like when they don't.
- why adults shouldn't talk about children in front of them.
- why an early childhood educator should be more like a mother hen than a drill sergeant.
- what stands in the way of continuity–an important ingredient in early childhood programs.
- why there is disagreement about focusing the early childhood program on the individual or on the community.
- that there is an early childhood culture that often differs from those of the families and even of staff members.
- a way to look at conflicting cultural priorities.
- that oppression affects child-rearing goals for some families.
- that culture is dynamic and always changing.
- why it is important that infants learn their home culture.
- the ways in which the early childhood culture is evolving.

The social-emotional environment of the early childhood program is harder to detect than the physical environment, but it is very real and affects everyone in it. As much care and consideration must go into planning for a healthy, safe, and nurturing social-emotional environment as for the physical environment. Although the social-emotional environment is not distinctly listed as an NAEYC Program Standard, it develops naturally when programs adhere to the other ten standards. Standard 3, called "Teaching," states that programs should use developmentally, culturally, and linguistically appropriate and effective approaches that enhance each child's learning and development in the context of the program's curriculum goals. The rationale is that teachers who optimize children's opportunities for learning use approaches that respond to the different backgrounds children bring, including interests, experiences, learning styles, needs, and capacities to learn. When teachers approach children's learning this way, they create a positive social-emotional environment. Standard 7, "Families," also relates to the social-emotional environment as a program recognizes the primacy of children's families and creates collaborative relationships with them. Sensitivity to diversity, language, and culture is part of this standard and thereby adds to the social-emotional climate of the program.

NAEYC Program Standards

Program Standard 3: Teaching
Program Standard 7: Families

QUALITIES OF THE SOCIAL-EMOTIONAL ENVIRONMENT

Many aspects of the social-emotional and physical environments overlap. How the physical environment is set up directly affects the qualities of the social-emotional environment. Some of the environmental qualities explored in Chapter 8 were cooperation, equity, authenticity, exploration, and aesthetics. This chapter will focus on the social-emotional qualities of respect, warmth, nurturance, acceptance, protection, responsiveness, and continuity.

Respect

What does it mean to create a respectful environment? The basic principle of many cultures' child-rearing practices is to teach children to respect adults. However, what does it mean for an adult to respect a child? Magda Gerber, a well-known child-care professional and founder of Resources for Infant Educarers (RIE) talked and wrote more about respect than any other concept. According to Gerber, respect is summed up by treating a child, no matter how young, as partner from day one and, thus, like a fully human person rather than like an object.

How do adults treat children like objects? Consider this scene: a caregiver with a baby in her arms is talking over the fence to a preschool teacher who is supervising four-year-olds playing in a sandbox. The caregiver is telling the teacher what a miserable morning she has had so far with the baby, who is whimpering softly as she holds him. She tries to perk him up by jostling him a bit. Then she tickles him. He squirms. Finally, she tosses him in the air and catches him, laughing at his screams. Are they screams of terror or laughter? She does not really know or care.

In the meantime, the teacher has moved over to where two children are beginning to argue. She comes up behind one and, without a word, abruptly picks her up and carries her over to the other side of the sandbox. The child looks startled.

What is wrong with this picture? Could you see the signs of disrespect? The caregiver was discussing the baby right in front of him. Maybe he could not understand what she was saying, but he probably suspected she was talking about him. She was acting as if he was not even there. Without trying to find out what the whimpers were about, she decided to distract him. She appeared not to care what he needed. Perhaps she was entertaining *herself* with this baby.

Tickling as a form of distraction is disrespectful. Would you ever think of tickling, say, a serious professor or a police officer issuing you a ticket? No. We only use tickling as a distraction with those who are helpless to stop us. Throwing the child in the air was also disrespectful–aside from being downright unsafe. "Throw balls, not children" should be the motto of every adult who lives or works with children.

Likewise, the preschool teacher did not tell the girl what she was going to do or why. She just picked her up like a sack of potatoes and moved her to another location. How would you feel if that were done to you? This same teacher probably sets up circle time by physically placing children in the chairs unless they quickly scramble in by themselves. She probably sees it as being expedient, but what she is really doing is treating the children like objects. It is likely this teacher also lines up children by pulling, hauling, and shoving them until they are situated just so.

Notice the lack of communication with the children in the scene. There was no attempt to observe what was going on or to try to understand what needs were being expressed. These adults did not listen to the children. They did not interact with them either, except in physical ways that were either distracting or demanding.

To summarize disrespectful adult behaviors:

- Talking about a child, even a baby, in front of him as if he did not exist
- Ignoring a child's feelings
- Distracting a child when she is unhappy
- Treating children like objects
- Refusing to listen to a child

Let us look at a different scene that shows how adults treat children respectfully. Kayla, an infant, is lying on a blanket fingering a soft ball. She looks up as she sees her caregiver approach. The caregiver comes at her from the front, not from behind. The caregiver does not want to surprise Kayla. She thinks it is more respectful to let Kayla anticipate what is about to happen. She says to Kayla as she leans down and holds out her arms, "I'm going to pick you up now. We're going outside." She pauses, watching to see if Kayla will respond. Kayla gives a little wiggle and lurches toward the caregiver. "Ah, you understood me, didn't you?" She smiles, reaches out for Kayla, and picks her up.

The director walks by and says to the caregiver, "How's Kayla today?" Instead of launching into a discussion with the director about how Kayla's feeling and what happened at home that morning, she looks at Kayla, including her in the conversation: "I think you're feeling okay, aren't you Kayla?" She has some information to pass on to the director but purposely saves it for later to avoid talking about Kayla and her home situation in front of her. She is treating Kayla like a human being worthy of respect, irrespective of her age.

Here is another scene that illustrates how respect is a learned skill: "Oh they're all so cute. They're just adorable!" gushes a new teacher's aide in a preschool program. The

TIPS AND TECHNIQUES

How Respect Relates to Clothing

Dressing children up in cute clothes as if they were dolls shows a lack of respect, according to Magda Gerber. Cute clothes are fine as long as they are comfortable and do not restrict movement. But babies who try to crawl in ruffled dresses are definitely hampered; crawling becomes virtually impossible when the knees get caught up in the skirt. Likewise, toddlers and preschoolers in slick-soled shoes cannot run, climb, or jump without fear of slipping, and pastel-colored clothing accumulates more stains than darker-colored clothes. Of course, there are times when children's dress is directed by the program. Consider the seasonal performance planned for parents. When children are asked to wear adorable but uncomfortable costumes that they do not want to wear, respect is also an issue.

Usually, however, the early childhood educator is on the other side of the clothes problem, trying to convince parents to dress their children for play in inexpensive clothes that nobody has to worry about. It is easy to inform parents at the intake interview and through parent information sheets or handbooks what the program considers appropriate clothing. Some parents already know that it is best to send their children in play clothes and are glad to comply. Other parents, however, are not so happy about dressing their children "down" for what they consider "school." Just as it is important that early childhood educators respect children, so it is important that they respect parents. Respecting parents does not mean issuing mandates about clothing but engaging in dialogue to understand the parents' attitudes. Then conflict-management skills come into play as the two parties try to resolve their differences.

teacher remembers when he felt that way about young children. But he has moved beyond those feelings and can see children as individuals and not talk about them as if they were monkeys in a zoo, there for his entertainment, nor do we want to value children for their extrinsic appearances. He has also learned other ways to respect children. He does not talk to them in a sweet-as-honeyed voice as if they were vacuous dolls. He relates to them on their level but with the same respect he shows peers and older people. Similarly, he tries not to praise them or have them seek out his approval by using phrases such as "good job" because phrases such as these do not convey enough information, nor do they acknowledge a child's feelings. Instead, one could say, "You did it!" or "I noticed that you used purple and blue in your drawing." It is important that children do things for their own pleasure, not for adults' approval. Accordingly, it is important to try to never label a child as being a "bad boy" or "good girl." Instead, caregivers should try to explain why a particular behavior may be inappropriate, and why. He vows to take some time in a staff meeting to help the new aide understand what it means to respect children. (See the *Tips and Techniques* box above for a further view on respect.)

Showing children respect is important for fostering healthy inward and outward social-emotional growth, and the best way to understand respect is to put yourself in a child's shoes. Would you feel respected if someone treated you the same way you treat children? Before we leave the issue of respect, it must be pointed out that the way to get children to respect you is to respect them. In that way, you both model respect and earn it, and there is a big difference between demanding respect and earning it.

WARMTH, NURTURANCE, ACCEPTANCE, PROTECTION, AND RESPONSIVENESS

Young children need to be handled with warmth and tenderness. They need a mother hen, not a drill sergeant. Just imagine a mother hen: those feathers offer a warm, cozy cover to little chicks as they scurry in at night to snuggle in the dark softness. But they do not go in just at night; they also scurry to the warmth of the mother hen whenever danger threatens. When they are scared or when they are cold, they receive what they need, when they need it. They receive warmth, protection, nurturance, and responsiveness.

Children need adults who respond in warm, understanding ways.
Stockbyte/Getty Images

Young children in an early childhood program may not need a mother hen per se, but they do need all those qualities of warmth, protection, nurturance, and responsiveness in their environment. When they are scared—because of real or imagined danger—they should be able to seek the figurative warmth of someone's protective wing. Whether it is a man or a woman offering the protection and nurturance doesn't matter; what matters is that he or she responds in a warm and understanding way.

As noted in Chapter 6, how a teacher responds to a child's needs indirectly teaches children, or models, unwritten social rules in regards to building relationships with others. By providing responsive and nurturing environments, children develop empathy and feel safe enough to explore novel activities and friendships. Indeed, research shows that children in good social and emotional health develop essential skills for later academic success, including confidence, emotional self-regulation, communication, and building close relationships with caregivers and peers.[1]

Again, it may help to see what the lack of these qualities looks like in order to understand their importance. Imagine a baby who isn't feeling well. He cries. Instead of a warm, understanding, responsive adult, a cold, uncaring one approaches him. Her goal is to stop the crying, so she tries repositioning him by turning him over in a rough, unfeeling way. This does not help. He cries harder. She picks him up and feeds him, even though he ate not too long ago. Suddenly, he spits up. The adult jerks him away from her body roughly. "Now see what you have done," she says harshly, trying to brush the spit-up off her clothes with the cloth she was carrying over her shoulder. She puts the crying baby down in his crib and goes to sponge herself off. When she finishes, she comes to attend to him. He is asleep. She sits down, relieved.

Here is another scene. Four-year-old Jenna misses her mother. She stands crying by the window where she last saw her. Ignoring Jenna, the teacher works around her, setting up the room and greeting other children as they arrive. When Jenna grabs for the teacher's apron as she passes by, her hand is roughly pried loose and she is told, "You'd better just get over this crying, Missy! It won't do you any good. Your mother isn't coming back!" No mother hen for Jenna in this adult. No warmth, understanding, or reassurance. No feeling of protection.

Still another scene. Jackson has waited what seems hours for the red ball, and now he finally has it. He starts to bounce it happily under the hoop, getting ready to toss it in when a whistle blows. "Time to come in now," the teacher announces in a loud voice. Jackson runs the other way, clutching the ball tightly. The teacher catches him by the back of his jacket. "No, you don't, young man! Didn't you hear the whistle? That's it for now." Jackson wrenches loose, throws the ball as far across the yard as he can, and flings himself on the ground pounding his fists and crying.

"That's enough of that. Time-out for you!" says the teacher sternly and she carries him to a chair just inside the door by the window. He sits slumped over and looks wistfully at the ball outside in the corner of the yard where it rolled. Tears roll down his cheeks. No mother hen for Jackson. No understanding of his frustration and anger. No warmth or snuggles. And no reassurance that he will have another chance to play with the ball.

What these three scenes illustrate is that when you tune in to what children are feeling, you can acknowledge and accept what they are trying to express and thereby respond in warm and nurturing ways. And who would not rather have the warmth and protection of a mother hen over a brash, unfeeling drill sergeant?

Continuity of Care

Warmth, nurturance, acceptance, protection, and responsiveness are all qualities that help adults establish relationships with young children. But one additional quality is needed–continuity. Children need consistency in the people around them in order to feel comfortable, safe, and secure in a program away from home. Continuity is a vital ingredient of the social-emotional environment.[2] Nevertheless, two factors work against continuity in early childhood education: high staff turnover rates and the frequent advancement of children from one classroom to another.

Continuity of care refers to the practice of assigning a caregiver/teacher to a child at the time of enrollment and continuing this familiar relationship throughout the duration of the child's time in a program. This practice supports the children and their relationships by limiting the number of changes they experience. This enables children to experience a stable, long-term relationship not only with their teacher but also with the other children in their peer group. Continuity of care has been shown to foster "staff and family cooperation and loyalty, creating a calmer, less stressful environment, and providing better opportunities for the growth and development of adults as well as children".[3]

High staff turnover is a common impediment to continuity of care, an important indicator of quality in early childhood programs. Skilled educators are essential for creating developmentally appropriate learning environments that optimize learning during a sensitive period of brain development for children. Quality programs, however, are labor intensive and expensive.

Outward pressures such as the fundamental quality mandates of low teacher-to-child ratios and minimum education requirements, along with funding constraints from both governments (public programs) and working-class families, keep wages low.[4] With the exception of public programs, which tend to offer higher wages, pay remains relatively low for the average early childhood education (ECE) educator.[5] The low wages, poor benefits, and more recently, high degree requirements sustain a revolving door in the field for ECE educators, with many leaving to teach kindergarten or transitional kindergarten (California) in public schools for better pay to comparable work.[6]

Let us look at the difference between two programs–one with a high turnover rate and one with a low turnover rate. The ABC child care center has always had a high turnover rate and high absenteeism among its staff. So when little Jamie arrives on a Monday morning, she is never sure who she will find there. She is getting to know teacher Dawn and likes her a lot and is hopeful that she will be there today.

It has taken Jamie a long time to let Dawn get close to her because of her experience with teacher Leanne. Jamie loved Leanne and always looked forward to seeing her. But suddenly one day Leanne was not there, and the director took her place. Jamie never knew what happened to Leanne, just that she would not be coming back. Jamie felt very sad and still gets upset when she thinks about Leanne.

Jamie started getting to know Dawn, who finally came to replace Leanne after a series of substitutes. Then just last week, Dawn was not there either. Jamie thought Dawn was gone too, but Dawn came to visit just before nap time and told Jamie that she had to be in the infant room. Jamie did not understand why. The problem was that although Dawn was assigned to Jamie's group, the director had to move her to the infant room last week because a caregiver was sick and the substitute the director finally found was not certified to work with infants. So the substitute replaced Dawn in Jamie's room.

mainstreaming A term that means placing children with special needs into programs that serve children who are typically developing. In some such programs support for the children with special needs may be minimal, so those who can't handle the mainstream may never feel they belong there.

integration The incorporation of children with special needs into programs with their typically developing peers and giving them the support they need so they really belong. Part of integration is giving attention to the interactions between the two groups of children. The goal is for all children to participate in the program to the greatest degree possible.

full inclusion A concept that goes beyond simply including children with special needs into whatever setting is the natural environment of their typically developing peers. Full inclusion means that such children, regardless of their disability or challenge, are always integrated into a natural environment and that services are as culturally normative as possible.

All this shuffling around is nobody's fault, but Jamie is suffering. She is having a hard time feeling connected to anyone in the program.

Now let us contrast the ABC program with the XYZ program. Steffan is a four-year-old in the XYZ child care center. He is well connected to several people in his child care program. Steffan came to the program as a six-month-old and had the same caregiver, Ms. Jones, until he was three. He adored Ms. Jones! Then he moved into the preschool classroom. There he became attached to teacher Janice. He could still visit Ms. Jones in the infant-toddler program, which was adjacent to his preschool room. This is an example where children within a close age group are moving together along with their teachers. Rotating of staff with the students helps maintain continuity for several years. For example, a young infant teacher becomes a toddler teacher and continues to move with the children as they grow and change classrooms.

For a while, Steffan was allowed to move back and forth between the two programs during a transition period. But the more challenging and interesting environment of the preschool classroom called to him, and before too long he was spending most of his time there. Janice knows him very well now–and he knows her. They have a close relationship. And he can still see Ms. Jones. He is glad of that.

Another example of continuity in an early childhood program is one where the environment stays the same and the staff modifies the environment to adjust to the changing needs of the developing children. Jamal has been in the same family child care home since he was a baby. He is now seven years old and considers this his second home. His provider, Barbara, has been in business for about ten years, and sometimes big kids come to visit who used to be in her care, giving Jamal the sense of a large family. Because he is the only child in his family, he has benefited from living in two worlds–the one at home and the one at Barbara's house.

A factor that works against long-term adult–child connections and continuity of care in child care programs is the fact that many early childhood programs follow a school model of grouping children together by grade levels. While this model may work well in primary and secondary school, moving young children too frequently as they reach various milestones during their first few years of life disrupts a child's ability to form the necessary attachments required to feel safe enough to explore and learn crucial social skills. Because grade school children move up every year, some professionals see that as a good way to run early childhood programs too. In fact, some take the idea even further by grouping children according to developmental stages (see Chapter 11). Thus, nonmobile infants may be in one room with one caregiver or set of caregivers. When they begin to crawl, they get moved up to a more sophisticated environment with other, appropriately trained caregivers. Walking is the next stage that requires a new environment and new specialists. After age two, the moving-up occurs annually. By then, the children have gotten so used to seeing adults come and go that they may well avoid feeling too attached to any one person.

FOCUS ON INCLUSION: A FEELING OF BELONGING

Everyone wants to belong, but children with developmental differences may have greater issues around belonging than other children. Some specialized vocabulary shows the history of helping move children with disabilities in the direction of belonging. First came **mainstreaming,** which means placing children with special needs into programs that serve children who are typically developing. In such programs support for the children with special needs is minimal. The main idea is to put such children into the mainstream of life with the hope that they will adapt. **Integration** takes mainstreaming a giant step further and means that attention is given to the interactions between children with special needs and their typically developing peers so that all children can participate to the greatest degree possible. **Full inclusion** goes even further and means that children regardless of their disability or challenge are always integrated into whatever setting is the **natural environment** of their typically developing peers. The idea is **normalization,** which means that services such as early care and education programs provided to those with special needs are based on circumstances that are as culturally normative as possible.

natural environment A setting (such as a home or early care and education program) where children with disabilities will find their typically developing peers. A natural environment can be defined by the fact that it will continue to exist whether or not children with disabilities are there.

normalization A term that means services such as early care and education programs provided to those with special needs are based on circumstances that are as culturally normative as possible.

In the words of Linda Brault:

> It's really about belonging. Children and families want to belong to their community, they want to be accepted and included regardless of ability, race, creed, or ethnicity. Placing children together is not enough. Settings that have had successful experiences report that they did not see the child with a disability as a guest or an outsider, but as a full member of the group. The staff and family worked together, sometimes with assistance from specialists already involved with the child, to adapt activities, modify the environment, and support the child in interactions with the other children.[7]

NAEYC Position Statement
on Inclusion

Inclusion is a right, not just a privilege, according to the **Americans with Disabilities Act (ADA).** Further, the NAEYC endorsed the Council for Exceptional Children's Division for Early Childhood's position statement on inclusion and created one of their own supporting integration through the stated belief that high-quality, developmentally appropriate programs should be available to all children and their families. Helping *all* children gain a sense of belonging through full inclusion programs benefits children with developmental differences. They have greater opportunity to belong to a community where they can observe and imitate other children becoming more independent and self-reliant. They learn to cope and problem solve while developing social skills, all of which helps them build a positive self-concept. Children with typical development also benefit by recognizing strengths in children with varying abilities and learning to become comfortable with children different from themselves. Helping others and learning to express caring, concern, and compassion can also be a benefit. Both groups benefit by exploring new aspects of friendship.

NAEYC Program Standards
Program Standard 8: Communities

Americans with Disabilities Act (ADA) A 1992 law (Public Law 101-336) that defines disability, prohibits discrimination, and requires employers, transportation, and other public agencies to provide access to the disabled in places of employment, public facilities, and transportation services.

Should the Program Focus on the Community or on the Individual?

Can you take a group of young children and turn them into a "community"? What does it mean to be a "community?" Can a community consist of one age group, or must it comprise a variety of ages? Can an early childhood program qualify as a community? Should it? The decision to focus on community or individualism is often determined by one's immediate social environment and one's cultural values and beliefs. The social-emotional environment can vary with each focus, often defining program curricula and how families approach program selection.

On the one hand, some early childhood educators believe the program should focus on the children as individuals and not stress their membership in the group—whether it be the children and adults in the classroom, the school, the surrounding neighborhood, or society at large. After all, young children are just developing their individuality. Why even introduce the idea that they must belong to a group? These educators worry that the idea of community pressures children to conform rather than to unfold in their own way.

On the other hand, some early childhood educators believe strongly that no child is too young to feel a part of something. The concept and reality of the "group" are important to bring into the picture. Children can be both individuals *and* group members.

If the primary goal of an early childhood program is to establish a sense of community, what kind of community should it be? Should the program become an extension of the community the children experience outside the center? Is the program located in the immediate community in which all or the majority of the children live? (Early childhood programs usually differ from neighborhood-based elementary schools in that they do not often serve the community in which they are located. Families commonly travel distances to take their children to the program that best meets their needs, taste, budget, and often eligibility circumstances.) Even if most or all of the children do come from the community in which the program is located, should it then be an extension of the community? What if the community has a high incidence of violent or drug-related crime?[8]

All these questions must be considered before deciding whether a program should incorporate the concept of community in its curriculum. Of course, there are other options besides focusing on either the individual child or the community. For example, it is possible to focus on the individual *and* community at the same time. Perhaps you want to fashion the program after a family model. In family child care and small center-based programs that serve mixed-age groups, a family model is easier to reproduce than in a program where the children are divided by age, the size of the groups is large, and the setting is more institutional than homelike. However, even in a larger, age-segregated program, a modified version of the family model can be created.

Regarding the family model, there are some questions to consider: do you want the program to replicate the children's homes? Should the program be as much like home as possible for each child, or should it purposely be different?

Managing Cultural Differences

The social-emotional environment of ECE programs is heavily affected by culture, as you will read in the examples below. Cultural differences can strain relationships between parents and practitioners, but children gain the most when programs implement approaches that increase tolerance and acceptance, reduce bias, and diffuse

Focus on Diversity

Normal Versus Different

Were you able to identify with either Rebecca or Joy? Did you think of the one you identified with as "normal" or "typical" and the other one as "different" or "atypical"? Because culture is so invisible, some people, especially those from the mainstream culture, think that they do not have a culture. They think of themselves as "regular," "ordinary," "normal." They may label others as "ethnic," but see themselves as having neither culture nor ethnicity.

But we all have tendencies toward ethnocentric thinking; in other words, we walk in *our own* shoes and see the world from our own ethnic perspective. Most of us are tempted to think our own ways are the *right* ones. We measure the world by our own cultural yardsticks.

conflicts through listening and dialogue. With today's increasingly diverse society, such approaches have never been more integral to a child's learning.

It is important to take a look at some basic ideas about culture before exploring the subject further. Culture is invisible. It has been said that one moves in one's culture the way a fish moves through water. The water is so much a part of the fish's experience, that the only time it becomes aware of the water is when it suddenly finds itself surrounded by air.

We are immersed in our culture the way the fish is immersed in water. We may be unaware of how much our culture influences our actions, our thoughts, our very perceptions. Culture determines everything we do–from our personal behavior (the way we sit, stand, walk, cross our legs, or gesture) to our interactive behavior (how close we stand to other people, what kind of eye contact we make, or how we send and interpret messages). Unwritten rules govern every aspect of our behavior–rules many of us do not even think about or notice until someone breaks them. Indeed, violations of our cultural rules may jar us, but they enrich our experiences and expose us to new ways of thinking and being.

Fish die out of water, but humans are luckier; when we find ourselves moving within a culture different from our own, we not only survive but we even grow from the experience. From learning about other cultures, we come to understand ourselves and other people better.

Children begin learning to be members of their own culture from birth. Let us look at how two mothers, Rebecca and Joy, socialize their babies: Rebecca holds her baby Emily in her arms. Emily smiles and coos. Rebecca smiles back and imitates her daughter's little noises. She makes her daughter laugh by clicking her tongue and opening her eyes wide with surprise. At her daughter's laugh, Rebecca bounces her slightly. Then they both laugh loudly. Rebecca gives Emily a big warm hug and nuzzles her neck. Emily squeals with delight. Rebecca holds her out at arm's length and jiggles her. Emily squeals some more. Both look happy and excited. Eventually, Emily's responses become less enthusiastic, and Rebecca takes this as a sign that it's time to quiet things down.

Joy, on the other hand, takes the opposite approach to her daughter Suzie. At Suzie's first smiles and coos, Joy responds warmly, smiling back and talking quietly to her. But when Suzie gets excited and starts to kick, Joy tones down her facial expressions and talks in even quieter tones. Instead of getting excited herself, she becomes more subdued. In contrast to Rebecca, who continued to heighten her response and stimulate Emily more until she had had enough, Joy calms Suzie down. She holds her still and makes soothing, murmuring noises. When Suzie starts bouncing, Joy holds her even closer and rocks her gently, humming a slow, rhythmic lullaby. She gently strokes Suzie's back in a way she knows will settle her down. It works.

These two mothers have very different intentions, but both behave in culturally appropriate ways. Rebecca purposely engaged her daughter; the more animated the baby became, the more delighted the mother was. Rebecca likes to see her baby excited; she thinks such stimulation is good for her. She enjoys lively interactions herself, so she promotes them. Rebecca is also sensitive to her daughter's signals that tell her she has had enough stimulation, and she stops before it goes too far. With her lively approach, Rebecca is helping her daughter become a member of her culture–a culture that values excitement, stimulation, and lively exchanges.

Joy, on the other hand, thought her daughter Suzie was in danger of being overstimulated. She worries about babies getting excited. She thinks calm, quiet babies are better off than noisy, animated babies. In her culture, tranquility and serenity are two highly valued qualities. Joy is teaching Suzie how to be peaceful. She is teaching her about equanimity.

Is one of these mothers right and the other wrong? You cannot answer that question without taking cultural context into consideration. See the *Focus on Diversity* box for further discussion on this subject.

Imagine that Emily is growing up to be the lively child her mother expected and that Suzie is developing into a calm, composed young girl. How might the two girls respond to preschool? What Emily finds boring and uninteresting, Suzie might find overstimulating, frightening, or upsetting. Of course, such a prediction is too simplistic. It is impossible to tell how two children will turn out just by knowing how they were socialized as infants. Genetics enters the picture. Also, there are individual differences within a culture. People who share one culture are not all the same! Rebecca's people are not all lively, and Joy's are not all placid. The variations may be infinite, yet there is still a cultural thread that holds it all together. Our individual values, manners, and ways of being are influenced by our culture. Culture is the framework on which all else hangs.

Certain aspects of culture are quite visible–food, dress, music, art, literature, holidays. Some aspects of culture show in behavior–including the way people raise their children. Child-rearing practices are often based on cultural beliefs about what children need, how they learn and develop, and even what their basic nature is.

In any early childhood program, there may be several cultures operating at the same time. There are the different **home cultures** of the children, as well as the home cultures of the staff members or providers (which may or may not come into play, depending on their training). For staff members or providers trained as early childhood educators, there is a third culture–namely, the **early childhood culture.**

home culture The family life of the child, which encompasses cultural beliefs, goals, and values—including how they play out in child-rearing practices.

early childhood culture The culture (largely unrecognized) that results from early childhood training. It is related to the dominant culture of the society but not exactly like it.

Children are individuals as well as members of their peer group, community, or culture.
Yellow Dog Productions/Photodisc/Getty Images

At present, the early childhood culture reflects, to a great extent, its European roots, but the heavy emphasis of European American culture has begun to wane slightly. In 1989, the NAEYC published *The Antibias Curriculum: Tools for Empowering Young Children,* which brought racial and cultural bias to national attention.[9] Since then, the NAEYC has remained a steadfast supporter of inclusion and cultural respect and diversity within communities. Through its various position statements, including a 2018 draft that focuses solely on advancing equity and promoting diversity in early childhood education, the NAEYC has encouraged practitioners to practice antibias approaches that promote diversity in all its forms. The most recent iteration of *Developmentally Appropriate Practice,* guides professional decision making based on three areas of knowledge: what is developmentally appropriate, what is individually appropriate, and what is culturally appropriate.[10]

NAEYC Position Statements on Developmentally Appropriate Practice and Responding to Linguistic and Cultural Diversity

National professional leadership and grassroots implementation are effecting changes at every level. The early childhood culture is expanding beyond its European American roots to increasingly reflect the diverse cultures of the many professionals and the families they serve. Teacher training programs are expanding their views, and early childhood educators are becoming more responsive to families' cultural differences and their varied child-rearing practices.

The Child's Home Culture

Children bring their culture to the early childhood program, but what they bring may or may not be recognized by staff members. Previous chapters have explored the differing views on promoting independence versus interdependence in young children, but the following section is designed to help you understand parental attitudes and behaviors in the context of a larger cultural picture.

Cultural Priorities: Independence or Interdependence? Marion Cowee tells about her experience with a mother who stressed interdependence more than she did. See the *Voices of Experience* box on page 261. Newborn babies are faced with two major tasks: (1) to become independent individuals and (2) to establish connections with others. The parents' job is to help their children with these tasks.

It seems logical that parents would focus on both, but they do not. Most parents either consciously or unconsciously pick one task or the other. Their choice depends on their goals, which are often determined by their culture. The result is that the child comes to define "self" as his or her own culture defines the concept. Two such definitions are

- The self is a separate, autonomous individual whose job it is to grow and develop into the best he or she can be in order to become part of a larger group. Personal fulfillment and/or achievement are all-important.
- The self is inherently connected, not separate, and is defined in terms of relationships. Obligation to others is more important than personal fulfillment or achievement.

What kinds of differences show up when one parent encourages independence and another one focuses on reinforcing connections? Parents who place a bigger value on independence are likely to encourage early self-help skills. They hand their baby a spoon when the baby first reaches for one. They teach their baby to sleep alone in a crib. Self-reliance, self-assertion, and self-expression are the goals of parents who focus on independence. The end result of reaching all these goals is self-esteem. Such children adapt well to early childhood programs that promote the same goals.

Independence-focused parents also teach their children to connect with others, but they put far less emphasis on developing such skills. "He will learn to share when he is ready," says the parent who believes his or her child must become an individual and understand the concept of ownership before he can learn to share his possessions.

Parents who are more concerned about their child's ability to maintain connections have a different view. They worry about their child becoming too independent, so they focus on creating interdependent relationships. An Asian woman in a conference once explained to me how dependence operates in a cycle: "First you are dependent on your parents and later they are dependent on you. That's the way it should be."

Sometimes parents who focus on interdependence downplay ownership and insist on sharing from day one. They may have little concern about self-help skills and continue spoonfeeding their children for several years–sometimes up to age four. Feeding times represent opportunities for the parent and child to connect–opportunities these parents are not in a hurry to give up. Some teachers would be shocked by a three-year-old wanting to be fed. Likewise, parents who expect their child to be fed at child care may be surprised at and disappointed with a program's policy on developing self-help skills.

Whereas parents who stress independence look down on "coddling" or "babying" children, the parent focused on interdependence sees nothing negative about such terms. Doing things for their children, even things they can do themselves, gives parents warm, positive feelings. They even go so far as to discourage independence when children start trying to assert themselves. There is a word in Japanese that

Keeping a Positive Relationship in the Face of Differences

I thought Yin was overprotective. She came every lunch time and spoon-fed her large-for-her-age, robust, four-year-old daughter Casey. I tried in gentle ways to convince Yin that Casey could take care of herself and that these self-help skills were important for her to learn if she was going to be successful in kindergarten the next year. Her mom continued to come every day to feed her. I told Yin that Casey was the only child whose parent came in to feed her. Yin said she was concerned that Casey wasn't eating all of her food and continued to come every day to feed her. I realized I was fighting a losing battle so I said to Yin jokingly, "you won't be able to do this once she's in kindergarten." We ended the school year with a positive relationship despite the fact that Casey was still being spoon-fed by her mom. A year later Yin and Casey came back to visit and I inquired about how lunch time was going at the elementary school. Yin proudly announced that she had gotten a job at the school as a lunchroom supervisor! She no longer spoon-fed her daughter, but she did make sure she got enough to eat. This experience taught me how powerful culture is; it is not something that can be willed away by logic or guilt. By a unilateral decree on my part, I could have forced an important part of a family's culture to be absent at my school, but would have destroyed the trust and joy that this family experienced.

–Marion Cowee

means "to graciously accept help." The idea is to teach independence-minded children to learn to let others help them even if they do not need the help.

Parents who promote interdependent relationships are less adamant about babies learning to sleep through the night alone. They are less anxious to promote self-expression, self-help, *even self-esteem.* In fact, any trait beginning with the prefix "self-" is suspect. These parents don't want to raise *selfish* children, they want to raise children who put others first.

The independence and interdependence approaches represent two different ways of looking at getting needs met. The parents who advocate independence teach their children that it is their responsibility to take care of their own needs as they become more capable, but children who grow up in "other-centered" families learn that the needs of others should be their focus, not their own needs. They will get their needs met, however, because while they are taking care of others, others are taking care of them. Needs are met in both types of homes, but the process in each is very different.[11]

Adults who are independence-minded and adults who are "other-centered" often have deep misunderstandings and disagreements with each other. The former will criticize the preschooler who stands and waits for her parent to put her coat on: "There's no excuse for a child to be so helpless." The latter will criticize the parent who excuses his preschool-age son for not properly greeting his grandparents: "There's no excuse for being disrespectful to elders." What to one is a minor issue is seen as a moral issue by the other.

Children who grow up in "other-centered" families learn to help each other rather than focus only on themselves.
Gaetano Images Inc./Alamy Stock Photo

Of course, most children, no matter how they are raised, do grow up to become both independent individuals *and* people who create and maintain relationships. Children accomplish both major tasks, even if their parents only focused on one. Indeed, parents expect their children to be both independent and connected, but they work harder on the trait they believe to be most important and leave to chance the trait they are less concerned about (or they work on it hit or miss). Most parents never make a conscious decision about what to focus on because it comes from a deeply embedded cultural value.

Oppression. There are some parents who do make a conscious decision to guide their children to become more independent or interdependent; their deliberate goal is to counteract the effects of oppression, such as racism. One parent focuses on firmly establishing her child as the member of a group, a part of her people, rather than glorifying the idea of being a unique individual; the parent sees the group as a buffer between the child and harsh reality. When children receive negative messages about themselves and their people from society at large, they need the group to validate a different reality about themselves. Parents prepare their children for the harsh reality of how the world will treat them when they leave the security of home by connecting them firmly with their own people. Their parenting methods relate to this goal of connectedness.

In contrast, some parents who are members of oppressed groups take the opposite approach. They teach their children to be independent and to value their individuality as a way of standing up to oppression. In either case, the parents' motive is not based primarily on culture but is rather a response to history and current social conditions.[12]

The Home Culture and How It Relates to the Early Childhood Culture. The challenge for the early childhood educator is to become more aware of what parents' motives are, how culture operates, and what else, such as oppression, might be influencing the parents' child-rearing practices. With this awareness should come respect for others, which brings us full circle in this chapter, which started with a look at respect. In addition to being accepting and respectful, early childhood educators should also help parents understand their own behavior better. Sometimes parents' motives, goals, and behavior are inconsistent. If so, the early childhood teacher can help the parent discover and deal with these inconsistencies.

For example, a new mother and her four-year-old son arrive in a center-based program. During the intake interview, the mother tells the director that her son needs strict discipline and that it is important to spank him when he misbehaves. The director respectfully explains that spanking is not allowed. The mother insists that it is a cultural issue with her and that the director must tell the teachers to spank her child or he will become wild and unruly and fail to respect them. In the mother's mind, this behavior is the worst of all sins.

The director, by taking her stance on spanking, is not only following her legal mandate but is also reflecting the early childhood culture, which she has taken on in addition to her own culture. Before entering the field of early childhood education, this director used to spank her own children. She believed in spanking; in fact, she thought it was a vital ingredient of behavioral control. Now, she looks at discipline as guidance instead of behavior control, and she takes a very different view on spanking. She has *become* a member of the early childhood culture.

As the intake interview continues, the director discovers that the parent has pacifist goals for her child. She takes a strong nonviolent stand. Gently, the director points out that the mother's use of spanking will work in opposition to her stated goals. The mother is not convinced at first, but the discussions continue with the director hoping the mother will come to see that spanking defeats her goal of teaching her child nonviolence.

The Dynamic Nature of Culture

Culture is not set in stone. Individuals change, families change, and so do cultures. They never remain static, especially when they bump up against other cultures. Culture is always evolving.

Through cultures change, early childhood educators continuously review the role the program takes in the evolution of a child's culture. Some children arrive with a home culture that differs from the culture they encounter at the early childhood program. Whether the children keep their own cultural perspective, expand or modify it, or lose it altogether and take on the program's culture depends on many factors, including the unspoken messages of the dominant culture and the media. Another factor is the children's age. The older the children are upon entering a culturally diverse program, the less likely they are to lose their own culture, particularly if they are well rooted in their home culture and language. It is the younger child who is more susceptible to the influence of the dominant culture.

Infant caregivers need to be especially sensitive to diversity issues. It is easy to overlook subtle differences in caregiver styles that may have a great impact on babies. Consider this example. Angel's big brown eyes stare at her caregiver Kimberly. Angel's chocolate-colored face contrasts with the paleness of Kimberly's arm, where she lies nestled. Kimberly speaks to Angel in soft tones. How does Angel experience this? She seems a little uncomfortable. Is she noticing that Kimberly's way of holding her is different from her mother's? Kimberly's voice and body language are different too.

Babies are adaptable. They are held different ways by different people. Why can we not just assume that Angel will get used to Kimberly and everything will be fine? Many early childhood professionals believe that early exposure to a multicultural environment makes a child more open later to diversity. In truth, we do not know exactly how babies become members of their culture. We do not know if early exposure to diverse cultures makes them more open or more confused. If indeed they become confused, we do not know if they will outgrow the confusion and become bicultural (or even multicultural)–that is not an inevitable result. Some children who do not receive strong cultural messages may grow up culturally inept. Commonly, children take on the mainstream culture and reject their own culture.

Here is another, more dramatic example of an infant responding to diverse practices. Michael is excited when he sees a bowl and spoon. He kicks his legs and waves his arms, but when Helen puts him in the high chair and places a bowl of finger food in front of him, he just sits there and makes no attempt to feed himself. He looks distressed at first. Finally, he slumps over in his seat with a glazed look in his eyes. His mother explains later that he was taught early not to touch his food with his hands. In fact, he used to be fed on his mother's lap, wrapped tightly in a blanket to prevent him from interfering with her. He obviously does not know how to respond to this new arrangement.

In Michael's family, independence is not a goal. This family believes that children are born with a stubborn independent streak (stronger in some than in others) that must be weakened. The parents' child-rearing practices are designed to create closeness and interdependence. They worry that their child will become too independent and no longer need them. Family closeness and interconnectedness are priorities, not self-sufficiency, which they think comes naturally anyway, whether you train children for it or not.

The ways in which a baby is cared for are a major factor in his or her ongoing socialization and the development of his or her future adult personality. Who the baby is (self-identity) and how he or she relates to others are influenced by the early months and years.[13]

It's also important to understand that parental child-rearing practices are designed to prepare children for adulthood in the family's culture.[14] Caregivers need to be aware of the possible effects they may have on infants whose backgrounds differ from theirs.

The increasing number of infant programs underscores the importance of considering the effect of an infant's exposure to cross-cultural settings. Infants are just beginning to learn their own culture. What happens to the infant–who has not incorporated all those unwritten rules and behaviors–who spends her days with people not of her home culture? Will she automatically become bicultural? Maybe. But many factors are at play, including unspoken messages by the members of the dominant culture. (Although this factor has a greater impact on older children, it is hard to say at what age children begin to take in unspoken messages.) If children learn to look at their home culture as inferior, they may take a negative view of themselves. Naturally, they want to be part of the "better culture," and they may reject parts of themselves and their families.

Of course, issues of cultural differences are not limited to infancy. So much attention has been paid to infancy because that is where it all starts. However, the issues of cultural differences and the need for cultural sensitivity prevail throughout early childhood education and beyond. For example, children of any age who come from a culture that regards pride as a great weakness, even a sin, will have problems in a classroom where personal pride is emphasized. How will these children handle such a double bind? Will they learn to feel pride at school and turn off those feelings at home, or will they choose one cultural view over the other? How many will take home the teacher's view and resist their family's teachings from then on? And if they resist this one aspect of their home culture, will they resist other aspects as well? Imagine what it does to children to be put in such a position.

For children in child care, cultural sensitivity and responsiveness can make a difference as to whether they remain firmly rooted in their culture, become more a part of the mainstream culture, become bicultural, or vacillate between cultures without feeling a sense of belonging.[15]

The Evolution of the Early Childhood Culture

Early childhood educators must learn to be respectful of differing practices yet still be professionals and share their expertise. It is important to recognize that families change when they come in contact with early childhood educators from different backgrounds, but the educators change too. Acculturation is a two-way process. Some old values and practices remain intact, some remain but are modified, and some are shed for newer ones.[16] This process opens up children, families, and early childhood educators to operate flexibly in two or more cultures.

Culture is not static. The early childhood culture is susceptible to change just like every other culture. If early childhood educators leave their minds open enough, sometimes they are the ones who are transformed when they bump up against different ways of thinking and being. Being an effective early childhood educator involves give and take. Sometimes we teach others, and sometimes we learn ourselves. When an early childhood educator embraces a different perspective and changes a program practice, he or she expands the parameters of the early childhood culture.

A Story to End With

I once attended a seminar at an institute for teacher trainers on cultural and linguistic diversity in early childhood education. The leaders talked about the "early childhood culture," but I was not so sure that such a thing existed. One exercise, however, proved to me and the group that, indeed, there *is* an early childhood culture that binds professionals together.

The leaders divided the participants into three groups. The first group's task was to focus on parents; the second, on teachers; and the third, on children: the directions were written on three cards and were quite simple. "List the characteristics of a good ___________ and those of a bad ___________." The blank spaces read "parent," "teacher," or "child," depending on the group.

The group that got the parent card went right to work. They did not have any trouble making their lists. The group with the teacher card also did not have any problems. It was clear to them what the distinguishing characteristics were of a good and a bad teacher.

I was in the group with the child card. We had a terrible time completing our task. First, we agreed that there were no "bad" children; it is not the child who is bad but the behavior. But then we argued about using the word "bad" at all—unacceptable behavior, unsocialized behavior, aggressive behavior, but not *bad* behavior.

We were feeling frustrated and indignant about this assignment. We had not yet figured out how to handle either our feelings or the task when we were notified that the exercise was over and it was time to report back to the larger group. We had to make a quick decision.

When it was our turn to share our two lists, we explained that we had refused to do the assignment because we agreed there is no such thing as a "bad" child. "All children are good," we said emphatically. The leaders who thought up the assignment nodded wisely. "See," they said, "That's the early childhood culture operating!" What we learned was that we agreed because we unknowingly shared a culture that arose from our professional training and experiences. If the group had been made up of individuals randomly picked off the street, they might not have been in such agreement that there are no "bad" children and not even "bad" behavior.

Summary

The social-emotional environment of the early childhood program should provide respect, warmth, nurturance, acceptance, protection, and responsiveness. But those qualities alone are not enough to create a healthy, safe, "growthful" environment; continuity is the final key ingredient. Two obstacles to continuity encountered in the early childhood field are personnel turnover rates and the yearly advancement of children from one classroom to another. Provisions must be made for children with special needs. Several cultural considerations must be taken into mind when planning the social-emotional environment of the early childhood classroom. The first consideration is whether to focus the program on the individual child or on the community—or even whether to fashion the program after a family model.

Another cultural aspect the early childhood program must take into consideration is the difference in child-rearing philosophies between families who focus on independence and those who value interdependence. The challenge is learning how to respect the home culture while fulfilling the professional responsibilities of the early childhood culture. Because culture continually evolves, the home culture of the families–and even those of staff members–will not remain exactly the same; nor will the culture of early childhood education.

REFLECTION QUESTIONS

1. Remember a time when you were treated disrespectfully. What happened? How did you feel? Remember a time when you were treated with respect. What happened? How did you feel? How can you use your own experience to help you treat young children with respect?
2. How aware are you of the social-emotional nature of the environments where you spend your time? Can you separate the effects of the physical aspects from the social-emotional aspects? What is an example of an environment's social-emotional aspect that has a negative effect on you? A positive effect? How does the social-emotional environment of this classroom affect you? How would you change it, if you could, to create a more positive effect?
3. Where do you feel most comfortable, safe, happy, and creative today? How much are your feelings a result of the social-emotional environment? Do you have ideas about how the physical and social-emotional environment interact to make you feel "at home"?
4. Name your culture. If you have difficulties, perhaps you do not think that you have a culture. You do. Everyone has a culture whether they are aware of it or not. Considering that you do have a culture, how do you think that you learned the rules of that culture?
5. How big a role do you think your culture plays in determining which social-emotional environments make you feel comfortable and at home?
6. What are your experiences of clashes between "community" and "individuality"? Are the goals of building community entirely compatible with learning to be an individual? Where do individual rights end and community rights begin? What are your ideas, feelings, and experiences on this subject?
7. Did your parent(s) and/or family of origin have strong independence priorities or stronger interdependence priorities? How do you know? What specific factors in the way you were raised tell you that one or the other was the top priority? Why do you think your parent(s) and/or family had the priorities that they did?
8. What are your experiences with oppression? What do your experiences lead you to believe about how to work with children to overcome the effects of oppression?

TERMS TO KNOW

How many of the following words can you use in a sentence? Do you know what they mean?

FOR FURTHER READING

Bohart, H., and Procopio, R. (2017). *Spotlight on Young Children: Social and Emotional Development.* Washington, D.C.: National Association for the Education of Young Children.

Bronson, M. (2012). Recognizing and supporting the development of self-regulation in young children. In C. Coppel (Ed.), *Growing Minds: Building Strong Cognitive Foundations in Early Childhood* (pp. 97-104). Washington, D.C.: National Association for the Education of Young Children.

Derman-Sparks, L., and Edwards, J. O. (2010). *Anti-bias Education for Young Children and Ourselves.* Washington, D.C.: National Association for the Education of Young Children.

Dischler, P. A. (2010). *Teaching the 3 Cs: Creativity, Curiosity, and Courtesy.* Thousand Oaks, CA: Corwin.

Epstein, A. S. (2012). How planning and reflection develop young children's thinking skills. In C. Coppel (Ed.), *Growing Minds: Building Strong Cognitive Foundations in Early Childhood* (pp. 111-118). Washington, D.C.: National Association for the Education of Young Children.

Espinosa, L. M. (2014). *Getting It Right for Young Children from Diverse Backgrounds: Applying Research to Improve Practice with a Focus on Dual Language Learners* (2nd ed.). Upper Saddle River, NJ: Pearson.

Jacobs, G., and Crowley, K. (2011). *Reaching Standards and Beyond in Kindergarten: Nurturing Children's Sense of Wonder and Joy in Learning.* Thousand Oaks, CA: Corwin.

Lally, J. R., and Mangione, P. (2011). The uniqueness of infancy demands a responsive approach to care. In D. Koralek and L. G. Gillespie (Eds.), *Spotlight on Infants and Toddlers* (pp. 7-13). Washington, D.C.: National Association for the Education of Young Children.

Quann, V., and Wein, C. A. (2011). The visible empathy of infants and toddlers. In D. Koralek and L. G. Gillespie (Eds.), *Spotlight on Infants and Toddlers* (pp. 21-28). Washington, D.C.: National Association for the Education of Young Children.

Tobin, J., Hsueh, Y., and Karasawa, M. (2011). *Preschool in Three Cultures Revisited: China, Japan, and the United States.* Chicago: University of Chicago Press.

10 Routines

Ariel Skelley/Getty Images

In This Chapter You Will Discover

- why caregiving is considered a part of the curriculum.
- what synchronous interactions are and how they create attachment.
- four rituals of care.
- what and how to feed children.
- how feeding issues vary according to the child's age.
- how to make diapering a valuable experience for the baby.
- the concept of readiness in toilet learning.
- that there is more than one way to toilet train.
- how resting issues vary according to the child's age.
- what cultural and developmental issues surround the grooming and dressing of young children.
- what a "transition" is.
- how to turn a transition from chaos to ritual.
- how to handle emotionally charged arrivals and departures.
- how cleanup time works in different programs.
- how group time works.

Daily routines form the framework for a young child's day; some children depend on them for a sense of security. Even adults grow familiar with and enjoy certain kinds of structure. Just as daily routines vary from person to person and home to home, so they also vary among early childhood programs. In some early childhood programs, the daily routines reflect a more open structure, and the general sequence of events moves with the rhythm and flow of the group; in other programs, the daily routines are precisely divided into time segments and regulated by the clock. (It's 10 o'clock–time for snack.) No matter what type of schedule the early childhood program follows, there are certain routines that should occur daily (see the *Tips and Techniques* box on page 273).

This chapter looks at four caregiving routines–feeding, toileting, resting, and grooming and dressing–and how they form an integral part of the early childhood program's plan for learning. We will also examine transitions and group time, two other routines that also function as important curriculum components. Most importantly, we will discuss how the various routines should be carried out in age-appropriate, interactive, and instructional ways.

As children leave infancy and toddlerhood, the emphasis on teacher involvement with their physical needs and care diminishes, but because programs for infants and toddlers are growing rapidly, everyone in early childhood education needs an overview of the essentials of their care and education. Not that this chapter does not also pertain to older children. Physical care and adult attention continue to be important throughout early childhood, even after toileting becomes a private affair; children continue to learn about eating and nutrition as they grow older, even if they no longer need to be held or spoon-fed. In fact, the nation's record on obesity shows that we must intensify our education of children on proper nutrition and exercise while children are still forming their habits.

Caregiving as Curriculum

The early childhood educator must recognize that children are *always* learning–even when they are engaged in eating, toileting, resting, grooming, and dressing. These caregiving activities take up a good deal of program time and can be considered a chore or a blessing depending on the point of view. They are a chore to those who fail to see their significance, yet they are a blessing to early childhood professionals who regard caregiving routines as a way to share an interactive and often intimate moment with a child. Caregiving routines provide built-in opportunities for interactions–interactions that enhance relationships and build attachment.

Although NAEYC's Program Standards don't link caregiving with curriculum, it's easy to make that link by juxtaposing Standard 1 on relationships with Standard 2 on curriculum. Reading the two standards with the idea that caregiving is goal-oriented behavior, it is easy to see that well-trained professionals can use caregiving times to create positive relationships with the children, as required by Standard 1, which is about promoting relationships. The rationale for Standard 1 reads, "Positive relationships are essential for the development of personal responsibility, capacity

TIPS AND TECHNIQUES

Elements of a Daily Structure

Whether an early childhood program follows an open schedule or a tightly structured schedule, the following elements should be a part of the daily routine:

- *Beginnings and endings.* There should be a start and a finish to the day or session, whether for the individual or for the group. Each child should receive some kind of greeting and farewell, even if arrival and departure times are staggered.
- *Need fulfillment.* Individuals–both children and adults–should have the ability to get their needs met with reasonable promptness without having to abide by an overly rigid schedule or inflexible group rhythm.
- *Balance.* There should be a developmentally appropriate balance among (1) individual freedom and consideration of the group, (2) challenge and security, and (3) stability and flexibility.
- *Choices.* Each day children should have opportunities to decide how to spend their time. This type of decision making is important practice for the future, gives children a chance to experience the consequences of the choices they make, helps children determine what they like and dislike, and, thus, empowers children.
- *Things to do.* The equipment, materials, and activities available to the children should be interesting, age appropriate, and culturally appropriate, and they should help children make cognitive connections and be deeply meaningful to them.
- *Varied opportunities.* There should be opportunities for outdoor and indoor play and work and for quiet and active play and work. Children should also have daily opportunities to play and work alone, in pairs or threesomes, in small groups, and, when appropriate, in large groups.
- *Emphasis on the whole child.* Children should have daily experiences that stretch their minds and bodies, recognize feelings, and facilitate social and emotional skills.
- *Relationships.* Helping children make connections with each other and with members of the staff should be a key focus each and every day.
- *Education.* The environment and the planned activities and projects should allow for meaningful and in-depth learning and development that relate directly to the children's lives and interests. In addition, adults should be on the lookout for and take advantage of unplanned learning and development opportunities. Adults should take care not to let teachable moments pass by.

for self-regulation, for constructive interactions with others, and for fostering academic function and mastery."[1] By the time you finish this chapter you should be able to see that Standard 2 also applies, because caregiving can promote learning and development in each of the following domains: aesthetic, cognitive, emotional, language, physical, and social. Even the rationale for the standard fits because caregiving can incorporate, as the standard says, "concepts and skills as well as effective methods for fostering children's learning and development. When informed by

NAEYC Program Standards

Program Standard 1: Relationships

Program Standard 2: Curriculum

teachers' knowledge of individual children, a well-articulated (caregiving) curriculum guides teachers so they can provide children with experiences that foster growth across a broad range of developmental and content areas." Sandy Baba shares her experience of creating an infant curriculum that fits NAEYC's program standards and made sense. See the *Voices of Experience* story.

I was taught by my mentor, Magda Gerber, that caregiving is curriculum. I eventually had the opportunity to travel to Budapest, Hungary, five times to visit the Pikler Institute, the orphanage where Magda received her training and derived her philosophy. That philosophy underlies this book and is being used by what is called Resources for Infant Educarers (RIE) to train early childhood professionals and parents in this country. Those trips to the Pikler Institute in Budapest opened my eyes to the deeper meaning of how caregiving becomes curriculum. The caregivers there are called nurses and they are trained to give focused one-on-one attention during caregiving routines. Magda taught about synchrony, but I never saw such synchrony as went on while the nurses dressed children or bathed them (or fed the younger ones). Because most of the time the children are in groups, and because this is 24-hour care, the nurses make use of the caregiving times to interact with each individual child, making language an important part of the interaction. You can see both the learning going on and the closeness of the relationship by watching the nurse carry out the caregiving routines. No other time of the day is as rich as those person-to-person interactions during caregiving. When caregiving is given the kind of importance that it

Family child care providers have to plan their space so it serves more than one purpose and responds to the ages of the children present.
FamVeld/Shutterstock

What Does an Infant Curriculum Look Like?

I vividly remember my struggle, during my second year of being an infant teacher, to create a classroom bulletin project. The classroom had six babies and my project was to post an infant curriculum on the bulletin board so that parents and families would know what their babies did in the classroom with the teachers during the day. Day after day I brainstormed on how the curriculum would look but just could not figure out how to begin the project. I researched at the resource library hoping to find a sample infant curriculum, but it was not much help. The questions and challenges I kept having were "So what is an infant curriculum? Why is it so difficult to create one?"

As a result of this difficulty, I started to reflect on the children's routines in the classroom. I started to think about Zoe, a four-month-old baby. She usually liked to play on the floor, babble to a teacher, and fall to sleep on her own, whereas Joey, a six-month-old liked to be held and read to before he took a nap. After reflecting on the children's individual schedules, how could I create a schedule to fit everyone? Suddenly, a light bulb went on in my head as I realized that there is no "set" curriculum for infants. Infants don't need a set schedule like older children, as they each have their own unique and personal timetable for napping, eating, playing, and exploring. Every moment of feeling the world was a learning experience.

When I look back now, I am glad I couldn't find anything to put on the bulletin board as an infant curriculum! However, I did create an infant routine flow chart to show parents what teachers did in the classroom to support their babies' development through the daily routines. Having gone through this professional "growing pain," I feel completely at ease with infants because I truly understand what babies need and can communicate to families what they can do to support their babies' development.

–Sandy Baba

is given in the Pikler Institute, it is easy to relate it to NAEYC's Standards 1 and 2 on relationships and curriculum. See *The Theory Behind the Practice* box on page 277 for a look at Dr. Emmi Pikler, the founder of the Pikler Institute.

After being in Budapest and seeing with my own eyes the contrast between a program that regards caregiving as a major feature of the learning plan and the average program in the United States, I am more convinced than ever this is an important chapter in this book. Routine is curriculum in infant care. If children are to be cared for in groups, even if it is not residential care, they need focused individualized attention, which happens during caregiving routines only if they are valued as important learning times.

Synchronous Interactions

The quality of adult-child interactions makes a big difference in building relationships. Children need a number of one-on-one interactions with adults every day. They need to have adults pay attention to them, respond to them, and respect who

synchronous interaction A coordinated interaction in which one person responds to the other in a timely way so that one response influences the next in a kind of rhythmic chain reaction that creates connections.

they are. Synchrony is what makes those interactions work. To illustrate, let's look at an example of a **synchronous interaction.**

Davy is a hungry baby! He just woke up, and now he's screaming hard. His caregiver, Jenn, is hurrying to get his bottle ready. She calls to reassure him she is coming. It's hard to tell if he hears her because he is screaming so loudly. Finally, she is there and ready. He quiets a bit as he hears her voice tell him that she is going to pick him up. She reaches out her arms, and he arches his back ever so slightly. She knows that he is responding to her.

A short time later, the two are seated comfortably in a little alcove by a window. Jenn nestles Davy's head in the crook of her arm and holds him so his head is raised slightly above his body. She picks up the bottle and touches the nipple lightly on his cheek. He turns immediately in that direction, and the nipple slips smoothly into his mouth. Davy squints his eyes into a tight line and sucks furiously on the bottle. He does not pause or breathe for what seems like a long time. Finally, he stops for a moment, lets the nipple go slack, and then clamps down and goes at it again.

Jenn does not talk to him or distract him in any way. She knows that he needs to focus on eating. Eventually, he slows down, and then Jenn starts talking to him. "You were really hungry," she says. "Wow, look at you eat!" He stops sucking for a minute, lets go of the nipple again, looks her in the eyes, and gives her a big smile. Then he turns his attention back to the task at hand—emptying the bottle and filling his stomach.

Can you see the synchrony in that interaction? Jenn was closely tuned in to Davy. She knew just what he needed and how to provide it. She even knew not to poke the nipple into his mouth. Instead, she touched the nipple to his cheek, triggering his rooting reflex. Such small actions empower children: Davy was the one who put the nipple in his mouth, not Jenn. She just set it up so that he could do it. Jenn knew when to talk and when not to talk. She managed to elicit a smile from him, which in turn made her smile and feel good. These two were in perfect sync!

It will not take many feedings like that one for Davy and Jenn to develop a close relationship with each other. Because Jenn is Davy's primary caregiver, she is the one who usually feeds him. Such consistency also helps promote their relationship.

What you just saw was a caregiver and a baby in a synchronous interaction. Now let us look at how a teacher and a preschooler interact synchronously.

Reba is standing by herself in the play yard and looking downcast. Teacher Tauheed approaches her and lays a hand on her shoulder. She looks up at him. He asks, "Is something wrong?"

"I don't feel good," says Reba quietly. Teacher Tauheed gets down on his knees and looks her in the face. A big tear rolls down her cheek.

"What is it?" Teacher Tauheed asks, as Reba suddenly throws herself into his arms, clinging tightly. He holds her close and feels her give a big sigh.

He is about to touch her forehead to check for a fever when she says, "I want my mommy." She breaks away from him and stands with her back turned and her

Emmi Pikler, MD

The Pikler Institute was established in 1946 by pediatrician Dr. Emmi Pikler. The city of Budapest commissioned Dr. Pikler to organize a home for infants and young children who were deprived of their parents until appropriate living arrangements could be found.

Dr. Pikler didn't follow the rules of standard institutions at the time, which followed strict hygiene guidelines and structure. Rather, she aimed to achieve conditions similar to the family, emphasizing the importance of attachment through primary caregiving and continuity of care. Dr. Pikler's concepts and practices included ideas on respecting the child as a person–where he/she is an active individual and is allowed to move freely in a safe environment.

Dr. Emmi Pikler's early research on gross-motor development showed the benefits of the baby's active role in his or her own development when allowed freedom of movement. She advocated never putting babies into positions they cannot get into by themselves, a rather drastic notion for those in the United States who are anxious to speed up development. Pikler went on with her research after founding the orphanage nicknamed Lóczy after the street on which it is located and now called the Pikler Institute. Pikler trained each caregiver (called nurses) to create a special kind of relationship between herself and a small number of the children in her care. The idea is that when each child has a particularly close relationship with one adult, caregiving becomes an educational and even therapeutic activity. Outside of caregiving times, children are given a good deal of freedom to develop independently of adults. To read more about the Pikler Institute and find further resources, read Myriam David and Genevieve Appell's *Lóczy,* published in Hungary in 2001 by the Association Pikler-Lóczy for Young Children (English translation by Jean Marie Clark and Judith Falk is available).

head down. He waits where he is. He does not reach for her. She turns around and says angrily, "They are so mean!"

"Mean!" repeats Teacher Tauheed, waiting to see what she will say next. She makes a movement toward him, so he holds out his hand. She takes it. He feels her shaking. "They said I couldn't play." She is tugging on him now, pulling him across the play yard to the sandbox.

"I can too play," she shouts at three girls digging in the sand. She holds on to Teacher Tauheed's hand tightly. "Come on, you can play," says one girl. "We never said you couldn't!" "Well, then I get the green bucket," shouts Reba. The same girl shrugs and shoves the bucket toward her. Reba lets go of Teacher Tauheed's hand and grabs the bucket. She plops down in the sand and starts filling it in great scooping handfuls.

Teacher Tauheed bends down close to her face. "Are you OK now?" he asks softly. She smiles back–a big smile. She does not need words. Her smile says it all.

"So you don't need me now?" he asks, moving away. She dismisses him with a regal gesture of her hand and goes on digging.

These interactions–between Jenn and Davy and between Teacher Tauheed and Reba–are like connecting links that eventually form a relationship. Given enough links, these children and adults may well become attached to each other–if they aren't already.

Attachment

attachment An enduring affectionate bond between a child and a person who cares for the child, giving the child a feeling of safety or security. Building a trusting secure attachment through consistency, responsiveness, and predictability shows children they can trust the caregiver to meet their needs and frees them to explore their environment.

Why is **attachment** important in early childhood education? At higher levels of education, nobody worries about whether students become attached to the teacher. Of course, it helps if students like the teacher, but feeling a close connection is not necessarily a prerequisite for learning. Early childhood education is different, however, because attachment is a vital part of the learning atmosphere. Young children feel secure when they are attached and are therefore freer to learn. Also, when adults and children have an attachment, they understand each other better; adults can be more effective teachers, and the teaching-learning process is enhanced.

How does attachment help mutual understanding? The younger the child, the more the adult must depend on nonverbal communication to understand what the child needs and where the gaps in learning are. Attachment is the closeness that facilitates the communication, understanding, and trust required for a child to feel safe and that his or her needs are met. Think of people you know well and are close to who send signals that you can read but that other people might miss. For example, I know a boy who tends not to express anger openly but instead twitches his face in a certain way and hums a little tune–always the same tune–when he is frustrated or upset. I also know a girl who twists her hair when she is tired and a baby who spits up when frightened.

Communication is difficult in a situation when you do not know the children and then have to determine what they need or what they are trying to express to you. When you cannot understand what they are saying or cannot read their signals, you are more apt to project your own needs onto the children than you would if communication between you were good. For example, you might overdress an active child because you felt cold from just standing around. Likewise, if you were feeling hungry or groggy, you might feed a baby unnecessarily or put him down for a nap prematurely.

Attachment not only facilitates communication and helps out the adult in the role of caregiver and facilitator of the teaching-learning process, but it is also vital to the mental and social development of young children as well. Of course, the most important attachment should be at home and usually is. Most children already have an attachment to an adult when they come into an early childhood program, so what occurs in the program is a secondary attachment. The aim is not to replace the primary attachment but to fill a need for close connection during the time the child is away from the family. Children need to be around adults who show that they care about them personally. Children need adults to be in tune with them to create the kinds of interactions that lead to close, lasting, ongoing relationships.

Early childhood programs can't simply educate and care for young children without also considering their attachment needs.

There are many models of caregiving, not just one, but there is also a limit to how many caregivers a child can relate to. For example, an infant-toddler program may have too many babies and too many caregivers and, thus, inadvertently prevent adult-child attachments. To help solve that problem, the program might use what is called a **primary caregiving system.**

primary caregiving system A caregiving system in which infants are divided up and assigned (in groups of three or four) to specific primary caregivers who are responsible for meeting their needs and record keeping. The goal of this approach is to promote closeness and attachment but not exclusivity. An important aspect of this system is for each child to know and relate to other caregivers as well.

However it is done, it is important that children, not just in infancy but beyond, develop ongoing relationships and feel connected to one or more adults in the program. Because such connections cannot be planned, the incorporation of primary caregiving systems in programs ensure that children can safely form the attachments required for healthy development.

PHYSICAL-CARE ROUTINES

This section focuses on how to carry out routines in four areas: feeding, toileting, resting, and dressing and grooming. It also explores some ways to make transition times easier as children move from one activity to another.

Feeding

Infants. During the first months of life, infants live on breast milk or formula alone. Welcome any mothers–and fathers–who are willing to take the extra time and effort to come and feed their babies periodically. For breast-feeding mothers, provision should be made for privacy and quiet. It may not always be easy on the caregiver because of timing (if baby is hungry and the mother is not there, that is a problem!), but it is worth the extra trouble if the baby can get breast milk and/or relate one-on-one regularly with his or her mother.

Bottle-fed babies also require privacy, quiet, and one-on-one care so they will not be distracted from the feeding process. Infants should be held when fed–*never* put down by themselves with a bottle. Infants put to bed with a bottle are a recipe for disaster: they can choke, and they are at risk for ear infections and tooth decay. Besides, they miss out on the intimacy of being held and attended to during the feeding process. Feeding can be a close, emotional experience–an experience that promotes relationships.

Remember the scene with Jenn and Davy? This example illustrated not only how a baby should be fed, but also how feeding contributes to closeness. Jenn and Davy were in sync with each other. Jenn demonstrated her knowledge of babies in general and Davy in particular, and Davy showed he knew how to interact with Jenn, starting with his first cries that let her know he was awake and hungry and ending with his big smile that drew her closer to him.

According to Erik Erikson, a major task of infancy is establishing basic trust. Caregiving routines contribute to the accomplishment of this task when they are done in a timely manner with warmth, sensitivity, and responsiveness.

Adding solids to the diet (in addition to breast milk or formula) usually starts around six months–give or take a few months. Solids may be held off until the child is a year old, depending on the family, the advice of the infant's pediatrician, or the program's policy. It is usually best to let the family take the lead about when and how to start solids. A general rule of thumb is to introduce one new food (pureed) at a time and to start with just a taste. Add another mouthful each day until, within a week or two, the baby is taking a reasonably sized portion. The reason for starting one new food at a time is to detect possible food sensitivities; if a baby had a reaction after eating a casserole dish with eight different ingredients, it would be hard to tell what the offending food was.

Do not add sugar, salt, or other seasonings. The natural flavor of the food is all that babies need. Take special care to avoid all artificial flavors and colors.

A task of the mobile infant is to explore, and food is the perfect medium. It is important to allow for messy meals and to give babies plenty of time to eat. Finger food gives younger children a chance to feed themselves (though some cultures discourage this). Spoonfeeding can also eventually be taken over by the baby, but be aware that cultural ideas about when this should happen vary widely.

Do babies have to sit in highchairs for feeding? Although it is common practice to use highchairs for convenience–to put the child at the adult's level–It is possible to run a program without a single highchair. Babies too young to sit can be held on laps. This gives more physical contact and closeness. Mobile infants can be seated at a low table in a chair they can get in and out of themselves.

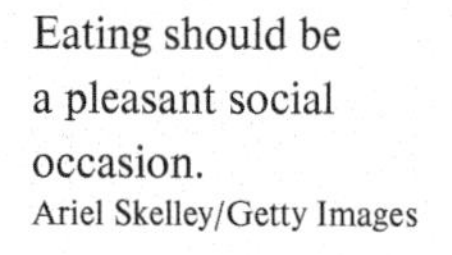

Eating should be a pleasant social occasion.
Ariel Skelley/Getty Images

This arrangement gives more autonomy and lets babies decide when they are finished. They don't have to wait for an adult to release them from any straps or belts.

Toddlers. The major task of toddlers is to try to do things for themselves and to show independence. If fostering independence is also the program's and the family's goal, children should be allowed to help not only with the feeding process, but also with the food preparation, insofar as they are able. Giving toddlers some limited choices also helps avert the power struggles that arise so easily. Toddlers need patience and understanding. It is also important to realize that children's appetites often drastically decrease when they reach toddlerhood. The baby who was once a ravenous eater may become a toddler who only picks at food and eats just a few bites at any given time.

Preschoolers and School-Age Children. Children at these ages are becoming competent eaters. Compared to an adult, they are still messy, but compared to a mobile infant or toddler, they are amazingly neat. They can be included in meal preparation and cleanup. Children of this age need to learn about nutrition so they can make wise food choices as they get older. Habits form early, and it is the responsibility of adults to help young children form good eating habits early in life.

The food pyramid should provide a nutritional guide. It outlines the types of foods and the appropriate proportions that constitute a healthy diet. Children–indeed, all of us–need far more grains, fruits, and vegetables than we need fat, oil, sugar, and meat. Getting enough protein is, of course, a concern for growing bodies, but protein comes in many forms–not just in meat and other animal products. Beans and other legumes are good sources of protein.

With obesity on the rise, early care and education professionals need to pay attention to their role in intervening with the kinds of patterns that are being recommended early in life. Although obesity comes from many causes–genetics, family patterns, and lifestyle–the main problem is that children who are obese eat too much and exercise too little. Even young children tend to eat large portions of processed sugary foods with high fat content. Regular doses of television also contribute to obesity. Lifestyles are established in the early years, so responsibility to help ensure that those lifestyles are healthy ones falls on early childhood educators.

Eating at any age can be a pleasant social occasion, so the environment should be set up to encourage interaction. A group of several children around small tables makes for better table conversation than masses of children at long picnic tables. If independence is a shared goal of the program and its families, children can learn to serve themselves from toddlerhood on. Self-service is empowering because children can make choices about how much or how little they want to eat. Some programs that serve only a snack make it a free-choice activity, involving a serve-yourself, and sometimes even a make-it-yourself, approach.

Toileting

Infants and Toddlers. Changing diapers is a routine, but it should be more than just a routine. Diaper changing should be highly individualized and interactive. Elsa Chahin, a lactation teacher and infant expert, trainer, and teacher of the RIE (Resources for Infant Educarers), writes about how to make diapering a personal connection and not a chore in the *Voices of Experience* story.

The way diapers are changed is important. To feel respected, children must be treated like human beings on the diapering table. It is not enough to offer a toy to entertain the top half of the child's body while dealing with the bottom half in a no-nonsense fashion. The "whole child" is present on the diapering table and should be acknowledged. Children should be incorporated into the process so they become part of a team, not just an object to be manipulated. Explaining what is going on and asking for their help and cooperation are important.

During diapering, a child learns about his or her own body and its sensations, processes, and products. The child also gets feedback about how adults feel about his or her body and its products. If adults teach shame, the child learns shame. If adults approach diapering with a natural and accepting attitude, the child comes to take toileting as a matter of course. It is hard to make diapering a meaningful experience when a child is distracted with a toy or some other means of entertainment during the diapering process.

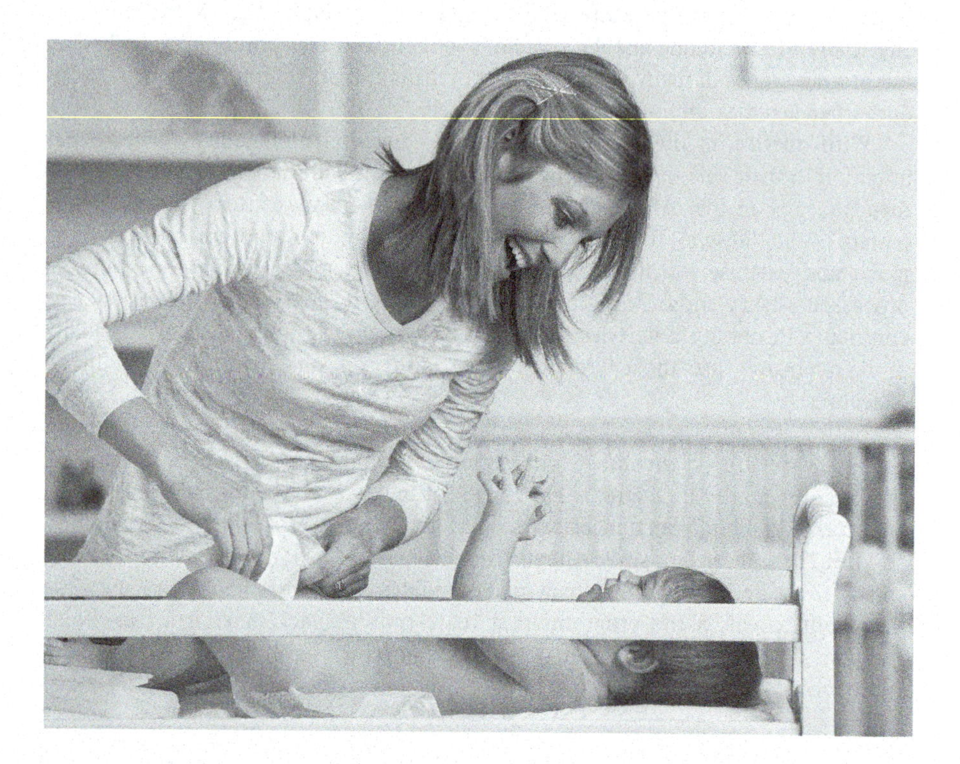

Children should be incorporated into the diapering process so they feel a part of a team–not distracted with other objects.
AVAVA/Shutterstock

Living Every Moment

I often get asked the same question by participants in my workshops: "How can I make diapering a pleasurable experience, especially when it is done so many times throughout the day?" I start by sharing a story I once heard about Napoleon preparing for battle. He was getting dressed by his very nervous valet who in his haste wasn't doing a proper job. Napoleon said to him "Slow down, I'm in a hurry." Sometimes in our urgency to get things done, we forget to be in the moment; we become less focused, less efficient, and we end up taking more time than if we had slowed down in the first place. And very possibly, in the end, we forget to enjoy the experience.

I ask the participants if the simple answer could be *to be present.* When one is in the moment, one can never be rushed because time doesn't exist. We can read the child's cues and be in synchronicity with their needs. We become responsive to what is dictated in that particular moment. I invite them to take their time to establish the connection with the child.

"Does it really matter to the child?" I often get asked. Children live in the moment; to a child nothing matters but the present moment. I share a story about a three-year-old boy, Leonardo, who was at home finger painting when his Dad, who had been away on a business trip, called on the telephone. Mom talked with him for a few minutes and mentioned that Leonardo was painting something special for Daddy. When Leonardo was handed the phone, Dad asked him 'What are you doing honey?" and without waiting for a response from his son, proceeded to say, "Mamma tells me that you're painting." Leonardo replied "No, Pappa." "What do you mean honey? Mamma told me that you're painting." Leonardo confidently responded, "No Pappa, I talking with you."

I can almost be sure that Pappa must have tapped his own forehead after his son's reminder of returning to the moment.

Every child has his/her own rhythm. When we slow down to match their natural pace and become open to what the present has to offer, we can witness the magic unfolding. The result is that the child can enjoy the diapering, the adult walks away with a smile for having shared the pleasurable experience, and both may be looking forward to the next oppor-tunity to establish this connection.

–Elsa Chahin

Diapering is a social experience as well as a caregiving routine. During diapering, adults and infants share an intimate interaction and build on their mutual relationships.

When adults do not value caregiving routines, especially diapering, they tend to set up an assembly-line process. Usually, the lowest-status person is the one who gets stuck all day at the diapering counter. But such an approach to diapering eliminates many of the benefits just described.

Like diapering, toilet training–or toilet learning–has to be done with sensitivity to the child and the family. It can't be governed by a blanket policy. It is usually best if the family takes the lead and the program follows that lead. See the *Points of View* box for two views on toilet training.

Preschoolers and School-Age Children. Most preschool-age children and school-age children are toilet trained. Except for an occasional accident, they require little from adults except the freedom to use the bathroom when they need to. Nevertheless, they may need reminders to flush and wash their hands afterwards. They may also need a little help tucking, zipping, and buttoning their clothes. And in case of accidents, it is a good idea to have changes of clothes available.

Some programs schedule a toileting time, especially programs whose bathrooms are not convenient to the rest of the setup. Other programs establish a toileting time based on the belief that children benefit by attuning their individual body rhythms to the group rhythm. Nevertheless, many early childhood educators frown on trying to make children all use the bathroom according to a schedule because it involves too much waiting time and puts too little emphasis on individual needs.

Resting

Infants. Rest time for infants is highly individualized. When they sleep, how they get to sleep, what signals they give caregivers to let them know they are tired, and even what coverings they prefer all depend on what infants are used to. Infants also favor different sleeping positions, but it's important for caregivers to know that young babies should be put to sleep on their backs or sides, *not* on their stomachs. Studies conclude that stomach sleepers are at higher risk for Sudden Infant Death Syndrome (SIDS, once called "crib death").

Cultural approaches to resting vary greatly. Although the common early childhood approach is to put babies to sleep in safe cribs in a quiet, darkened room away from the activity area, not all children are used to sleeping that way. Some babies never spend any time separate and alone in a crib; instead, they spend their first months constantly in someone's arms or bed. They eat, sleep, and nurse in the midst of daily activity, creating their own rhythmic pattern. A person who believes in schedules, private space, and alone time, even for babies, may disagree with this approach to infant care.

Toddlers and Preschoolers. Whether to schedule specific nap times depends on the program's philosophy, if not outside regulations. Some adults feel perfectly comfortable getting children as young as two onto a schedule that fits into the program schedule. Nap time comes after lunch, and children learn to pace themselves so they can wait. Napping lasts a certain amount of time; children who cannot sleep the entire time learn to lie still and rest until the group is beginning to wake up. This approach works best when the adults believe firmly that a napping schedule is in the child's and the group's best interests.

However, not all adults see nap time the same way. Some adults want children to learn to read their own body signals and to rest when they are tired, not when the clock says it is time. They believe it is important for children to learn to take care of their own needs instead of subordinating those needs to a time schedule.

Toddlers and preschoolers rest on mats or cots. They may have favorite blankets or "snugglies" to sleep with. These "transition objects" comfort children who miss their families more at nap time, when the room is quiet and there is no activity to

POINTS OF VIEW

Two Views of Toilet Training

Determining when a child is ready for toilet training depends on whether one believes in encouraging independence or interdependence. There is no one correct approach, only a difference that is grounded by one's social and cultural values. The early childhood culture agrees with child-rearing experts: Toilet training is really "toilet learning" and is part of an independence-promoting curriculum. Toilet learning cannot begin until the child is old enough to have independent urges. The child must also be ready in three areas. First, the child must be physically ready (meaning he or she can "hold on" for a period of at least an hour or more; several hours is a better indicator of readiness.) Second, the child must be intellectually ready (i.e., he or she understands when and how to use a toilet). Finally, the child must be emotionally ready (in other words, he or she must be *willing* to use the toilet). When the three areas of readiness occur together, toilet learning is usually easy and rather quick!

In contrast to this approach, some people believe that toilet training should start at birth or soon thereafter. By trying to read the child's signals, the adult catches bodily wastes in a potty or toilet. Eventually, the child gets good at sending signals and the caregiver at reading them. In addition, the adult may prompt the child to let go at a signal–usually a *shhshhshh* sound.

Advocates of toilet learning often respond to the second technique sarcastically: "Oh, it's the adult who's trained, not the baby. It's impossible to train babies." They believe that independence is the goal of toilet learning, not interdependence. The advocate of the second approach, however, sees things differently: "Why wait and deal with all those diapers? Besides, I get satisfaction from helping children, even if it's an inconvenience to me." For this individual, toilet training is a joint process that establishes interdependence. In addition, it provides the adult and child a different kind of opportunity to work and learn together.

distract them from their feelings. Also, going to sleep makes some children feel vulnerable. Letting go of control in order to go to sleep is hard for some children, especially in a new place. Do whatever you can to help children feel safe, secure, and comfortable.

Some programs have a going-to-sleep ritual that is identical every day. The caregivers change the environment at nap time to signal to the children that it is time to be still and to rest; they may darken the room, store toys out of sight, play or sing soothing music, or even rub some children's backs. Nap time is often preceded by a quiet time. The teacher or provider might read a story to relax the children before they go to their cots. The more signals the children get that tell them what's about to happen, the easier it is for them to settle down.

School-Age Children. School-age children may have a rest period but seldom take an official afternoon nap like preschoolers do. Nevertheless, it is important for school-age children to have places to go to rest when they are tired or need to be alone.

Being in public all day every day is hard on some children; they need time apart. Setting up the environment to provide semiprivate, quiet spaces is important. One program had a large closet full of cushions and books where children could choose to go whenever they wanted a peaceful place to unwind or to read.

Grooming and Dressing

Infants. Infant grooming usually occurs after diaper changes and meals and right before going home. Remember to always use a clean washcloth and towel with each child. Change infants' clothing when it gets wet or soiled.

Toddlers, Preschoolers, and School-Age Children. Most children in half-day programs don't require much grooming, except to learn to wash their hands after toileting and before and after meals. It is the young children who stay all day and take naps that need the most grooming attention. At the end of nap time is a good opportunity to assess each child's need for grooming. Most children need their hair combed when they get up (some parents may be upset if they arrive and find their child uncombed).

Children can, of course, learn to groom themselves and should be encouraged to do so, especially preschool- and school-age children. Be aware, however, that opportunities for individual care and warm interaction are sometimes missed when caregivers stick to a firm rule of never doing anything for children that they can do themselves. Some children enjoy having their hair brushed, even when they can do it themselves. It simply feels good. Even children who can tie their own shoelaces may just want a little attention and special care. Be sensitive to children's needs, and be sure to offer one-on-one attention to even the most competent children.

Then again, there may be children who are able and willing to take care of themselves but come from families that stress interdependence. The families may frown on self-care for young children. Be aware of these cultural differences and try to understand the family's and the child's perspectives.

Dressing children is time consuming! Going outdoors in cold climates, for example, takes substantial time and effort on the part of children and adults. When you look at dressing and undressing as part of the curriculum–a chance to build relationships, a time when children get one-on-one attention, when they get the touching they need–it may not matter so much that it is time consuming. It is not a waste of time–it is part of the curriculum. The more the children can help themselves and each other, the easier it is on everyone and the less time they have to wait to go outside.

FOCUS ON INCLUSION: ADAPTING ROUTINES FOR CHILDREN WITH SPECIAL NEEDS

Feeding. Some children with special needs may require help at mealtime. It is important to get information from parents on particular ways to feed their children. Creative solutions can help alleviate some feeding obstacles. For example, a child with weak lip control may need to use a straw instead of drinking directly from a glass. Some children may need to have their cheeks or throats massaged to help them chew and swallow. A gum massager can help "wake up" the mouth and make the child more aware of oral sensations in preparation for eating. Moreover, some food textures are problematic for children with neurological problems; pureed food may be the answer, even if the child is beyond infancy. Slowly moving from pureed food to food with more texture may give the child a chance to develop chewing and swallowing skills.

NAEYC Position Statement
on Inclusion

No matter what age or what special circumstances, eating should be a pleasurable, healthful, social experience. It should be regarded as an integral part of the early childhood program.

Toileting. Age or stage may have nothing to do with the need for some aspects of physical care. For example, some children take much longer than others to gain control of bladder and bowel muscles, and some may never manage it. This should not keep them out of natural settings, settings where you would find their typically developing peers. Indeed the law says that children cannot be rejected because they are in diapers, even if the teachers have no experience in such matters, which is why this chapter is relevant for *all* early childhood professionals, not just infant-toddler specialists. It is important to take care of diaper changing or toileting in a warm and matter-of-fact way, not grudgingly. If it seems that a child needs toilet learning or training, but is not getting it, talk to the family and the specialists who may be working with them. Find out what the plan is for toilet training, their timeline, and their expectations. What is happening at home? What do they expect of you? As with all other matters, the teamwork approach is best!

NAEYC Program Standards
Program Standard 8: Communities

Resting. Although the information about SIDS makes it sound imperative that all infants should sleep on their backs, nevertheless, it is important to understand and weigh the risk factors related to positioning infants with special needs. Find out from the family what is the safest sleeping position and if they do not know or are not sure, have them ask their pediatrician or other specialist who knows their child. Some infants should not be on their backs, and you need to know which ones.

Make sure that all children's needs for rest are recognized and responded to. Some children, whether they have identified special needs or not, have less stamina than other children. Some are medically fragile and sufficient rest is important to keeping their immune systems in the best possible shape. Also be aware of differing responses to stimulation. A lively noisy classroom may be exhausting for some children and yet they may have a hard time settling down to rest without a quiet place to do so. It may be challenging to meet all children's needs, but that must be the goal.

Grooming and Dressing. It can help to know from the start how much and what kind of help each child needs. You also want to know the family's position on self-help skills, which may be a cultural issue with them. When considering self-help skill development, it is important not to put the same set of expectations on every child regardless of family goals. Knowledge of the typical skills associated with developmental stages can mislead you to expect more than a child is capable of in some cases. Some specific approaches to dressing can be helpful when it comes to positioning. You want children in positions where they feel secure and have the most freedom of movement. Some very specific techniques can be learned from family members or their specialists. For example, you need to know how to put this particular child's arm in a sleeve in ways that keep the muscles relaxed. You also need to know how to put pants on this particular baby without making his legs scissor.

Other Routines

This last section looks at different kinds of transitions, both those that occur between activities and those that coincide with the daily arrival and departure of children. Cleanup time is a special kind of transition. Finally, we will take a look at group times–another routine found in most programs.

Transitions

transitions The passages between one place and another or one activity and another. Examples of transitions include arrivals and departures, cleanup time, and going outside. Transitions occur as often as children change activities either as a group or as individuals.

Transitions are the passages between one place and another or one activity and another. Included under the category of transitions are arrivals and departures to and from the program day or session, as well as cleanup time. In some programs, all transitions are accompanied by a cleanup period; in others, cleanup is done only once or twice a day or session.

Transitions occur as often as children change activities, either as a group or as individuals. In most programs, the day is full of transitions that bridge the time and space between activities or routines. Depending on how the program is set up, the children as a group may experience transitions every few minutes or go several hours without one; moreover, the flow of the group and adult needs may influence the number of transitions. Some adults enjoy a set schedule that moves at a brisk pace. They like change. Others are willing to let things unfold more naturally and move at a slower pace, which results in fewer transitions.

Think about rhythm and pace. The children's needs should be the primary consideration. Do young children need a lot of changes, or is adult misconception of boredom driving the rhythm and pace? A fast-paced program may be based on the belief that children have short attention spans–in fact, it is often adults' attention spans that are limited. Children's attention spans are influenced by a number of factors, not the least of which are adult pace and adult expectations. When adults expect children to get bored, they quickly move children on from one thing to another without allowing them to go deeper into their explorations, activities, or projects; they send the message that nothing stays the same for long. Is that what we really want for children? A fast-paced program with rapidly changing scenes is a little like a music video–full of momentary blips that merely leave impressions rather than lasting concepts. While fast paced, activity-filled, programs may seem like a solution to preventing boredom, children are natural explorers whose learning and interest are more enriched when given ample time to play in any given activity. Today children commonly receive indirect cues from home (e.g., modern media and

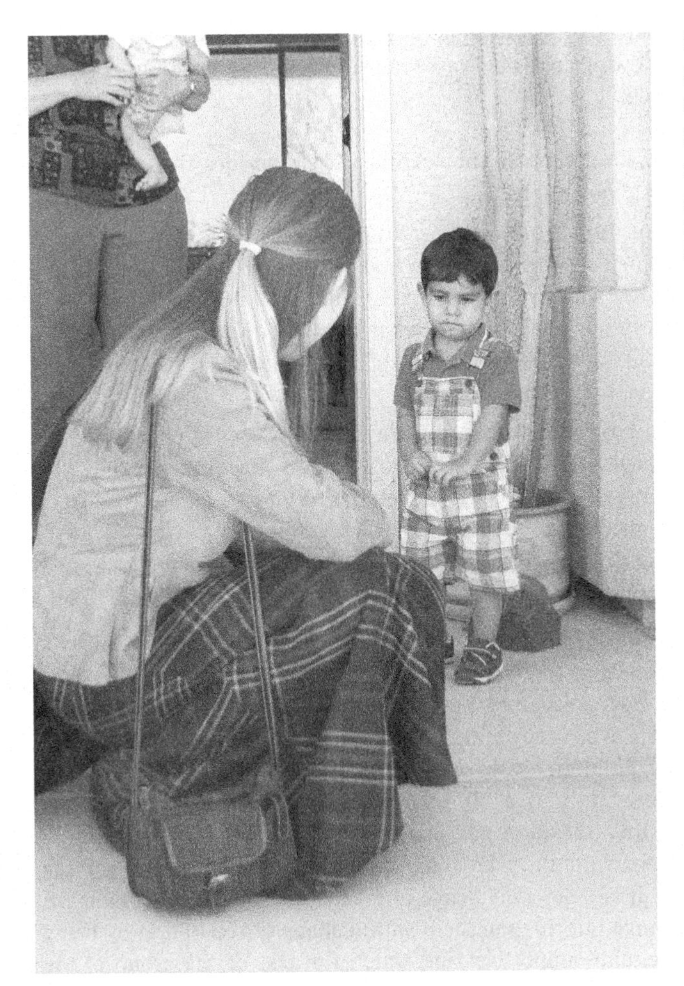

Separation anxiety can arise at arrival time. Children may feel insecure–and parents may feel ambivalent as well.
Lokibaho/iStock/Getty Images

fast-paced routines from busy families) to shorten their attention span, but a quality program can counter such cues through aptly paced schedules and flexible curricula that give them the freedom to explore their interests.

Transitions should be approached as rituals, but they sometimes feel chaotic, which is why thought and planning need to go into how they are handled. Although most of the ideas here relate to children of all ages, in some cases transitions for infants and toddlers are slightly different. To make transitions easier, make the world as predictable as possible. When children know what is going to happen, they can cooperate easier than if adult actions come as a surprise. Even very young infants, if they know what to expect, learn to respond in ways that help the adult. When cooperation becomes a habit from infancy, those uncooperative periods that are part of most children's development pass by more smoothly and quickly. Adults do not get angry because they know that the cooperative spirit is still there, and they appeal to it.

Transitions may be hard or easy, depending on the group and the adults' skill in handling them. The following are some ways to ensure smooth transitions between activities for older children:

- Warn children ahead of time so they can adjust to the idea that the current activity is about to end.
- Try to arrange transitions so there is as little waiting as possible. Problem behavior often coincides with waiting periods. There is usually a way to prevent lining up the whole group to wash hands. If children are called a few at a time, the others can continue what they are doing so they do not have to wait around. In some programs, children wait up to one-fourth of the day!
- Eliminate chaotic, crowded areas in the classroom by adjusting the schedule or rearranging the physical environment. If, for example, the cubbies are all grouped in the same part of the room and everyone heads for them at once, there may be shoving and crowding.
- Do not wait to start the next activity until every single child is present and attentive; instead, begin as soon as the first children arrive so something is already happening when the others get there.

Arrivals and Departures. A specialized kind of transition occurs when children arrive for the day and again when they leave. These moments are the biggest transitions of all, because the children leave one environment and set of people for another. Separation anxiety can play a big role at arrival time. Not only children may feel insecure during this transition but also parents.

Arrivals can set the tone for the whole day and create a juggling act for the early childhood educator. It is quite a challenge to greet children and parents, make them feel welcome, exchange information, help with feelings of separation by being understanding, reassuring, and supportive, and watch the rest of the group at the same time. A well-prepared environment helps a lot to entice children to interesting materials, toys, and activities and to make the separation easier.

If crying, protesting, or anger is part of the transition, it is important to accept the feelings and proceed slowly. Let both the child and the parent know you care about their feelings, and show confidence that the child will be fine. For some children, the parent's presence for a while eases the transition; for others, it is agony.

Some children are comforted by a transition object, such as a favorite blanket or toy. It may be something they bring from home or something they look for each day in the program. Some children are comforted by a transition ritual. For example, one child's transition ritual was to enter with her mother, be greeted by her teacher, find her name tag on the entrance table, say good-bye to her mother by giving her a kiss on the cheek and waving her name tag at her, run to her cubbie to put her things away, run back to the window to give a final wave of the name tag to her departing mother, stand at the window until the car disappeared, and then go to her teacher to have her name tag pinned on. The girl was fine on the days she was able to carry out her ritual. Any change, however, upset her, and she needed special support to leave her mother and to make the transition into the classroom.

At the end of the day or session, departure routines may also be emotional. Some children are delighted to be picked up, and others are upset. Sometimes the timing matters. The first child to go may not feel ready. The last child to go may feel deserted.

It is important to understand why a child might ignore the parents, run the other way, fall apart, start behaving in unacceptable ways, and protest about leaving when the parents arrive at the door. Do not automatically assume that the child does not love his or her parents or is afraid of them. (Inexperienced early childhood educators sometimes suspect abuse when they see a child resist going home.) There are a number of other reasons:

- The child may be involved in an activity that he may not want to leave.
- The child may resist *all* transitions, even going home.
- The child may be angry at the parent for leaving her all day; the parent's reappearance at night may trigger this feeling and cause the child to snub the parent as a form of punishment.
- The child may feel so relieved to see the parents again that all the stress he has been holding in all day suddenly comes out when he feels secure enough to fall apart.
- The child may test to see what will happen if two powerful adults are in the room and she behaves in an unacceptable way. Will they be so busy talking to each other that they will not notice? Will they be unclear about whose job it is to guide the child in this situation? Some children just want to find out what will happen.

It is surprising but not unusual that a child who complains and protests in the morning about being left does the same thing all over again when it's time to go home. See the *Tips and Techniques* box on page 292 for an example of this phenomenon.

Cleanup. Cleanup is a special kind of transition, and if adults look at it as a chore, so will the children. However, cleanup that becomes a habit is more of a ritual than a chore. For example, one program taught children that every activity has a cycle and that cleanup is part of that cycle. At first, the children needed to be reminded to finish the cycle when they abandoned toys or materials or got up from the table without clearing their plates and cups. Eventually, however, they began to think the same way as the teachers, and they had an unfinished feeling if they did not complete each cycle.

That is one way to look at cleanup time. Here is another: the scene is a preschool. A number of activity centers are set up in two rooms that open into each other. Above each activity center is a chalkboard with a piece of chalk hanging from a string, high enough out of reach so that only teachers can use it. On these chalkboards, the teachers write the names of the children seen using the different centers

Tips and Techniques

What's Wrong with Christopher?

"What's wrong with Christopher?" a student teacher in my practicum class once asked me. "He cries when his mother or father leaves him at day care in the morning and again when either of them comes to take him home in the afternoon."

"Maybe he has a transition problem," I replied. "Are there other times that he protests or cries? Does he have trouble at school following established routines, for example, sitting down for circle time or coming inside when it's time?"

"He does have all those problems you mentioned. It's hard to get him to quit doing something and start doing something else. His mother reports the same problem at home."

I gave this student teacher the following five suggestions:

- Do not push or prod. This creates resistance. Understand that this child takes longer to accept change and that transitions cause unhappiness. If you are a person who moves quickly through transitions without any difficult feelings, it may be very hard for you to understand someone who does not.
- Create a consistent routine. A predictable flow of activities might help him anticipate schedule patterns and develop coping skills and habits.
- Prepare the child for transitions by letting him know before they occur. For example, give warnings before cleanup, or alert him a few minutes before his parents' arrival.
- Allow plenty of time for transitions, but don't be too upset if that does not always work. Sometimes resistance to a change of activity lasts exactly as long as the time there is to resist.
- Most of all, try to be patient and understanding about this problem. Christopher may never come to *enjoy* transitions, but rest assured, the crying will eventually stop!

during the morning; when cleanup time comes, the teachers hold the respective children responsible for the area they played in. On the surface it seems like a clever system, but the result is that children try to beat the system.

Watch three little boys as they move through the classroom. They settle in the block corner and are about to build a structure when one spies a teacher heading their way. They jump up and run off before the teacher notes that they were playing there. Their timing is exquisite. She misses them completely. Then they are off to the dress-up clothes. They spend their time by the racks, discarding clothes on the floor after trying them on–always on the lookout to be sure they do not get caught. And they do not. They are long gone by the time someone notices the mess. Finally, a quick pass by the finger paint table allows them enough time to gush around a bit, but they do not stay long enough to get their name on the chalkboard.

These boys are very clever. They know what they are doing. By the time cleanup time comes, the rooms are in disarray, but nobody holds the boys responsible for anything. They grin triumphantly as they head for the outdoors.

Obviously, the policing approach illustrated in this scene is ineffective, but it is not just the approach to cleanup that matters. The physical setup of the environment can also facilitate the cleanup process. An environment that is orderly and has a place for everything is much easier to clean up than one that looks almost as chaotic when clean as it does when it is messy.

Some early childhood educators create a consistent signal to let children know that cleanup is about to start. Some sing a little cleanup song while the process is occurring to let children know that this is a special time reserved for one activity–cleanup. Other teachers get children to clean up as they go along, instead of making it a special period. One program eased into cleanup by tidying up areas the children were not playing in before the official cleanup time started. They then hung "closed" signs on the areas.

The "closed" sign is an example of how cleanup can be a "cognitive" activity. When children match toys to signs, symbols, or pictures indicating where they belong on the shelves, they are playing a type of matching game. When they sort out the housekeeping area, putting baby blankets in one drawer and dishtowels in another, they are categorizing. When they hang up kitchen utensils on a pegboard set up with hooks and silhouettes of the specific objects, they are working on visual perception and symbolic development. Counting parts of games or puzzles to be sure that all the pieces are there is another cognitive activity. These are only a few of the cleanup activities that contribute to emerging literacy and mathematical concepts. If you look closely, you will find many others.

Group Time

Group time (sometimes called "circle time") is often the highlight of a program's day–a favorite routine that becomes a ritual. Most programs, with the exception of infant programs and some toddler programs, have at least one time in the day when children are gathered into a group (or several groups). Group time may be upon arrival. In child care situations, however, children do not always arrive at the same time; group time may occur mid-morning–either before or after playing outside–right before or after lunch, or at the end of the day (or even several times throughout the day).

group time A period during which children come together to participate in some specially planned activities, such as singing, storytelling, movement activities, or discussions. The content, duration, and frequency of group time (sometimes called "circle time") vary according to the children's age and developmental needs.

The goal of group time varies among early childhood programs. If the focus of the program is to promote a feeling of belonging to the group, the children may come together at least once or twice a day or session.[2] Some early childhood educators see group time as preparation for kindergarten. They believe that children need practice in sitting, listening, and participating in group activities. Other professionals see it primarily as a teaching or discussion time. Still others design their group time experiences to relate to the "whole child": they offer music and movement activities, cognitive and perceptual games, and story time (both storytelling and reading aloud from children's books).

An antibias curriculum might use group time as an opportunity to talk about issues relating to culture, race, gender, and abilities.[3] Teachers use "persona dolls," to discuss real or pretend incidents and ask children to solve problems and engage in value clarification: "How would you feel if you were Maria?" or "What should

Group time is sometimes the highlight of a program's day.
DGLimages/Shutterstock

Marco do about that situation?" or "Is that fair?" Each of the persona dolls has particular attributes (of race, culture, ethnicity, gender, class, or abilities). Sometimes teachers make up ongoing stories about the dolls to expand the children's horizons.

Infants. Some infant programs have a "circle time" with singing and simple movement games or activities, but such group times are more for adults' enjoyment than for the infants' entertainment or learning. Infants benefit most from group experiences that arise spontaneously, as they find each other and adults while they are on the floor. Nonmobile infants are dependent on others coming to them to create the "group experience."

Toddlers. Toddlers benefit from short-term group experiences if they are free to come and go and if the group is very small. It is important not to cxpect groups of toddlers to sit still and listen for more than about 10 minutes. In fact, many programs wait until preschool to start a regular routine of calling a group together.

To keep toddlers present and attentive requires some powerful techniques, some of which may send negative messages to children about their natural urges to explore, leave and come back, touch each other and the things around them, and move around. Toddlers should feel positive about what they do naturally, and curbing their instinctive behaviors is a disadvantage of trying to have a group experience that resembles a preschool circle time. Indeed, many group times occur spontaneously in toddler programs: a child brings a book to the teacher. When she sits down to read, she suddenly has a lapful of children all anxious to see the book. Most will not stay long, but some may stay to the end of the book and beyond.

Preschoolers and School-Age Children. Some early childhood educators use group sessions for talking, problem solving, and discussing. Nevertheless, it is important to have appropriate expectations for younger children. Most three-year-olds cannot sit still in group sessions much longer than two-year-olds can. Four- and five-year-olds do much better in discussion sessions than younger children. Of course, most school-age children have the maturity and the experience to be good group members and to get a lot out of group times.

To be successful, a group time should be

- Adapted to the age group.
- Suited to the attention span of the group.
- Planned with age-appropriate, well-timed activities that all the children have an opportunity to participate in.
- Conducted in an area big enough for the children to sit comfortably apart either on mats, chairs, or a rug and so they can all see the teacher.
- Planned ahead of time so that the children do not have to wait for the teacher to locate and organize supplies or musical and visual aids.
- A choice, ideally–according to the children's needs and developmental abilities.

Group time is preceded by a transition during which children move from whatever they were doing to join the group. Skilled teachers begin with some preliminary songs, finger plays, or activities that involve and welcome children as they arrive individually or in clusters to the group. It is easier for children to join a group that is already engaged in something interesting than it is to sit down until the entire group is convened and wait for something to happen. It is also easier for children to behave in socially acceptable ways in a group that is actively involved rather than in one that is waiting.

A Story to End With

I am a person with a transition problem, so I understand those little kids who never want to stop what they are doing to start something else–least of all a meal or a nap! And even when nap time is over, getting up is just as trying as settling down to sleep.

It has taken me a lifetime to even begin to enjoy getting up in the morning or to look forward to going to bed at night. I hate getting ready for a vacation, and I have difficulty getting back to my normal life when I return home. I put off taking a shower, but once I am in, I do not want to get out. In other words, I have trouble getting started and I have trouble stopping. I am not good at transitions. Even my birth reflects this theme of resisting change; I put off being born as long as I could. Two weeks past the due date I finally made my entry into the world. Talk about dragging your feet!

I still struggle with change: letting go of a relationship, moving, leaving one job for another. Those are big changes, and, like most people, I find them painful. But I also suffer over little transitions. Even getting up from the table when I am through eating is something I put off doing.

My transition problem makes me look like a "dawdler." Because I take so long making changes, I am often late when I need to be somewhere. To an objective observer, it might appear that I just need more time to get going. Starting earlier should help. But it does not. Because I am resisting the *change,* I resist for exactly the amount of time I have to resist. I can give myself a whole extra hour, but I am still late because I use the extra time for getting ready to put off leaving. Getting me started sooner does not get rid of my dawdling; it just prolongs it.

I am hopeless. But the advantage to having this problem is that I understand perfectly children who have the same problem. We just do not like change!

Summary

Caregiving is part of the curriculum, not something that has to be gotten out of the way to get to the curriculum. Caregiving routines provide opportunities for synchronous interactions, which lead to attachment. Attachment is important in early childhood programs for the security and well-being of the children and to enhance their opportunities for learning. The early childhood curriculum includes four physical-care routines–feeding, toileting, resting, and grooming and dressing. As always, cultural variations enter into the administration of these routines. Two other types of routines are transitions and group times. Transitions are those bridges between one place and another or between two activities. The most emotional transitions and the ones that deserve special consideration from both parents and caregivers are arrival and departure times. Group time is considered as preparation for kindergarten by some; for others it becomes a time for discussion and the presentation of new experiences.

Reflection Questions

1. Are you a by-the-schedule person? Why are you the way you are? How well do you think you would fit in most early childhood programs? How do you feel about your answers to these questions?
2. Remember a time when you experienced synchronous interactions–when you were in sync with someone. Describe it. How did that feel? How did you benefit? Can you use your experience to help you recognize synchrony in caregiving?
3. What are your personal experiences around feeding issues? How can you use your experiences to help in your work with young children?
4. Are you a napper? Were you always a napper? How do your sleeping habits relate to your understanding of the sleeping needs and program demands the early childhood educator may have to juggle?

5. Do you remember your own toilet training? Do you think you have any issues left over from it? What are your ideas and feelings about toilet training other people's children?
6. Are you a person who has "transition problems"? Can you understand a child who does?
7. How do you feel about order? To what degree should order be kept in young children's environments? To what degree should the children themselves be held responsible? What were your own childhood issues about "picking up" (if any)?

Terms to Know

How many of the following words can you use in a sentence? Do you know what they mean?

synchronous interaction 276
attachment 278
primary caregiving system 279
transitions 288
group time 293

For Further Reading

Balaban, N. (2011). Easing the separation process for infants, toddlers, and families. In D. Karalek and L. G. Gillespie (Eds.), *Spotlight on Infants and Toddlers* (pp. 14-19). Washington, D.C.: National Association for the Education of Young Children.

Butler, A. M., and Ostrosky, M. M. (2018). "Reducing Challenging Behaviors during Transitions: Strategies for Early Childhood Educators to Share with Parents," *Young Children, 73.4* (September 2018). Retrieved from www.naeyc.org/resources/pubs/yc/sept2018/reducing-challenging-behaviors-during-transitions.

Collins, M. C. (2010). ELL preschoolers' English vocabulary acquisition from storybook reading. *Early Childhood Research Quarterly 25* (1), 84-97.

Crawford, M. J., and Weber, B. (2013). *Early Intervention Every Day: Embedding Activities in Daily Routines for Young Children and Their Families* (Baltimore, MD: Brookes Publishing; 2013).

Crawford, M. J., and Weber B. (2016). *Autism Intervention Every Day: Embedding Activities in Daily Routines for Young Children and Their Families* (Baltimore, MD: Brookes Publishing; 2016).

Duncan, S. (2011). Breaking the code–Changing our thinking about children's environments. *Exchange 33* (4), 13-17.

Elliot, E., and Gonzalez-Mena, J. (2011). Babies' self-regulation: Taking a broad perspective. *Young Children 66* (1), 28-33.

Florez, I. R. (2011). Developing young children's self-regulation through everyday experiences. *Young Children 66* (4), 46-51.

Kersey, K. C., and Masterson, M. L. (2011). Learn to say yes when you want to say no to create cooperation instead of resistance. *Young Children 66* (4), 40-44.

Kovach, B., and Patrick, S. (2012). *Being with Infants and Toddlers*. Tulsa, OK: Laura Briley.

Mardell, B., Rivard, M., and Krechevsky, M. (2012). Visible learning, visible learners: The power of the group in a kindergarten classroom. *Young Children 67* (1), 12-19.

Thelen, P., and Klifman, T. (2011). Using daily transition strategies to support all children. *Young Children 66* (4), 92-98.

Vesely C. K., and Ginsberg, M. R. (2011). Strategies and practices for working with immigrant families in early education programs. *Young Children 66* (1), 84-89.

11 Developmental Tasks as the Curriculum: How to Support Children at Each Stage

What Children Need: A Broad View

Developmental Stages

Young Infants

Mobile Infants

Toddlers

Two-Year-Olds

Three-Year-Olds

Four-Year-Olds

Five-Year-Olds

School-Age Children

A Story to End With

In This Chapter You Will Discover

- why this chapter is not called "getting ready."
- what children can do and what they need at eight different developmental stages.
- why developmental charts alone are insufficient for determining what children need.
- two other bases of knowledge that determine what children need at what stage.
- how developmental expectations differ from culture to culture.
- how children with special needs are children first and children with disabilities second.
- how to help a young infant develop trust.
- what to do to support mobile infants' exploration urges.
- how to expand toddlers' horizons.
- what increasing autonomy looks like in two-year-olds.
- how to aid the three-year-old increase in competence.
- how to support a sense of initiative in four-year-olds.
- how to expand the world of the five-year-old.
- how to champion the school-age child as a learner.

In this chapter, we will look at the sequence of a young child's development in terms of eight unique stages. The stages encompass the years from birth through age eight: (1) young infants, (2) mobile infants, (3) toddlers, (4) two-year-olds, (5) three-year-olds, (6) four-year-olds, (7) five-year-olds, and (8) school-age children. This chapter will explain what children need in terms of adult support and resources at each stage in their development. Research in child development is comprehensive, and from it we know what to expect from children and how to meet their needs at any given stage.

NAEYC Program Standards

Program Standard 1: Relationships
Program Standard 2: Curriculum
Program Standard 3: Teaching
Program Standard 4: Assessment

The National Association for the Education of Young Children (NAEYC) places great value on developmental perspectives. The Program Standards for the NAEYC Accreditation use the word *development* (or *developmental*) in all but one of their standards, either in the standard itself or in the rationale for it. The only exception is Standard 5, Health. Further, the word *develop* (or *developmental*) occurs 19 times in the remaining nine standards. The words we use show our values and our perspectives. The same is true for organizations. Remember that everyone in the world does not use concepts of development to explain how children grow and change. This does not invalidate NAEYC's standards; it just reminds you that respecting diversity means honoring multiple perspectives on even the most basic well-researched concept, such as developmental patterns.

This chapter could be called "Getting Ready." A major question parents of preschoolers have for their children's teachers is, "Are you getting my child ready for kindergarten?" Parents of kindergartners worry about whether or not the program is getting their children ready for first grade. Parents of first-graders have the same worry about second grade, and so forth. The developmentalist's idea of "getting ready" may be different from that of the general public. What the preschool teacher is doing is giving the child a good preschool education because that is what is needed for kindergarten readiness. Four-year-olds are different from five-year-olds. They do not need a kindergarten program; they need a preschool program. The same goes for kindergarten. Getting ready for first grade does not mean doing first-grade work. It means doing kindergarten work.

trickle-down effect The result of expectations and approaches appropiate for older children appearing in programs for younger ones in the name of getting them ready for what is to come.

A **trickle-down effect** occurs when expectations and approaches appropriate for older children begin to appear in programs for younger ones. In other words, children are taught to do things they are not ready to do in the name of getting them ready, similar to expecting infants to learn how to talk or walk before they can barely coo or crawl. Obviously, such an activity would be ridiculous for infants. Yet for some parents, making toddlers or two-year-olds sit still through a long circle-time "calendar session" does not seem ridiculous. The fact that some children may be learning to recite the days of the week does not mean much when you consider that children that age do not even have a clear concept of the difference between today, tomorrow, and yesterday. Those squares and numbers on the calendar, no matter how cute and attractive, do not mean a thing to them.

James Hymes, a pioneer in early childhood education, emphasized often that the way to get a child ready for the next year is to encourage them to do what they need to do right now. Each stage of development has a particular set of characteristics and limitations. Instead of pushing the child toward the next stage, urging the crawler to walk, for example, celebrate the crawling. When children have done thoroughly what it is they need to do at each stage of development, they will automatically move on to the next stage. Early childhood educators should not jump ahead but instead give children a good grounding in what they can do and understand. Pushing children into realms beyond their abilities disregards information we have about the differing needs of each developmental stage. This chapter will teach you how to prepare children for what is ahead by very thoroughly meeting their needs at their present stage of development.

This chapter does not take the place of a child-development course. Every student who prepares for a career in early childhood education should study child development. In a child-development course, students learn about each of the stages of development and the particular characteristics that each stage has as well as its limitations and challenges. The goal of this text is to examine the role of early childhood education in responding to those characteristics and challenges.

What Children Need: A Broad View

Developmental stages are important for the early childhood professional to grasp and incorporate in his or her work with children. Every early childhood professional should be able to answer the question, "What do four-year-olds need?" for example, with some specific developmental information. However, you should never rely on developmental charts alone to determine what the individual child needs at a given stage. Developmental information is important, but it is limited because it represents only one body of knowledge. You also have to take into account individual differences, familial, cultural, and societal context, as well as the immediate context in which you ask the question.

When you ask the question "What does this child need?" and consult your knowledge about the individual, the answer you come up with may differ from the answer that relates to the child's developmental stage alone. Maybe this child just gained a baby brother and he is exploring becoming a baby again himself, or–an even more traumatic example–maybe his mother just died and he is acting more like a two-year-old than the four-year-old that he is. He will not need what all four-year-olds need right now. His needs are individually determined by the circumstances.

While taking the *individual* into consideration, you must also look at cultural information. You cannot make decisions about what a child needs without looking at the family context. What is the culturally appropriate way to look at this child's needs? How do family goals, perceptions, and beliefs influence the information we have about ages and stages?

To summarize, the three bodies of knowledge you must always consult to make decisions about what children need are

Developmental ages and stages

Individual differences

Cultural context[1]

NAEYC Position Statement
on
Developmentally Appropriate Practice

To show how these three bodies of knowledge interact, let us consider a few examples–the first two of which will look at two children who encounter similar situations in two different cultures.

Imagine a cautious, shy, fearful three-year-old girl. She lives in Italy, and today is her first day at preschool. She arrives outside the door with her eyes downcast. She does not look at the teacher, who is sitting by the door, ready to greet her. The shyness will be taken into account, but the girl will still be expected to greet adults in a socially accepted way; greetings are considered an appropriate developmental expectation for a three-year-old. Shyness is an individual variation in this particular child but is not considered an excuse for being disrespectful to elders. As she enters the room, either her parent or her teacher–or both–will make sure she goes through the motions of a proper greeting.

Now imagine that the girl is a cautious, shy, fearful three-year-old living in the United States. This is also her first day at preschool. She too enters the room with her head down, but instead of being stopped to greet the teacher, she walks by her. The teacher in this scene will say hello but not expect or push for an answer. She will excuse the girl in her own mind, both because the child is shy and because she is young. The mother will likely say something to reassure the teacher that when her daughter feels more comfortable and is ready she will naturally respond to greetings in her own way.

Finally, let us explore another example of the American early childhood cultural context. Let us say the child isn't a shy three-year-old, but a lively, outgoing four-year-old. She waltzes into the classroom and, without a glance, runs right by the teacher, who is crouched down by the door ready to greet her. She immediately joins a conversation with two children at the play dough table. Most likely, no one will make a fuss about the girl's ignoring the teacher. The teacher and parent will probably credit the child with being an independent individual who demonstrates initiative.

As the last two examples illustrate, respect for elders takes a second place to other values in American early childhood culture.[2] The *Focus on Diversity* box on the next page contains more information about how the cultural context affects adult perceptions of developmental stages and behavioral expectations.

As you can see, developmental stages by themselves provide an incomplete picture of the whole child; they are only a rough guide based on averages. They say nothing about individuals–and each child is an individual, including (but not limited to) children with special needs. Children with special needs are children first and children with disabilities second. It's important for those who work with children who have special needs to understand the normal sequence of development and to emphasize the *child* in all children. But remember to use developmental norms only as guidelines, not as a strict ruler. Avoid comparing the children to norms. Professionals with special training in assessment can use developmental norms to chart and plan for progress, but it is important for a beginning teacher not to worry about where a child "should be." Instead, learn to see what that child can do, and help him or her develop those abilities. Work on immediate needs rather than push toward a distant developmental goal.

FOCUS ON DIVERSITY

Culture and Developmental Timetables

Culture affects three aspects of child rearing: beliefs about what children can and should do, parental values and goals, and child-rearing practices. Combined, these three aspects affect adult expectations about when and what children should do.

Research shows that mothers' beliefs about developmental timetables are influenced by their culture.[3] Mothers expect their children to show behaviors their culture values earlier than behaviors their culture does not value. For example, a group of mothers in San Francisco expected their children to become verbally assertive and to cope with their peers much earlier than they expected them to show emotional control, courtesy, and obedience. In contrast, mothers from Tokyo had very different ideas about how early emotional control, courtesy, and obedience should appear; because they valued these behaviors, they expected them to appear much earlier than did the San Francisco mothers. The Tokyo mothers did not put such a value on verbal assertiveness or coping with peers; they expected those behaviors to develop later.

Another example of contrasting values involved Italian mothers from a city near Rome and a group of mothers from Boston. The Italian mothers expected crawling, self-feeding, and unsupported sitting to occur later than the Boston mothers expected these behaviors. The first group seemed to discourage the behaviors, while the second one seemed to encourage them.

When teachers from two cities in different countries (Amherst, Massachusetts, and Pistoia, Italy) were tested on their expectations of developmental timetables, it was clear that both experience with children and training in child development shaped their perceptions. Culture was a factor too, but not to the extent that it was with parents. The two sets of teachers agreed on developmental sequence; only the timing was different on some specific items. For example, the American teachers thought that the skill of cutting with a knife occurred around four years, while the Italian teachers thought that it occurred more than an extra year later. The American teachers expected social skills with peers to develop earlier than did Italian teachers, while the Italian teachers expected social skills with adults to develop earlier than did the American teachers. In addition, the two cultures held different values regarding these skills. The American teachers placed greater value on independence and individuality, while the Italian teachers saw family solidarity as something important to emphasize.

Developmental charts may or may not be a help in understanding an individual child with special needs, especially one who does not fit into any category. The sequence of development may be the same (though not always), but the ages may vary greatly. Also, the family of a child with special needs is like any other family in that they are raising their child in a cultural context that must be taken into consideration. A child's skills and progress can be different in a family who values independence and self-help skills from those of a child whose family values interdependence among family members. Whether the family regards the situation as a blessing or a curse can also make a difference in how they respond to the experience of having a child with special needs. The family who sees lifetime dependence as a curse may work harder toward promoting independence, or, if that same family believes the child will never be completely independent, they may give up and institutionalize the child. Conversely, the family who sees dependency as a blessing might not push for independence or feel the need for permanent care for the child away from home. It is important to remember that there are no rights or wrongs, only differences in perspective that require understanding from early childhood professionals.

Developmental Stages

The rest of this chapter is concerned specifically with the body of knowledge about developmental stages–a key tool for understanding how to respond to children's needs. Let us start with infants.

Young Infants

Young infants are defined as babies who are not yet mobile and range in age from newborn to somewhere between 5 and 11 months. Infants of this age are dependent on others to fulfill their needs. Typically their development is rapid, as they go from barely being able to lift their heads to eventually rolling over and getting up on their knees to crawl around.

cephalo-caudal development The developmental pattern of human beings that proceeds in a head-to-foot direction.

proximal-distal development The developmental pattern of human beings that progresses from the middle of the body out to the extremities.

Physical Development and Learning. Infants' development is from head to feet and is termed **cephalo-caudal development;** that means infants first gain control of the muscles located in and around the head and then work their way down. Development is also **proximal-distal,** which means from the middle out toward the extremities. You can observe these two progressions easily: infants lift their heads before their chests, and they control their arms before their hands and their arms and hands before their feet.

From birth to about the first six months of life, infant movements are reflexive; their muscles automatically respond to certain stimuli. Arms wave and hands grasp and hold on without conscious intention. The Palmer grasp reflex, for instance, commonly occurs when an object is placed in an infant's open palm, causing the hand to reflexively close on the object. Infants, however, have no control over their reflexes. Although it is tempting, don't put rattles into the hands of very young infants! Until the grasp reflex releases, they are stuck with the rattle, like it or not, and may end up banging their heads. They have no control.

Toward the end of the first four to six months of life, many reflexes begin to fade and infant movements become more intentional as they respond to objects of visual interests in their environment with the maturation of their body and sight. They begin to reach and grasp objects at will, a progression in the development of their respective gross- (i.e., large physical movements) and fine- (i.e., smaller coordinated movements) motor skills. Indeed, infants are sensory learners, particularly through their mouths, as theorized by Sigmund Freud (see Chapter 1). Educators

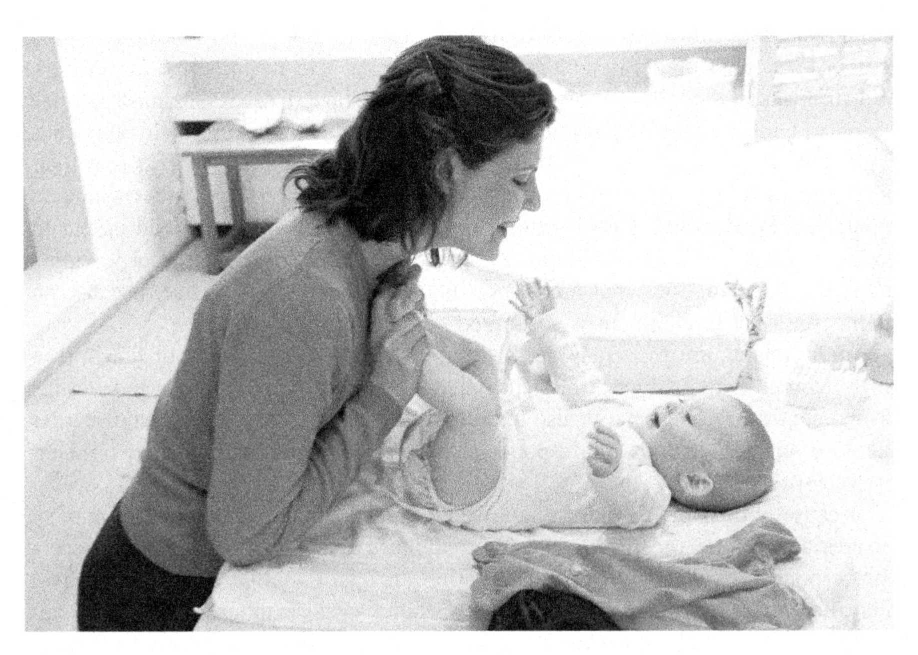

An adult-infant "conversation."
Eric Audras/Onoky/Corbis

and caregivers can encourage gross- and fine-motor development by making safe and age-appropriate objects of varying shapes, sizes, and textures available and accessible in the nursery. Expect everything to go into their mouths. And remember, eating is one of their greatest needs and pleasures. It should be a prompt and satisfying experience because it not only responds to physical needs but also helps them learn trust.

For sleeping, infants need a cozy, comfortable crib or bassinet, although it is important to understand that cribs are not necessarily required baby equipment in all cultures. In this country, however, it is generally agreed that infants need cribs that meet federal safety standards for safe, secure, and undisturbed sleep. Keep in mind that cribs are designed for sleep. For cognitive stimulation and development, infants learn best when given the freedom to move and explore in a developmentally appropriate play space.

Newborns prefer being tightly bundled in a blanket or swaddled and tucked in warm arms or a small confined space whether awake or asleep. But as they mature, they need more room to maneuver and a wider horizon to view. When awake, young infants should spend more time in a playpen or on a mat or blanket located in a clean, safe space so they can feel free and move unencumbered. In the Pikler Institute in Budapest, at three months of age babies are considered ready for larger spaces. They are placed for periods in a large playpen with other babies. The surface is carefully considered–it must be wood and hard enough that the baby can feel the effect of gravity. The hard surface covered with a light cloth gives the baby the ability to move freely and though they do not go anywhere at first, they do move their bodies. Soft rugs, foam pads, and quilts hamper movement.

Social-Emotional Development and Learning. Healthy social and emotional development for young infants is contingent on nurturing and responsive relationships. As Erik Erikson theorized in his first psychosocial stage, trust versus mistrust (see Chapter 1), infants' most paramount need is to form attachments to their caregivers. The health of such relationships helps them determine the level of trust that their needs will be met in their environment. This will affect how they will respond to people and uncertainty as they grow. This means that caregivers should attend to feedings and other caregiving duties promptly with affection, while also providing ample warm interactions that are respectful, sensitive, and responsive, because an infant's physical development is so limited.

Young infants are extremely nearsighted in the first few months of life, with a range between 8 to 15 inches of sight, but they respond well to faces and objects of high contrast. Young infants also learn to discern higher-frequency sounds earlier than low-frequency sounds and respond well to motherese or heightened-pitch infant-directed speech.[4]

Because of their physical and cognitive limitations, infants learn trust by making gestures and noting the responses of the adults in their environment. Caregivers who respond promptly with sensitivity within an infant's visual and auditory field build the foundation for trust and forming healthy social responses. Back-and-forth responses may continue for several minutes, and such interactions help infants learn the effects of their gestures and fortifies their bond with their caregivers.

Infants also respond well to talk, particularly from those to whom they are attached to. Motherese, or infant-directed speech, in response to infant's smile, coo, or cry further strengthens the infant's trust in his or her social environment by providing comfort and reassurance of safety, particularly in strange situations such as the introduction of new faces or transitions.

It is important to note that while infants may smile and make eye contact, smiling is a reflex in the early weeks of life. Intentionally social smiling typically emerges first during the sixth week of life and helps young infants start to learn how to generate positive responses from those in charge of their care. Moreover, infants can be overstimulated or tired easily during interactions. Be mindful of the physical signals for rest such as fussiness, sleepiness, or visual avoidance (e.g., averting eyes or turning the head away).

At about six months of age, infants' visual and auditory acuity will have improved, widening their options for exploring their surroundings; their willingness to do so will be partially affected by how safe they feel in their environment. Over time, the depth of the trust and responsiveness in their relationships will help infants distinguish the people in their environment. Infants will develop a natural wariness of strangers, and caregivers should provide consistently safe and responsive care to cultivate healthy social emotional growth as they transition through their developmental milestones.

Cognitive Development and Learning. As previously implied, infants are sensory learners, so much of their cognitive development is connected to their social-emotional and physical development. Infants rely heavily on their relationships with their primary caregivers to not only fulfill their basic needs, but also to provide the resources and responsiveness required to explore the people and objects around them. For instance, during the early months of infancy, cognition lies in the muscles and senses, a developmental period that Piaget called **sensorimotor cognition.** Because young infants have a need and desire to explore the physical world, their physical positioning is important. Placed on their backs, babies have a wider view of what is around them and can hear with both ears at the same time. Lying face up also gives them more freedom to use their arms and hands. Simple objects of high contrast with viewing distance, such as a mobile, encourage young infants to explore.

sensorimotor cognition A cognitive stage described by Jean Piaget that occurs from birth to age two. During this stage, children's cognition develops through movement and sensory exploration of the physical world.

While young infants need sensory input and a physical world to explore, excessive sensory stimulation ignores their developmental needs. Avoid toys that move too fast, make discordant noises, or have many bright lights. Human contact is one of the best sensory experiences for the early weeks and months. It is also never too early to talk, read, or sing to infants; during the first six months of life, they are constantly listening and are statistically analyzing the sounds of human speech of both native and non-native language to create a nascent mental model of the language environment.[5] Language development begins long before an infant can utter his or her first word. As the first year progresses, infants will gradually develop the physical and cognitive dexterity to reach out to toys and other objects that they will enjoy manipulating.

Infants will also start the process of developing object permanence, or the ability to recognize that objects and people exist even if they cannot see them, a concept first introduced by developmental scholar Jean Piaget. Infants who have yet to fully develop it often show surprise and delight when adults "disappear" and "reappear" in games like "peek-a-boo." Contemporary research, using more nuanced techniques, have challenged Piaget's time frame for gaining object permanence, suggesting that some infants have some understanding of object permanence as early as 3.5 months as opposed to 18 months.[6] Nevertheless, it is a crucial milestone. It suggests that children can conceptualize objects in their mind, a precursor to other advancements in cognition.

Attachment also relates to cognitive development (as well as physical and social-emotional development). Treating a parent or primary caregiver differently from others is an attachment behavior that signals cognitive development. Stranger wariness is another. Both behaviors show that babies can differentiate familiar people from unfamiliar people.

Mobile Infants

A whole new world opens up to infants once they become physically able to move around. Their mobility differentiates their needs from those of younger infants. The age range for beginning mobility is from 5 to 11 months, and the average is about 7 months, when the baby starts moving in some fashion, be it crawling, creeping (the most common), rolling, scooting, or whatever other means the child invents to go forward at will. This stage lasts until the infant can walk alone, which usually happens anywhere from approximately 9 to 17 months (this varies greatly from infant to infant).

Physical Development and Learning. Mobile infants, more commonly called "crawling babies," need space to move. A standard-sized playpen is too limiting and, therefore, an inappropriate environment for them. Because mobile infants love to explore, provide them with plenty of interesting objects and, most importantly, a safe environment for their explorations. Intensive childproofing is called for, especially in mixed-age settings, where crawling babies risk getting hurt. If the opportunities present themselves, they will pop small things in their mouths or put pencils up their noses (or in their eyes), so keep all small and sharp objects out of their reach. They will also poke things in electrical sockets and chew on electrical cords, so don't ever leave a socket unprotected or an extension cord plugged in the wall with one end loose. More than one baby has seen a cord dangling and popped it into his mouth, resulting in a terrible shock. It is not enough to simply monitor these things; *they must be covered up, put up high, or removed from the area.*

You may even think that you can teach babies not to touch things and that everything will be okay. Please reconsider. Mobile babies are in the "touching stage"; that is what they are programmed to do. If you have to inhibit that impulse in the interest of safety, the effects may be long lasting. Some children who are taught early not to touch or explore lose their desire–even their capacity–to experience the world in new ways, use their senses, and try novel experiences. Infants learn best when given the freedom to reach out and try things. Set up an environment so these natural urges are encouraged, not curtailed.

Also consider the need mobile infants have for upright exploration. Most will pull themselves up eventually and begin to cruise by hanging onto whatever is available. Plan the environment to include low rails, couches, tables–whatever provides safe support for children who are anxious to get upright but cannot yet do it on their own.

Furthermore, set up the environment so that infants do not constantly rely on adult help. Their developmental urges push them to learn new skills; they do not need an adult to teach them how to stand or walk. (See the *Tips and Techniques* box on the next page for a view on this subject.)

Social-Emotional Development and Learning. The mobile infant should be showing signs of attachment to a primary caregiver and using that person for a secure base from which to explore. Fear of separation begins at about nine months, and mobile infants may protest when being dropped off at the early childhood program. In addition, depending on staff turnover and how well the infant care team works together, mobile infants may also protest when their primary caregivers leave the room.

Infants at this stage need opportunities to become self-assertive and to practice beginning self-help skills. They also need adults to interpret the effects of their actions on others, to help them express their separation fears, to accept their feelings, and to assist them in developing coping skills. Adults should express honest feelings and provide good models for children.

Because mobile infants are sometimes willing performers, adults like to teach them tricks that they think advance social development such as waving bye-bye or playing patty-cake. But here it is important to remember the principles of respectful interactions. Do not ask infants to show off cute tricks, even if they seem to enjoy it because that is exploiting children for adult entertainment. Because adults commonly interact with infants this way, the last statement may come as a surprise to you. But think about it. Always try and place yourself in the child's shoes, ask yourself: "Would you like to be treated like a trained monkey?"

TIPS AND TECHNIQUES

How Much Help Do Babies Need to Learn to Move Around?

Magda Gerber, an American infant expert, says, "Don't put babies into positions they can't get into by themselves."[7] Babies should not be sat up before they can get to a sitting position on their own. They should not be put into a standing position until they are ready to get up on their feet without help. Never walk a baby around by holding her hands; let baby's first steps be on her own.

Gerber bases her ideas on research from the 1930s done by Emmi Pikler, a Hungarian researcher and pediatrician. Gerber was trained by Pikler, and from the methods and approaches that Pikler used, Gerber created a philosophy of infant care that is widely known in this country. At a time in the 1960s and 1970s when infant stimulation was the major approach to infant care, Gerber was one of the first to describe how to treat infants with respect. Her ideas, which were considered radical at the time, are now widely accepted in the early childhood profession.

"Never push development," says Gerber. Let babies develop in their own time and their own way. She feels particularly strong about milestones and developmental time tables. She urges parents and caregivers to be concerned about the *quality* of development in each stage, not the age at which it appears.

Cognitive Development and Learning. Mobile infants can begin to remember games, toys, and people from previous days. They can also anticipate people's return. They are beginning to work on what Piaget called **object permanence**–that is, they are coming to understand that things still exist when they are out of sight. They pull the cover off of something they have seen hidden. Playing peek-a-boo helps develop this ability. Peek-a-boo gives babies control over making people disappear and come back; as such, it is a game of power, even if it only looks like plain fun.

object permanence A cognitive milestone described by Jean Piaget that occurs in later infancy, when babies develop the understanding that objects and people continue to exist when they cannot be seen.

Mobile infants need an environment that allows for problem solving, exploration, and interesting experiences. Crawling babies enjoy moving freely around the floor and stopping any time they want to investigate a variety of interesting–and safe–objects. They can solve manipulative problems and should find plenty of opportunities to do so in their environment. For example, mobile infants enjoy taking things out of containers and putting them back in. They especially like real-world objects, such as pots and pans, wooden spoons, and so on. Infants also progress in their understanding of object permanence, particularly during this stage when they are able to manipulate objects on their own. For instance, an infant may hide objects in containers or use a blanket to hide himself to elicit a reaction from his caregiver.

Indeed, they are now interested in discovering the consequences of their actions, or cause and effect. Intentionality has set in; mobile infants try things out by conducting experiments, like throwing food on the floor, to see what happens next. Patience and understanding are necessities with children this age.

Mobile infants pay attention to conversations and may respond to words. Some older crawlers may carry out simple commands and use words such as "mama" and

"Checking In"

Watch Tasha. She's 18 months old and has been enrolled in an infant center for 6 months. When she and her mother arrive for the day, her mother sits down on the floor in the middle of the playroom for a few minutes to get Tasha settled in before she leaves.

Tasha is excited to be there. She rushes over to her caregiver, pulls on her pants, then continues on to check out some dolls that are enticingly arranged along one wall. All of a sudden she stops. Is her mother still here? She looks around the room. Yes, there she is. She runs over to her mother and throws herself on her lap. But only for a minute.

She gets up from her mother's lap and toddles over to see what is new in the playroom. She picks up a purse and is carrying it on her arm when another child approaches her and grabs the purse. That action knocks her over, and she sits down hard on her padded bottom. She looks surprised, then distressed. Right away, she starts scanning the room. Oh, there is Mom! She gets to her feet unsteadily and wobbles over for a pat and a hug. That is all she needed. Then Tasha is off to play again.

But wait a minute! Mom is getting up. She's saying good-bye. Uh-oh! But sweet relief–her caregiver comes over to her. Tasha raises her arms as if to say, "Pick me up." The caregiver does. "Bye-bye Mommy," the caregiver says for her. Tasha waves a little reluctantly and struggles to get down. She stands at the window, with tears coming to her eyes. Then she turns around and looks for the caregiver. There she is on the floor, near where her mother had been sitting. She runs to the caregiver and climbs onto her lap.

"dada." They can use intonation and may repeat a sequence of sounds. It is important that you talk to babies and read books with them.

Finally, because of their mobility, interaction with their peers is a new experience for crawling babies. They see each other as some of the "interesting objects" in their environment. And one of your responsibilities is to help them learn how to interact without hurting each other.

Toddlers

Infants become toddlers once they start walking. They may start as early as 9 months or as late as 17 months. The average age is 11.7 months. But remember, developmental stages are not exact; some children reach milestones earlier or later, outside the boundaries set by staged norms.

Physical Development and Learning. Toddlers need to move. At first, practicing walking takes all their focus. It may be hard for them to get up and down, so they do not spend as much time exploring objects on the floor as they used to. However, once they are adept at getting back down to the floor from a standing position, they continue on as before, exploring everything they encounter. Sometimes they like to carry things around with them.

When they first begin to walk, toddlers waddle, spreading; their feet to provide a wide base of support. After mastering walking, toddlers begin to run and can even climb stairs by holding on to the next riser or a handrail; going down, however, is harder for them. They need a variety of toys on low shelves to choose from, such as play people and animals, dollhouses, containers filled with objects, measuring cups, spoons, and so on. The environment should be predictable but also interesting; put out toys on a rotating basis to maintain their novelty.

Toddlers need plenty of opportunities to exercise, such as running, tumbling, and climbing, in both indoor and environments. Although some adults believe such activities can only be done outdoors (which may be true for older children), toddlers should be allowed to run inside, ride small, wheeled toys, and play on plastic slides. Learning to use their bodies and their rapidly emerging skills is extremely important for strengthening gross-motor skills and cultivating confidence and self-control. They cannot just sit around indoors and wait to go outside.

The environment should also be set up to encourage fine-motor development. Examples of fine-motor activities are stringing large beads, manipulating large LEGO blocks, and eating with their hands or a spoon. Taking off clothes is another fine-motor activity that some toddlers can do; however, putting clothes on may be too difficult for them.

Social-Emotional Development and Learning. Early toddlers are beginning to explore their newly discovered independence. With encouragement, toddlers begin to recognize that what they do can affect changes in the people and objects in their environment. While they might begin to ease their stranger anxiety, they still crave security from their primary caregivers. (See the *Tips and Techniques* box at left for an example of a toddler "checking in.") Attachment, both at home and in the early childhood program, allows them the freedom to explore. A sense of security is a vital requirement for exploration and growth.

Toddlers, with budding physical and cognitive abilities, depend on adults to provide necessary limits and guidance because they have yet to be able to recognize common dangers as they explore. The environment for young toddlers requires careful thought and planning because it is the third teacher, a concept originated from the Reggio Emilia approach. They understand more than they are able to say because their minds are attuned to absorbing language but their vocal structures have not fully matured. With a limited, albeit budding, vocabulary, they often use one word to represent an entire sentence. You may notice that toddlers play alone and along-side, but not with, other toddlers. They can follow instruction, though they may show signs of resistance quite often because independence is new and exciting.

Emotions during toddlerhood run high and quickly go from one extreme to another. Fears also weigh heavily in the life of a toddler and can create sleeping difficulties, separation problems, and other complications. Their major fears are associated with losing their parents. Hide-and-seek and "chase" games that they invent themselves give them some control over getting people back when they want them. Hide-and-seek helps reassure them that things do not disappear permanently.

And when they play chase, they prove to themselves that the adult wants them when he or she chases after them.

Keep in mind that toddlers, like infants, also tire from too much action but lack the ability to say so, which may result in outbursts and tantrums. Being responsive in such cases requires educators to recognize the signals of toddler fatigue; help toddlers wind down with a soothing story or song.

To facilitate social-emotional (and language) development, label feelings and emotions during challenging interactions that involve adults and/or children. This not only models healthy emotion regulation, but also helps toddlers begin building vocabulary required to express their feelings and talk through challenging times that might otherwise become outbursts and tantrums as they increasingly become more verbal.

Another major worry for toddlers is the functioning of their own body. They try to understand how it works and seek to control its functions. They also worry about losing body parts.

Cognitive Development and Learning. Self-directed play continues to be the optimal means of learning for toddlers. There are many types of play, but at this cognitive stage, toddlers are beginning to pretend by mimicking simple actions made by the adults in their world. This type of play will form the basis of more complex play that occurs between 24 and 36 months, such as symbolic play, or the ability represent an object with another unrelated object (e.g., banana as a phone), and sociodramatic play, or role-play with peers. Because they are gaining new abilities as they grow and mature, they also continue to remain physically active and curious, exploring, experimenting, and solving simple thinking problems–sometimes in ways that challenge your patience. Toddlers follow their curiosity, often without intentional goals, and, they can get easily distracted as something more compelling comes along. Toddlers have attained Piaget's object permanence: they know a thing exists even if they cannot see it. Because of the continuous advancements in memory and mental representational abilities from infancy to toddlerhood, you cannot as easily remove a forbidden object, give them a toy to play with, and assume they will forget all about the forbidden object.

Talking may slow down when walking begins, but not for long. Soon thereafter, toddlers rapidly add new words to their vocabulary, or what experts call the "word spurt" or "language explosion." Most love books and not only want to be read to but like to "read" themselves, holding the book, saying words or making sounds, and turning the pages. They will point to pictures and sometimes name objects in the picutres.

They imitate actions from memory, showing that they are progressing from infancy, when they could only imitate in the presence of a model. Thinking is still action oriented (i.e., children still need to use their bodies for problem solving most of the time), but they are beginning to build a store of symbols and images in their minds to prepare them for eventually being able to work out problems in their head. They learn a lot from solving problems, so it is important not to do everything for them. Indeed, educators best support learning through scaffolding, a guidance approach that nudges children toward a new solution or understanding through indirect means such as environment design or hints. Scaffolding is a concept related to the zone of proximal development (ZPD), a model of learning developed by Lev Vygotsky (see Chapter 1). The ZPD is the gap between what children can do on their own and what they can learn to do with a little help from a teacher or a more expert peer. Sometimes it may be tempting to rescue children from their problems, but by being aware of developmental milestones and individual abilities, educators best facilitate learning by encouraging toddlers to work through problems on their own and provide help when they cannot move beyond their frustration.

Two-Year-Olds

Two-year-olds are at an "in-between" age in early childhood education. Some early childhood programs are able to group two-year-olds together by themselves, which

Young children appreciate opportunities to explore and manipulate objects.
Ariel Skelley/Getty Images

is the most ideal setup. Other programs, however, often group them with infants and toddlers or with three- and four-year-olds. When they are grouped with younger children, they have to deal with an infant-toddler environment that doesn't have enough challenges for two-year-olds; they are simply too old for the younger group. Nevertheless, when they are grouped with preschoolers–the most common arrangement–they often get involved in power issues. Two-year-olds are already dealing with power issues that relate to their stage of development, and being lowest in the pecking order only compounds their struggles. Unless staff members or providers are extremely vigilant, some two-year-olds enrolled in preschool programs end up being aggressive four-year-olds; when (after a couple of years of being dominated) they are finally big enough to stand up for themselves, they go overboard.

Physical Development and Learning. Two-year-olds need space to walk, run, hide, chase, climb, crawl, tumble, push things, and ride wheeled toys. They may not be ready for pedals, but they like to scoot around. They need things to explore, throw, take apart and put back together, dump, and refill. With these kinds of needs, it is easy to see why they do not fit into a preschool environment so well. They would rather dump a puzzle to hear the clatter on the floor than sit patiently and put it together. A two-year-old might investigate the things on the science table quickly with the magnifying glass; but when he is finished, he is likely to take a big, old purse from the housekeeping corner, empty the science table into it, and take it outside and dump it into the sandbox. This is normal two-year-old behavior.

Two-year-olds are developing a major interest in how their bodies work. Bladder and bowel control eventually become issues. They are in Freud's anal stage and Erikson's stage of autonomy (see Chapter 1). The trick is to make toilet training an autonomous activity rather than a power struggle. When you can play a partnership role instead of trying to dominate the child, toilet training becomes toilet learning and proceeds more smoothly. Some time before they are three, most children are able to keep dry during the day. However, accidents happen, and you must have a forgiving attitude toward them.

Social-Emotional Development and Learning. Two-year-olds are best known for their growing sense of autonomy. They discover that they can assert themselves and that they don't have to go along with everything. Some say no every chance they get, especially if they hear no from the adults around them. Sometimes they do not mean it; they are just trying out their power. "Me do it" is another two-year-old expression of wanting to be an independent individual. And then there's the famous "Me! Mine!" which shows that two-year-olds are beginning to see themselves as possessors. Until they thoroughly understand that they can own objects, they have a hard time with the concept of sharing. They may also struggle with waiting and turn-taking. Two-year-olds need understanding adults in their lives who help them feel powerful and independent, yet still protect and treasure them.

Fears may affect two-year-old behavior, especially fears about separation. Fears about controlling body processes may also arise, especially when children are prematurely toilet trained–before they have the physical abilities they need. Some two-year-olds worry about gender differences: "Why doesn't she have a penis?" and "Will mine disappear?" Fear of the unknown affects everybody, but for two-year-olds, this fear may become gigantic. They try to figure things out, but without mature reasoning abilities, they end up with mistaken conclusions. They need you to help them sort things out and give them clear explanations. They need opportunities to play out their anxieties as well. Children at this age start to engage in sociodramatic play as a way to process their experiences, particularly if they feel scared or anxious. Props in a dramatic play area encourage them to set up a pretend situation that they can control.

If the two-year-old is dethroned as the baby of the family by a newborn, her behavior may be affected. Sometimes it takes until the baby begins to walk for the threat to be real, but often it is right in the first weeks or months that the two-year-old begins to act out. Regressing, or returning temporarily to an earlier stage of development, is a common response to dethronement. Perhaps the two-year-old's limited reasoning goes like this: "If I'm a baby again, maybe I'll get the attention my baby brother is getting now."

Two-year-olds need caregivers and teachers who have worked through their own control issues. Otherwise, the adult and child will butt heads as the adult tries to control the child and the child tries to take power into his own hands. Empathy begins to emerge, but two-year-olds are naturally egocentric; they have yet to develop the capacity to understand the world through the perspective of others. Two-year-olds and other preoperational children, according to Piaget, assume that everyone see the world as they do. Nevertheless, they begin understanding emotions and start to recognize distress in others; they may try to offer comfort.

Two-year-olds can play with other children, but they often engage in parallel play (see Chapter 4) rather than interact. In parallel play, two or more children play alongside each other talking to themselves at the same time, affecting each other's play. Parallel play is something like two students in a computer lab, each working on her own computer and dealing with separate problems. People at computers often talk out loud to themselves, and they sometimes connect to what their neighbor is saying and doing without interacting directly. Just as students in a computer lab need identical equipment, so two-year-olds need duplicate toys to aid in their parallel play. Having several of the same kinds of toys is better in a program for two-year-olds than having a wide variety of toys.

Cognitive Development and Learning. The reasoning of a two-year-old is based not on logic but on personal fears and wishes. They think wishing makes it so. If they are angry at baby brother and he gets sick, they worry that they caused the illness. They need adults who understand how their reasoning works to help sort out what really happened.

Two-year-olds' cognitive development is growing rapidly. Multiple physical and mental skills are blossoming. Their ability to use symbols and images to think is increasing, and they can make comparisons and see the relationships between two objects. Their vocabulary is growing rapidly every day and despite their short sentences, they can communicate effectively; and again, they understand even more than they can speak. Continue to label, read, and sing to encourage this vocabulary growth. Music, in particular, can cultivate culture and emergent literacy skills such as phonemic awareness (essential for decoding) through fun and play; word games, nursery rhymes, chants, songs, finger plays, and rhythmic activities can promote learning.[8]

Two-year-olds also need structure and a predictable routine. They need adults who know that the daily routines of eating, toileting, and grooming are important opportunities for children to learn about adults, the world, and their own abilities to control their bodies and their behavior. They need adults who (1) understand that play is learning and (2) provide appropriate objects and materials for them to play with.

Two-year-olds also need a predictable indoor and outdoor environment. They enjoy books of all kinds, flannel board stories, songs, finger plays, and art projects that allow them to explore various media without models, instructions, or any emphasis on the "product." Food as an art material in toddler and two-year-old programs is an issue because children of this age need to learn to distinguish between food that goes into the mouth and other objects and materials that don't. And although scented play dough, felt pens, and finger paints are popular, they can confuse some children under three who are not sure yet what can be eaten and what can't.

Three-Year-Olds

Three-year-olds have now conquered many of the challenges they faced as two-year-olds. They are increasingly independent without having to be so assertive about it. They may be easier to get along with and more cooperative than they were at two. They may also have increased confidence but still not the judgment to avoid unsafe behavior. If three-year-olds are grouped with older children, they may want to

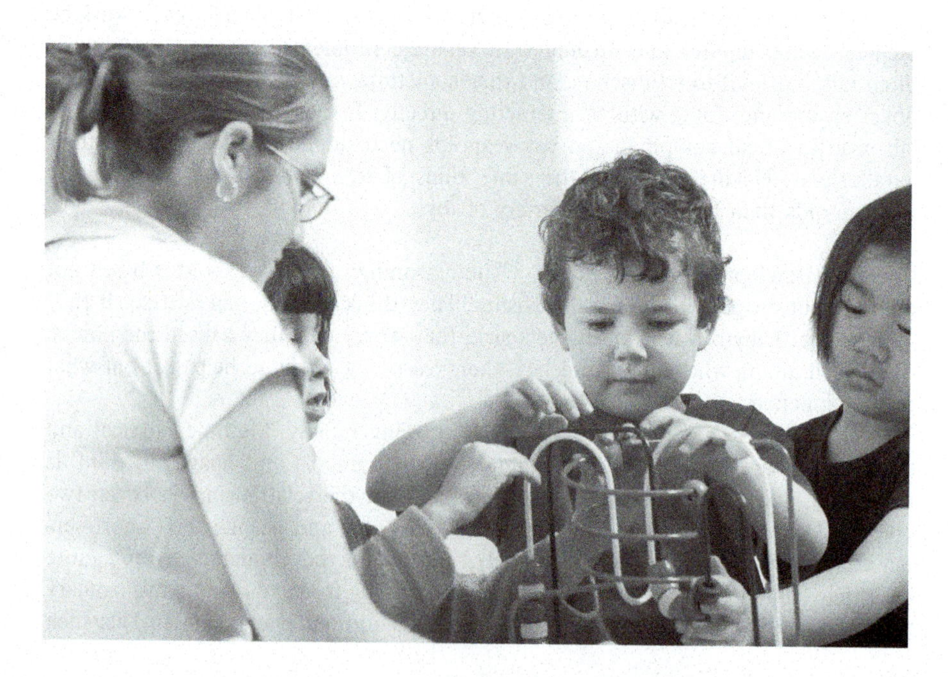

Three-year-olds are working on fine-motor skills. This child is very focused on moving those beads along the wire.
IT Stock Free

imitate the older children's behavior; they need to be closely watched for their own protection. They also need help to cope with the frustration of not being able (or allowed) to do what they want. It takes sensitivity to help them maintain their self-confidence *and* keep them safe at the same time.

Physical Development and Learning. Three-year-olds are working on fine-motor skills. They can put together a three- to six-piece puzzle and not be compelled to throw or dump the pieces. They begin to snip with scissors. Their self-help skills are blossoming! They can feed and dress themselves (but not yet handle buttons, zippers, snaps), and they need plenty of different opportunities to use these developing fine-motor skills.

Their gross-motor skills are increasing as well. They can pedal a tricycle a short distance, swing on a swing when someone gets them started, walk up stairs using alternating feet, catch a ball with two hands, and even turn a forward somersault. They need and want to practice these skills, so the environment should be set up accordingly.

Most three-year-olds are toilet trained in the daytime and beginning to be at night time as well. Toileting facilities should be conveniently located and available to the children on an as-needed-basis–not according to a schedule. They may still need some help with clothing, especially when they have urgent bathroom needs, so help should be readily available.

Social-Emotional Development and Learning. Three-year-olds are beginning to understand the concept of sharing and taking turns; they are not always willing, but they can, if they choose to. You should encourage sharing and taking turns, but do not always expect the child to comply. Three-year-olds have become more cooperative, and most comply with an adult's requests more often than not. They relate to other children and sometimes play with them, but at other times they may engage in parallel play. Although they can sit for longer periods and listen to a story or book, it is best to limit group time to small numbers of children and to short periods of time. Three-year-olds enjoy group activities most when they know they are free to come and go at will.

Three-year-olds have become slightly less egocentric, but they still have trouble relating to how others feel if they are not feeling that way too at the moment. They may take pleasure in destructive acts because they don't see them from any point of view but their own: squashing a sowbug has no more emotional meaning than cutting a piece of paper in half.

A sense of initiative is growing in three-year-olds. They have the imagination to dream up great enterprises, many of which won't work or must be limited or forbidden. It is important not to curb them too sharply or restrict them unnecessarily, as three-year-olds are beginning to discover who they are and what they can do. They need the freedom to choose and try to implement the projects they dream up.

Although there are growing signs of maturity in three-year-olds, they can easily revert to younger behavior; when they are upset, they may cry, suck their thumb, or lash out. Be tolerant of less mature behavior, and never rebuke them harshly or ridicule them.

Cognitive Development and Learning. Three-year-olds are growing in their ability to classify and name. Most know at least 10 body parts. They can arrange objects into categories and name the categories. They can distinguish big and little, long and short, heavy and light. In a rich environment with plenty to choose from, children are able to practice and increase these skills. They need objects and materials, other children, and adults who support them and facilitate their learning. They need adult support for their natural curiosity and opportunities to explore cause-and-effect relationships.

Three-year-olds do not usually go by rules because they do not understand or remember them. As such, guidance approaches should not depend on rules. (See Chapter 5 for more on this subject.)

Language skills are growing too. Three-year-olds can talk about the past, present, and future, and they can explain things in sequence. The classroom or family child care home should be a lively place with lots going on. Children should not be expected to keep quiet but should be encouraged to talk as much as they want to.

Four-Year-Olds

Four-year-olds are becoming more mature, but they still learn by doing. They need a variety of opportunities to interact with the external world in order to understand it in increasingly complex and rational ways. Although four-year-olds are moving toward kindergarten, it is important not to impose a kindergarten-readiness curriculum on them. Readiness approaches that are highly academic in nature ignore the children's needs in the here and now.

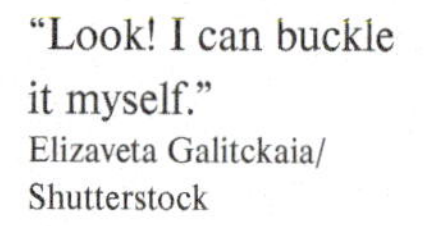

"Look! I can buckle it myself."
Elizaveta Galitckaia/Shutterstock

Do not make four-year-olds sit and listen in groups for long periods or do worksheets because that is what you think kindergarten will require. Instead, give them the opportunity to be a four-year-old–no more. Meeting the developmental needs of four-year-olds is the best way to prepare them for becoming five and going on to kindergarten.

Physical Development and Learning. Four-year-olds continue to perfect the skills they began acquiring as three-year-olds. They can use scissors with more control and can cut out simple shapes. They can work more complex puzzles, construct with blocks, and string beads, and they are beginning to draw recognizable pictures. Their gross-motor skills are also accelerating. They can run with ease and change direction; climb up and go down stairs using alternating feet; jump; stand on one foot; pedal a tricycle around corners and through mazes, even back it up into a parking place; and throw, bounce, and catch a large ball. They need plenty of opportunities to do all of these activities and more. There should be a good balance between gross and fine-motor activities, and, whenever possible, children should be able to move between indoors and outdoors on a regular basis daily.

Self-help skills are also growing. Four-year-olds need encouragement and opportunities to practice doing things for themselves. They can take care of many of their own needs skillfully, such as buttoning and unbuttoning their own clothing and zipping their own jackets. At meals and snack times, they can serve themselves and use knives, forks, and spoons. Most stay dry at night or get up to use the bathroom.

Social-Emotional Development and Learning. Four-year-olds also need encouragement to do things for other people. With their increasingly competent fine-motor skills, they can be a big help to adults by buttoning sweaters and zipping jackets of younger children. They also need encouragement and opportunities to cooperate. Adults who model and teach prosocial skills advance children in their abilities to relate to others in positive ways. Positive guidance techniques help children stay within the limits and gain a good self-concept at the same time. As with any age child, appropriate behavioral expectations are important.

If four-year-olds have success when they ask for help, they are encouraged to seek assistance when they need it. They can carry on conversations with adults and other children. They have increased attention spans; it is not unusual for them to work at something for 20 or 30 minutes or longer. They are more likely to ask for something they need or want from another child than just grab it. They understand when they have done something wrong and sometimes offer an apology without being told to. They take turns much more easily, especially when they have been taught to do so. They engage in cooperative activities and projects with other children. Socially acceptable behavior is a concept that has meaning for them.

Four-year-olds can spend more time in larger groups than three-year-olds, but most of their time should be spent as they choose–either alone or in small informal groups of their choosing. Children of this age start to prefer playing with peers. Children may be excluded or rejected from a peer group, and educators should use guidance strategies to help such children process negative feelings and think of coping strategies. Also, keep in mind that large-group, teacher-directed instruction is still developmentally inappropriate.

Cognitive Development and Learning. Four-year-olds have acquired a firmer grasp of number, shape, and size. Most can recognize and name familiar colors. They can retell a story and keep the facts and sequence straight. They remember the words to songs and can sing them. They are gaining an understanding of where objects are in relation to each other and can differentiate between "beside" and "on top of," for example, depending on their experience. Most can understand and carry out a series of three directions (though they may not always be willing to do so).

Although four-year-olds are showing increasing cognitive abilities, continue to provide them with concrete learning experiences, not abstract pencil-and-paper activities. The environment should encourage free choice most of the time. Also, provide opportunities to increase cognitive skills by encouraging small groups of children to work on ongoing projects.

Nurture and even stimulate their curiosity by tapping into your own childlike curiosity. Everything about the social-emotional environment should encourage curiosity at every stage.

If they are in a print-rich environment, some four-year-olds will show outward signs of emergent literacy (more about this in Chapter 13). Their emerging literacy skills should find many opportunities to continue to develop. They need to see how reading and writing are useful through exposure to meaningful literary experiences, not through mechanical or rote instruction. Books, stories, field trips, thank-you notes, signs, and writing down children's stories, poems, and chants are all a part of the emergent-literacy curriculum. Four-year-olds need plenty of opportunities to play with writing through painting, drawing, copying, and inventing their own way of putting symbols on paper. Talking and listening are also part–in some ways the most important part–of the emergent literacy curriculum.

Five-Year-Olds

Five-year-olds are in transition as far as program needs go. Sometimes it seems they don't fit well anywhere but in a program set up just for their age alone. They may not find enough to do in an environment set up for three- and four-year-olds, and they may be unable to hold their own in an environment set up for school-age children. Being either the youngest or the oldest of a group can be a problem.

Five-year-olds are social beings and enjoy friendships and group activities. Their physical and cognitive skills are increasing; they like to get better at what they can already do as well as take on new challenges.

Physical Development and Learning. Five-year-olds usually enjoy showing off their motor skills. "Watch me!" they say as they walk a balance beam or low wall, try to turn somersaults, or work on learning to skip. Tricycle riding is easy, and the practiced five-year-old can back into a tight space with a trailer on the back. If two-wheelers are available, some five-year-olds have an interest in learning and an ability to ride them. Balancing on one foot, jumping, and hopping come easily now.

Fine-motor skills allow some five-year-olds to begin making intricate designs and patterns, work on invented writing, and reproduce shapes and letters. They can handle buttons and zippers, and some can tie their shoes–though for others this skill comes later.

They need plenty of opportunities to practice old skills and challenge themselves with new ones. They are gaining in judgment and need to practice risk taking within limits. Also, their world is expanding, and they need more room to move as well as opportunities to get out into the community.

Social-Emotional Development and Learning. Five-year-olds can be cooperative. They are becoming more social-minded. They may even have generous impulses. They play with others in increasingly complex ways, especially in their dramatic play, where they weave a reality all their own about who is what and which events will unfold in what order: "Pretend that you're in a boat and I'm in the water, and you have to rescue me before the bad guys come." A rich environment with plenty of freedom and choices and adult resources and support helps five-year-olds go ever deeper into their play and projects.

Empathy is a feeling five-year-olds are beginning to develop, and they can sometimes put themselves in another person's shoes. They are becoming better at sharing and taking turns, and they may be affectionate and caring, especially toward younger children. They need opportunities to learn about other perspectives and relate to people in positive ways. Adult teaching and facilitation are useful for developing social skills.

Their sense of humor is progressing, and some of their jokes begin to make sense, although five-year-olds still find some things funny that adults do not understand. It is important that you demonstrate your own sense of humor to encourage the children to explore their own. Also, model curiosity to encourage children to be inquisitive.

Five-year-olds have better emotional control and are learning the difference between feeling an emotion and acting on that feeling. They still need help from adults to find ways to express feelings without hurting anybody or anything.

Cognitive Development and Learning. For some five-year-olds, cognitive skills are beginning to lead to easily recognizable literacy skills. Children this age begin to distinguish letters, numbers, and words; read familiar signs; and play at invented writing and spelling. Their concept development is blossoming: they demonstrate an understanding of "same" and "different"; "size," "shape," and "color"; and "more" and "less" to name a few. Five-year-olds can put a small number of objects in order by size and by length. They can tell which is first, second, or last in a series. Some five-year-olds are beginning to tell time, and they know what a calendar is for. Many programs carry out daily "calendar sessions," which make sense to five-year-olds, though most younger children find them a mysterious ritual.

The kinds of games, activities, and projects five-year-olds are drawn to further develop and refine their various cognitive skills. It is important for you to be a good observer of children to know what challenges to present that will further their unique interests and help them continue in their development.

All children should spend their days in a print-rich environment, but by five, they must have plenty of opportunities to discover and play with print in various forms, such as stencils, stamps, and magnetic letters. They should be exposed to print used for many purposes–in signs, books, notes, and labels. Although most five-year-olds lack the cognitive development to read at the skill level of primary school children, they are moving toward it. When they are truly ready, they will have also developed the ability to sit still for longer periods. At five, they need to be allowed–perhaps very gently encouraged–but *not pushed* into developing conventional literacy.

Watch out for competition. Children are exposed to this societal value in many aspects of their lives, and they may bring their competitive drive to their early childhood program. But do not encourage this behavior. Young children gain far more when a program stresses cooperation over competition.

School-Age Children

This stage comprises children from age six through age eight. Because this is an introductory book, its scope is broad. As such, any reader intending to work with school-age children will need to look elsewhere to learn more about the specific developmental characteristics and needs of each of these ages. For the purposes of this text, however, we will look at the commonality of needs of children ages six through eight. The following discussion applies to children in developmentally appropriate primary programs and in child care. Many children in this age group come to the early childhood program only for surround care–the hours that precede and immediately follow standard primary school hours. However, whether they attend the program all day or part of the day, the principles are the same.

Physical Development and Learning. If they have had practice, school-age children are usually good at meeting their own needs. They have developed the fine- and gross-motor skills they need for primary education. Their fine-motor skills allow them to use scissors in more intricate ways and to draw pictures that are increasingly recognizable (if they so desire, but abstract art may also be part of their repertoire if given the choice). Most school-age children have the fine-motor skills for at least rudimentary writing and reading, although the degree of competence varies by child.

Adults don't need to worry about physical care needs as much with children this age as they do, say, with infants and toddlers. The main issue is to allow them the freedom to take care of their own physical needs. Toileting should be an individual affair as the need arises, not something that has to wait for "recess." In fact, in a developmentally appropriate program for children this age, physical needs and skills are well integrated into the curriculum and there is no "recess" from learning, even though outside time may provide different kinds of experiences.

Six through eight-year-olds continue to increase their coordination skills, which is no easy feat considering the rapid changes in their bodies as they grow. Children in this age group can control their movements and put them together into sequences such as skipping, turning somersaults, and pumping a swing. They can throw, kick, and catch a ball. They need plenty of opportunities to be outside and do all this! And with their increased cognitive skills, they can play games that entail rules. They can also cooperate enough to play team sports. Children with special physical needs should be accommodated so they can participate to the best of their capabilities. Children motivated to take risks should find opportunities to do so safely, with some adult help to prevent dangerous situations. Above all, do not stress competition, because the goal is to help children feel good about who they are and what they can do rather than compare themselves to each other.

Social-Emotional Development and Learning. As with younger children, the school-age curriculum should integrate the whole child; the social aspects of the program should not be separate from the emotional, physical, and cognitive aspects. Children of this age are still "learning to learn," and they should feel good about themselves.

One of the skills that they should have been working on all along is resolving conflicts. In a classroom where the teacher tells children how to behave and sets up punishments for misbehavior, resolving conflicts may be skipped over entirely. In order to live in the world, children must know what to do when someone disagrees with them and both sides want their way.

The emotional overlay of each learning experience is an integral part of the whole. This is true at all ages but is stressed here because sometimes professionals regard school-age children as being ready to study subject matter or focus on academic skills regardless of whether they feel good and see themselves as growing as a learner or not. It is very important, as in the earlier years, that children not be compared to each other or to a standard that identifies some as superior and others as lacking. This is why competition should be avoided. All children should have a chance to see themselves as learners–no matter what their achievement level. Having positive feelings about themselves helps them become good learners, whereas negative feelings get in the way and undermine their confidence.

Cognitive Development and Learning. School-age children need to move forward in their abilities to think logically. Recognize, however, that they still think in terms of the concrete world. They will not be capable of abstract thinking that involves sophisticated reasoning, taking into account variables, and coming up with propositions until they are older and reach Piaget's stage of formal operations. In the meantime, children of this age are firmly rooted in the stage Piaget calls **concrete operations.**

They are gaining more skill in symbolic representation beyond the mental images they once used for pretend play; they can store and retrieve memories of actions to incorporate into their play and creations. Now they can use symbols for reading and writing to a greater or lesser extent, depending on the individual.

School-age programs (as well as early childhood programs for younger children) should include children with special needs. If children with special needs arrive already seeing themselves as learners and leave with the same attitude about

concrete operations
A cognitive stage described by Jean Piaget during which young children (ages seven through eleven) can use what Piaget calls "mental operations" to reason about the concrete world. At this stage, children are not yet capable of purely abstract thinking that involves taking variables into account or coming up with propositions.

themselves, the program will be highly successful. In order to accomplish this worthy goal, these children must be regarded as unique individuals with particular developmental patterns and timing. They must be allowed to move at their own pace and not be pushed.

Perhaps the hardest skill not to push is reading. So much is made of children being "at grade level." Nevertheless, it is important to realize that not all children can be expected to read by six. A few learn earlier, some take a little longer, and others take much longer. Early readers are not necessarily better readers, though they may be at first. Eventually, reading skills level off.

Math, social studies, art, music, drama, dance, and science are pertinent subjects for children of this age–and younger–but, again, they should not be taught separately; instead, they should be carefully integrated into the curriculum.

Social studies, in particular, can be looked at as belonging under the heading of social-emotional development; the early childhood program becomes a human relations laboratory in which children learn hands-on from immediate experience how to be an individual and still get along with a group. They practice relating to each other while they work on social skills, explore values, and have opportunities to plan, share, and work cooperatively with others.

A Story to End With

As the parent of a baby born prematurely who had developmental delays, I had qualms every time I went into the diapering area of the infant program we attended together. What stirred my feelings was a developmental chart plastered on the wall right above the diapering table. I hated that chart and tried not to look at it. I did not want to measure my son's progress against the norms outlined on that chart.

I never did quite "place" him, because I could not diaper him and do the calculations to adjust for his prematurity at the same time. He was born at 27-weeks' gestation–three months early. I never quite knew where he lay on that chart, but I knew he did not measure up. Of course, my experience in the diapering area was totally irrational. I knew that children with special needs do not fit neatly into developmental timetables or age norms. My head told me to forget the chart, but my feelings would not let me.

The constant assessment by the various experts in the program was a different story. My son was followed by a speech pathologist, a physical therapist, an occupational therapist, and a developmental specialist. We had all those experts working with us and helping me see what he needed. In addition, I had a caring and supportive social worker. These professionals were always looking for signs of progress, and they found them. Dealing with them was different from dealing with that cold, old chart in the diapering area. I felt anger every time I saw the chart, which seemed to be pointing an accusing finger at him and at me.

I think this experience made me a more understanding early childhood professional. I am now well aware that although charts don't tell the whole story, it is hard to convince parents otherwise. I am cautious about using ages and stages as measuring sticks and timetables. I emphasize that they are tools to help us understand what children need, not ways to judge children or their parents.

SUMMARY

Developmental information based on what is known about ages and stages is only one of three bases of knowledge for making decisions about what children need; the early childhood educator must consult (1) developmental knowledge, (2) knowledge of the individual child, and (3) knowledge of the cultural context. Developmental expectations differ from culture to culture. Regarding the individual is also important for children with special needs because they may not progress according to norms or schedules.

The early childhood program serves children from birth through age eight. Broken down into eight stages, young children's developmental needs are as follows (in brief): (1) young infants need a trusting relationship, (2) mobile infants need freedom to explore safely, (3) toddlers need freedom to expand their horizons, (4) two-year-olds need freedom to become more autonomous, (5) three-year-olds need opportunities for expanding competence, (6) four-year-olds need support for developing a sense of initiative, (7) five-year-olds need support to deal with their expanding world, and (8) school-age children need encouragement to regard themselves as good learners.

REFLECTION QUESTIONS

1. How important is it for a young child to learn to properly greet an older person and learn other similar ways of showing respect? Do you think your answer relates to your culture?
2. Were you ever compared with a norm and found lacking? If so, what effect did this have on you? How do you think it affects children to be given the message that they somehow don't measure up?
3. If you were a parent of a child who had special needs, do you think you would consider it a blessing, a curse, neither, or both?
4. Which stage of your own development do you remember best? Why?
5. Do you remember learning to read? If so, what do you remember and how can you use your own experience to work more effectively with young children?
6. Considering that all of us have "special needs," even if they do not show, what special needs do you have?

TERMS TO KNOW

How many of the following words can you use in a sentence? Do you know what they mean?

For Further Reading

Bohart, H., and Procopio, R. (2017). *Spotlight on Young Children: Social and Emotional Development.* Washington, D.C.: National Association for the Education of Young Children.

Copple, C., and Bredekamp, S. (2009). *Developmentally Appropriate Practice in Early Childhood Programs Serving Children Birth through Age 8* (3rd ed.). Washington, D.C.: National Association for the Education of Young Children.

Copple, C., and Bredekamp, S. (2006). *Basics of Developmentally Appropriate Practice: An Introduction for Teachers of Children 3 to 6.* Washington, D.C.: National Association for the Education of Young Children.

Copple, C., Bredekamp, S., and Gonzalez-Mena, J. (2011). *Basics of Developmentally Appropriate Practice: An Introduction for Teachers of Infants and Toddlers.* Washington, DC: National Association for the Education of Young Children.

Galinsky, E. (2010). *Mind in the Making: The Seven Essential Life Skills Every Child Needs.* New York: HarperCollins.

Healy, J. M. (2011, March-April). Brain Readiness: Impacting Readiness–Nature and Nurture. *Exchange,* 33 (2, serial no. 198), 18-21.

Katz, L. (2012). Developing Professional Insight. In G. Perry, B. Henderson, and D. R. Meier (Eds.), *Our Inquiry, Our Practice: Undertaking, Supporting, and Learning from Early Childhood Teacher Researcher(ers)* (pp. 27-132). Washington, D.C.: National Association for the Education of Young Children.

Leong, D. J. and Bodrova, E. (2012). Assessing and Scaffolding: Make-Believe Play. *Young Children 67* (1), 28-35.

Mardell, B., Rivard, M., and Krechevsky, M. (2012). Visible Learning, Visible Learners: The Power of the Group in a Kindergarten Classroom. *Young Children 67* (1), 12-19.

12 Observing, Recording, and Assessing

DGLimages/Shutterstock

Observing

Recording

Anecdotal Records

Running Record Observations

Incidents Reports

Journals

Photographs, Sound Recordings, and Videos

Checklists and Mapping

Time Samples

Assessing

Assessing the Children

Assessing the Program

A Story to End With

In This Chapter You Will Discover

- how observation, record keeping, and assessment are intertwining processes.
- how to use a series of incidents reports to understand behavioral patterns.
- how a journal can be kept by more than one person.
- the value of photographs, sound recordings, and videos as records.
- what checklists look like.
- how to use "mapping."
- what time samples can show.
- what portfolios are and what they can contain.
- approaches to assessing children's learning and development.
- some ways to assess program effectiveness.

FOCUS ON DIVERSITY

Culturally and Linguistically Appropriate Responses to the Needs of Latino Children

A study[1] sought to examine the kinds of responses given to the unique education and linguistic needs of Latino children. They came up with culturally and linguistically appropriate practices related to language development and early literacy learning, child assessment, approaches to support equity and diversity, and parental involvement. These approaches reported by 117 state administrators of early childhood programs represented ones that were recommended or being used in early education and intervention programs that served Latino families and their children. All of the administrators generally agreed on the importance of preserving children's home language–important for ensuring that children do not become alienated from their families and communities.

One of the ways early childhood education (ECE) educators facilitate the development of young children is to measure their progress. However, unlike primary school children, who are tested and ranked with graded, timed exams, young children are indirectly assessed primarily for improving various aspects of the program. ECE educators use the tools of observation, record keeping, and reflective assessment. The results are used not only to shape teaching, curricula, and the environment for individualized development in the classroom (and possibly at home when results are shared with parents), but also to understand the child better in the context of his or her family and school environment. Look at the *Focus on Diversity* box above to see a study about the needs of a particular group of children–Latinos.

NAEYC Program Standards
Program Standard 4: Assessment

In the words of NAEYC's Program Standard 4, "The program is informed by ongoing systematic, formal, and informal assessment approaches to provide information on children's learning and development. These assessments occur within the context of reciprocal communications with families and with sensitivity to the cultural contexts in which children develop. Assessment results are used to benefit children by informing sound decisions about children, teaching, and program improvement."[2] You won't be doing formal assessments right away, but the late Patricia Nourot, shows one assessment strategy anybody can use in her *Voices of Experience* story.

What do early childhood educators do with what they learn about each child and the children as a group? They take what they learn as a resource for goal setting, for implementing the teaching-learning process. Planning for development and learning entails making opportunities for a variety of experiences, creating projects, providing resources, and analyzing, adding to, and/or rearranging the environment. Besides focusing on children, early childhood educators also reflect on their own effectiveness as well as that of the program. An early childhood program is about learning and development–everybody's learning and development, adults and children alike.

Through stories and examples, this chapter explores effective assessment that includes observation and record keeping (sometimes called documentation). You will also learn about portfolio development, a major, ongoing assessment tool for early childhood educators.

To get a picture of how observation, record keeping, and assessment are really one synchronous process, let us look at the following scenario. Cathy is a teacher of four-year-olds. Every morning she and her co-teacher have a group time called "morning news." She starts with a good-morning song that incorporates each child's name. This beginning ritual not only lets the children hear their own name repeated by the group, it also reinforces Cathy's memory of who is present and who is absent. Later, she will fill out the attendance report and write some anecdotal records about each child. Right now, though, she is getting a feel for the mood of the group and of each individual.

The news session helps Cathy peek further into the minds and hearts of the children, as she listens to each child report what is new in his or her life. Cathy writes some notes on a large piece of paper as the children talk, putting into print before their eyes what they are saying. Later, she will recopy, expand on, and hang up the "news notes" for the parents to see and comment on when they return. She will also make notes for herself based on what particular subjects seemed to hold energy for the children.

Morning news is one of many sources of curriculum ideas and directions for Cathy and her co-teacher. The two will get together at nap time to go over the notes and look for project ideas to carry the children's interests further and deeper. Their planning is dependent on their observation and recording of the children's interests.

Observation, documentation, and assessment are complimentary tools that improve the teaching–learning process. The results help administrators and educators improve various aspects of the program, including the layout of the various play areas in the environment.

Later in the morning, Cathy gets a small group of children ready to go outside for a different kind of observation-and-recording activity. Last week, they planted a pole in the courtyard on the day of the vernal equinox. The children now gather chalk, paper, and pencils to record their new observations of the shadow cast by the pole.

They hold a preliminary study session indoors before they go out. Cathy asks the children, "Where do you think the shadow will be, and what do you think it will look like?" Some children tell her, and others draw their idea of what the shadow will look like. One of the children goes to the bulletin board where a number of pictures of previous shadow observations hang. She studies them carefully. Cathy listens and writes down the children's answers to her question. Some ideas are far-fetched, but Cathy listens to them, glad that they are giving her clues to the way they think. She does not correct them. She knows that they are "constructing" knowledge, and she does not want to interfere with that process.

On the first day, Cathy and the children studied the movement of the shadow during the program hours, going out to check every hour and drawing its changing shape and position; they leave different-colored chalk traces on the ground of what they are observing so they can study those traces later. Now they are studying the changes over a week by going out at the same time each day. It takes careful planning on Cathy's part to get them outside at exactly the right time. She asks the children to help keep track of the clock so they are ready to observe and draw at exactly 10 o'clock.

The hand is at five minutes to ten when they pack up and go outside. "What color chalk shall we use to mark today's shadow?" asks Cathy. Two children have a brief debate about which color to choose; they each want a different color and begin to argue about it. The argument is settled quickly by a third child, who points out that both the colors mentioned have already been used. Purple is the color they all decide on. The time is perfect. The child who drew yesterday chooses someone to draw today, and the other children watch carefully to be sure the drawing is accurate. When the drawing is finished, Cathy takes a picture of it.

"What do you think?" asks Cathy. "Is the shadow different from yesterday?" Upon close examination, the children decide it is not. Cathy reviews the previous days with them. When they compare the chalk traces over a week's period, they can see a slight difference. They set to work recording their observations and speculating about what makes the shadow move. Cathy records their conversation. As they are working, a cloud moves in front of the sun. A little girl notices the changing light and suddenly jumps up from her drawing saying, "What about the shadow now?" She gets down on her hands and knees to look closer. "Look what happened to it!"

Cathy asks them to compare how the shadow looked before and how it looks now. She suggests they draw the new version. They set to work. One boy draws a picture of the sky with the cloud hiding the sun. Cathy takes two pictures–one of the sky and the cloud and one of the pole and the ground to compare to the shot she took in bright sunlight.

Assessing Through Interacting

One way to be with children is to join in their play as a "parallel player." By jotting down notes about what I observe during our play together, I am able to assess the children's learning and development. The pitfall, of course, is to make sure that the child's agenda, rather than my own, leads the play, and that the questions I ask are "authentic," showing genuine curiosity about the child's thinking and feeling.

For example, three-year-old Tamara is playing at a low table with large connecting blocks. I observe her as she puts two blocks together to form an upright tower and then puts two more across the top horizontally. I am curious about her block play but do not want to interrupt, so I sit down next to her and begin my own block structure, imitating what she has made. She turns to me, her face alight with pleasure. "Look, Pat, we both made 'T's'!" She traces the shape of the letter "T" on her structure and then on mine.

"So we did," I respond. "T is for Tamara," she observes. Tamara then names all the colors in her structure and mine. "We have yellow and blue and red," she notes. We touch the colored blocks together saying in unison "Yellow, blue, red."

I begin to make a pattern with my blocks, red, blue, red, blue, and Tamara soon follows suit. We build together quietly for a few minutes and soon Tamara links the two lines of red and blue blocks we have made. "Red-blue, red-blue, red-blue" she repeats as she touches each block. I am briefly tempted to ask her how many there are or launch into a discussion about creating "a-b patterns," but instead I say, "What do you think about our red and blue blocks?" Tamara responds, "We're building a bridge for the train to go and when the red and blue are friends together, it makes a magic bridge."

"Thanks for playing blocks with me," I say as we respond to the clean-up bell by putting blocks away in the basket. "Sure, anytime," she smiles, as she walks to circle time.

I write down this anecdote on the Post-it notes I carry in my pocket, and later stick the dated note on a card in her file. I also make a note to myself that Tamara is using patterns, can make the letter "T," and can name colors. I think about how I might extend the concepts she has revealed in her play through some patterning activities with rocks and shells the next week and how we might as a small group make the beginning letters of children's names with the connecting blocks. My curriculum plans evolve from Tamara's play.

–Patricia Nourot

There are lots of observation and documentation going on here. The children and Cathy make records of the process the children are engaged in as they explore the movement of the sun and its effect on the pole's shadow. Cathy keeps the records they produce and makes them available to the children as they continue their shadow project. The drawings and written records boost their memory and guide their understanding.

Through this kind of observation, record keeping, and assessment, children make connections and gain continuity in their mental processes and in their activities. Likewise, adult observations and record keeping allow for ongoing assessment, by which adults gain insights into the way children's understanding works and where their interests lie.

Now that you have seen a big picture, let us look at each element of observing, recording, and assessing from the early childhood educator's point of view. We will focus first on observing.

OBSERVING

Observation as a skill was introduced at the beginning of this book and has been discussed throughout the chapters. Examples abound of how professionals use observing, reflection, and recording to improve their work with children. Sometimes, you may even have to go beyond observation and reflection and test out your hypotheses. Following is an example of a family child care provider using observation. Veronica is always on the lookout for hearing problems in the children she cares for. Her own child had a hearing loss that was not detected until she was three, when a pediatrician discovered the problem and went to work to correct it. Veronica will always remember the day her daughter came in excitedly from the garage and shouted, "Mommy, Daddy's radio talks!" Her daughter had apparently never before been able to hear the radio her father listened to while at his workbench.

Veronica has concerns about two of the children currently in her program. Neither seems as responsive to noise as she thinks they should be. One, a baby, can sleep through anything. She shows his mother today when she arrives by clapping her hands loudly near the crib. Without even flinching, the baby sleeps peacefully on. The mother tells Veronica that the baby is just used to noise. He is an adopted child who lived for a long time in a foster home where, according to the social worker, the noise level was extraordinarily high. The mother promises, though, to get it checked out. It turns out that the baby's hearing is fine. The mother is right. He tunes out noise.

The other child Veronica is worried about a four-year-old who often fails to respond when she talks to him. Because this unresponsiveness is new, Veronica wondered if something was wrong at home or if he was going through something in her program. But then she noticed that he responded to her if he could see her face when she talked to him. She expresses her concern to the parent, who also agrees to get the child's hearing checked. It turns out that he has fluid behind his eardrums that is affecting his hearing. The problem clears up with treatment, and the child becomes his old responsive self.

Veronica is a good observer. She also reflects on what she observes. Furthermore, she has problem-solving skills. Because of her assertive approach, she prevents possible language delays that sometimes result from auditory impairments.

Here is a different example of how observation and reflection pay off. This example occurs in a center setting. Brittany is a small, slender, rather delicate three-year-old. She has been in preschool for a month now and seems to like it very much. She talks a lot about her friend Giovanni, whose family just moved here from Mexico. She looks forward to her program every morning and climbs happily into the car.

But suddenly things have changed. Before leaving home, she protests and says she does not want to go to school. She has to be urged to get into the car and out again. Sometimes she cries in distress when her mother leaves, but she gets over it in a hurry. Brittany's mother expresses her concern to the teacher and asks if she understands what is happening. The teacher explains that this is a common pattern. Sometimes separation issues do not hit children until after the first weeks. It is quite possible that Brittany is merely expressing delayed separation anxiety. The mother accepts the information and leaves. The teacher, however, decides to observe Brittany more closely to see what else she can learn. And, indeed, she learns a lot.

The next day, when Brittany arrives, the teacher greets her as always but keeps an eye on her, just to see what happens when she gets to school. Brittany shows signs of tears, but they do not materialize. She is not crying, but she looks apprehensive. She leaves the teacher for the play dough table, and as she sits rolling out a ball absentmindedly, she periodically glances at the door. Thinking about her mother, the teacher decides.

Brittany keeps an eye on the door as she moves to the puzzle table. The door opens, and in walks Giovanni. He spots Brittany immediately and barrels his way over to her. She gets up from the puzzle table and backs away from him. He looks excited as he continues to pursue her. He has his arms out as if to give her a big hug, but when he meets up with her, his clumsy attempt to hug her ends up knocking her over. He then sits on her, jabbering away happily. He is twice her weight, and she can barely move. She manages to wriggle out from under him just as the teacher reaches her to help. Brittany seems to understand that the greeting is over and she does not have anything further to worry about. The two then settle in at the puzzle table together and look happy.

The teacher has her own puzzle to put together. She continues to observe Brittany and Giovanni. What she sees are two three-year-olds who enjoy each other's company, even though they do not speak the same language. After another morning of watching, a pattern begins to emerge. Giovanni arrives after Brittany and gives her another enthusiastic welcome that ends exactly the same way. Giovanni knocks Brittany over and sits on her.

What is happening is that Giovanni does not have words to greet his friend, so he uses his body. He is big and strong and does not realize that throwing his weight around is not the same as giving a hug. He does not understand the effect of sitting on her. The teacher sets out to change this pattern, to teach him words and gentle actions that convey his message without upsetting Brittany. She also helps Brittany understand that Giovanni does not mean to scare or hurt her. It does not take long to change the pattern and clear up Brittany's anxieties. It was not a separation issue after all. It was Brittany's fear of Giovanni's daily greeting that changed her attitude about coming to school. As illustrated by these preceding examples, without observation or reflection, ECE educators may fail to notice or misread problems that can uniquely hinder a child's ability to play and learn effectively. Observation and reflection are invaluable methods that help ECE educators both recognize and maintain the conditions that enable each child to thrive.

RECORDING

Observations are helpful for noticing apparent progress or problems. Documentation is another essential, complementary tool for educators because it can reveal patterns of development that typically go unnoticed with observation and reflection alone. Record keeping/documentation tracks activities that may seem ordinary in isolation. Yet the process of reviewing a series of records captured over a period of time can reveal hidden insights. Record keeping can be helpful in assessing children's behavior and thinking processes (as individuals and as groups) and in assessing the effectiveness of the environment as well. The following sections describe several record-keeping methods.

Anecdotal Records

anecdotal records A documentation method that briefly describes an activity, a snatch of conversation, a chant, and so on. Anecdotal records can be based on reflection or written on the spot.

Anecdotal records are a means of creating brief written descriptions of an incident or something that stands out about a child. Make sure each is dated, because when you look back on them you may be able to see a pattern or a progression.

Anecdotal records can be based on reflection–remembering things that happened that day. It's also sometimes possible to jot down things as you see them happening; keeping a small pad of paper or 3-by-5-inch cards and a pencil in a handy pocket can help you make such on-the-spot records. Use the paper or cards to jot down quick snatches of conversation, little poems the children make up, and chants they sing on the swings. You may learn something from reading these notes later, or you can use them as language samples or examples of the child's creativity.

Running Record Observations

running record observation A method of documenting that gives a blow-by-blow, objective, written description of what is happening while it is happening. A running record can include adult interpretations about the meaning of the observed behaviors, but it must separate objective data from subjective comments.

While anecdotal records are mostly reflections or brief notes of something worth capturing on paper, a **running record observation** is a lengthy detailed description of what is happening while it is happening. The goal of running records is to capture all the behavior as it occurs, ideally with such detail and description that someone reading it would "see" just what the observer saw. It works best when focusing on one child, though, it is also possible to record interactions as well. The problem is getting everything down. It takes skill to write a good running record observation. You cannot write whole and complete sentences during the observation or the action is over before you have got it recorded. You are still looking at your paper when the child has gone on to something else. Quick brief notes are the answer to the problem, but they need to be written up right away after the end of the observing period. Otherwise you risk forgetting what the notes mean and the details that go with them.

Here is an example of a running record observation. The setting is mid-morning in a preschool classroom. The child is a five-year-old with special needs who is part of an inclusion program. The observer has done informal observations on this child at different times of the morning for the past three days. All she knows about the child is what she has seen. She knows nothing about the child's background. She tries to record only what she sees without any interpretation because she does not feel she understands the child well enough to guess at the meaning of the behaviors she is observing. The purpose of the observation is to get input from the specialist who visits regularly to help the staff work with this particular child.

10:00 J. is walking slowly in front of a window, running a wooden spoon along a radiator just below it. The spoon makes a series of clicking noises, and the child is laughing but shows no joy on her face; in fact, she has a blank expression. Her eyes are fixed on the end of the spoon. She comes to the end of the radiator and turns abruptly on one heel and starts back again. She repeats exactly the same action and the same laugh. Her arm bumps up and down as the spoon clicks on the radiator. Her feet move mechanically along the floor. Her expression does not change. When she reaches the other end of the radiator, she turns in the same way, spinning on one heel, and starts back again. She is now on her third repeat when a teacher comes up to her and holds her arm. Reaching out with her other hand, the teacher takes the spoon away, saying gently, "Would you like to come sit down now for snack?"

10:02 J. does not protest as the spoon is removed from her grasp. She stops moving, stands perfectly still, and gazes upward at the ceiling, her eyes wide and staring. The teacher holds her arm and slowly moves her to the snack table. J. goes without resisting but never looks at the teacher or ahead to where she is going. She is staring into space.

10:04 J. sits stiffly in a chair at the table already half filled with children. She continues looking at the ceiling, with her head tilted back and her hands in her lap. She does not move or make a sound. Her face is expressionless. A child comes and sits next to her. She pays no attention. She seems unaware of what is happening around her, or, if she is aware, she gives no indication. She does not respond in any way until the teacher puts a plate of pretzels and little round crackers in front of her. She looks down at the food, staring briefly. A slight smile crosses her lips and disappears. Then she raises her right hand to her plate and picks up a cracker, using a mechanical movement. She puts the cracker in her mouth, chews, and swallows.

10:06 Her tablemates are chatting with each other and with the teacher, who is seated opposite J. J. pays no attention but rivets her focus on the process of picking all the crackers off her plate and eating them one by one.

10:08 J. has finished her crackers. She shifts her position slightly and puts her left leg under her. She still looks stiff. She picks up a pretzel, holding it between her thumb and forefinger and puts it on the table beside her plate. Then she methodically picks up another one and places it on top of the first. She continues stacking pretzels one by one. She watches what she is doing but does not seem to show any real interest.

10:10 The girl next to her begins stacking her pretzels too. J. pays no attention. Another child also starts stacking pretzels, and the two talk to each other and to J. about what they are doing. J. does not look at them or respond but continues to stack the pretzels until her plate is empty and she has a rather tall, leaning tower.

10:12 The teacher returns to the table with a pitcher and begins handing out juice glasses. She sees the pretzel stacks and reminds the children not to play with their food. "If you aren't going to eat it, get up and throw it away," says the teacher.

10:14 "Up, up and away with TWA" says J. in a singsongy voice, semi-mimicking the teacher. She slouches back in her chair. Her muscles are looser now than they have been up to this point. Her head tilts back again, and she stares at the ceiling, repeating, "Up, up and away with TWA." The child next to her reaches over and knocks down her tower, but J. never notices. She seems totally absorbed in looking at the ceiling and repeating the phrase.

Incidents Reports

incidents reports (or event sampling) A method of documenting a particular type of repeated occurrence from beginning to finish. Sometimes called "event sampling," incidents reports focus on one of a variety of behaviors, such as aggressive incidents or parent–child separations.

Sometimes called event sampling, **incidents reports** narrate a particular type of repeated occurrence from beginning to finish. For example, you might observe aggressive incidents involving the group or a particular child. Observe and record everything that happens before, during, and after each incident; by doing so, you may be able to see patterns that suggest why the incidents occur.

In the case of aggressive incidents, perhaps there is a child who lashes out whenever he reaches a certain frustration level. When you recognize the pattern, you can then step in before the child reaches his coping threshold. Or perhaps there is a child who hits for no apparent reason—with no show of emotion or warning beforehand. You do not know why she is doing this, but by recording a few incidents, you might begin to realize that she hits to get what she wants—attention. In fact, now that you see the pattern, you are surprised at how much attention she receives for hitting. One quick slap and a yelp from the victim brings at least one and sometimes two adults running over to talk, touch, and in other ways acknowledge the existence of this child. Never mind that the adult's focus is on scolding; for a child in serious need of attention, any kind will do.

Of course, it is hard to both record incidents and respond to them—especially in the case of aggression. You would not just sit back and continue writing if you saw a child was about to get hurt. Writing incidents reports requires that another adult be present to handle the problems so that the observer does not have to change roles and be the mediator.

Journals

Some early childhood educators find journal writing an important recording device. Journals may contain all sorts of writing: anecdotal records, running records, incidents reports. They may also contain drawings or photographs.

A journal can be a two-way tool—that is, both parents and the early childhood educator can contribute to it. Some programs keep the journal with the child; it goes home in the evening and returns to the center in the morning so that everyone gets a chance to write in it. Used this way, the journal becomes an important means of communication with parents. From the journal, parents know a little of what went on while they were gone and the early childhood educator can keep up with what went on at home.

Incidents reports can give insights into patterns occurring in the classroom or in individual children.

For infants and toddlers, the journal can be used to record such specifics as diaper changes, feeding times and the amount consumed, and sleeping times. It can also go beyond the mundane and give accounts of the child's moods, interactions, and activities. Journals for preschoolers and school-age children usually focus on activities, interactions, interests, and incidents. When children are able to write themselves, they can contribute to the journal and get some good practice writing.

Photographs, Sound Recordings, and Videos

Capturing the processes and products of children at work and play can be useful for teachers and parents alike. Teachers can help children delve further into their explorations of various subjects by recording interviews and conversations about those subjects, as Cathy did with her four-year-olds in the opening scene of this chapter. Capturing interactions or group conversations can provide valuable records if it is done often enough so that children aren't distracted by the process. It is important to note here that because children are underage, consent forms by parents should be collected before obtaining any kind of visual documentation that photographs a child.

Checklists and Mapping

developmental checklist A method of documenting and assessing a child's development. A developmental checklist might be broken down into specific categories, such as physical, psychomotor, cognitive, social-emotional, and language.

Developmental checklists can be useful ways of looking at progress as long as they are not used as "report cards." Although an experienced professional with a good developmental background can informally assess children without a checklist, most early childhood educators have difficulty keeping track of specifics without some kind of structure. A checklist provides that structure. Checklists are usually divided

FIGURE 12.1 Sample developmental checklist for four-year-olds

Motor Skills

Gross-Motor Skills	Fine-Motor Skills
Child has been seen to:	
❑ pedal a tricycle	❑ feed self without help
❑ hop on one foot	❑ hold crayon/felt pen between two fingers and thumb
❑ balance on one foot	❑ stack 10 blocks
❑ jump over obstacles	❑ serve water or milk from a pitcher
❑ walk on a balance beam	❑ hit nails with a hammer
❑ catch a ball with both hands	❑ string small beads
❑ throw a small ball overhand	❑ button and unbutton own clothing
❑ walk up and down stairs alternating feet	❑ use knife for spreading

into "domains" of development–for example, "physical" or "psychomotor," "cognitive," "social-emotional." Sometimes "language" is a separate category, or it may be included under "cognitive."

Figure 12.1 illustrates a portion of a sample developmental checklist. You will notice that this checklist includes several self-help behaviors under the fine-motor-skills heading. Self-help skills vary according to cultural perspectives on independence versus interdependence. If the child being observed comes from a family who values interdependence over independence and individuality, such self-help skills might not show at age four. The child may have never had an opportunity to use a knife, button his own clothes, or feed himself. In this case, it is important not to judge the child's self-help abilities since they have no meaning for him at this particular time in his life. When using checklists–as in every other aspect of the early childhood program–remember to be aware of cultural contexts.

environmental checklist A method of documenting the setup and/or use of the environment in an early childhood program. An environmental checklist can be used to assess a specific child's use of the environment, or it can be used to assess the effectiveness of the setup itself.

An **environmental checklist** is another documentation method that records the use and effectiveness of the physical setup. For example, by keeping track of who is where every few minutes, you can see some patterns of individual and group behavior. Analyze Figure 12.2. Why has Maria never entered the dramatic play area? She has never been in the block area either. In fact, after looking at the checklist, you can see that Maria never left the art table during the time she was observed. Does Maria have a great interest in art, or is something else going on? If subsequent observations show that Maria never spends free play time anywhere but the art table, you would want to find out why, in such an interesting and exciting environment, Maria has limited her options so drastically. Does she need to broaden her experiences, or should you be feeding her artistic interest?

Look again at Figure 12.2. What does it tell you about Chris? He moves around a lot, sampling a new area every 15 minutes. Why is that? Is he rejected by other

FIGURE 12.2 Sample environmental checklist

Checklist

How Children Use the Indoor Environment

Observer ______________________ Date ____________________

Time of Observation No. 1 10:15 No. 2 10:30 No. 3 10:45

No. 4 11:00 No. 5 11:15 No. 6 11:30

Enter the number of the time of observation in the blocks of the different areas for each child

Name of child	Kai	Sonia	Kelly	Maria	Chris
Blocks	1 2 3 4 5	1 2 4 5	1 2		
Art table		3		1 2 3 4 5 6	5
Easels			3		1
Dramatic play			4 5		6
Manipulatives	6	6	6		2
Music area					
Book corner					3
Science table					4

children playing in the area? Is he restless? Is he new to this program and just trying out everything? Of course, you cannot answer these questions by reviewing this one chart (representing only a little over one hour), but you can see that he has been in six different activity areas in that short period and that some further observation and documentation may be necessary.

What about Kai, Sonia, and Kelly? Kai spent a whole hour with the blocks. Sonia was there most of that time, except for a quick trip to the art table. She went back to the blocks and stayed there until Kai left, and then she went to join him and Kelly in the manipulatives area. The three started out together at 10:15 in the blocks and ended up together in the manipulatives at 11:30. Is this a pattern? Have they become a threesome? Why might you want to know that?

Mapping serves a similar purpose to an environmental checklist by helping you understand the specifics of how a child functions in the environment. Mapping is particularly useful with crawling infants and toddlers—children who tend to move around a lot. Through mapping, you can see where they go and what the

mapping A method of documenting how a specific child functions in the early childhood environment. Using a map of a room or area, the recorder plots the path of the child and records such activities as interactions with other children or with adults. Start and end points are notated as well as the duration of the observation. Mappings can also be used to assess the use and effectiveness of the environment itself.

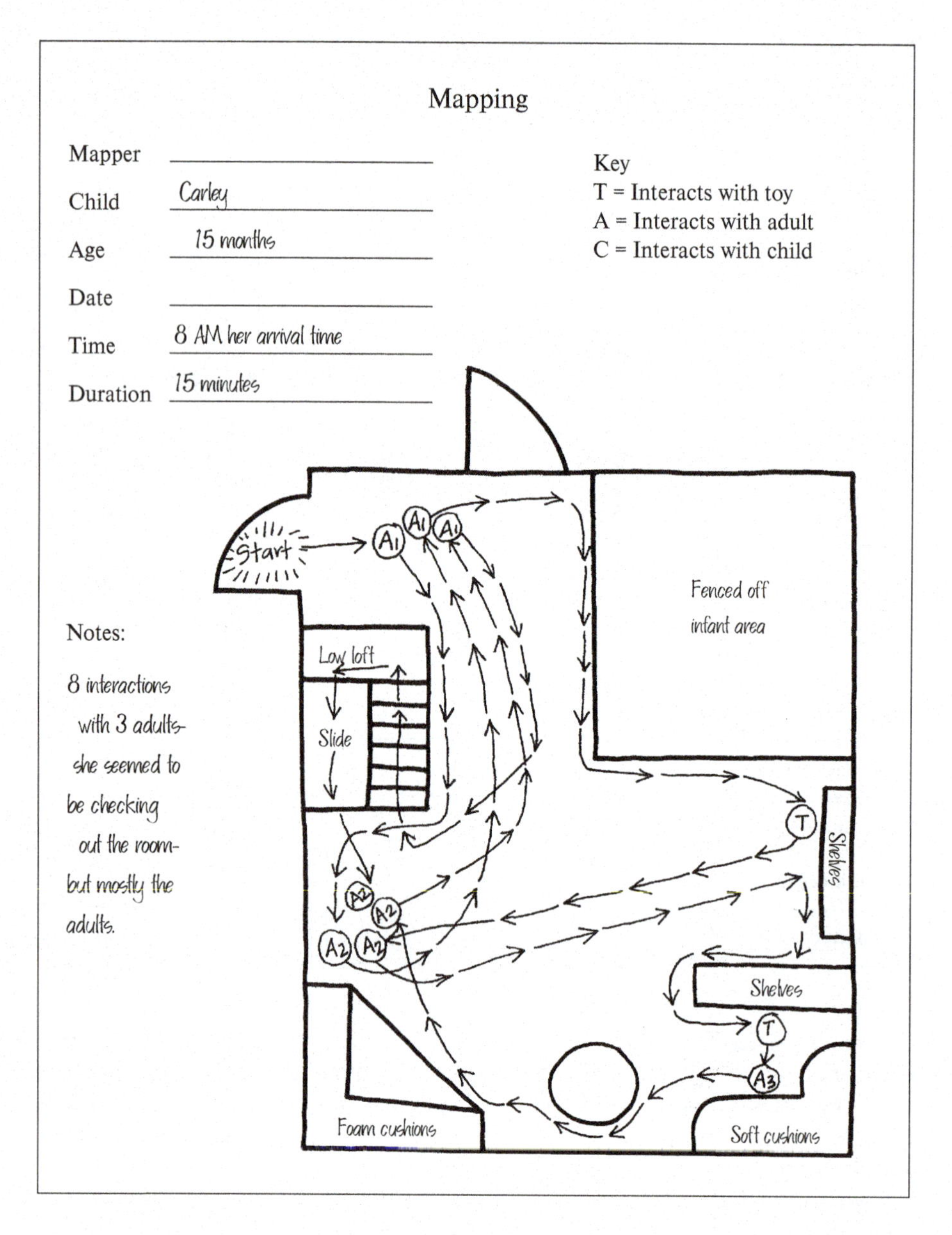

FIGURE 12.3
Sample mapping form

patterns are. Do you see how Carley in Figure 12.3 interacts only with adults—never with children? Is it because this mapping was done first thing in the morning? Maybe a different pattern would emerge at a different time of day. Contrast Carley's map with Blake's in Figure 12.4. He never interacts with a soul and not often with toys either–at least during this mapping period. He seems to wander aimlessly. Maybe you would have noticed that without the mapping activity, but putting the picture down on paper makes it easier to see and review with others.

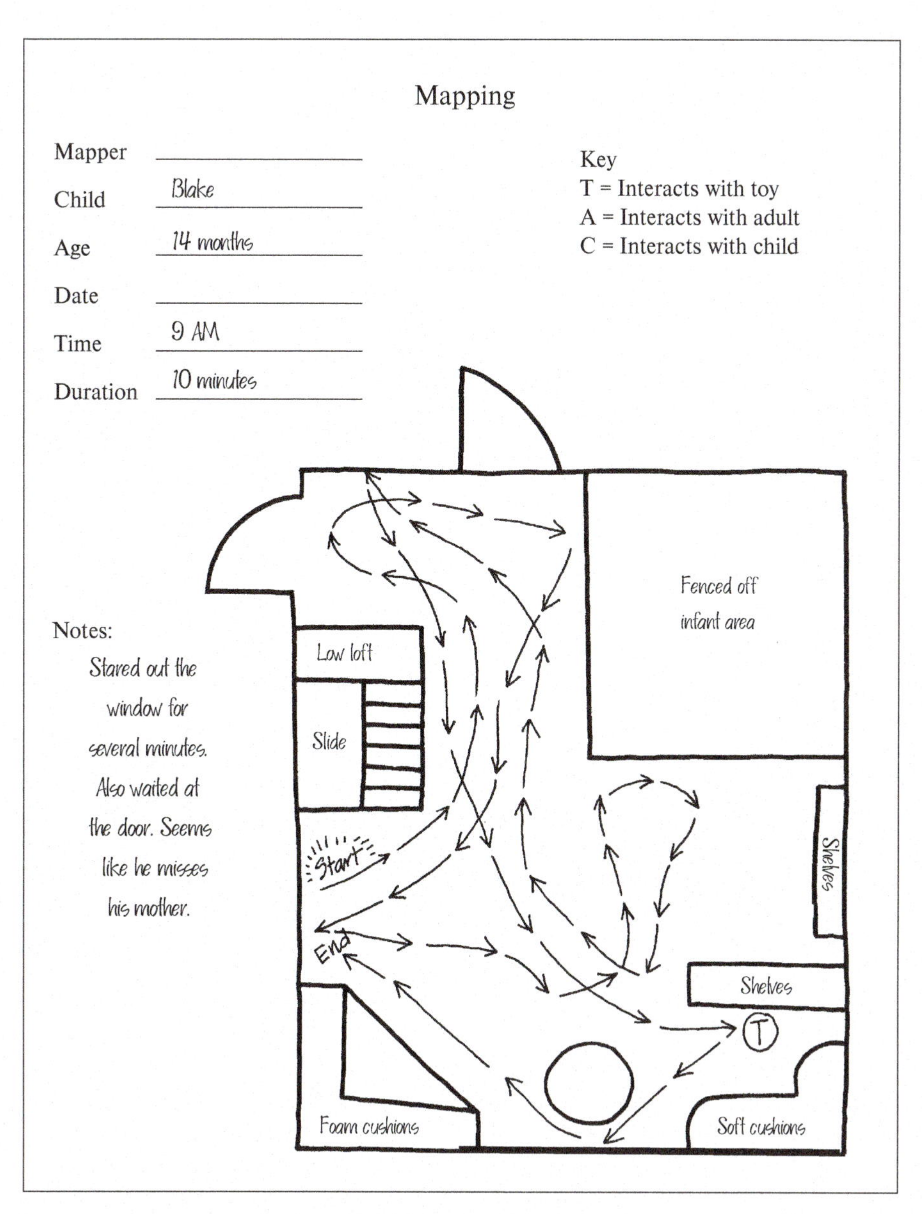

FIGURE 12.4
Sample mapping form

Time Samples

Collecting **time samples** of a few targeted behaviors of small groups of children is another way to learn about individual and group patterns. How many of the three-year-olds in the sample in Figure 12.5 were attentive and actively involved in circle time? How many were not? What might such an observation tell you about the children or about the way circle time is being conducted? What might it say about the appropriateness of this particular activity for three-year-olds?

time sample A documentation technique that involves collecting samples of targeted behaviors of small groups of children within a specific time frame as a way to learn about individual and group patterns.

FIGURE 12.5 Time sample form

Time Sample

Observer ______________________ Date ______________

Time duration 24 min Time increment 3 min

Description of setting 3 yr old circle time

Time	Child 1	Child 2	Child 3	Child 4	Child 5
10:00	A	A	P	S	M
10:03	A	I	P	S	T
10:06	C	A	S	T	M
10:09	C	I	I	I	S
10:12	C	S	I	M	I
10:15	S	M	T	C	M
10:18	C	A	A	A	S
10:21	C	A	C	S	M
10:24	A	P	P	P	A

CODE:

A = attending to what is going on
P = participating in what is going on
C = compliant but not attending or participating
I = interacting with another child – not attending
S = involved with self – not attending
T = trying to get the teacher's attention – not attending
M = moving away from the group – not attending

It might help you better understand this time sample if you knew what was happening in each of the time segments. At 10:00 the teacher started with a finger play. Only one child was participating, but two more were attending. Child 4 wasn't bothering anybody; he was playing with his shoelaces. Child 5 was trying to go outside, but a teacher brought him back. Things went downhill from there. The teacher got out a calendar at 10:03. From then until 10:09, she lost most of the group. She had a hard time getting their attention back when she was ready to read a story. She read the story roughly between 10:10 and 10:20. The children's attention was wandering during most of the story. She got all their attention at the end, however, when she had them pretending to be jack-in-the-boxes. Three of the five jumped as directed, and the other two watched. If you were this teacher and realized how little the children were participating and attending to your circle time, what would you do?

ASSESSING

We have looked at various methods of observation and record keeping. And because it cannot be separated from observation and record keeping, we have also begun to explore assessment. This section further discusses how to assess children's learning processes and their development as individuals and as a group.

We will also look at the role of assessment in evaluating the program in terms of its environmental effectiveness and its ability to meet the needs and goals of the children and their families, the administration and staff, and the community at large. To begin, we will look at assessing children.

Assessing the Children

Diagnosis and Prescription Versus Authentic Assessment. Let us compare two programs that have very different assessment approaches. Program 1 uses a standardized developmental checklist, which is administered every six months. If the staff members have seen a child demonstrate a specific skill on the checklist during the natural course of the program, they give the child credit for "passing" the item. If, however, they have not observed the particular skill, they informally "test" the child. They set aside a week for the testing procedure so they can be clear about what exactly each child can and cannot do.

The idea is to understand where the children are in their development and to pinpoint areas of weakness. They then address those weaknesses by creating activities and exercises for individuals and for the group throughout the year. Twelve months later, they reassess to see how far each child has come. They also note which children fall below the norm in particular skills and report both progress and areas that need improvement to the respective parents at a developmental-assessment conference.

This method is based on the same diagnostic-and-prescriptive method that a medical doctor uses to determine what's wrong with a patient and what treatment is needed. See the *Points of View* box on page 346 for a parent's view of how this assessment approach works.

Program 2 has a different approach to and purpose for assessment. They use a variety of methods on an ongoing basis throughout the year, including a lot of observation and documentation. They regard what they do in the name of assessment as part of the teaching-learning process, not something separate. They do not use their assessments to "grade," categorize, or compare the children to each other or to a norm. They never talk about "pass" or "fail."

The point of their assessment approach is to find out more about each individual child and the group, as well as document process and progress. They are more likely to get a true picture than Program 1 because assessment is ongoing rather than periodic. They seek to understand the child in a natural context, not how well he or she performs in a test situation. They use what they learn to pick up on individual and group interests and plan an emergent curriculum that responds to interests and needs. Teachers also do a lot of self-reflection as they assess their own effectiveness and that of the program.

POINTS OF VIEW

One Parent's View of the Diagnostic-and-Prescriptive Approach to Assessment

Here's my experience with my child's preschool and the way they evaluate my child's progress. When they explained how it worked, it seemed OK to me. This was my first child in preschool, and I just assumed there would be a "report card" at the end of the year, as there is in elementary school. It turned out not to be a report card but a conference with the teacher.

I felt OK about it the first year, except when I looked at the checklist they showed me in the conference, I saw that my daughter failed a couple of things I know she's good at. I don't know if she was just having a bad day when they did the assessment or what.

The second year, I volunteered in my child's classroom and got to see how the evaluation process really works. And I don't like it. They took the checklist and highlighted all the areas my child was behind in. Now I see that she was lucky that she failed some items that she is good at because they encourage her to keep doing them, and that makes her happy. She happens to be a little girl who likes to sit quietly and do crafts or play indoors with dolls.

Her gross-motor development, they told me, isn't up to par, so they are working on it. I know from looking at that checklist that she failed a lot of items, so I guess that means she is way behind the other kids. They were working to correct her problems. They insist now that she spend more time outside, and they push her to use the slide, which she hates. She likes the swings OK, but every time she gets on them, some teacher is there bugging her to learn to pump her feet. She never did like outdoor time very much, but now she is beginning to hate it with all this emphasis on what they call "skill building." I feel frustrated that they focus on her weaknesses so much and tend to ignore so many of her strengths! I don't think that approach makes her feel very good about herself.

authentic assessment A method of assessing children according to what they know, can do, and are interested in, which can then be applied to ongoing curriculum planning. Authentic assessment avoids comparing children to a norm or grading them. It also avoids standardized testing, which measures isolated skills and bits of knowledge out of context.

How are these two programs similar? Both programs have a developmental focus, and they both assess "the whole child" by looking at the three domains of development: physical, cognitive, and social-emotional. Both have the children's best interests at heart. In spite of the similarities, however, they have very different views of how to assess the children and how to apply their findings.

Program 1 uses a diagnostic-and-prescriptive approach that is too narrow and focuses more on weaknesses than on strengths. It represents a one-shot approach rather than a continual process. It has a pass-or-fail orientation, which can be disheartening to parents and children alike.

Program 2 uses what's called **authentic assessment** which is a much broader way of looking at children's progress and how it relates to curriculum goals. Instead of focusing only on weaknesses, authentic assessment zeroes in on what the children know and do and are interested in and then uses the results for ongoing planning. Authentic assessment gives teachers insights into children's thinking processes—insights that reveal gaps and misconceptions. All this information provides teachers with input for creating projects that will advance the children's strengths, carry their interests deeper, and help them close gaps in understanding and clear up misconceptions. Children are not "graded down" for what they do not know or cannot do but rather given opportunities to continue learning. Authentic assessment avoids measuring isolated skills and bits of knowledge out of context. Authentic assessment contrasts sharply with the kind of standardized tests that overlook the child's everyday life and are neither developmentally appropriate nor measure items related to the actual curriculum.

Portfolios. As you can see, authentic assessment is used for a variety of purposes. One of the tools of authentic assessment is the **portfolio.** Portfolios are collections of samples of the children's work; they document both process and product. Portfolios can be used as ongoing assessment devices, as well as a way to document the child's best work, serving as an ending record of what the child accomplished.

portfolio One of the tools of authentic assessment. Portfolios are collections of samples of children's work; they assess both process and product. Teachers, children, and parents can all contribute to portfolios in order to broaden the assessment to reflect developmental progress in the home as well as the early childhood setting.

When Program 2 first learned to do authentic assessment, they started making portfolios for each child. All of a sudden, they stopped the long-standing practice of sending home all the children's work at the end of every day. Because the children were used to the idea that they made things to take home, at first they resisted leaving anything overnight at the program. But the teachers were clear that a change was needed. With everything going home every day, it meant that there was little ongoing work and that each project was necessarily small and self-contained because it had to be completed in one session.

Program 2 began to introduce more ongoing, longer-term projects that extended over a period of days. This change made it easier to say to the child, "This isn't finished yet, so leave it here and you can work on it tomorrow." The teachers also created collaborative projects involving several children, so it was not clear who

Portfolios may contain photos or videos of processes such as the cooperative work these two children did on this block structure.
FatCamera/E+/Getty Images

should take the project home. The children began to see that things were different now, and they were more willing to leave the center at night without carting away every single thing they had worked on that day.

At first, the program collected and saved the week's work and then on Friday asked the children to select one piece to keep for their portfolio, keeping the children involved in the selection and self-evaluation process. The teachers paid attention to why the children chose certain pieces. They wrote down their words, which became part of the portfolio as well. The adults learned more about the children, and the children learned more about themselves. As a result of saving a piece of work for the portfolio, the children sometimes wanted to repeat something–to do it differently or elaborate on it. The teachers encouraged their inclinations and came to see the value of revisiting ideas. Now collecting for the portfolio is an ongoing process with children *and* adults deciding together what should go into it.

When a program decides to use a portfolio approach to assessment, it must decide what the collection should reflect and what means will be used to assess it. Portfolios can be a record of one or two areas of development–such as physical and cognitive development, for example–or they can cover all areas of development. Portfolios may contain children's drawings, scribbles, and invented or conventional writing, as well as lists of books read by the child or to the child to record that part of the child's emergent literacy. Photos or videos of processes (such as the building of a block structure or the creation of a carpentry project) may also be included, perhaps accompanied by the comments of adults and children to give a more complete view of the process than just the visuals themselves. The words add a lot. Likewise, the finished projects should be documented by pictures and words. What did Kyla say about her completed LEGO sculpture? What did the children who built the system of waterways and dams in the sandbox have to say about it? Furthermore, have children draw pictures of their processes and their products. Drawings, together with the child's comments about either the process or product or both, can be very useful for revealing how the child is making sense of the world. Recordings of children's interactions and conversations, as well as interviews, are also valuable. Whatever is added to the portfolio should be labeled, with a date and explanation of the setting, the circumstances, and any other pertinent information about the item.

Portfolios are wonderful additions to parent conferences because they provide concrete examples of the child's work and play. Parents can *see* how the child is progressing; the portfolio takes the conference out of the theoretical realm and into real life. In addition to the teachers and children, parents can also contribute to the portfolio. Just as the journal is used as a two-way record in some programs, so the portfolio can be used as a reflection of the child's life in the program *and* at home. Parents can contribute drawings, interviews, comments, and other kinds of documentation about what goes on at home. When this is done, the parent conference becomes a reciprocal conversation, rather than a one-way meeting in which the teacher does all the reporting and the parent is a mere recipient. Assessment then becomes broader and reflects developmental process and progress in more than one setting. It also helps to bridge the home-school gap–one of the primary goals of the early childhood program. It is not just what happens when the child is in the program that matters, it is also what goes on at home.

Self-Assessment. Children should have a part in their own assessment. Self-reflection and self-assessment are important skills to learn early, and adults help develop such skills when they ask children for their opinions: "What do you think about what just happened?" or "What do you like best about your work?"

Assessment is not constructive when there is pressure to perform and succeed according to some standard. When a teacher, whether consciously or unconsciously, emphasizes "right" answers by rewarding them, children learn very early to act as if they understand things they do not. Questioning children or probing to see what they know and what they do not know is a delicate skill. We need to help children say with confidence, "I don't know. I don't understand. Tell me more." There should be no shame attached to such statements, yet how many of us admit that we do not understand something? Most of us learned early to hide our ignorance. Let us be careful that we do not do the same to the children in our care. We cannot evaluate and further learning if we deny our own areas of ignorance or hide what we do not know from ourselves and others.

Assessing the Program

Assessing the Environment. The recording devices used to assess individuals can also be used to assess the program. Figure 12.6 shows a pattern of use of the indoor

FIGURE 12.6 Sample environmental checklist

Checklist

Use of the Indoor Environment

Observer ______________________ Date ________________

Time	9:30		9:40		9:50		10:00		10:10		10:20		10:30	
Boys Girls	B	G	B	G	B	G	B	G	B	G	B	G	B	G
Blocks	4	2	3	0	4	0	4	0	3	0	4	0	3	0
Art table	2	2	1	0	2	3	1	3	2	2	0	0	0	0
Easels	0	1	1	1	0	0	2	0	1	1	0	0	2	
Dramatic play	0	3	0	3	0	4	0	4	0	2	0	4	0	6
Manipulative area	2	2	3	3		1	2	2	1	1	2	3	3	2
Music area	1	0	0	0	3	2	1	0	1	3	2	0	0	0
Book corner	0	0	0	0	0	0	0	0	0	0	0	0	0	0
Science table	1	0	3	2	0	1	0	1	1	2	3	1	2	2

environment at free play time by a class of four- and five-year-olds. This checklist illustrates which activity centers are neglected and which ones attract more boys than girls, and vice versa. You can see in Figure 12.6 that boys never go near the dramatic play area, girls seldom go to the blocks, and nobody goes to the books. During the staff meeting, the teachers will have to brainstorm on how to change the environment so that both genders use all the areas. Will small figures in the block area get the girls interested? Will water in the play sink in the dramatic play area attract the boys? Also, what can we do about the neglected book area? (It turned out the book area was located in a raised area in a back corner and teachers seldom went there. When it was moved closer to the action and staffed by an adult, the children began to flock to it.)

Mapping can also tell you about how the environment is being used. For example, a comparison of maps like the one illustrated in Figure 12.7 was made to look at a problem in an infant-toddler program. Toys were constantly being scattered around–never left in the area where they belonged. Several toddlers' movements were mapped, with particular attention to their carrying behavior. They found out that the favorite activity of these toddlers was to pick up toys and carry them from one place to another and then drop them. Figure 12.7 shows that in 15 minutes one child dispersed six toys to new spots.

After the staff members in this program discovered this pattern, they had a meeting to talk about it. One suggestion was to change the behavior by using the behavior-modification technique of rewarding desired behavior. However, they reconsidered this plan after deciding that the behavior of these children was typical for toddlers. In other words, picking up, carrying, and dropping are "walking activities" for toddlers, not indications of a misbehavior. They decided to drop the rule that toys belonged in their own areas and were not to be removed to other areas. By changing the rule and their own expectations, they relieved a lot of their own frustrations about always keeping the environment neat and orderly. They were also able to engage the children in putting things back by using the fact that they liked to carry things from one place to another.

Assessing the Program from Different Perspectives. From Chapter 1, this book has discussed the value of self-reflection. Professional early childhood educators not only ask questions about the environment and about the children as a group and as individuals, but they also assess themselves–not to grade themselves–but to learn more, to upgrade their skills, to discover their own strengths, and, ultimately, to use all this knowledge to make the program better.

Reflection is also a key element in a comprehensive scheme for program assessment recommended by Lilian Katz, an early childhood researcher.[3] She suggests looking at the program from many angles. If you start from the bottom, you see the program from the children's point of view: What would it be like to be a two-year-old in this program? a four-year-old? a six-year-old? Is this a place where each individual gets needs met? Is this a warm and friendly place with interesting things to do? Is this a place to stretch and grow and feel good about oneself?

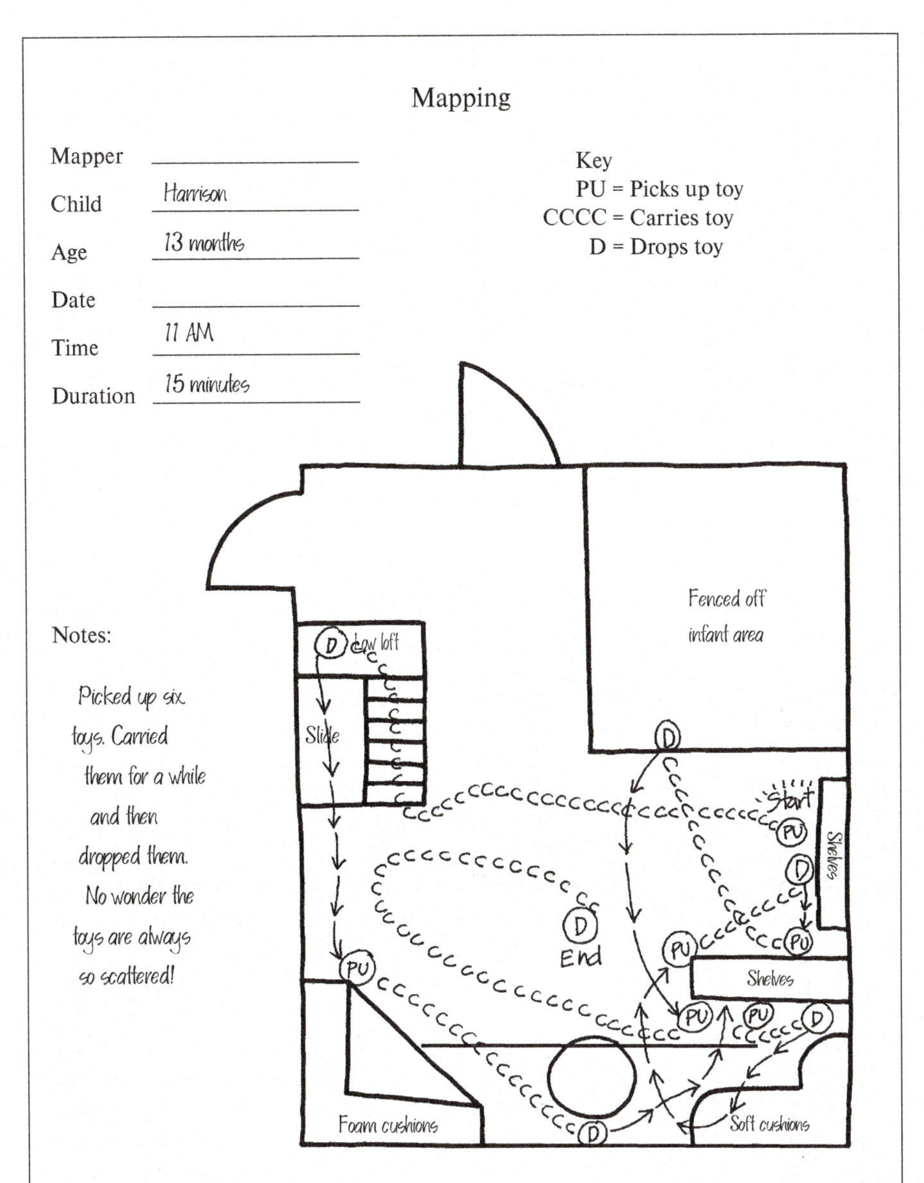

FIGURE 12.7 Sample mapping form

You can also look down from the top: How is this program viewed by the administrators, by the funders, by the board of directors? (This is often the prevailing view because the individuals at the top are usually the ones most involved in program assessment.) How does this program fit the regulations and meet the standards? What are the strengths and what areas need improvement?

Look at the program from the staff's viewpoint: What is it like to work here? Is this a place where you find support for yourself and your ideas? Is this a place where you can be creative? Can you stretch and grow? Does your work give you satisfaction and a sense of accomplishment?

Try to see the program through the parents' eyes: is this a comfortable, welcoming place where they feel at home? Does this program promote a team approach that includes the parent? Does this program fit cultural values even if it has to stretch and grow to do so? Is this a program where all families are helped to feel that they "fit"?

Finally, imagine how the community views the program: is there public understanding of what the program does and how it benefits the community? Is there a connection between the program and the community? Does the program reflect the community in various ways?

A Story to End With

For me, the hardest subject to observe is myself. And the hardest time for me to observe myself is when I am caught in the grips of my own feelings. I get caught when children push my buttons.

Button pushing works this way: a child exhibits some behavior that has been tried before and has proved to be effective. I then respond in a completely predictable fashion. As soon as a child learns that a particular behavior will get a reaction out of me, that child has a bit of power that he or she did not have before. Once we discover we have it, power is something we all like to use. Children are no exception.

Before long, a pattern is firmly established. When a child needs to feel powerful, he or she just finds me and does a little button pushing. Voilà like magic, I get emotional. Of course, I am a professional, so I do not carry on at school the way I might at home. Nevertheless, even under my professional veneer, button-pushing children can always find the real me—the emotional human being.

It is not that I do not want children to know me as an emotional human being, but I would rather relate human-to-human without the device of button pushing. That's why I have worked so hard all these years to change some patterns.

The secret to taking the power out of button pushing is to be *inconsistent*—to respond in a surprising way. Yes, I know, we have all been taught how important consistency is. But in this particular case, inconsistency is what is called for. The problem is, it is easy to think and talk about changing one's response when no buttons are being pushed. The difficulty is remembering to step back and observe oneself the moment of the button pushing.

It is possible to rehearse beforehand a number of alternative responses. I could ignore the behavior and just walk away. I could calmly redirect it. And if I were really good at being inconsistent, I could just laugh. That would surprise the button-pushing child. What is clear to me is that all it takes to break a pattern is to make some small change in my typical response. It is simple, but it is also astonishingly hard to step back from an emotional situation long enough to remember that there are alternatives.

Someone once said that freedom lies in the gap between the stimulus and the response. That is what I am working on—to recognize and use that gap. It takes a lot of self-awareness, it is not easy, but by practicing self-observation, I can now stop and ponder the moment button pushing occurs instead of automatically reacting.

SUMMARY

A major purpose of observation and record keeping is to gather information to facilitate curriculum planning, which entails creating projects, providing resources, and analyzing, adding to, or rearranging the environment. Early childhood professionals need to become good observers and must learn ways to record what they observe. Through careful observation and record keeping (sometimes called "documentation"), they seek to know the group and each individual in it. The tools early childhood professionals use to record information include anecdotal records, running record observations, incidents reports, journals, photographs, recordings (both audio and visual), checklists, mappings, time samples, and portfolios. They use the knowledge they gain from using these tools to plan long-term and immediate goals, keeping in mind that goals represent values that must take family and cultural contexts into consideration. Early childhood professionals also assess themselves to determine their own effectiveness and that of the program. A comprehensive program assessment involves evaluating a program in terms of its ability to meet the needs and goals of the children and their families, the administration and staff, and the community.

REFLECTION QUESTIONS

1. Are you a good observer? Explain your answer.
2. Can you remember a time when you observed something that other people had failed to notice? What was it and why were you the only one who noticed?
3. Can you remember a time when someone was observing you? What does it feel like to be observed?
4. Have you observed a child and noticed that he or she was uncomfortable with your watching? What can you do to make a child more comfortable in that situation?
5. If you were the teacher of the circle time described on page 343 what would you do if you noticed the children weren't paying attention?
6. What are your own experiences with being assessed in an educational setting? Can you relate these experiences to what you are learning about assessment in this class?

TERMS TO KNOW

How many of the following words can you use in a sentence? Do you know what they mean?

anecdotal record 336
running record observation 336
incidents report (or event sampling) 338
developmental checklist 339
environmental checklist 340
mapping 341
time sample 343
authentic assessment 346
portfolio 347

For Further Reading

Bates, C. (2014). Profile: Digital portfolios: Using technology to involve families. *YC Young Children, 69* (4), 56–57.

Bruce, Tina (2011). *Learning Through Play: For Babies, Toddlers and Young Children.* (2nd ed.). London: Hodder Education.

Carter, D. R., Norman, R., and Tredwell, C. (2011). Program-wide positive behavior support in preschool: Lessons for getting started. *Early Childhood Education Journal, 38* (5): 349–55.

Caspe, M., Seltzer, A., Kennedy, J. Cappio, M., and DeLorenzo, C. (2013). Infants, toddlers, and preschool: Engaging families in the child assessment process. *YC Young Children, 68* (3), 8–15.

Curtis, D., and Carter, M. (2012). *The Art of Awareness: How Observation Can Transform Your Teaching* (2nd ed.). St. Paul, MN: Redleaf.

Elicker, J., and McMullen, M. (2013). Infants and toddlers: Appropriate and meaningful assessment in family centered programs. *YC Young Children, 68* (3), 22–27.

Ferguson, C., Green, S., and Marchel, C. (2013). Kindergarten and primary grades: Teacher-made assessments show children's growth. *YC Young Children, 68* (3), 28–37.

Guss, S., Horm, D., Lang, E., Krehbiel, S., Petty, J., Austin, K., Bergren, C., Brown, A., and Holloway, S. (2013). Toddlers: Using classroom quality assessments to inform teacher decisions. *YC Young Children, 68* (3), 16–21.

Hancock, C., and Carter, D. (2016). Preschool: Building environments that encourage positive behavior: The preschool behavior support self-assessment. *YC Young Children, 71*(1), 66–73.

Kaiser, B., and Rasminsky, J. S. (2016). *Challenging Behavior in Young Children: Understanding, Preventing and Responding Effectively* (4th ed.). Upper Saddle River, NJ: Pearson.

Laski, E. (2013). Preschool and kindergarten: Portfolio picks: An approach for developing children's metacognition. *YC Young Children, 68* (3), 38–43.

Liewra, C., Reeble, T., and Rosenow, N. (2011). *Growing with Nature: Supporting Whole-Child Learning in Outdoor Classrooms.* Lincoln, NE: Arbor Day Foundation.

Pikler, E. (2006). *Unfolding of Infants' Natural Gross Motor Development.* Los Angeles: Resources for Infant Educaters.

Shabazian, A. (2016). Birth to grade 3: The role of documentation in fostering learning. *YC Young Children, 71* (3), 73–79.

Wylie, S., and Fenning, K. (2015). *Observing Young Children: Transforming Early Learning Through Reflective Practice* (5th ed.). Toronto, Canada: Nelson.

PART

3

Planning for Learning and Development by Enhancing Children's Curiosity, Joy, and Sense of Wonder

In some ways the last three chapters in this book are both the most traditional and at the same time the least traditional—depending on your background and perspective. Chapter 13 closely follows early childhood tradition, but Chapters 14 and 15 fit into the disciplines used and understood by the general population and by teachers of older students. Early childhood educators do not label the subject matter in the same way. Instead, we tend to categorize learning and development into the areas of physical, cognitive, and social emotional—what we call the "three domains of the whole child." We make sure that our planning for educational experiences covers these domains without thinking if they are called social studies or science.

Other traditional ways to categorize learning and developmental experiences are by "centers" or "areas." These are places set up in the classroom, and sometimes outdoors as well, where children can have hands-on, interactive experiences with materials, individually or in small groups. Many times the centers are open-ended and the children explore and use the materials as they choose. Other times the centers may also have "task cards" with suggestions, or even assignments, of what to do. Some of the centers may have the same names as the subjects of Part 3, for example: the art area, music area, science table. But some are quite different. Take the "block corner," for example–which does address the three domains but does not fit into any single discipline understood by the general public. How about the "dramatic play area"–another traditional center in an early childhood environment? And do not forget the play dough table. These areas were explored in the earlier chapter on the physical environment.

In keeping with the theme of diversity, I have chosen to explain approaches to curriculum in a number of different ways. This last way is designed to help both you and those who do not know about early education see that learning is a continuum. The courses taught in later grades and even in college have their roots–their foundations–in the experiences offered to the very young in early childhood programs. We do not have language classes or even language lessons, but language development and learning are a vital part of everything. Literacy is a language tool–a way to use language more remotely and record it more permanently through visual symbols or tactile ones (in the case of Braille). A program may have a formal approach to teaching math, such as addressing math concepts during small group activities or during circle time. Many programs have a "math table" with manipulative materials on it. Those obvious math activities, however, should not take the focus off the mathematical opportunities of many other experiences. Unit blocks are highly mathematical, for example. The children may seem to be using the blocks only for imaginative purposes, yet they are gaining concrete math skills at the same time, when they substitute two short blocks for one long one or put two triangles together to make a square. Other areas with great potential for mathematical learning include the sandbox and the water table.

Science teaching goes way beyond a hands-on science table display, science projects, or developmentally appropriate science experiments. The teacher who can discern science in everything can bring out the scientific concepts in any experience. The key to effectively facilitating the teaching-learning process is to stay within the realm of developmental appropriateness.

The last chapter looks at art, music, and social studies. When public schools' funding is cut back, the first things to go are art and music, because compared to academics, they seem extraneous. Yet, in early childhood, art and music are essential for developmentally appropriate, holistic learning. Social studies is almost never called that, but if you examine most early childhood education approaches, you'll see social studies. There is usually a starting focus of self-study–perhaps a unit called "all about me," with the goal of increasing self-awareness, self-image, and self-esteem. The movement in the "social studies curriculum" is from self to others, and you'll see that every day in the way teachers guide and facilitate interactions. Community is also a part of the progression of social studies, both living in community and studying it.

Early childhood exposure to experiences in art, music, and social studies are just as crucial to future outcomes as math and reading. Such experiences bolster the soft skills such as effort, initiative, and self-control that are normally not measured but are correlated with future job stability and better health. Indeed, early child measures of academic performance fade by primary school, but the returns on soft skills persist into adulthood.[1]

Although it is not important to label the subjects with the names of traditional disciplines of the adult world, using names that are easily recognized by the general public helps the image of early childhood education. We are a valid part of the education system and need to be recognized by a society that has the tendency to think of anyone who works with children younger than kindergarten as "babysitters."

The text will close with a direct focus on the holistic theme that has underscored the entire book. Chapter 15 will demonstrate how early childhood education (ECE) educators use tools such as the topic web to create a holistic curriculum.

So now you are ready for the final section of the book. Enjoy!

13 Enhancing Children's Learning and Development Through Language and Emergent Literacy

Ingram Publishing

Diversity and Language

How to Facilitate Language Development

Facilitating Language Development in Infants and Toddlers

Facilitating Language Development in Two-Year-Olds

Facilitating Language Development in Three-, Four-, and Five-Year-Olds

Facilitating Language Development in School-Age Children

Emergent Literacy

A Reading-Readiness Approach Versus an Emergent-Literacy Approach

Emergent Literacy for Infants and Toddlers

Promoting the Development of Emergent-Literacy Skills in Three-, Four-, and Five-Year-Olds

Promoting the Development of Emergent-Literacy Skills in School-Age Children

A Story to End With

In This Chapter You Will Discover

- what language allows children to do.
- how language must be learned in context.
- some issues surrounding bilingual education.
- how some children risk losing their home language when they come to an English-only early childhood program.
- how adults facilitate language development.
- what role pretending plays in language development.
- why a young child might say "he goed."
- what role arguing plays in language development.
- how an emergent-literacy approach differs from a reading-readiness approach.
- how an emergent-literacy approach is applied in infant-toddler programs.
- how teachers and providers facilitate emergent literacy in programs for preschoolers and kindergartners.
- how teachers and providers facilitate emergent literacy in programs for school-age children.

NAEYC Program Standards
Program Standards 2 & 3
Curriculum and Teaching

One cannot write a book about early childhood education without mentioning language development and learning in every chapter. This book is no exception. Language and communication underlie all the aspects of early childhood education discussed so far. Language learning is dependent on cognition. We use language to think, and we think when we use language. Many people consider the two processes as one or at least as being closely related. Cognition, that is, gaining knowledge through the senses, experiences, and by thinking, whether specifically mentioned or not, is embedded in these last three chapters.

Remember, this book is based on a holistic approach. Although this chapter focuses on language, it is impossible to discuss language without considering the whole child. Language involves physical, perceptual, mental, and social skills. Emotion figures in too because feelings affect language development and language is used to express feelings.

Let us start with some definitions. The word *language* (whose Latin root means "tongue") can be defined as "the formation and communication of information, thoughts, and feelings through the use of words." Language development and learning also eventually entail producing and understanding the written word; in other words, language is talking, understanding, reading, and writing. Language is both active and receptive–that is, we transmit it (talk) and receive it (understand). By studying language development and learning from the beginning of life, we know that understanding comes first, then talking. The same holds true for learning a second language (at any age); most beginning learners can understand more than they can say–at first.

emergent literacy
The ongoing, holistic process of becoming literate—that is, learning to read and write. Emergent literacy contrasts with a reading-readiness approach, which emphasizes teaching isolated skills rather than allowing literacy to naturally unfold in a print-rich environment.

Emergent literacy is defined as the ongoing process of becoming literate, that is, learning to read and write. Emergent literacy is included in the holistic package of language development, and it starts at birth. Remember, the purpose of language–whether oral or written–is communication.

What does language allow children to do? It helps them make cognitive links, clarify their needs, gather information, and label objects and experiences and store them symbolically so they can remember and talk about them later. Language development also involves categorizing and classifying on an increasingly complex level. It allows children to plan as they organize and order their experiences. It also increases their ability to cope. Eventually, language allows children to reason.

Let us look at how the skill of classifying develops: "Doggie," says the caregiver, pointing to a picture. "Doggie," imitates the child. The next day, driving down the highway, the child sees a cow. "Doggie!" he says, pointing. "No, not doggie, Cow!" says the caregiver. At this point the child may or may not accept the new label. If he classifies all four-legged animals under the label "doggie," it may take a while for him to understand that this is a different category of animals. Eventually, he will get it straight and come to see that there are hierarchies of classes; that is, "animals" is the larger category under which "dogs" and "cows" fit as separate subcategories. And then there are smaller classes of dogs and cows, like Labradors, Cocker Spaniels, and Bassett Hounds and Jerseys, Herefords, and Angus. Conversely, the category "animals" falls under an even larger category, say, "living things."

It takes years for children to sort all this out, but eventually they do. And they come to discover that one thing or being can belong to many categories at the same time. An adult can be a teacher, a mother, and a daughter. A child can be Chinese, American, Christian, and San Franciscan.

Just learning that beings and things have labels is a big step forward. At first, adults provide the labels, but before long, the child is pointing and asking "What's that?"

As mentioned earlier, it is hard to separate language from cognition; we know what a child is *thinking* by listening to what he or she is *saying*. Jean Piaget and Lev Vygotsky created their theories of cognitive development by observing children and listening to them.

Language is caught, not taught.[2] You do not need to set up language lessons to teach language. Language is learned in context. Children use whatever language they have to talk with adults and other children, and when the other person responds, they try to understand the message and answer in ways that keep the conversation going.

Diversity and Language

What language should early childhood programs emphasize? For many readers, that question may seem puzzling. If your answer is "English, of course!" you may not be aware of some of the complex issues that surround this seemingly obvious question. You may also not be aware of the National Association for the Education of Young Children's position statement on responding to Linguistic and Cultural Diversity. The 1995 statement is clear: "For the optimal development and learning of all children, educators must **accept** the legitimacy of children's home language, **respect** (hold in high regard) and **value** (esteem, appreciate) the home culture, and **promote** and **encourage** the active involvement and support of all families, including extended and nontraditional family units." The NAEYC has not updated this statement; the 1995 version is still the most current on its website.

NAEYC Position Statements

NAEYC/IRA Position Statement: Learning to Read and Write NAEYC Position Statement: Responding to Linguistic and Cultural Diversity

As has been stressed throughout this book, the statement emphasizes that early childhood educators acknowledge and respect children's home language and culture. When they do, ties between the family and programs are strengthened, which gives increased opportunity for learning because young children feel supported, nurtured, and connected.

But what about the child whose **home language** is English? There's probably no reason that the English-speaking child cannot learn a second language (or a third, if the child already has two). Children all around the world learn two, even three languages starting at birth. It is possible for children to learn two languages in early childhood programs, but there may be unintended consequences to such instruction.

home language The language spoken at home. For many children that language is English, but for many others, it is a language other than English. The term can also be used for a particular way of speaking English that differs from what is called "standard English."

Whether a child should learn two languages at once in the early years depends on the risk factors associated with losing the home language. If the child comes from an English-speaking home and the family wants the child to learn a second language, the risk factors are minimal. Given that the second language is taught in natural and meaningful ways by someone who is a competent speaker of that language, the child may become bilingual in a few years and even biliterate if

How Will Noemi's Language Unfold?

Noemi is four years old and has some developmental delays related to her diagnosis of Down syndrome. She is enrolled in both a special preschool class in the mornings and an inclusive child development program in the afternoons. Noemi has an 11-year-old sister, Margarita, who attends the local middle school where classes are held in English. Margarita is a star student, having won several academic honors over the years. Both of Noemi's parents are Spanish-speaking and are currently enrolled in English as a second language (ESL) classes at the community college. Noemi's mother cleans homes for a living and her father works in construction. Due to the high cost of living in their town, the family lives in a very tiny one-bedroom apartment. Their dreams for both of their daughters are for them to be contributing members to society, to go to college, and to have a happy life.

Noemi and Margarita's mother prides herself on spending time with each of her daughters to support their learning. She has magnetic alphabet letters on the refrigerator, memory and matching games, puzzles, an art easel, and a few children's books at home. She knows that books are important so she takes them to the library twice a month to pick out new books. Before Margarita gets home from school, Noemi and her mother walk to the neighborhood park and her mother labels objects, places, and people they see along the way.

One of the teacher's assistants at the afternoon child development program is bilingual and uses Spanish with the children from time to time. The bilingual children in that program are comfortable using Spanish when they are playing with each other. Noemi is much more vocal in this afternoon program and her teachers actually see her as one of the class leaders. The teachers in the morning class see Noemi as a shy, withdrawn child. As a result, many of the goals on her Individualized Education Plan focus on language and communication. When Noemi gets home, she shows her boisterous and active side. Her mother says that Noemi "sings" along to ranchero and popular music. She knows about 10 words in Spanish and uses them quite effectively by varying her intonation or loudness and pairing her speech with gestures. She is popular with the other young children in the apartment complex, all of whom use Spanish when playing together.

Noemi's parents worry because they are not sure how her language development will unfold as she grows up. All the specialized services offered to Noemi and her family thus far have been in English, including speech therapy. Although they would like for Noemi to learn English well like her sister has, they also do not want her to lose her Spanish.

–Rebeca Valdivia

exposed to print in both languages. However, because preschool enrollment of children from culturally and linguistically–only classroom versus one that accommodates a second language in its teaching. For many decades, early research erroneously suggested that bilingualism was an impediment to learning. According to Fred Genesse, there are many myths associated with bilingualism–such as the one that young children lack the capacity for learning more than one language at a time.[3] Another myth is that delays are associated with learning two languages. Around the world children have grown up in bilingual and trilingual homes, and their families have never considered the learning of more than one language to be a problem. Those families are well aware of the advantages of competence in more than one language. Moreover, research suggests that dual-language classrooms are both socially and cognitively beneficial for all children.[4] When compared to their monolinguistic peers, bilingual children can have advantages that persist throughout their childhood and well into adulthood, from the earlier development of theory of mind to protections against dementia and cognitive decline.[5]

Before embarking on a bilingual goal in early education, it is important to understand what a particular family wants for the child and what the child needs. Emotional issues figure in here. Will being in a bilingual setting add to the child's sense of self and feelings of security or not? Does the program have staff members who speak the child's home language? If not, how comfortable will the child feel if there is no one around who speaks or understands the language? In the *Voices of Experience* box on page 362, Rebeca Valdivia, a bilingual education expert, shares Noemi's story, which relates to some of these questions.

In the United States, there are **language-immersion programs** with bilingual goals where the child is exposed only to the target language. The approach seems to work for most English-speaking children; however, CLD children who are in English-only programs often struggle due to the lack of support and cultural empathy. Often they are in immersion programs where the goal is not bilingual education. Rather, they are expected to learn English quickly without any regard to the possible loss of their home language. This approach is subtractive rather than additive and has grave implications for the child's bilingual potential, self-concept, self-identity, and connections with the family and other members of the home culture.

In the United States, it is less punitive for an English-speaking child to learn in an immersion program in another language than for a non-native English-speaking child to learn in an English-only classroom. The risk factors for losing the home language are far less for the English-speaking child; he is surrounded by English at home and in the greater world and will retain his home language in spite of learning another language. However, the Spanish-speaking child in an English-only program, for instance, may turn her back on her home language if she inadvertently gets the message that English is better than her own language.[6]

Language-immersion programs have existed for a while, but changes have produced improvements over time. Today, many immersion programs with bilingual goals are showing effectiveness for both native English speakers and for English learners whose primary language is other than English.

It is important to observe how a child is doing in a language-immersion program. Does the child feel isolated and alone? Some children in early childhood programs feel scared and lonely, even when surrounded by people who speak their language. Separation from their families is a major issue for some children that can be compounded when combined with communication problems. Imagine what it would be like to spend long hours in an environment where no one spoke your language–even

language-immersion programs Have as their purpose the learning of a language that is not the child's home language. When children who are at risk for losing their home language are put into English language immersion programs, the result is often the replacement of their home language with English. An approach that has had more positive results in the United States is called two-way language immersion programs or dual language immersion programs. In this approach, half the children have English as their home language and the other half come from a different language group—such as Spanish. In this situation each group learns the language of the other as instruction occurs in both. Children are more likely to end up bilingual in two-way language immersion programs.

worse, where no one understood your language either. Of course, plenty of children, along with adults, have been in such situations and have overcome their initial insecurities in a relatively short time. Still, such an experience has different effects on different people; for some people, the effects can be long lasting and harmful to their self-identity and self-esteem.

A better model for the beginning years is a bilingual or multilingual setting, where the home languages of the children are spoken by one or several staff members. Children may or may not hear English in their first years, but they definitely need to hear their home language if they are to develop in it, feel good about themselves, and feel firmly attached to their family.[7]

Perhaps you are under the impression that most children in the United States come from English-speaking homes, but it is not true. (Even if it were true, bilingual education would still be valuable for any country that wanted to produce bilingual citizens capable of functioning effectively in more than one language setting.) Twenty-two percent of children will speak a language other than English at home, a rising rate that widely varies by state, from a high of 44 percent in California to a low of 2 percent in West Virginia.[8]

Home-language preservation is important! Bilingual education must be sensitive and appropriately timed. Think what we lose when these children switch from their home language to English, which commonly happens when no provisions are made for valuing and preserving the home language; not only do children lose cognitive- and emotional-development opportunities, but we as a society lose bilingual citizens as well. For these reasons, the NAEYC advocates cultural and linguistic responsiveness and sensitivity in its position statement.[9]

Laurie Makin and her colleagues in Sydney, Australia, study and advocate for the preservation of home language in early childhood education programs. In their book *One Childhood, Many Languages,* they list the following guiding principles for early childhood educators:

- Families are key participants in early childhood language learning.
- The languages children bring to early childhood programs should be maintained and developed.
- Early childhood programs should be culturally and linguistically relevant.
- In language-rich environments, all children can explore other languages as well as their home language.
- Bilingual children have specific language needs and individual approaches to learning languages.
- Being bilingual is beneficial for all children.[10]

How to Facilitate Language Development

So suppose the first question has been answered—What language should early childhood programs emphasize? The next question is, How do adults teach language—whether one or two (or even three)—to children in early childhood programs?

As mentioned earlier, language is "caught" (not taught) through interactions with other people. Language learning is both an internal and external process and depends on the child's developmental level and needs as well as the people in the environment and the kinds of language interactions that occur. Vygotsky took an interactionalist view of language development and described how adults move children forward in their language development. By using an assisting method called "scaffolding," adults move children into what he called the "zone of proximal development." When adults are aware of not only the child's present level of linguistic skill and conceptual development but also of what is likely to come next, they can provide appropriate input to enhance the child's language learning and understanding.[11]

Language development is facilitated when relationships are formed and when the people involved have common interests to talk about. But the question is, What should the early childhood educator talk about with children to facilitate their language development and create attachment and form relationships? Lilian Katz has some ideas about adult-child conversations for children preschool age and beyond. She has observed early childhood classrooms for many years and claims that U.S. teacher-child relationships focus mainly on the routines and rules of the class or on the children themselves–their conduct and performance. None of those subjects make for particularly inspiring conversations.

In contrast, Katz's observations of Reggio Emilia schools in Italy reveal that the relationships and therefore the conversations in these programs focus on projects that reflect the deep interest of both the children and the adults. Katz writes, "Both the children and the teachers seem to be equally involved in the progress of the work, the ideas being explored, the techniques and materials to be used, and the progress of the projects themselves."[12] Adults and children work together on these projects with the children playing an apprentice role. Such collaboration is different from a teacher instructing and directing children in a one-sided monologue. It's also different from talking to them mainly about rules and proper conduct or praising them for their performance.

What other types of conversations are appropriate? Some conversations can be playful and meaningless, such as when children explore language and sounds and adults respond lightheartedly. This kind of exchange is fun and encourages creative language production; it also promotes adult-child relationships.

Facilitating Language Development in Infants and Toddlers

In infancy especially, loving and playful exchanges provide the basis for language development, as babies learn conversational turn taking through nonverbal interactions at first, then through vocalizations, and eventually through verbalizations.[13] It is important to use verbal language to converse with infants as early as possible. From the day they are born, their senses are attuned to linguistically rich and loving interactions to spur brain development. Therefore, infants learn from day one and gain an understanding of their environment long before they can speak their first word. Watch adults and infants and you will see that conversational turn taking involves more than just language exchanges. Adults mimic

Focus on Diversity

Cultural Contrasts in Communication Styles with Infants

Some cultures value nonverbal infant communication more than verbal interactions. For example, in cultures where babies are constantly carried or kept close by their caregivers, verbal exchanges are not essential to communication; when a baby cries, the caregiver is immediately present to give a warm caress, squeeze, or jiggle to reassure the baby. In other cultures, however, babies sleep in cribs in separate rooms and ride around in strollers, rather than being carried. When these babies cry, their caregivers–who may be in a separate room or unable to see them over the hood of a stroller–must use the words ("I'm right here. Don't worry, I'm coming.") to reassure the infant that care and attention are on the way.

Classic studies comparing Japanese mothers with European American mothers reveal differences in how they communicate with their babies.[14] Think about two mothers, Rebecca and Joy, relating to their babies in very different ways. Joy is far less animated in how she responds and talks to her baby than Rebecca. Like Joy, Japanese mothers tend to use indirect communication. They value intuition; they are just as empathetic as European American mothers, but they do not feel the need to put their thoughts and feelings into words. Nonverbal empathy is the goal. In contrast, although not all European American mothers are as animated as Rebecca, many value direct communication and use it to connect with their babies as well as stimulate them.

babies, and babies mimic adults. How much or how little the adult talks may be dependent on his or her temperament, personality, and even culture. (See the *Focus on Diversity* box above for a discussion of cultural differences in adult-infant communication styles.)

Adults make it easier for children to learn language when they use scaffolding–that is, when they prompt them and use other ways of subtly helping them to understand what is being said. Most adults also naturally employ labeling in their conversations, and they tend to use simple, short sentences that refer to what the child can see or do. Repeating what the child says and putting it in correct form and pronunciation seem to come naturally to many adults.

To illustrate, let us look at an exchange between a caregiver and a toddler: the caregiver says, "Go get your coat." The child looks blank, so the caregiver says, "We're going outside. It's cold out there." He pauses, looking at her. Her expression doesn't change. "You need your coat." He shows his own coat to her. "I have my coat on; you need yours." The child looks questioning. "Your coat is on the hook over there." The child still stands there, looking from caregiver to hook. "Go get it," he says. "Coat?" asks the child. "Yes, get your coat–the blue one." He points to a blue coat hanging on a hook. "Go outside?" says the child starting for the coat rack. "Right, we're going outside now," says the caregiver.

"Expanding" is another valuable method of facilitating language development at any age. It may not come naturally to every adult but can easily be learned. Here is an example of expanding: "Mommy!" says a child, looking wistfully at the door. How the teacher/provider/caregiver responds depends on whether it is 8:05 a.m. or 4:15 p.m. At 8:05, the adult says, "Yes, you miss your mommy. She went to work. She'll be back this afternoon." At 4:15, the adult says, "Yes, almost time for Mommy. She'll walk in that door!" The child may not have any concept of "this afternoon," or "almost time," but by hearing those phrases in context, eventually he'll develop the concepts "past" and "future" to add to his understanding of the here and now.

As you can see, conversations with children up to two are mostly about the present or the immediate past or future. Caregiving routines, for example, are legitimate subjects of conversation, but conversation also depends on being responsive to the child's initiations and responses.

Facilitating Language Development in Two-Year-Olds

Two-year-olds' conversations usually contain words that have to do with daily experiences and that name things and actions. They can put into words what they are doing, and they are capable of using their imagination and language in pretend play. Following is a scene that illustrates a two-year-old's language and symbolic development.[15]

A. pulled her shoe off her foot with one hand, turned to her caregiver, smiled, and said, "Take shoe off." She lifted her shoe to show her caregiver. "Yes, I see, you took your shoe off by yourself," replied the caregiver. A. put that shoe down and took off her other one. "Now you have both off," said her caregiver. A. stared at her shoeless feet for two seconds and then crawled over to a ball lying nearby.

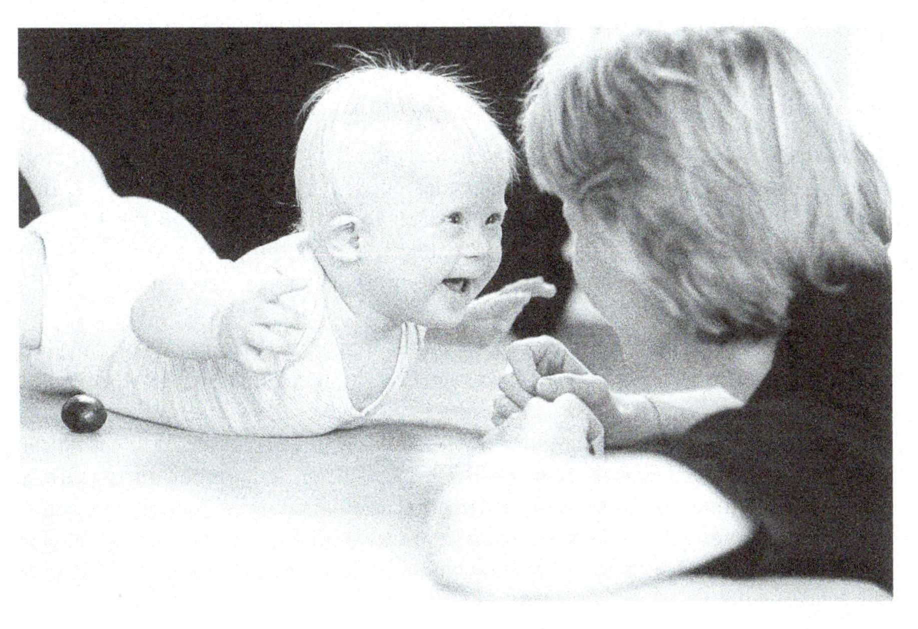

Language is caught, not taught.
Ursula Markus/Science Source

She picked it up and put it back down. It rolled away. She looked at the ball and said "No," pointing at it with her finger. Then she looked at her caregiver and smiled. The caregiver expanded on what she said and put into words what she perceived the child was feeling: "You didn't want the ball to roll away." A. crawled over to the ball and picked it up and held it next to her cheek. She dropped the ball and laughed and squealed. Then she ran across the room to get a baby doll. She picked up the doll and brought it back toward the caregiver. As she walked, she looked at the doll and said, "baby" several times. She walked by the caregiver and went to sit down by the window with her back to the room and laid the baby doll in her lap. She said, "Baby go nigh-night" to no one in particular. Then she kissed her doll and said, "Go nigh-night. Sweet dreams." A. closed her eyes and pretended to snore.

What we see in this scene is a little girl with a caregiver nearby who joins in her play by verbally responding to what the child initiates. Notice how at first the caregiver responds by expanding on what A. says. By the end of the scene, however, A. goes into her own world and no longer involves the caregiver, who then keeps appropriately quiet. A. talks about what she is doing. She moves from a realistic mode to a pretend mode: she talks to her caregiver, to the ball, and to the doll. She says "Nigh-night" to the doll the same way an adult would say it to her. She symbolizes sleep by snoring. She is a good imitator.

Facilitating Language Development in Three-, Four-, and Five-Year-Olds

Taking a project approach is a way that preschool and kindergarten teachers facilitate language development in their students. Three-, four-, and five-year-olds are more likely than younger children to have conversations that revolve around projects. Their projects are deeper and longer term than those of toddlers, and the verbal exchanges that grow out of them can be as interesting to the adults as to the children. While the children are talking about what they are doing and learning, they can also be encouraged to symbolize it in various ways, through drawings, diagrams, three-dimensional models, and, of course, dictation and, eventually, their own writing (captions, journal entries, stories, poems, and so on). Again, adults should take a holistic approach to oral language and the language of other kinds of symbols. Recognizing, valuing, and encouraging a variety of ways that children can use symbols are important to language development in this age group.

What are other ways adults can facilitate language learning? Providing language models is one way. (The *Focus on Diversity* box illustrates how modeling language differs among cultures.) Adults should also allow and encourage conversations of all sorts. Most children enjoy talking to each other. When children are actively engaged in conversations with each other, it is important to refrain from interrupting. Furthermore, be sure that the schedule of the various activities and routines is not so brisk that it prevents the children from settling in and talking to each other. An appropriate pace can be an important factor in language facilitation.

FOCUS ON DIVERSITY

Communication Styles in High- and Low-Context Cultures

As discussed in the *Focus on Diversity* box on page 366, communication styles differ among cultures. One way they differ is in their varying degree of dependence on context versus language to communicate with others. For example, mainstream American culture fits into the low-context end of the communication continuum; this culture shows a high dependence on words to communicate meaning–rather than on context. **Low-context cultures** value language greatly. In an exchange, a low-context person aims for good, clear communication through verbal means. The goal is to be articulate. And even though spoken language does not carry all the meaning in an exchange, low-context people place more importance on it than on the unspoken context and nonverbal communication (such as body language).

In contrast, a person from a **high-context culture** derives information from the context and gains only minimal information from the language of the message. The high-context individual pays close attention to pauses, body language, feelings, relationships, and a history of shared traditions with other members of the culture. The Japanese are an example of a culture at the high-context end of the continuum.[16] Communication for a high-context person is not direct and is not expected to be by other members of the same culture. Many things can be left unsaid because the meaning comes through the context.

As you would imagine, low- and high-context cultures approach language develop ment in early childhood differently. Low-context cultures place a great emphasis on encouraging children to talk. In contrast, communication in high-context cultures depends greatly on traditional knowledge passed from the older to the younger generation; if that knowledge is absent, communication suffers.

Because early childhood education is based on the perspective of a low-context culture, it is important for professionals in the field to be sensitive to the needs of high-context children. These children need to be well grounded in their own culture; otherwise, their interpersonal communication skills with other members of their culture will suffer. For these children, it is not just a matter of knowing their language, but also of being skilled in the communication style and of understanding the whole context.[17]

Talking is only part of communication; listening is the other half. Again, modeling is the best way to teach listening skills. Think what happens when we model the following poor listening skills:

- When we vaguely hear but do not really listen
- When we listen but do not pay close attention
- When we listen selectively

low-context culture A low-context culture is one that depends a great deal on words to convey messages rather than emphasizing context.

Think of how different those "listening lessons" are from showing a child that we are giving our whole attention to what he or she is saying! The *Tips and Techniques* box describes four levels of listening. As you will see, level 4 listening is what

TIPS AND TECHNIQUES

Four Levels of Listening

Check out these four levels of listening. Are you aware when you are listening at each of these levels? How often are you listening at level 3?

Level 1 listening is not really listening at all; it is simply *hearing*. Hearing is a sensory process associated with listening. For hearing to occur, the sound waves only need to be received, but their meaning does not have to be perceived by the brain. In other words, we hear, but we do not listen; we take in sound, but we do not process the message. The ability to hear is often necessary for listening to take place, but hearing alone is not enough for communication; indeed, people with hearing impairments have the ability to listen.

Level 2 listening is a step above hearing. The words register but only slightly. Level 2 listening is like being in a holding pattern: we wait for the other person to stop speaking so that we can begin speaking. We spend the waiting time rehearsing our own message. We listen just enough to loosely connect with what the other person is saying. In a level 2 conversation, both parties may have their own agenda, but they may not be able to relate very much to the other person's interests. A conversation at level 2 is similar to the way children engage in parallel play.

Another type of level 2 listening is not conversational but occurs when we simply take in information without processing it. The information goes in but is not sorted, analyzed, or interpreted. At this level, the listener, like a computer, is indifferent to the source of the information.

Level 3 listening is true listening. We register and receive what is being said. We listen intently with a particular purpose in mind and screen out whatever is irrelevant to that purpose. At level 3, we listen offensively, as a lawyer listens to the other side's witness–trying to trap her with her own words. We also listen defensively, like a person listening skeptically to a sales pitch over the phone, trying to figure out what the catch is. Arguments typically move back and forth between offensive and defensive listening.

Level 4 listening is wholehearted, open listening. The goal is to understand the message being sent. At level 4 listening, we are actively involved with what the speaker is saying. At level 4, we are receiving more than the actual words themselves. At the same time we receive and listen to the entire message, we have the ability to interpret the meanings behind the words (by noting semantic nuances and inflections in the voice) and to analyze body language and other contextual cues. At level 4, we do all that while using our own verbal and nonverbal responses to indicate we understand and have a feeling for the other person's meanings and emotions.

high-context culture A culture that depends more on context than on spoken or written language to get messages across.

early childhood educators should practice in the interest of facilitating language development in young children of all ages.

But besides modeling language and listening skills, what else can adults do to facilitate language development in preschoolers and kindergartners? Refrain from correcting them. Children will correct their own errors as their perceptions and memory improve and as they gain more experience in the world. The child who says "he goed" or "I have two foots" is just beginning to learn the structural rules of the language; he has overgeneralized a past-tense rule or a pluralization rule. But without anyone ever correcting him, he will eventually start saying "he went" and "I have two feet" when he hears that is what other people say.

Help young children expand their vocabulary. As children move through the preschool years, the number of words they use increases astonishingly and sentences get longer and more complex. Adults help this process by giving children plenty to talk about while doing projects and studying topics of interest. As they talk, they come up with ideas, plan, run into problems, evaluate solutions, and begin to predict outcomes and consequences. They can do some of this on their own, but having an adult nearby to scaffold adds depth and breadth to the thinking process and the language that expresses it.

Interactive and collaborative play, such as that which goes on in the block and dramatic play areas, presents rich opportunities for learning and practicing language. Children may or may not need an adult to facilitate this kind of play, depending on how good their communication and social skills are. Let us listen to two skilled boys having a discussion about pretend playing in the block area.

"Brian doesn't like to play with action figures anymore. He thinks he's too old," said five-year-old Tim to his friend. "That's because he doesn't know how to play the big-kid way," said Paul. "He just knows the 'bang-bang' baby way."

"Tell me about that," said a teacher passing by who just happened to overhear this conversation. She was curious, and, besides, she had a paper due on dramatic play for a class she was taking. "Well," responded Paul, "when you're really little, you just go, 'bang you're dead.' But when you grow up, you start making the figures talk. When you're really older, you make up whole stories. That's what we do, at least when we don't get too busy building bases and stuff for the figures. The story part is what Brian doesn't know about."

The teacher was delighted. A five-year-old had just outlined the sequence of development of dramatic play with small figures. She would write her paper about how children gain a sense of power by creating their own small worlds. They transform reality and practice mastery over it through the use of language. In addition to a sense of personal power, the teacher had observed how the children develop sophisticated communication skills through pretend play; they deal with several levels of communication, as the figures themselves interact and the players who control them also interact. The children also gain social skills, as they practice negotiation and cooperation on a real and a pretend level. Furthermore, they acquire significant intellectual skills by constructing mental images and learning to deal with the world in a symbolic way, as well as practicing story sequences.

All this was going through the teacher's head as she continued to watch the boys. They were putting the finishing touches on a complex creation of blocks and figures. She was thinking how her paper was practically writing itself when, suddenly, the fire alarm rang.

Paul jumped up and went straight to the door the minute he heard the bell, but Tim stood his ground, protesting. "Oh no, not now," he shouted. "We can't have a fire drill now!" He looked extremely agitated.

The teacher hustled him toward the door and tried to reassure him. "You can pick up where you left off when we get back." "Sure," he said, grumbling. "*If* I remember where we were." He turned back to look at the figure hanging from a tower by one foot and at another one hiding behind a box. The boys had created two impressive structures, one of blocks and another of words and actions. The blocks would remain, but the words would likely be lost. Nobody could solve that problem for him. The story was interrupted.

As he walked out the door, he came up with his own solution. "Oh well," he said looking at the teacher, his face bright now. "If I forget, then when we get back we can just start another episode!"

register A particular style of language or way of speaking that varies according to the circumstances and the role a person is filling at the moment.

Children's language development is truly amazing. They not only learn all the skills already mentioned, but they also learn different ways of talking for different occasions. Linguists use the term **register** to describe these different ways of talking. Children speak one way when they talk to each other, another way to parents, and still another way to teachers. The following scene demonstrates how children shift registers.

Four-year-old Erin is on the play phone. She perfectly imitates the tone and conventions of telephone talk. "Hello, is Daddy there? Hi, Daddy. I'm having fun playing. What are you doing? I have to go now. Good-bye." Then she gets down on the floor and crawls over to Charlene, who is standing at the sink. "Pretend I'm the baby," Erin says, using the word "pretend," which helps Charlene understand Erin's shift in register. Erin's voice and facial expression change, as she says, "Go pee-pee, mama!" she tugs on Charlene's blouse.

Puppets facilitate language development as children use their voices to give them different personalities and roles.

"Stop that, baby!" says Charlene sternly. "It isn't time to go pee-pee! It's time to eat. Get in your chair." Erin obediently climbs into a chair at the table, and Charlene serves her some pretend food and then sits down herself. Once the meal is over, the girls decide to switch to something else. The signal word is again "pretend."

"Pretend we're doctors and we have to operate," says Erin, no longer a baby. She hands Charlene a doll and then gets one for herself. Soon, both girls are doing surgery on dolls using the kitchen table and play knives from the cupboard. They have changed register again. Now they are talking as doctors, not as mommies or babies.

When the teacher tells them it is time to clean up, they switch to yet another register and ask for more time to finish what they are doing in their best preschool-age manner. Besides demonstrating an ability to shift register, these two girls, like the boys in the block area, use language to establish and negotiate roles.

Interactive play, especially dramatic play, presents rich opportunities for learning and practicing language.
Image Source/AGE fotostock

Facilitating Language Development in School-Age Children

How is language development different for children who are six, seven, and eight years old? Perhaps the most noticeable difference is that children in the primary grades add written language to their oral skills. Reading and writing help expand their vocabulary, and by the time they are eight, they have approximately 20,000 words at their command.[18] Conversations run deeper and go on longer, especially with adult support. Children gain such language skills as manipulating thoughts, understanding perceptions, and trying to influence the thinking of others. Moreover, arguing, which annoys some adults, is nevertheless a step up from physical fighting and is also part of language development. Instead of just trying to tolerate arguments, adults can help children get better at expressing their points of view clearly.

One notable difference between school-age children and preschoolers is their use of humor. They now tell jokes that are funny, and they love puns, riddles, tongue twisters, and secret words. Language grows in its function as a useful tool to add to the pleasures of life.

To facilitate language development, children should have plenty of opportunities to manipulate real objects and learn through experience. Even though they are now old enough to sit still, they still learn best through active engagement with real people and things. They *construct* knowledge, so they need many challenging opportunities to develop and practice the thinking skills that show up in their language use. Playing is still important to learning and should be encouraged.

Modeling good listening skills is just as important for school-age children as for those under five. To contrast good listening with poor listening, look at the following two scenes.

Seven-year-old Jessica comes into the family room, throws her backpack on the couch, and then flops down beside it, moaning, "I hate that old teacher! She gave me a test that was impossible, and when I couldn't do it I just wanted to die!"

The adult reacts emotionally to what she sees and hears instead of trying to truly understand it. Glaring at the backpack on the couch, she responds, "Oh come on now, don't be so dramatic. If you failed at a test maybe it's your own fault for not studying hard enough. And don't say you hate someone. That's not nice! You know where that backpack belongs, now go put it there, young lady, and next time study harder!"

Can you imagine Jessica slouching off with a pout and leaving her backpack on the couch? The adult's response cut off the conversation by using criticism and by giving orders, admonitions, and advice.

Let us run the same scene again. This time the adult is a family child care provider who has good listening skills. Watch her open up communication instead of shutting it down.

Jessica comes in, throws down her backpack, and moans, "I hate that old teacher!" "Which teacher?" asks the provider, ignoring both the word "hate" and the backpack. "My P.E. teacher," answers Jessica. "You're really mad at her," says the provider, putting Jessica's feelings into words. "Yes, she gave us this physical fitness test today, and it was too hard." Jessica looks as though she's about to cry.

"You feel bad because you didn't do well on the test," says the provider. "I sure do. I was the first one, and everyone was watching me, but I just couldn't do it! Some of the kids laughed at me." The provider commiserates with Jessica: "It's awful to be laughed at." "I think they were just nervous because they were going to have to do it too. But they didn't have to because the teacher discovered she was giving me the third-grade test. No wonder I couldn't pass it. I'm not a third-grader!" A tear rolls down Jessica's cheek.

"Your teacher made a mistake and gave you the wrong test . . ." says the provider, rewording the information. "Yeah, it was a pretty bad mistake." Jessica takes the tissue the provider offers. "I tried and tried, but I just couldn't do it! I felt awful." Jessica looks as though she is starting to feel better now. "I can imagine how you felt," says the provider in a sincere voice.

Getting up abruptly from the couch, Jessica ends the conversation with a big sniff and the words, "Well, I have to see Julie. Is she here? I want to find out how her kittens are doing." Picking up her backpack, Jessica runs off, satisfied that she has been heard.

Notice how different the second conversation was from the original version! The provider focused on listening to what was behind Jessica's words instead of asserting her own agenda. She encouraged the conversation to continue so that the problem became clearer. She did not jump to conclusions, nor did she try to solve the problem for Jessica. Sometimes children's feelings are so painful for adults that they quickly dismiss problems without working through them. What a relief it must have been for Jessica to be heard instead of criticized and ordered about. Jessica did not need to do anything more about this problem other than talk about it. Just expressing her feelings soothed her.

Emergent Literacy

In 1985, the Early Childhood and Literacy Development Committee of the International Reading Association published a statment about literacy practices in preschool and kindergarten that continues to remain relevant today.[19] In summary, the statement expressed concerns about children under six being subjected to rigid, formal prereading programs with little attention to developmental appropriateness, individual development, or learning style. It was the association's belief that pleasure in reading should be a primary goal, but it gets lost when the focus is on isolated skill development or abstract concepts. When literacy development is not integrated with oral language and writing, children are often exposed to activities that stress right answers and suppress curiosity, critical thinking, and creative expression.

In 1998, the NAEYC and the International Reading Association came out with a new Position Statement on "Learning to Read and Write." The rationale in the statement can be summarized as

NAEYC/IRA
Position Statement:
Learning to Read and Write

- Learning to read and write is critical to a child's success in school and later in life.
- The early childhood years–from birth through age eight–are the most important period for literacy development.

- The primary purpose of this position statement is to provide guidance to teachers of young children and others who are in a position to support early literacy.

The International Reading Association made some recommendations in 1985 that are paraphrased and summarized in the following list:

- Build instruction on what the child already knows about oral language, reading, and writing.
- Respect the language the child brings to school, and use it as the base for further language and literacy activities.
- Focus not on isolated skill building but on meaningful experiences and meaningful language.
- Integrate reading experiences with communication in general (including talking, listening, and writing) and with other areas of study, such as art, math, and music.
- Encourage children's attempts at writing without pressuring them for the proper formation of letters or conventional spelling.
- Encourage risk taking and experimentation with talking, listening, writing, and reading.
- Model the use of language and literacy.
- Read regularly to children.
- Use evaluative processes that are developmentally and culturally appropriate for the children being assessed (see Chapter 12).

NAEYC/IRA Position Statement: Learning to Read and Write

The recommendations are still valid today. In addition, the 1998 Joint Position Statement states that children need relationships with caring adults who engage in many one-on-one, face-to-face interactions with them to support their oral language development and lay the foundation for later literacy learning. The statement then lists sample important experiences and teaching behaviors for the different age groups. Two of these experiences for infants and toddlers are talking to them with simple language while being responsive to their cues, sharing cardboard books with babies and reading to a toddler on the adult's lap or together with one or two other children. For preschool children the first suggestion is to form positive, nurturing relationships with them and engage in responsive conversations with individual children, model reading and writing behavior, and foster children's interest in and enjoyment of reading and writing. Six more ideas for experiences and teaching behaviors follow the first one. The word instruction does not appear until the practices recommended for kindergarten and the primary grades. Instruction includes but is not limited to daily experiences of being read to, independently reading and writing many kinds of texts such as stories, lists, messages, poems, and reports.

We should strengthen our resolve to ensure that every child has the benefit of positive early childhood experiences that support literacy development. Likewise, administrators and teachers of all levels have the responsibility to educate every

Painting provides practice in eye-to-hand coordination and increases children's ability to write when they are ready.
Mike Watson Images Limited/Glow Images

child, regardless of his or her ability to learn, even if resource-intensive interventions are required. With support and encouragement, every child can learn.

A Reading-Readiness Approach Versus an Emergent-Literacy Approach

Compare the following two approaches to helping pre-first-grade children increase their reading and writing skills. It is easy to tell which one follows the NAEYC and International Reading Association's recommendations.

In program A, the children are taught reading-readiness skills. Three- and four-year-olds are given exercises to strengthen their left-to-right hand movements–even crayoning and finger painting must be done left to right. Some of the children are Chinese American, and the writing they see at home flows from top to bottom, not left to right; nevertheless, this inconsistency is overlooked because literacy in Chinese is not valued by this program. In addition, the children are learning the names and sounds of the alphabet in a daily half-hour circle time. The three-year-olds especially have a hard time sitting still that long, even though the teachers attempt to make the lessons entertaining. After circle time, the children are sent to work stations to fill out phonics worksheets and practice writing the letters they are learning. The teachers sit with small groups and attempt to keep the children on task—in spite of continual resistance.

After they either finish the task or sit for 20 minutes at the table, the teachers correct their errors with red pencils and dismiss them to go outside to play. All the children get a smiling-face stamp on their work when it goes home, but under the smiling face, the number of errors is immediately obvious to both children and parents. Many of the children do not care, but a lot of the parents do. In the name of developing reading-readiness skills, the children are given crayons and coloring-book sheets each day and instructed to stay in the lines. The teachers explain to the parents that this is an important exercise for developing fine-motor skills and eye-hand coordination.

Program B's approach is quite different. Here, literacy experiences for three- and four-year-olds are embedded in the program, not created as distinct activities and lessons. The children naturally develop eye-hand coordination by using the materials set up in the many activity areas of the classroom. Materials that are often available include beads for stringing, pegboards, and puzzles. Activities set up in the art area include finger painting, easel painting, free drawing with crayons and felt pens, cutting with scissors, and pasting collage pieces.

In Program B, teachers read books and tell stories often during circle time and also during free play periods. At almost any time of the day you can find some children in the book area, snuggled into couches and cushions looking at books. At least one child is able to read to himself and sometimes reads to other children. There is almost always an adult with a child or two nearby or on his or her lap sharing a book together. And if you look closely, you can see that the languages of the children's families are reflected in the book selections. Some of these books are handmade. In fact, several times a week, children take turns dictating stories to a teacher, and some of these stories are made into books for use in the classroom.

One area of the classroom is always stocked with writing materials and some basic art materials. Children are encouraged to play at writing, which usually starts as scribbling and eventually transforms into invented writing and finally **invented spelling.** Children's work is not corrected. They correct themselves as they grow more aware of conventional spelling and begin to ask, but until that time, they manage to create ways to spell words without ever having a formal phonics lesson.

invented spelling The way children spell when they first begin to write, going by the sounds of the language more than by conventional spelling rules. In other words, they invent their own spelling.

Oral language is considered an important part of this literacy program, and children are encouraged to talk—to each other and to adults. Listening is also encouraged. Literacy is seen as only one part of a much larger curriculum area called "communication."

When we compare the two programs, we see that program A is teaching a variety of isolated skills that make no sense to the children. Furthermore, the concept of reading readiness is disconnected from meaningful oral language. The children's writing attempts are corrected rather than encouraged, and there is no creativity or experimentation in any of the exercises they perform. Little about this approach to literacy is developmentally appropriate.

Program B, however, is taking a true emergent-literacy approach by building on what the children already know about oral language, exposing them to a print-rich environment, and encouraging them to experiment with reading and writing. Everything they do is *meaningful.* There is no pressure to perform, only encouragement to use the wealth of materials available, some of which are obviously connected to literacy and others not so obviously. The teachers read to the children, thereby acquainting them with books, and they tell stories, which exposes the children to storytelling conventions.

Emergent Literacy for Infants and Toddlers

What about infants and toddlers? Does emergent literacy begin as early as the infant and toddler years? Yes. All language experiences are considered part of emergent literacy. As children increase in verbal skills, they are moving toward learning to read. Believe it or not, caregiving is an important time for early literacy development. During those essential activities of daily living, if caregivers use language in context, infants learn vocabulary, such as body parts, without ever having "lessons." When caregiving activities are performed as intimate exchanges between two human beings, relationships grow. Warm relationships provide feelings of security and are important to brain development, which in turn relates to early literacy.[20]

Books are part of the process; children learn the conventions of reading, such as turning pages and holding the book right-side up. Perhaps most importantly, as small numbers of infants, toddlers, and adults cuddle together over a book, children learn to associate reading with pleasure and closeness. When pleasure is the goal, adults are not compelled to read the book from front to back or to do all the reading themselves; young children like to grab books and do their own "reading" and page turning. They should be allowed to handle books and pretend to read and point at pictures and name them. Forbidding this kind of active involvement in the hopes of turning tots into good "listeners" can harm children's enjoyment of books. Patience is required. Eventually, they will come to understand that adults have the keys to meaning, and they will start listening–that is, until the time they are ready to start reading themselves.

Early writing skills begin to develop in infant and toddler programs when children first pick up a Cheerio and later try using crayons or felt pens. Beginning scribbles are valued as early stages of both writing and drawing. At first, the marks themselves are not as important to the young child as the sensation of using her body. It takes a while for the child to focus on the product of her effort. Eventually, though, the child will pay attention to what happens when she uses arm and hand movements in conjunction with a writing tool on paper. Progress continues as the child reaches the stage of naming or explaining drawings or writing. Eventually, those marks become recognizable to adults as well, though that accomplishment may take five or more years.

In the name of early literacy, some people stick printed labels on everything in sight with the hope that babies will begin to absorb the connection between the object and the printed word. This can give the room a cluttered look and be distracting. It is important to recognize that babies are learning symbols and reading signs appropriate to their developmental stage, but not necessarily printed words. Here are the kinds of signs they "read." The squeak of the door opening draws the baby's attention to the fact that her father might appear in the doorway to take her home. Her coat in the care teacher's hand signals that she is about to get dressed to go outside. The smell of her mother's clothing comforts her when she sleeps with it. (This is a clever device some caregivers use. They ask the mother to bring in a piece of clothing she slept in to provide comfort to a baby who is having separation issues.) Note that these examples are not exclusively visual but relate to a variety of senses, as is appropriate to infant and toddler development. All these kinds of "readings" are precursors to focusing on visual recognition of print. They are also more basic (as they should be) than understanding that printed words stand for things and printed letters stand for sounds. In the name of school readiness, some adults feel pressure from the message that earlier is better and thus jump ahead to attempt to teach babies skills that they cannot understand.

Promoting the Development of Emergent-Literacy Skills in Three-, Four-, and Five-Year-Olds

During the first five years, without any formal lessons, children accomplish a good deal:

- They begin to understand the value and functions of print.
- They learn that written words carry meaning.
- They connect written words with sound. It is an exciting moment when a child realizes that spoken sounds can be represented with symbols on paper.
- They begin to recognize environmental print and can pick out the names of their favorite fast-food places, read stop signs and other road signs, and recognize logos and brand names on familiar products, like cereal boxes.
- Eventually, they come to distinguish between drawing and writing and between letters and numbers.
- They learn to associate books with reading and reading with pleasure.

What can adults do to promote this path of development in young children? Help them make connections between symbols and objects. Put up pictures on the shelves where toys and games go so the children can match the picture to the object. Draw outlines of block shapes on the shelves so children can match the shape to the object. Use symbols of all sorts. Make tags with houses on them for children to wear when they play in the housekeeping area.[21] Teach universal symbols, such

Writing is a natural activity for young children, if they see it modeled.
Corbis/VCG/Getty Images

as the slashed zero for "forbidden" or the male and female figures on public restroom doors. Eventually add words alongside the symbols so that children can make verbal associations as well.

Expose the children to their names in writing. Betty Jones and John Nimmo, in their book *Emergent Curriculum,* describe the excitement of a preschool girl, Althea, during an open house as she shows her family the many places where her name appears in the classroom: on the helper's chart, on the alphabet that circles the room (A is for Althea), on a ladybug painting that she made and on which she wrote her own name, on a body tracing, on a book she wrote about her family, on her portfolio, and on every page inside her portfolio.[22]

Create a print-rich environment so that words and phrases abound. Try to make the print relevant to something. Do not just post it for the sake of having print around. Notes on bulletin boards at a child's eye level have meaning. A note stuck on the door that says the class has left on a field trip has meaning. A written reminder to the cook about tomorrow's picnic has meaning. Thank-you notes have meaning. Invitations have meaning.

Capture the children's own language in print–dictated stories, poetry, rhymes, chants. These can be short or long, illustrated or not.

Writing is a natural activity for children. They start as toddlers with crayons and felt pens and eventually move on to pencils to create invented writing. Having a particular area in the classroom equipped with writing materials will encourage children to write and support them in their attempts. Blank books (paper stapled together) sometimes provide incentive to create a story. Magnet letters, stamps, an alphabet chart, and a picture dictionary are other good resources for the writing area.

POINTS OF VIEW

Marble Painting and Emergent Literacy

Marble painting is a popular preschool activity. The traditional process goes like this: children roll marbles in paint and put them on a clean sheet of paper placed in the bottom of a shallow box or tray. By tilting the container at different angles, the marbles roll around, leaving interesting trails. The children get a sensory experience from placing the marbles in paint, picking them out, and putting them on the paper.

It was surrounding this innocent art activity that a debate arose on the Internet. On one early childhood LISTSERV, someone described a couple of variations of marble painting. One was to use paint-covered beads and small jewelry boxes to create tiny designs; another was to use paint-covered plastic eggs (with things inside them to make sound as they roll around) and a huge refrigerator box–the added benefit of this activity being that it would take several children cooperating and coordinating their movements to tilt the box. (Someone even suggested using hard-boiled eggs rather than plastic eggs, which brought up yet another debate–regarding the appropriateness of using food as an art material.)

The Internet marble-painting controversy arose when someone questioned its value. This person suggested that although some physical skills were involved, children didn't get much out of the activity and, furthermore, that it was not creative art and certainly not a cognitive activity. Several arguments came back over the Internet in response. One message discussed the significance of leaving traces. According to this individual, leaving traces is a basic literacy concept. Snails leave traces from which we can see where they have been. Paint-covered marbles also leave traces, and we can follow their trail as they move across and around the bottom of a container. When children walk barefoot through paint or water and then across paper or cement, they also leave traces. Writing, too, is a way of leaving traces. From writing, we can follow the trails of history.

Although young children cannot make the symbolic connection from marble painting to writing, they do benefit from seeing many different ways to leave traces. Eventually, when they learn to write, they come to under-stand some of the many functions of reading and writing.

Encourage children to read everything in sight, not just words. They can read their paintings, people's facial expressions, and signs, like footprints in the snow. (The *Points of View* box above provides another example of reading tracks and traces.)

Promoting the Development of Emergent-Literacy Skills in School-Age Children

What about school-age children? As children reach five and six, their literacy skills begin to resemble what the general public would recognize as reading and writing. They can usually print their names and other words; some are becoming proficient

Children eventually come to see themselves as readers. Adults have a big influence in this process.
Wavebreakmedia/Shutterstock

at invented spelling. Some are beginning to realize that conventional spelling also exists. If they have been in an early childhood program or home where an emergent-literacy approach is taken, they are good at exploring, experimenting, and playing with language in creative ways–both oral and written. They do these explorations without fear of failing or making a mistake.

Sometime in the next few years they will recognize themselves as readers. They will also discover that reading opens many doors and provides access to new information, which will motivate them to increase their skills.

Adults have a big influence on this process. When they value the ability to read, provide resources and a quiet place for reading, and read themselves–both to children and silently–they promote reading skills in the children. Adult models are important. Although they may not read books to themselves in front of children, they demonstrate the importance of literacy by reading aloud directions, recipes, notes, and letters. And, of course, there is no person too old for storytelling or reading books aloud. Being read to can be a lifelong pleasure. Kitty Ritz, a first grade teacher, has a unique way of making literacy a family affair. See the *Voices of Experience* story.

Writing progresses in the primary grades, as children are encouraged to write daily. At first, they produce mostly expressive writing–that is, they write about themselves, their experiences, their feelings, and their ideas. They also begin to write stories that have literary elements such as beginnings, endings, plots, and characters. They begin to pay attention to the rhythms and sounds of language and attempt to capture them in print. Eventually, they will move to expository writing–that is, writing to argue, persuade, direct, and explain.

Family Reading

It is 8:16 in the morning, and my first-grade classroom is full of readers, from infants and toddlers to grandparents. A bright yellow sign on the open door proclaims that it is Family Reading Day: everyone is welcome, and you can read anything you want anywhere you want.

Betsy is reading a board book, "Read to Your Bunny," to her baby sister and mom in one corner. Andy, Larry, and Karl each have a big book atlas lined up in a row. They excitedly point to different parts of the maps and charts with large pointers made out of dowels, "claiming" natural resources or countries. Anne and Brigit sit on either side of Anne's mom, singing "Today Is Monday" and "Chicka Chick Boom Boom" (for the zillionth time this year). Jack reads out loud surrounded by his mother, father, and four-year-old sister who holds her own book, mimicking all the reading behaviors of her adored older brother. Daisy and her father crouch over a book about butterflies, then make a trip to the butterfly house where the caterpillar is munching away. Mr. Brown is discussing a book on biomes with his son Mark. He calls me over to find out whether I'm familiar with a term. I bring over a children's dictionary where we find the word and a definition. Today Toby's grandparents are in town from the southern part of the state, so they have dropped in to see what first grade looks like. Toby is giving them a tour of the walls, his journal, and the library. Karl rushes over to show me an example of a compound word in the animal book he is reading, knowing that we discussed this kind of word recently. Juan and his mom help each other sound out the English in a beginning reader, speaking briefly in Spanish to make sure they comprehend the story.

Out of the corner of my eye, I see Ethan and his father reading by the computers. Ethan is beaming because this is the first time he has had a parent attend Family Reading. He and I had tried everything we could think of to entice one or both his parents into the classroom: special invitation letters, reminder notes, phone calls, and conferences. Finally Ethan can proudly show his father what it means to be part of Room 28 on Family Reading Day.

When I ring the chime at 8:30, everyone quiets down to listen for news and announcements, about upcoming events, homework, and special recognition of home reading. Then it is time for kisses, hugs, and goodbyes. The parents, siblings, and grandparents file out while the first graders return books to baskets, bins, and shelves. Ethan, full of new pride, sits up front this morning next to his friends. I smile at him and wink. We both know his dad will be back for more Family Reading.

–Kitty Ritz

Children should be encouraged to draw and write about projects and other activities they are doing. A message center can also be a feature of the environment, where adults and children leave messages for the group or for individuals.

Fluency is still more important than mechanics in these first years as writers. Eventually, editing, revising, and rewriting can be taught, as children learn the need to polish a piece of writing for public use. Children should be encouraged to help each other with the editing process.

A Story to End With

In a college parenting class one day, I was telling my students how important it is to put needs and wants into words. I was adamant about how adults must model assertive communication for children. I asked the class how children will ever learn to express themselves clearly if the adults around them do not model clear communication. "Don't expect people to read your mind," I said firmly. "Tell them what you are thinking and feeling. Then children will learn how to put their own needs into words."

A student's hand rose timidly. She said shyly that she did not understand what I meant. In an apologetic voice she explained that she did not think children should have to put their needs into words. After all, she said, her mother always read her mind. She never had to tell her mother what she needed because she just instinctively knew. She thought that it should be the goal to get beyond words. When two people were really close, they did not need to talk to communicate in words, she said. This student was from Japan and had come to the United States for a year's study abroad.

Another hand rose. This time a woman from Mexico spoke. "I don't see anything wrong with mind reading. It's a sign of close attachment." She gave an example of how her sister always managed to call her at the exact moment she was thinking about something to tell her sister.

I realized then that I had been speaking from my own culture and ignoring the communication styles and goals of other cultures. I come from a low-context culture, where we like to put everything into words. The two students who disagreed with me both come from higher-context cultures, where indirect communication is important and messages delivered without words have great value.

Summary

Language allows children to make cognitive links, clarify their needs, gather information, label objects and experiences, categorize and classify, store information symbolically for later retrieval, plan, organize, and order experiences. The early childhood program should be sensitive to the diverse language backgrounds of its children. Preserving home language is vital, and bilingual education is desirable for all children, even those from English-speaking homes.

Children learn language from carrying on conversations with adults and other children. The content of these conversations develops with age. Infants' and toddlers' conversations tend to be playful exchanges, discussions about caregiving routines, or dialogues about events in the here and now. By two, children have usually added imagination to their talk as they begin to pretend. Preschoolers' conversations revolve around projects. Their language skills are becoming more sophisticated; they can communicate on different levels, shift registers, and develop story lines. School-age children's conversations run deeper and go on longer. They enjoy humor that is understandable even to adults and also engage in arguments, which can be considered language practice. The primary role of the adult in facilitating language development is to be a model of both language and listening skills.

This book takes an emergent-literacy approach to reading and writing. An emergent-literacy approach is different from a reading-readiness approach because it is more holistic and integrated. It builds on language skills the children already have instead of teaching isolated skills out of context. The adult role in emergent literacy is to encourage conversation and language exploration in a rich environment that contains meaningful print, including books. Writing and drawing tools should be available on a daily basis to encourage children to explore and experiment with writing.

REFLECTION QUESTIONS

1. Do you speak a language in addition to English? What have been your experiences with–and feelings about–learning English?
2. Have you had experience with bilingual education?
3. Were you read to as a child? If so, what, if any, are your memories of the experience?
4. What (if anything) do you remember about learning to read?
5. Are you a good reader? Why or why not?
6. Do you speak the language of your ancestors? If not, why? How do you feel about that?

TERMS TO KNOW

Can you use the following words in a sentence? Do you know what they mean?

emergent literacy 360
home language 361
language-immersion program 363
low-context culture 369
high-context culture 369
register 372
invented spelling 378

FOR FURTHER READING

Burman, L. (2008). *Are You Listening: Fostering Conversations That Help Young Children Learn.* St. Paul, MN: Redleaf.

Burns, M. S., Johnson, R. T., and Assaf, M. M. (2012). *Preschool Education in Today's World: Teaching Children with Diverse Backgrounds and Abilities.* Baltimore, MD: Brookes.

Garcia, O., Kleifgen, J. A., & Cummins, J. (2018). *Educating Emergent Bilinguals: Policies, Programs, and Practices for English Language Learners* (2nd ed.). New York: Teachers College Press.

Helm, J. H., and Katz, L. G. (2010). *Young Investigators: The Project Approach in the Early Years* (2nd ed.). New York: Teachers College Press.

Hullinger-Sirken, H., and Staley, L. (2016). Preschool through grade 3: Understanding writing development: Catie's continuum. *YC Young Children, 71* (5), 74–78.

Johnson, J., and Dinger, D. (2013). *Let Them Play: An Early Learning (Un)Curriculum.* St. Paul, MN: Redleaf.

Lindfors, J. W. (2008). *Children's Language: Connecting Reading, Writing, and Talk.* New York: Teachers College Press.

Michael-Luna, S. (2015). What parents have to teach us about their dual language children. *YC Young Children, 70* (5), 42–49.

Norton-Meier, L., and Whitmore, K. (2015). Toddlers through grade 2: Developmental moments: Teacher decision making to support young writers. *YC Young Children, 70* (4), 76–83.

Passe, A. S. (2012). *Dual-Language Learners Birth to Grade 2: Strategies for Teaching English.* St. Paul, MN: Redleaf.

Pilonieta, P., Shue, P., and Kissel, B. (2014). Preschool: Reading books, writing books: Reading and writing come together in a dual language classroom. *YC Young Children, 69* (3), 14–21.

Pradis, J., Genesee, F., and Crago, M. B. (2011). *Dual Language Development and Disorders: A Handbook on Bilingualism and Second Language Learning* (2nd ed.). Baltimore, MD: Brookes.

Valdes, G., Capitelli, S., and Alvarez, L. (2010). *Latino Children Learning English: Steps in the Journey.* New York: Teachers College Press.

Wessels, S., and Trainin, G. (2014). Kindergarten and grade 1: Bringing literacy Home: Latino families supporting children's literacy learning. *YC Young Children, 71* (5), 74–78.

Wohlwend, K. E. (2011). *Playing Their Way into Literacies: Reading, Writing, and Belonging in the Early Childhood Classroom.* New York: Teachers College Press.

14 Fostering Joy in Developmentally Appropriate Experiences in Math and Science

Hero/Corbis/Glow Images

The Constructivist Approach

What Do Children Learn?

How Do Children Learn?

Math

Infants and Toddlers and Math

Preschoolers and Math

School-Age Children and Math

Concepts of Time and Space

"Real-World Math"

Games

Science

A Constructivist Approach Versus Formal Science Lessons

Physics and the Project Approach

Chemistry and the Project Approach

Two Basic Science Concepts

Nature Study

Transitions Projects

Basic Equipment and Materials for Math and Science Learning

A Story to End With

In This Chapter You Will Discover

- what a constructivist approach to learning entails.
- what kind of math- and science-related knowledge children "construct."
- what math has to do with infants and toddlers.
- how preschoolers and school-age children construct math knowledge.
- how young children develop their concepts of time and space.
- what "real-world math" means.
- how playing games such as dominoes and board games helps children learn math.
- why a constructivist approach to science is better than formal science lessons for young children.
- how children learn physics through a project approach.
- how transformation and representation relate to a science curriculum.
- how children learn chemistry through a project approach.
- how children learn about nature.
- some examples of "transitions projects."
- some basic materials and equipment that belong in every early childhood program.

NAEYC Position Statement

NAEYC/NCTM Position Statement 2010: Early Childhood Mathematics: Promoting Good Beginnings

How do adults teach science and math to young children? The question itself is misleading. We do not teach these subjects as they are taught in upper grades; instead, we observe, note the children's interests, set up the environment accordingly, ask sensitive questions at just the right time, and allow the children to explore and experiment. The purpose of this approach is to give children opportunities to *construct* knowledge about the physical world and explore ways to represent their findings symbolically.

THE CONSTRUCTIVIST APPROACH

constructivist approach A view based on Jean Piaget's work that suggests that children do not passively receive knowledge through teacher-led instruction but rather actively construct it themselves.

physical knowledge One of three kinds of knowledge described by Jean Piaget. Physical knowledge involves an understanding—in concrete rather than abstract terms—of how objects and materials behave in the physical world.

logico-mathematical knowledge One of three kinds of knowledge described by Jean Piaget. Logico-mathematical knowledge comes from physical knowledge and involves an understanding of relationships between objects through the use of comparison and seriation.

While the **constructivist approach** has not been named directly before this chapter, it is the basic approach to learning that has underpinned this book. The constructivist approach stems from the work of Jean Piaget, and can be applied to every aspect of learning, not just math and science.

Piaget described three types of knowledge children gain in their early years: physical, logico-mathematical, and social.[1] **Physical knowledge** develops as children gain concrete experiences and ask questions about the world. Children gain physical knowledge when they discover that they can stack small blocks on a bigger one but that the reverse does not work as well. Physical knowledge is also gained when they learn that marbles roll downhill but not uphill. Children construct this knowledge by interacting with the physical world, as well as working with peers and adults.

Building on their physical knowledge, children proceed to develop **logico-mathematical knowledge,** which is about relationships between objects: this cup is bigger than that one, but smaller than that other one. When children explore and manipulate objects, they begin to understand these relationships. This knowledge develops within their minds as they interact with the real world.

Social knowledge is knowledge about the physical world that can only be learned socially, such as labels. For example, a child can play with blocks for weeks and construct a good deal of physical and logico-mathematical knowledge, but the child cannot "construct" the label "blocks." The label is socially determined. The child has to learn from another person (usually an adult) that these wooden items are called "blocks."

As you become familiar with Piaget's work, you see that math and science are inseparable from the other aspects of curriculum. Although this chapter handles math and science separately, the constructivist approach does not. As you continue reading, you will notice that some science activities look more like art activities and that math activities can come in the form of pure play or even a cooking project. Like reading and writing, math and science cannot be separated from each other or the other various aspects of early childhood learning.

What Do Children Learn?

What do children learn about math and science during early childhood? Young children deal with concepts of space, time, physical property, motion, and evidence. They also learn about change–what causes it and what forms it takes–and about estimation and prediction. What happens to a Halloween jack-o'-lantern if it sits on the shelf through November? What is this fuzzy, gray stuff growing on the pumpkin? Why is it caving in? How many days will it take to flatten?

Children also learn about measurement. How long a board will it take to bridge this gap? Anything can be used as a measuring tool, both standard and nonstandard objects. Children may be interested to know how long the board is in terms of their own feet. Is the board the same number of "feet" for each child? They can also learn standard measurement using rulers, yardsticks, and tape measures. Furthermore, children can learn about other sorts of measurement instruments, such as thermometers, odometers, speedometers, and clocks.

social knowledge One of three kinds of knowledge described by Jean Piaget. Social knowledge relates to knowledge about the world that can only be transmitted socially, such as labels for objects.

Young children learn about money. One child care center director let the children count the change in the fund-raising jug on certain afternoons when just a few children were still present. They loved it. The younger ones sorted, and the older ones counted, stacked, and recorded. There was something for everyone to do, and the activity held their interest because they all knew that money is something important to grown-ups and that it belongs in the "real world."

Children also begin to learn about nature in the early years. Nature is especially interesting because it too belongs in the "real world." Nature is best learned by getting out and exploring it.

How Do Children Learn?

Young children learn math and science in a number of ways. Certainly many early childhood materials, toys, and equipment invite children to explore, experiment, solve problems, interact, and, ultimately, construct knowledge. Early childhood educators also contribute to children's construction of knowledge by setting up the environment for exploration and experimentation, posing provocative questions at the right time, pointing out intriguing inconsistencies, and helping children pursue areas of interest, often through project work. (See the *Points of View* box on page 392 for two contrasting perspectives on how children learn.)

Questions, problems, and issues arise in the normal course of a day. Some problems or questions come up during free play periods such as measuring a board to bridge a gap. Nature studies arise spontaneously when, for instance, a bird builds a nest in the play yard, a spider spins a web on a structure, or a snail leaves a trail on the cement. They can also be planned, such as taking a field trip to a local marsh or bringing in a special animal visitor.

As you read this chapter, keep in mind that science and math are closely related subjects. The following sections are divided into separate categories, but this is done merely for convenience to explain a variety of concepts in an organized fashion.

MATH

When many people think of math, they think of counting or maybe adding. However, math in early childhood education encompasses an enormous range of thought

POINTS OF VIEW

Two Perspectives on How Children Learn

One perspective on learning suggests that adults transmit information to children. Long ago, on the cover of a *Psychology Today* issue was a picture of an adult with a faucet for a mouth who was bent over a small boy with a funnel in the top of his head. The faucet was open, and a stream was pouring into the child. That picture represented the child as an empty vessel to be filled with adult knowledge. What that picture depicted is quite different from the perspective on which this book is built.

In the constructivist view, children are born with enormous potential for building knowledge, but they cannot do this all by themselves. Instead, they need to interact with the world–with the objects and the people in it. Children are well equipped to do just that; from birth, they are capable of attracting others to interact with them. They have built-in motivation to reach out and grasp, to poke and prod, to shake and bang, to pull things apart and discover how they work. The constructivist-oriented adult will take advantage of children's natural motivation and provide numerous opportunities for them to explore, invent processes, experiment, discover concepts, and build theories–in other words, to construct knowledge of the world around them.

processes and activities far more basic, as well as more sophisticated, than counting and addition.

NAEYC Program Standards
Program Standards 2 and 3 Curriculum and Teaching

Math is more than numbers and symbols; it is a particular way of looking at the world, understanding it, analyzing it, and solving certain kinds of problems. The symbols and processes of math are simply tools. We teach math backwards when we teach mathematical symbols before children have constructed some basic knowledge and have started to regard the world from a mathematical perspective.

Infants and Toddlers and Math

When babies notice that the round shape fits in one part of the puzzle and the square shape in another, they are doing math. Discriminating one shape from another is an early geometry skill.

When toddlers pour sand in the sandbox or water at the water table, they are learning about quantity, mass, and volume. Eventually, with enough experience, they will come to see that equal amounts of sand or water remain equal even when poured into different-shaped containers. This may seem like such a straightforward conclusion to most adults that we often forget children must first learn, or "construct," this concept. Piaget called the concept "conservation" because it involves "conserving" or holding onto the idea that although appearances may change, amount or number will not change unless something is added or extracted. Eight ounces of water poured into a baby bottle looks different from eight ounces poured into a cat dish, but the amount is still eight ounces of water. Adults know that logically, but young children have to acquire this understanding. Until then, they are easily deceived by perceptions.

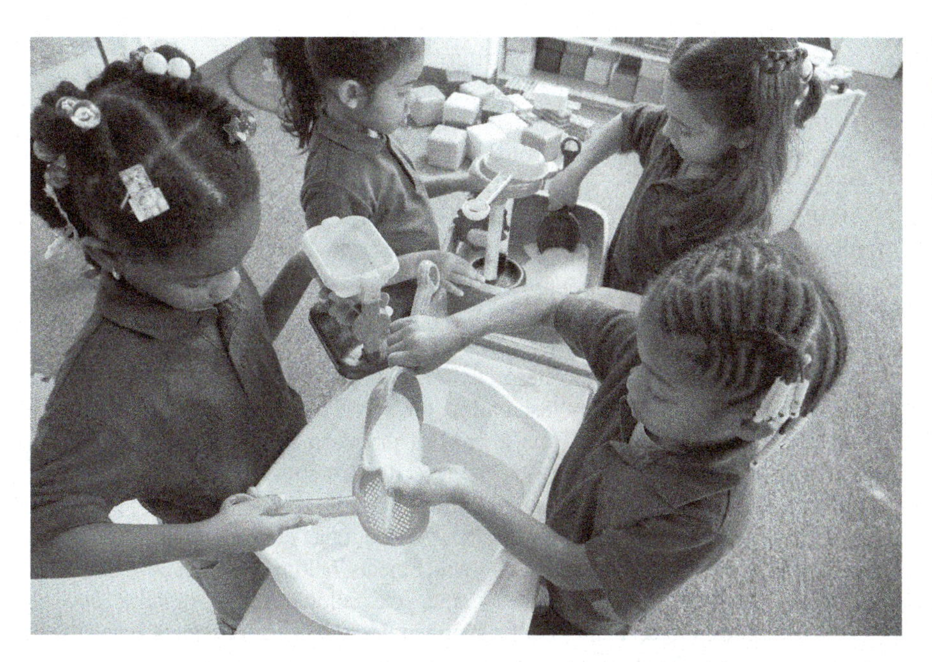

When toddlers mix and pour sand in the sandbox, they learn about quantity, mass, and volume.
Juan Silva/Getty Images

Yet even adults can overlook logic and be deceived by what they see. Have you ever thought about the shape of cereal boxes? The front and back of the box are tall and wide but the sides are shallow. When they sit face-forward on the grocery store shelves, they appear to contain a lot of cereal but they are designed and displayed to deceive. The first time I poured a big box of cereal into a paper bag (to get the prize from the bottom of the box). I was astonished at what a small amount of cereal that big box actually held.

Adults may have lapses of logic and occasionally be deceived by their senses, but young children rely only on their senses–their perceptions. I know in my head that eight ounces is eight ounces. I also know that three apples, whether whole, sliced, or made into applesauce, remain three apples–unless something is added or taken away. Young children, however, are not able to "conserve" volume or number, so they can never be sure whether three apples remain three or magically turn into four or two when they change form. Until children move beyond Piaget's preoperational stage, when they are neurologically limited to perception to process their thoughts, they cannot deal with abstract math concepts.

Preschoolers and Math

Although preschoolers rely less on perception than toddlers, math still needs to be very concrete for them. Four-year-olds are learning math when they play with wooden blocks and manipulative materials. When they set out the napkins on the table to match the number of chairs, they demonstrate the ability to make what is called **one-to-one correspondence,** a skill that is a prerequisite to true

one-to-one correspondence The understanding that counting involves assigning one number to each object or person being counted. This form of counting differs from reciting numbers by rote.

counting. Children need many opportunities in the concrete world to explore one-to-one correspondence before they realize that counting involves assigning one number to each object or person being counted. Until they can do that, counting is done from memory and has no more meaning for the child than nonsense rhymes.

Math materials should be manipulative for preschoolers. For example, snap-together blocks help children learn on a body level how many short blocks it takes to make one long block. They may not talk about it, but they are learning. Eventually, they may notice quantity and even begin to count, if it seems interesting or useful to them at the time.

Sorting and classifying are other activities children are naturally drawn to. The old-fashioned button box holds fascination for some children or the hardware jar full of screws, bolts, and nuts. Even a bowl of dry corn, beans, and peas exercises young fingers and minds. Sorting objects according to some obvious physical attribute is a simple math task that many children enjoy.

Classifying can be even more complex. Encourage children to designate the categories and decide for themselves which objects belong together and for what reasons, such as grouping according to shape, size, color, material, or other obvious physical attributes. Objects can also be grouped according to function (these all are fasteners), location (these are all found in the kitchen), or other more creative classifications. Physically manipulating the objects while classifying them makes the experience concrete and therefore interesting for young children.

Graphing, charting, and voting are other useful activities for young children. "What's your favorite color?" is a question a group of children can use to survey the class. With a little help, they can chart the answers. Eye color is another survey that lends itself to charting. Have children vote with their bodies: "Everyone who wants to go with Anthony's mother on the field trip stand by the window. Those who want to go with Sally, the cook, stand by the kitchen door." Adults can then show the children how this physical voting method can be symbolically represented.

School-Age Children and Math

Although school-age children are ready to learn symbols, they still need concrete experiences to understand math. Many of the preschool activities mentioned are also appropriate for school-age children. Older children can take those same activities, such as classifying and graphing, to a deeper level.

Concepts of Time and Space

Measuring time is a skill young children take years to fully develop. The lessons start when adults help children notice various sequences during the normal course of the day: "We wash our hands before we put bibs on; we put bibs on before we eat." "After story time comes nap." "Your mommy comes right after teacher Christina leaves." Before children learn to read clocks, they develop a sense of time passing, and adults help them gain this sense.

Some teachers help children record the passage of time in creative ways. An ordinary kitchen timer is one tool to demonstrate the passage of short periods of time; with enough patience, the children can see the timer progress around the circle. An old-fashioned sand-filled egg timer is an even better visual aid, showing the passage of time as the sand flows.

One preschool had a bake sale project. The children signed up for shifts to help sell the goodies by gluing a picture of themselves on a piece of paper. The teacher then wrote the starting time of the child's shift and drew a clock, showing how the clock's hands would be positioned at that time. When the "schedule" was complete, it was posted for all to see.

A different time measurement exercise was conducted in another program. The children made snowballs and stored them in the freezer for three weeks. Once a week, the children removed the snowballs and traced their gradually decreasing sizes onto pieces of paper. Another program created a calendar that identified personal events from the children's lives rather than regular national holidays. And yet another program used a calendar format to distinguish between the days they were in school and away from school. They used red squares to indicate the days school was in session and blue squares to identify holidays and weekend days.

By elementary school, children are learning to read both clocks–digital and face–and calendars. Learning time on this symbolic level in preschool helps children understand the concepts of "today" and "tomorrow" and have an idea of how long an hour takes to go by–ideas that most adults have a firm grasp of but children have to learn.

Children can learn about the passage of time by measuring the evaporation of a snowball.
Blend Images/Image Source

Part of understanding time is learning that it is linear and cyclical—that is, it runs forward and also in circles. The seasons and the cycles of the moon are examples of the cyclical nature of time. Children tend to be interested in both topics, which are themes found in many children's books.

Children also need to gain experience with space. They learn to judge what objects will fit in certain spaces when they put toys and materials away after using them. They pull tricycles into parking places. They build block enclosures and then discover what will fit inside them. Spatial concepts are part of real-life learning. As children draw, they gain further experiences with spatial relationships.

"Real-World Math"

Young children develop mathematical thinking when they deal with mathematical problems in the real world. For example, teachers can introduce mathematical concepts while preparing for group time: "If we have six children in this group, how many chairs do we need for everyone to have a chair?" Children also create their own **real-world math** experiences. Once they begin to understand quantity, they start counting, estimating, and comparing: "Hey, she has more grapes than me" or "It's not fair. He has lots of Legos, and I have just a few."

real-world math Math activities that relate directly to problems in the child's own world—as compared to theoretical problems that have nothing to do with the child's reality. Real-world math is sometimes called "authentic math."

Cooking is a great real-world activity that entails numerous mathematical concepts. Recipes call for exact amounts of ingredients (three tablespoons of this and half a cup of that), specifically sized pans (pour into a greased 12-by-12-inch pan), and precise temperatures and cooking times (bake at 350°F for 20 minutes).

Games

Children also move forward mathematically by playing games. Playing dominoes requires one-on-one correspondence to match up tiles with like numbers of dots. Dice games require young children to match up the dots on the dice to the squares on a playing board; when two dice are being used, they tend to move the number of spaces on one die and then deal with the other. As children become more proficient at adding, they say, "Oh, there's a five—and another five." So they move five and then count off six, seven, eight, nine, ten, matching the squares to the numbers they are counting off on the second die. Eventually, they learn the sums by heart: "Oh, yes, five and five—that means I can move ten squares."

SCIENCE

NAEYC Program Standards

Program Standards 2 and 3 Curriculum and Teaching

The constructivist view, on which this text is based, looks at the child as a scientist. It makes sense if you think about it. Both children and scientists are curious, and they have a natural tendency to find out more about the things that intrigue them. Children have a sense of wonder and excitement about the world. Scientists manage to keep alive those same childlike qualities. Scientists seek to construct new knowledge. So do children. Both actively explore and experiment. Both seek to test and retest their theories.

Anyone who sets out to study something is a scientist–adult or child–whether he or she is studying something "scientific" or not. When children construct knowledge, they act just like scientists, even when they aren't studying something "scientific." Whenever children (and adults as well) explore, observe, reflect, describe, and categorize, they are acting as scientists. The adult's major role in helping child scientists is to provide resources, ask provocative questions, and assist them in describing, categorizing, and especially recording findings.

A Constructivist Approach Versus Formal Science Lessons

As you will see from this chapter, taking a constructivist approach is very different from setting up formal science lessons in which children are taught labels, concepts, and processes through unrelated displays, activities, or teacher-controlled experiments. Children need other children to interact with in order to construct knowledge. When they encounter physical problems in the environment in the presence of other children, they discover that not everyone sees things the same way. Each child may have a completely different perspective on an issue or take a different approach to solving a problem. Adults cannot substitute for children because they have too much knowledge and authority. It is through peer interaction that children gain certain insights that they would not gain otherwise.

Adults can, of course, share their knowledge with children, but when they do so, they sometimes stifle children's problem-solving abilities. A useful adult role when a child is faced with a problem is to ask a leading question, not to provide an answer. "How can I get these two milk cartons to stay together?" asks the child. If the adult says, "Try tape," the child will run to the tape dispenser rather than use her own mind to think. If the adult says instead, "Hmmm, I wonder what we have here inside or in the play yard that you could use," the child will have to do her own thinking rather than rely on the adult's pat answer.

From all sides comes a good deal of pressure to improve education through memorization and direct instruction, which focuses on "right answers." The constructionist view emphasizes exploration and problem solving rather than memorizing answers given by an authority (the teacher). Peter Johnston, in his book Opening Minds: Using Language to Change Lives, shows how the attitude of teachers and the words they use make a big difference in outcomes. When the goal is to create lifelong learners, the teacher doesn't give an answer or a solution but encourages the children to use their brains and go to work on the problems they encounter.[2]

She may come up with solutions the adult never thought of. "Will egg white work?" she asks herself as she spies a group of children in the other room beating eggs for meringue. She wanders over to the dress-up area and finds some belts. "Can I lash the milk cartons together?" Then she gets a new idea. "Maybe there's a stick outside that I can poke through the two cartons and hold them together that way." "Would tinker toys work?" "Of course, there are glue sticks over in the art area, and they may work." Obvious solutions are not necessarily what she will come up with first. A child's mind is too fresh and creative to merely reach for standard adultlike solutions.

Physics and the Project Approach

The following scenario depicts a program that takes a project approach to physics learning: two boys are swinging and dragging their feet in the sand under the swings. When they get off to examine the trails they have made, the teacher notices their interest. He gets off the swing and stands watching his friend still dragging his feet, continuing to make trails in the sand. Then the friend gets off, twists the swing, gets back on, and tries to watch his feet make a spiral path. He gets off, dizzy, and looks at the marks. In the meantime, the other child has begun to run around in circles, dragging a shovel. He too examines his path.

The teacher notes all this activity around making trails. Later in the day, when the boys are sitting together near the writing area, the teacher suggests that they try drawing their track-making observations. They take the suggestion and make some drawings. The teacher labels them and writes down what they say about them.

The next day, the teacher introduces to the boys the idea of creating a tracking project. Three girls are also interested in joining them. The teacher talks about ways to represent the tracks and then takes the group outside to the swings, armed with pencils and paper-filled clipboards so they can do their recording on the spot. When the other children come out, the teacher puts up a sign "swings closed" so the project workers will not be interrupted. The other children know about doing projects and respect the project workers' rights to have temporary, exclusive use of the swings.

A little later, the teacher takes the closed sign off the swings and invites the five children to come inside. There they discover a plastic jug full of sand hanging from the ceiling by a rope. Under it is a tarp. When the teacher takes a piece of tape off the bottom of the jug, the sand begins streaming out onto the tarp. "Let's see what kind of tracks you can make with this pendulum," she says to them. They take turns moving the pendulum and watching the trail. "I want to draw it," says one girl. She finds her clipboard and adds a drawing of the back-and-forth pattern displayed on the tarp. Before she is finished, the boys remember the circles they were making yesterday. They find their drawings pinned on the bulletin board and take them down to study them. They are anxious to see if they can make circles with the sand in the pendulum. It is not easy. It takes a few tries to move the pendulum in a circle.

transformation and representation Two processes that distinguish the constructivist approach from other teaching–learning approaches. Transformation involves processes of change, while representation portrays change in the form of traces. Activities of transformation and representation facilitate children's symbolic thinking.

Just imagine how many different forms this project could take. How about a paintbrush on a string? The medium could be paint on paper or water on cement. The paint tracks would be permanent reminders–the water, only temporary.

What we are seeing in each of the preceding experiments is what George Forman and David Kuschner call **transformation and representation.**[3] Children act on objects to change them–in this case, make them move. Representations are the kinds of traces left by transformation. The boys discovered they could make marks in the sand by swinging and dragging their feet. They created representations of the swing's movement. The teacher took it a step further by inviting them to make representations of the representations–ones they could preserve. The teacher took yet another step by helping them explore swinging with a pendulum.

A week later, we tune in to the same program and see something going on in the block area. Three of the four children from last week, plus two other children, are down on the floor conducting a physics experiment involving a pendulum. This pendulum is a tennis ball attached to a string hanging from the ceiling. These children are dealing with both math and science concepts on a concrete level.

Two of the children are bringing blocks over to stack under the tennis ball, which is in the hand of another child. Three blocks are stacked when the boy with the ball says, "That's enough." He lets the ball go, and it swings way above the stack of blocks. "Missed!" says the boy who let the ball go. "The string is too short," says another child. "The block stack isn't high enough," says a third. "I think the blocks are in the wrong place," remarks a fourth child quietly. No one hears her.

Finally, after two more tries and misses, another child begins to see that the stack is not placed at the lowest spot of the ball's swing. He and the girl who first mentioned the problem try to talk to the boy who is holding the ball. They want him to let the ball go so they can see where to build the stack. He refuses. Then they figure out that they can move the stack directly under the place where the string is attached to the ceiling. They move the blocks. "It needs more blocks," says the girl with the soft voice.

"How many more blocks do you think it will take?" asks the teacher. "About a hundred eleventy" answers the child holding the ball. "Not that many," says the second. "Three," says the girl.

"Let's see if three will do it," says the third boy, who is already putting three more blocks on. The next swing knocks off the top block. The children clap delightedly. The boy who stacked the blocks immediately sets up another experiment by taking the top two blocks off and standing the third one on its end. "What about this?" he asks the others. The first boy swings the ball, which knocks the block to the floor.

In the meantime, the girl is saying to the teacher, "I want to see what happens if we make the string longer." The teacher shows her how to change the string's length, and she is ready for another experiment. To finish up today's work on the project, the teacher will help the children reflect on their observations. Eventually, she will take some photos of the various pendulum experiments, organize the drawings, and put them together up on the walls, so the children will continue to have access to the record of what they did and the results. This kind of documentation can be used in the future to help children revisit their observations as well as apply them to new experiments. Some samples of their work will go into their portfolios, along with more photos taken by the teacher.

Chemistry and the Project Approach

What are other kinds of "science" experiments young children perform? Following is an experiment put together by a child on her own.

It is a warm, sunny day, and Rosie has found a puddle in the corner of the play yard. She is busy hauling a bucket of sand from the sandbox to the puddle. When she pours the sand into the puddle, it immediately disappears. Rosie returns to the sandbox for another bucketful, which she dumps near the puddle but not in it. Then she holds a handful of sand over the muddy edge of the puddle and slowly releases it. A breeze catches the stream of sand, and Rosie is transfixed by its movement. She looks down at the sand on the mud. "Sugar," she announces, noticing the sand's white grains against the dark, oozy mud. She looks at her dusty hands and claps them. Then she puts one hand on top of the sand in the mud and pushes. Her hand and the sand disappear into the ooze.

Rosie is allowed to continue experimenting on her own; she is enrolled in a program that views each child as a scientist. To her teachers, Rosie is busy constructing knowledge, not a mess. But this type of experiment is not one every teacher would encourage; the *Points of View* box lists several reasons why.

Of course, Rosie's experiment could be translated into other formats. An adult worried about mud play but dedicated to allowing children to follow their passions and construct knowledge might set up the experiment in another part of the yard in a more controlled way. It could even be brought indoors and set up on a table for greater control. Christine Chaille and Lory Britain describe a related experiment in their book *The Young Child as Scientist.*[4]

A boy, Lucas, is seated at a table in front of containers of flour, salt, sand, water, and oil. First, he experiments by adding water to a mound of dry flour. Initially, he uses a spoon to stir the ingredients, but then he uses his hands. Eventually, he washes off his hands. Then he begins on the next step of his experiment, which is to add salt to water with his hands; he feels the salt dissolve in his hands. The scene continues and ends happily without a big mess.

To follow up on Rosie's self-initiated experiment, her teachers might set up an array of materials similar to the ones described by Chaille and Britain. The experiment might not end up as neatly as Lucas's, but any teacher who sets out that particular combination of substances must be prepared for the eventuality of mess making (in the name of science). And, of course, if this type of exploration is considered legitimate science, all the elements of project work could be applied. The adult's role is to observe and provide for further and deeper learning, as well as to help the children document and record their findings and reflect on them.

Two Basic Science Concepts

You don't have to have a degree in physics or chemistry to teach physical concepts to young children. You can create a whole line of questioning based on two concepts–movement and change. Review the experiments discussed so far in this chapter. All of them have been based on two simple questions, "How do things move?" and "How do things change?"

behavior contagion
A phenomenon that occurs when children are influenced by each other's behavior. It is most noticeable in its negative form, when children are doing something they aren't supposed to do.

According to constructivist researchers Constance Kamii and Rheta DeVries, children learn "physical knowledge" by observing and creating movements and changes in objects.[5] What will happen to this sand when I pour water on it? What will happen when I roll this ball down the slide?

Kamii and DeVries list four factors that facilitate the construction of physical knowledge. First, there should be a connection between what the child does and how the object or substance responds. Pouring water on sand and feeling the sand is better than just watching an adult demonstrate the effect of water on sand.

POINTS OF VIEW

Rosie's Mud and Sand Experiment

"The sand stays in the sandbox" is a familiar statement heard in many early childhood play yards. Indeed, sand costs money and has to be conserved–and even protected from neighborhood cats. Sand also has to be kept away from grass, or it will eventually kill it if dumped in large enough quantities.

A rule to keep sand in the sandbox would have put an end to Rosie's experiment, but sand is not the only issue here. What about Rosie's clothes? She is bound to explore that puddle further, and mud can leave permanent stains. Will her clothes be a problem?

"Yes," say some adults. From a teacher's or a family child care provider's perspective, eventually having to clean up Rosie will make it difficult to keep an eye on the other children. Furthermore, there is the issue of **behavior contagion;** it will not take long for the other children to join in and turn the experience into a free-for-all. Some early childhood educators would not frown on a mud free-for-all for the children, but they know that some parents might be upset.

What might a parent's perspective be on mud play and experimentation? Here are several views:

- "I send my child to school dressed nicely. We are a decent family and have a deep respect for education. When my child is dirty, it reflects on our family. We don't want to see her covered in mud!"
- "The stereotype of my people is that they are dirty, and I want nothing to do with promoting this stereotype. When my daughter has mud on her clothes and under her fingernails, it looks as though no one cares for her. I want her to be and *look* well cared for in this program. I want the world to know that she has a good mother who chooses quality child care."
- "Hey, I pay a fortune for my child's clothes. I don't want them muddy. She can experiment with other things besides mud holes."
- "I purposely send my child in old clothes so she can explore and experiment to her heart's content. I save her good clothes for those few special occasions when it matters how she looks. At school I want her to feel free."
- "I don't care if my sons get dirty, but little girls are supposed to be neat and pretty."
- "I want my daughter to have equal opportunity. How can she develop fully if she is restricted from some of the experiences that boys have–just because she might get her clothes dirty?"

Even in the name of curriculum, teachers, caregivers, and providers cannot afford to ignore parents' views. Just as adults must help children understand and respect each other's perspectives, it is important to do the same with parents. When we acknowledge, respect, and talk about differing perspectives, we can then take a collaborative approach, which, in the end, is richer and more satisfying than pushing a single perspective.

Second, the child must be able to try things different ways. Without variation, there is no experimentation. After the child rolls the ball down the slide a couple of times, she will be anxious to try something new: "How will sand on the slide affect the ball? What about water? Will a block go down the same as a ball? What about a toy car?" Third, the more the child can see how the object or substance reacts,

Adults and older children use binoculars for studying nature. This child is making a pair of play binoculars out of cardboard tubes. She can use her home-made binoculars for focusing in on things and practice being an observant scientist.
Index Stock/Alamy Stock Photo

the better. Some processes are invisible, such as air circulation, electricity, and gravity. Indeed, children can learn about such processes by observing their effects, but they'll get more from observing processes that react directly to their own actions on them. And fourth, the more immediate the reaction, the better, especially for younger children. Adding tempera paint to shaving cream shows immediate change whereas planting and caring for a seed is a process that takes longer and requires more patience. Older children can learn to observe over a period of time, but younger children appreciate fast results.[6]

Clear plastic objects are ideal for making certain processes visible. Sending toy cars, trains, or balls into mailing tubes at different slants helps children learn about the relationship of speed to angle of descent. By using a clear plastic tube, the children can see the process happen before their eyes. With a cardboard tube, the cars simply disappear into the tube and then suddenly reappear, thereby limiting the observation of the process.

Nature Study

Children cannot explore biology and ecology in the same way they explore some of the other branches of science, such as physics. Experimenting with inanimate objects, such as a pendulum, is one thing, but nature consists of living creatures and organisms. Children must learn a basic respect for nature. We do not want them to get the message that it is okay to take eggs out of a bird's nest to study them and then to pull the nest apart to see how it is made.

As they construct knowledge about biology and ecology, we want children to understand the connections between nature and ourselves–our interdependence on each other. The web of life is an important concept. We cannot teach nature study without teaching values.

Adults' attitudes make a big difference in what children learn. When an adult is an interested, curious, and respectful observer, children are more likely to approach observation the same way. When the teacher models interest in learning more about what the children are studying, that's also a powerful message. Teachers can model the use of resources to expand knowledge by seeking out books and other aids to take an avenue of inquiry further.

Children today lack the kinds of outdoor experiences in nature that previous generations have had. One thing children are missing out on are the many sensory experiences that nature provides. Children who spend their days in a world of plastic do not even know what it is like to be in natural surroundings. Educational practices do not focus nearly enough on nature. Because of pressure on teachers to produce outcomes, teachers see learning as something children must sit indoors to do.

We want children to be gentle and respectful of nature, but we also want them to construct knowledge, which they need to do in a hands-on way. Therefore, the challenge is to serve the children's needs while teaching them values at the same time. They cannot perform hands-on experiments with animals in the same way they can with blocks, balls, and pendulums.

There is also much to be learned from nature inside as well when teachers bring natural objects inside for children to observe and study. Instruments such as magnifying glasses (and for older children microscopes) augment observation with the naked eye. Some teachers have a "nature table" where they either set up exhibits themselves or let children bring things to put on it.

Transitions Projects

Children can, however, observe transitions in nature. Change is a theme that runs through all branches of science. Some transitions are a lot slower than others, especially those that occur in nature. Watching plants grow and eggs hatch takes what seems like forever–nothing as immediate as running a toy train down a plastic tube or mixing water in sand.

One program found a variety of ways to study transitions after an exciting event triggered interest in the children: some chickens were nesting in the ground cover outside the child care center. Hens and roosters had been part of the scenery for a long time, but the children had never seen a chick until one spring day, when little yellow fluff balls were noticed following their mothers around. It was an exciting day. The classes took turns coming out in small groups to observe the chicks, and several projects arose as a result of the children's interest.

Two-Year-Olds Studying Transitions. In the two-year-old room, a teacher decided to explore eggs with the children. Here is the scene. Yasmin, along with three other children, is listening to the teacher tell a story. The teacher is showing the children pictures from a book. On one page is a picture of a large bird; on the other page, some decorated eggs. Yasmin is listening and watching intently.

When the teacher finishes the story, she clothespins the pages of the book open and lays it on the table. She then says, "I'll show you some real eggs now." She goes over to the counter and picks up a wicker basket. She brings it back and shows the children four brown and two white eggs. Yasmin reaches out with her index finger and gently touches the eggs in the basket. She looks at the open book and then back at the eggs in the basket and says, "her mama, her mama," to the other children. Then she looks at the teacher and repeats, "her mama." The teacher answers, "Yes, Yasmin, eggs come from birds." The child then touches the picture of the eggs in the book, tracing the outline of the eggs with her index finger. She reaches into the basket and touches a real egg, again using her index finger.

The teacher asks the children if they would like to make a picture. Yasmin claps her hands excitedly and says, "Yes!" The teacher gives the children white paper and several felt pens. Yasmin grabs a brown pen and circles round and round on the paper. Then she takes a blue pen, makes zigzag motions across the picture, and then announces, "All done." It seems as though she has drawn the eggs she has seen in their natural color and then has added a blue design to them, like one of the decorated eggs pictured in the book.

The teacher takes a brown egg out of the basket and asks Yasmin if she wants to hold it. Yasmin stands up and says, "Yeah!" She puts her hands together, palms facing up, and sits back down. Two other children put down their felt pens and come to stand by Yasmin. The teacher sets the egg in Yasmin's hands. She cups the egg very carefully and stares at it. After holding the egg for a few minutes, Yasmin is asked to pass it to Daniel, the child sitting next to her. She stands up and turns with her entire body, holding her cupped hands out. Daniel grabs the egg from her cupped hands. Yasmin sits back down.

Daniel shakes the egg quickly and then bangs it on the table top. The egg cracks. Yasmin jumps up, saying, "No!" Then she says to the teacher, "Him broke it." "Let's see what's inside the egg," says the teacher, taking the egg from Daniel and putting it into a bowl.

Whether the project was originally designed to examine the inside of the egg or just the outside is unclear. In any case, the teacher was flexible and continued the lesson by observing the cracked egg and its contents. She obviously knew she was taking a chance passing eggs around to two-year-olds.

The next day, the teacher brings a carton of eggs and helps the children crack them to make scrambled eggs. The day after that, they hard-boil some eggs and have a long session of peeling them once they have cooled. While they are peeling and eating, the teacher helps them reflect on their previous experiences with eggs and compare uncooked and cooked eggs.

Preschoolers Studying Transitions. The preschool teacher went a different direction with interest generated by the hatched chicks. She brought in silkworm eggs–tiny specks of gray–and placed them on a tray lined with clean paper. She put magnifying glasses nearby for the children to examine the tiny comma-shaped eggs.

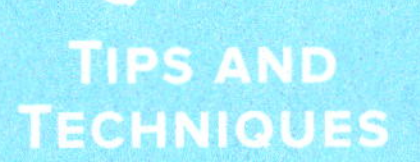

Skills Scientists Need and Sample Science Themes

- Examples of skills scientists need: observation, asking questions, investigating, categorizing, documenting, reflecting, coming to conclusions.
- Examples of tools used in scientific investigations: magnifying glasses, binoculars, flashlights, measuring instruments such as tape measures and scales, clip boards, paper to document findings, writing instruments, trowels for digging, boxes for collecting, books and field guides.
- Examples of science themes:

1. Air: learning what it can do and how we know it exists using kites, fans, balloons
2. Water: exploring the properties of liquids, learning about flow and what sinks and what floats, transforming water into ice and steam
3. Ice: watching it melt and discovering what objects or substances melt the ice faster than others
4. Steam: seeing the transformation of water by heat
5. Weather: observing and charting changes in the weather, measuring temperature and rainfall
6. Rocks and minerals: observing differences, comparing weights, learning names
7. Living things: planting seeds and watching them grow, identifying plants, studying insects, learning about animals
8. The senses: becoming aware of the five senses and using them (smell boxes, taste experiences, identifying objects in feely bags, matching shaker sounds)
9. How things work: using simple tools and taking apart simple appliances, fixing toys
10. The human body: listening to the heart with stethoscopes, drawing "big mes" and labeling body parts, studying skeletons

It did not take long in the warm room for the gray specks to begin to wiggle. At this point, the teacher transferred them to the bottom of an empty, paper-lined terrarium. Then began the daily ritual of gathering fresh mulberry leaves and throwing out the old ones. The children shared the tasks, some doing the feeding and housecleaning and others taking pictures and recording the silkworms' progress. The teacher also created a schedule outlining who would get to take them home each weekend.

The silkworms ate many leaves and grew surprisingly fast. If the children were especially quiet, they could hear the steady crunching as the silkworms devoured the mulberry leaves. The children took pictures twice a week to document their progress. They also kept records, named the silkworms, and even dictated stories about them.

It was an exciting moment when the first silkworm, now about four inches long, started spinning. The teacher put sticks and folded cardboard into the terrarium so there were lots of nooks and crannies to which the silkworms could secure their cocoons. One after another, the silkworms curled up and wrapped themselves in cocoons–some in white and others in yellow, soft, cozy balloons of silk.

Can You Make a Set of Blocks?

One day, a beginning early childhood professional asked for advice over the Internet on making a set of unit blocks out of wood scraps. She explained that she did not have the money to invest in a real set, but that she wanted the children to have some kind of blocks to play with. Some of the responses cheered her on and gave advice about how to measure carefully, sand sharp edges, and preserve the wood so it would not splinter. Others said that homemade blocks could not be produced to accurate enough measurements to do the job they need to do; the blocks would not fit together well and could not be used as a mathematical tool.

Indeed, both perspectives have validity. Homemade blocks are not the same as hardwood unit blocks. Yet, wood scraps by themselves, even without being made into block shapes, can be used by children to explore and experiment with size, shape, balance, fit, and a variety of other physical concepts. The fact that they are cheap or even free may also loosen the restrictions on their use; a good set of hardwood blocks is a big investment, and most adults are uncomfortable seeing them get hammered, painted, dunked in water, or hauled outside, as is apt to happen during creative experiments.

The children's work was done. Now all they had to do was wait. They checked off the days until the first moth emerged. From then on, things happened quickly, as the moths mated and then the females laid hundreds of tiny yellow eggs. The eggs were stuck everywhere–some neatly lined up in rows and others clustered in random patterns.

Suddenly, the flurry of mating and egg laying stopped. Everything was still. The moths had died, but the children were not sad. They clearly understood that it was the end of one cycle–and indeed, the beginning of another, for the new eggs were turning from yellow to gray. The eggs would now be stored in the refrigerator and brought out again the following spring.

This project, however, was finished for the year. Everything was well documented by the teachers and children. The walls and portfolios were full of charts, pictures, and other kinds of information about silkworm development. The book corner contained three new books about silkworms–one made by a child, one produced jointly by the teachers and children, and one published commercially.

Teachers who love science can teach science any time, any place, as well as plan for materials and active experiences children can learn from. See the *Tips and Techniques* box for ideas of how to expose children to scientific concepts including the skills they can acquire and themes they can study.

School-Age Children Studying Transitions. In the school-age room, yet a different project evolved from hatching of the chicks. This teacher brought in fertile eggs to incubate. The children helped set up the incubator and created a system for checking on the eggs, turning them, and marking off the days until they were expected to hatch. The egg-hatching project was slower and less exciting than the silkworm project, but the children were old enough to be patient. They charted each other's guesses about which egg would hatch first and when.

When the chicks finally began hatching, children from the other classes were invited in to watch. The older children took turns videoing. Afterward, they had fun watching the tape on fast forward.

The school-age children had researched beforehand how to care for the newly hatched chicks and went right to work creating a pen for them. Going to the feedstore was an interesting project in itself. Journal writing afterward reflected the different experiences the children had on the feedstore field trip. The children kept careful records of the costs involved and made plans to sell the eggs the hens would eventually produce in order to recoup some of their expenses.

Basic Equipment and Materials for Math and Science Learning

By now, you have an idea about the types of materials that are useful for various math- and science-related experiments and projects. Following is a listing of the most basic materials and an explanation of their uses.

- *Hardwood unit blocks.* These blocks are made to precise measurements so that they are all exactly the same width and thickness. They are called "unit blocks" because the basic unit is the square. Two squares equal one short rectangle, two short rectangles equal one longer rectangle, and so forth. Most early childhood professionals consider a good set of hardwood unit blocks to be essential early childhood equipment. See the *Tips and Techniques* box for another perspective on investing in hardwood blocks.
- *Sand.* Sand is a versatile material that should be present wherever young children spend their time. Sand is useful all by itself and becomes even more interesting when things are added to it. Digging and filling tools such as buckets, clear containers, funnels, scoops, and sieves are traditional sand implements, but children often come up with their own ideas of what to use in the sand. They enjoy natural objects, such as sticks, leaves, and stones. A water source adds another interesting dimension to sand play. Children find their own ways to get the water to the sand and vice versa, even if the water source is some distance away.
- *Water.* Water is infinitely interesting. Children can spend hours exploring water's properties. Hoses, drinking fountains, sinks, water tables, buckets, dish pans, and plastic wading pools are all types of equipment that can be used for water play. Like sand, water is interesting all by itself, but add objects to it–or add water to various substances to watch them transform–and the children's fascination grows.

Clay is responsive to young hands–and provides a sensory experience as well as enabling children to explore processes of change.
Will McIntyre/Science Source

- *Play dough.* Made from various recipes of flour and salt (sometimes cooked, sometimes not, sometimes with oil or cream of tartar) play dough is a standard early childhood material. Making it is as much an exercise in transformation as is playing with it. Play dough is very responsive and can be used alone or with tools and toys.
- *Clay.* Like play dough, clay is responsive to young children's actions. It can provide a unique sensory experience, especially when combined with water. Some early childhood educators feel strongly that young children should have several years of pure exploration and experimentation with clay before being taught how to make things; they believe that the process of exploring the properties of clay should be emphasized over the product. Others feel that both play dough and clay are mediums that lend themselves to symbolic representation. Sculpting, like drawing, can be a means of self-expression; according to one point of view, sculpting and drawing are two of "the hundred languages of children."[7]
- *Writing, drawing, and painting equipment.* Paper of all sorts, pencils, felt pens, crayons, tempera paint, and brushes, as well as other miscellaneous

supplies, such as string, tape, glue, and staples, should be available for children to use creatively as desired.

- *Loose parts.* Wooden boxes, ladders, boards (cleated and plain), and loose parts of all sorts give children plenty of opportunities to build and explore. From these items, children may create ramps, bridges, and slides for climbing up, walking across, crawling under, or sliding down. Of course, adult supervision is required to ensure safety.

This is just a starting list of very basic early childhood equipment and materials. Use your imagination to come up with dozens of other things that belong in the environment to teach math and science as well as all the other aspects of the curriculum.

A Story to End With

I remember the day a science experiment got out of hand. I did not set out to teach science; I only wanted to make play dough. I gathered the ingredients and some three- and four-year-old children to help me. Then came the big question–What color do we want to make this play dough? The children looked over the available containers of tempera powder. When they discovered the jar of black powder, they got excited. "Black," they all agreed. So black it was.

Black was popular then for several reasons. Lately, our class had been talking about darkness. It was fall; the days were getting shorter. Halloween was around the corner. One child started talking about the black spiders she was going to make.

So we went to work mixing the dry ingredients. Into the big bowl went the flour, followed by the salt, and then a whole lot of black tempera powder. I had not intended for so much to be used, but my back was turned when two children dumped it in. I did not really mind, because the stuff was ancient and, besides, maybe the play dough would turn out really black instead of just dark gray. I had visions of licorice-colored play dough.

The big mistake came when I turned my back at the wrong time again, and two children added water–twice the amount called for. I was still a beginning teacher at this point and had not learned how to assert a little more control over "cooking" projects. I was not thinking about exploratory science at the time either.

The children dived in with spoons and hands and started mixing. What a mess. The play dough was pure liquid–black liquid. "More flour!" I announced, as I marched toward the cupboard where the flour was kept. Two children followed me, dripping what looked like runny tar onto the floor. I scooted them back to the bowl.

I reached into the cupboard–no more flour! In the meantime, while I was searching the shelves, one child took the container of black powder and dumped the rest in. Unfortunately, it did not thicken the mess. It just made it a shade darker. I frantically searched the cupboard. There *must* be some thickening agent hiding somewhere, I thought. There was not. The children did not mind, though. They were having fun playing in the black soup they had created.

The other teacher popped her head in from outside and announced that it was almost cleanup time. Her eyes grew wide when she saw what was happening. Cleanup time–fine with me. I thought we had better get started since it was going to take a while to clean up the mess.

What I had not planned on was the fact that the sink was across the room from where we were working. Two children traipsed over to it, black ooze running off them the whole way. The others watched their trail and thought it was cool. They too abandoned the bowl and started across the room, adding their own drips to the ones already spotting the floor. Soon, the path to the sink was full of black footprints as well as drips.

We were demonstrating transformation and representation all right. The children noticed how the room was transforming, but I did not comment on it because I was too busy trying to keep my cool. I should have worked with them on reflection and documentation but had no enthusiasm for it.

When the children tried to help, the mess got even worse. That black paint just would not go away. Even diluted with water, it was still mighty black. I have a feeling if I visited that room today, I would still find traces of that transformation project that got out of hand! After that project, I learned to be more organized. I also learned to keep a tight lid on the jar of black tempera powder.

SUMMARY

Adults do not *teach* children about math and science; they give them opportunities to *construct knowledge.* Their role is to observe, to note the children's interests, to set up the environment accordingly, to ask sensitive questions at just the right time, to allow the children to explore and experiment, and to provide resources when needed. This method of teaching is based on a constructivist approach to learning. According to the theories of Jean Piaget, the constructivist approach helps children learn physical knowledge, social knowledge, and logico-mathematical knowledge.

Early childhood educators contribute to the construction of knowledge by creating projects that help children pursue areas they are interested in. Project work facilitates the kinds of interactions, explorations, and problem solving that result in knowledge construction and conceptual development. Children also learn math concepts (such as time and space) through "real-world" activities and by playing games.

Taking a constructivist approach to science is very different from setting up formal science lessons and teaching children labels, concepts, and processes through unrelated displays, activities, or teacher-controlled experiments. Transformation and representation are two basic science concepts that young children explore as they learn early lessons in physics and chemistry. Three primary questions children ask themselves are "How do things move?" "How do they change?" and "What traces do they leave as a result of their movements or changes?"

When children learn about nature, especially biology and ecology, they cannot explore and experiment the same way they can explore the inanimate world. Children need to develop a respect for nature and an understanding of our interconnectedness with nature. Adults' attitudes make a big difference in what children learn. When an adult is an interested, curious, and respectful observer, children are likely to approach nature study the same way.

The early childhood math and science curriculum depends on some basic materials: unit blocks, sand, water, play dough, clay, writing, drawing and painting materials, and assorted loose parts. But this represents a bare-bones list; using their imagination, early childhood educators can add endless pieces of equipment and materials to facilitate all areas of learning.

REFLECTION QUESTIONS

1. What do you remember about learning math when you were a child? Think about both formal learning in school and informal learning outside of school. How can you use your own experience to help you understand what young children need?
2. What do you remember about learning science? Think about both formal learning in school and informal learning outside of school. How can you use your own experience to help you understand what young children need?
3. How did you learn about nature? Did you have nature study in or out of school? Did you have someone who loved nature who shared that love with you?
4. How do you feel about children getting dirty in the name of exploration and learning? What is your perspective on messy activities? What would you do if you worked in an early childhood program where the director and staff had the opposite view of messes from yours?
5. How would you feel about a child discovering that by holding a magnifying glass over a bug at just the right angle to the sun, he can kill the bug. Would you allow the child to continue the experiment in the name of science?

TERMS TO KNOW

How many of the following words can you use in a sentence? Do you know what they mean?

constructivist approach 390
physical knowledge 390
logico-mathematical knowledge 390
social knowledge 390
one-to-one correspondence 393
real-world math 396
transformation and representation 398
behavior contagion 400

FOR FURTHER READING

Brooks, J. G. (2011). *Big Science for Growing Minds: Constructivist Classrooms for Young Thinkers.* New York: Teachers College Press.

Copley, J. V. (2010). *The Young Child and Mathematics* (2nd ed.). Washington, D.C.: National Association for the Education of Young Children.

Editors of Teaching Young Children. (2014). *Exploring Math and Science in Preschool.* Washington, D.C.: National Association for the Education of Young Children.

Gelman, R., Brennerman, K., Macdonald, G., and Roman, M. (2010). *Preschool Pathways to Science: Facilitating Scientific Ways of Thinking, Talking, Doing, and Understanding.* Baltimore, MD: Brookes.

Helm, J. H., and Katz, L. (2016). *Young Investigators: The Project Approach in the Early Years* (3rd ed.). New York: Teachers College Press.

Heroman, C. (2016). *Making and Tinkering with STEM: Solving Design Challenges with Young Children.* Washington, D.C.: National Association for the Education of Young Chilldren.

Johnston, P. H. (2012). *Opening Minds: Using Language to Change Lives.* Portland, ME: Stenhouse.

Moomaw, S. (2011). *Teaching Mathematics in Early Childhood.* Baltimore, MD: Brookes.

Pollman, M. J. (2010). *Blocks and Beyond: Strengthening Early Math and Science Skills Through Spatial Learning.* Baltimore, MD: Brookes.

Ranck, E. R., and Anderson, C. (2010). Blocks: A versatile learning tool for yesterday, today, and tomorrow. *Young Children, 65* (2), 54–56.

The Early Math Collaborative–Erickson Institute. (2014). *Big Ideas of Early Mathematics: What Teachers of Young Children Need to Know (Practical Resources in ECE).* London: Pearson.

Wilburne, J. M., Keat, J. B., and Napoli, M. (2011). *Cowboys Count, Monkeys Measure, and Princesses Problem Solve: Building Early Math Skills Through Storybooks.* Baltimore, MD: Brookes.

15 Integrating Art, Music, and Social Studies into a Holistic Curriculum

In This Chapter You Will Discover

- how early childhood education deals with the traditional subject areas of music, art, and social studies.
- how an "art" activity can sometimes look like a "science" project and vice versa.
- how children develop physical and cognitive skills through inventive and creative art experiences.
- how children express feelings through art.
- how to talk to children about their art.
- what adults can do to facilitate the artistic process.
- why children benefit from an open-ended exploratory approach to music and art.
- some of the different ways children make music.
- what a "music center" should offer children.
- how music can be a focus of group time.
- how the study of self, others, and the community fits into a social studies approach.
- what role adults play in passing on biases and negative images to young children.
- what children learn when they study "community."
- how early childhood professionals create a holistic approach to early education.
- how early childhood professionals can show parents a balanced approach.

Unlike higher education, early childhood education cannot be divided neatly into distinct subject areas. When you look at children busy in a preschool classroom or a family child care home, for example, you could try to define what they are doing using traditional discipline labels, but you might be wrong. Take art, for instance. In the last chapter, we saw science projects that looked a lot like art; however, their goal was to construct physical knowledge. When children sit down at the "art table" to use play dough or clay, are they involved in "art"? What if one child is using the play dough or clay for creative expression while another is using it to represent some concept or feeling? Indeed, some children enjoy pounding play dough or stroking clay mostly for the sensory experience—for the pure joy of feeling it.

The same is true for drawing. When documenting a project by drawing wildflowers, for example, are children being "artistic" or are they "recording data"? When toddlers scribble with a felt pen, are they engaged in an artistic, sensory, or emergent literacy experience?

Think of music. Perhaps music is considered a subject in itself, but children incorporate elements of music into many daily experiences—when they hum a tune while tricycling, beat a rhythm on the edge of the sandbox, or make up a song or a chant as they go to sleep.

And then there is social studies. One could even say this whole book has been about social studies. Certainly, teaching children prosocial behaviors has been an ongoing thread woven throughout every chapter. Young children are almost never admonished to "do their own work." Children's "work" is hard to distinguish from play, and it is almost always collaborative. So you would have to say that practically everything that happens in an early childhood program concerns "social studies."

Social studies starts with self-knowledge and moves to learning about others. The most important learning involves getting along with others, resolving conflicts, cooperating, sharing, and communicating. Children also learn firsthand about community as part of their social studies curriculum. An important social studies goal is respecting diversity—that is, appreciating cultural, racial, class, gender, and ability differences.

The goal of this last chapter is to celebrate creativity and community in the three areas of art, music, and social studies and to bring home the idea that every aspect of the early childhood curriculum is overlapping and interconnected. Language, emergent literacy, math, science, art, music, and social studies occur naturally as children explore a rich, well-planned environment, follow their interests in project work, collaborate with each other, and get their needs met.

ART

What is art anyway, and what does it look like? Art is self-expression through the use of various mediums. This creative process involves some or all of the following: color, form, line, shape, and texture. To qualify as "art" in the mind of many early childhood professionals, an activity should be open-ended so that children can

A Journey Begins with a Single Step

When I first entered the field of early childhood education in 1983, most of the information and training were centered on themes, precut activities, and patterns. We were encouraged to have the children put cotton balls on white construction paper in the shape of clouds during "weather week" and watercolor precut fish shapes during "under-the-ocean" week.

Very quickly I realized that I was bored with the choices and knew that the children deserved more–but what? I began asking questions and attending workshops, seminars, and conferences that were geared to what was being called the "process."

There was a whole new world of concepts that I had not known before. The simple question, "Is it process or is it a product?" started an awesome journey toward the discovery of respecting the real abilities of children and allowing for open-ended creative experiences that would enable them to explore the world around them and much more.

To my amazement when I offered the children opportunities to do print art with potato mashers, plungers, dog and infant toys, combs, and pairs of shoes with paint, clay, dough, and water, we both discovered the detailed patterns, graphs, and intricate designs in everyday objects. This developed into discovering patterns on plates, serving trays, tiles, and walls.

As the process of art became an everyday occurrence, the children's abilities to understand math, reading, language, and problem-solving concepts deepened. It led us to profound experiments, discoveries, and conclusions about the classroom environment and the world around us.

I sometimes look back at the teacher I used to be and feel sad. However, I quickly remind myself that most of the greatest works of art took years or even decades to complete. A great teacher is one who recognizes that the profession is a journey and continuously evolves her teaching styles.

–Kisha P. W. Champion

decide for themselves both process and product. Art in this sense differs from preplanned projects that children complete according to directions or a model, such as gluing precut petals onto an outline of a flower. Although art may involve creating or arranging elements in an aesthetically pleasing manner, most early childhood educators agree that it is the process, not the product, that is important. Art is a form of communication that conveys impressions, ideas, perceptions, and feelings. A student, whose name I have long forgotten, once gave me her very broad definition of art: "Art is the way of living in which you appreciate beauty in its infinite forms. Art is everything you do, every choice you make."

NAEYC Program Standards

Program Standards 2 and 3
Curriculum and Teaching

Art should not be encouraged in only one area of the classroom or family child care home. Art can be anywhere. Children can illustrate their writing, draw sketches to go with their science experiments, or create maps of a room, the play yard, or even the neighborhood. Artistic endeavors may be seen as projects in themselves, incorporated as parts of a project, or isolated to an individual activity that is not obviously connected to anything else.

Activities in early childhood education that fall under the category of art tend to be either flat or three-dimensional. Flat art includes brush painting, finger painting, drawing, and collage. Three-dimensional art can be collage as well, depending on the materials used. Other three-dimensional art activities include creating sculptures out of play dough, clay, wood scraps, cardboard, and papier-mâché. In the *Voices of Experience* box on page 417, Kisha P. W. Champion shares the evolution in the way she thought about art for young children.

How Children Benefit from Art Experiences

What do children get out of what we call "art"? As children experiment at the art table, they improve fine-motor skills. Their perceptions sharpen as they become more experienced at noticing similarities and differences, color, shape, size, and textures. They begin to organize their perceptions into thoughts and actions that then translate into patterns and symbols. Art experiences present children with opportunities to invent and create while getting to know themselves and their world better. Through art, children express feelings they can't talk about, but they can put them down on paper, smear them around on the easel, or pound them into play dough. (For two contrasting perspectives on the goals of art for young children, see the *Points of View* box.)

Through drawing children gain fine-motor skills and learn to express feelings that are hard to talk about.
Liquidlibrary/Getty Images

How Free Should Art Be?

Some people believe that children should be encouraged to represent their perceptions, ideas, and understandings through various artistic mediums. When children use art this way, their concept development expands at the same time as their creative skills. By working regularly in a representational mode using different mediums, children sharpen perceptions and increase their physical skills, including fine-motor control and eye-hand coordination. Their powers of communication also broaden when they learn to use more than oral language to express their ideas and thoughts.

A contrasting viewpoint argues that "representation" is not art. When adults urge children to represent things, they move them away from self-expression and free exploration–two of the basic characteristics of true art. Children should be given materials and allowed to use them in whatever way they want, *without* adult input. When they are asked to express their understandings of the world through artistic mediums, children are forced into a cognitive process rather than an artistic one.

Let us take a look at how art is integrated into a preschool setting. This scene illustrates what started as a project and ended as a potential art career. One spring morning, four-year-old Bo Pik and several other children discovered snails on the cement walk. At first they were interested in just observing how they moved. When that fascination wore off, the teacher helped them identify and look for snail trails. They found them everywhere!

Then the teacher helped them bring some snails inside in a paper-lined cardboard box to study for a while. She was careful to explain the importance of treating the snails with respect. After watching the snails move around on the paper, the teacher put the snails back outside. Then the children sprinkled powdered paint onto the paper to highlight the trails.

Bo Pik suggested having some snails walk on black paper to see if their trails would show up without the powdered paint. Back outside they went for some more snails. It was not as easy to find them the second time because the sun was high and the snails had sought cover. After some searching, the children found a few snails and carefully placed them on the black paper. They were excited to see that the trails did show up as shiny lines.

Over the next few days, the children came up with different ideas: having the snails move through food coloring to create their own colorful trails, putting the snails on glass to watch their movement from below, and sending them through an "obstacle course." They discovered in this last experiment that the sharper the object the snails crawled over, the more slime they laid down.

Some of the children recorded their findings, using paper and pencils. Bo Pik did a detailed drawing of a snail and then went outside to look at the plants where the snails had been found. She said she wanted to draw the snail in its home instead

Art, language, and emergent literacy all come into play as this girl illustrates her own book.
FatCamera/Getty Images

of as part of an experiment. She added plants to her drawing. Then, Bo Pik went from the snail project to the easel and painted what she called a "snail picture," showing several snails under the bushes surrounded by plants. The project continued for a week, and Bo Pik continued to record, making sketches on her clipboard before turning them into paintings.

Every day, Bo Pik found new ways to put snails in her artwork. For several days she created play dough snails, carefully rolling out "snakes" and then spiraling them into snail shells. Even after the snail project had ended, Bo Pik kept on representing snails in various mediums, stylizing her artwork over time. Eventually, the snail became her trademark. Every one of her paintings and drawings incorporated a snail somewhere.

Then one day, without explanation, she moved on to rainbows as her trademark. The spiral shape, however, still reappeared in her paintings, drawings, and sometimes in her sculptures. It is hard to say if Bo Pik was more interested in snails or in their symbolic representation. The spiral that so caught her fascination is an ancient symbol as well as a shape that regularly appears in young children's scribbles. The mandala, a crossed circle, is another shape that seems to be timeless and universally interesting.

Facilitating Art Experiences

What do young children need from adults to facilitate the art process? First, manage expectations for both the adults and the children. Children should feel that what they will do will be respected and valued. Likewise, educators should have age-appropriate expectations to appreciate both the process and final product. A toddler's scribbling, for instance, is art that aligns with his or her age-related abilities.

Second, you does not have to display every piece of artwork, but if you choose to display some pieces, you should make sure that each child's work is represented. When only "choice" art pieces are displayed, children learn quickly who has talent and who does not. Likewise, when the bulletin boards are covered with teacher art, children get the message that theirs is not decorative, or when the walls sport 30 identical cookie cutter art projects, the children learn that creativity is not valued. To demonstrate that real art is valued, a classroom should display a variety of children's creative work, drawings, paintings, and three-dimensional pieces. To demonstrate that art is integrated into the whole curriculum, displays should include project work, children's comments (that they have written or dictated to the teacher), and photographs of the process.[1]

Third, be mindful of what you say to children while they are engaged in the art process. (Sometimes you do not have to say anything at all. Avoid distracting a child who is in deep concentration.) Instead of asking, "What is it?", it is better to ask if the child wants to tell you about what he or she is doing. As noted in the *Points of View* box on page 419, children's art can be representational or a form of self-expression. Furthermore, avoid making a judgment when commenting on children's art; do not compare, criticize, or suggest. Furthermore, watch out for praise. Praise distracts children from their inner sense of satisfaction. Rather than gush ecstatically over every picture or play dough creation, it is better to comment on the process of the child's sense of expression. Talk about what you see without judging. You can focus on the process itself by commenting on what the child is doing and how you perceive he or she feels about it: "I see how you are covering every piece of white on that easel paper. You're being very careful not to miss a spot." or "It looks like those big wide sweeps of your arms feel good!" You can also talk about form, color, line, balance, or whatever else you see in what the child is making.

Fourth, provide a variety of materials and the freedom to use them as the children choose. Time is another important factor. When time is divided into small segments, children are forced to move rapidly from one activity to another. To become deeply involved in art (or anything else), children need plenty of time. Likewise, setting up a brand new art activity every day and putting away yesterday's materials encourages children to dabble rather than creating and exploring on deeper levels. Of course, children appreciate change, so children may lose interest if the selection of materials never changes. Nevertheless, there is something to be said for making the same interesting combinations of materials available over a period of days instead of just one short "art period."

Fifth, cultivate the social side of art. Design the environment to encourage children to work together. Large tables discourage interaction, but smaller, narrow tables allow children to talk back and forth and share materials. Also introduce communal projects; do not limit art to individual projects. Put easels side by side, and allow children to work on each other's paintings–as long as everyone is agreeable. Encourage children to work together on murals, joint collages, and other collaborative

body tracing An example of an activity that requires cooperation. The child lies on a piece of paper on the floor that is at least as long as he or she is tall and someone traces an outline of the child's body.

projects. **Body tracing,** for example, is an activity that takes two people–the tracer doesn't necessarily have to be an adult. Other joint art activities include sidewalk chalk art, quilt making, and large weaving projects.

MUSIC

Music should take the same open-ended, integrated path as art in an early childhood program. Children should be encouraged to make music with their voices, bodies, and instruments in creative ways. Exploring music, like exploring art or science, uses physical skills, involves feelings and social relationships, and enhances cognitive development by opening up opportunities for honing problem-solving skills.

NAEYC Program Standards
Program Standards 2 and 3
Curriculum and Teaching

Ensure that music is an open-ended creative experience rather than a formal lesson or a "performance." As in art, realize that although talent helps, it is not required to create music. Music is inherent in a child's natural rhythms and can emerge in everyday experiences–if given a chance. Everyone is "musical"; given the appropriate encouragement and time, anyone can learn what talented people naturally come by. It is a mistake to give children the idea that they cannot sing, for example. Far too many adults were told as children that they could not sing or were not musical, so they gave up actively participating in music and just became listeners. For young children, creating music is just as entertaining and amusing as listening to it. Both processes naturally encourage the development of important cognitive and social skills. As with visual art, it is the process, not the product that makes the most significant impact in learning.

Facilitating Music Experiences

Joan Bell Dakin is a music teacher in an elementary school. She shares her love of music and talks about how she works with children in the *Voices of Experience* box on page 423. Be appreciative of children's spontaneous efforts to create music, whether melody, rhythm, or beat. Also encourage them to respond to music with movement. Show interest in what they are doing, but do not judge. To facilitate their musical creations and responses, set up a music center where children can create music as well as listen and respond to it. Such a center should have apparatus for playing recorded music, as well as rhythm instruments (drums, shakers, triangles, and so on). Props, such as play mikes and filmy scarves, add an imaginative dimension to the listening experience. If the area is large enough, children will have room to do creative movement. Mirrors enhance the children's experience by adding a visual element to music and movement, and tape recorders and camcorders allow children to listen and see themselves create and respond to music. Be sure that the music center is away from quiet areas–such as the library corner.

Group Time: Singing

Group singing is a part of early childhood programs everywhere. It may be a spontaneous occurrence or part of a regular circle time–or both. Group singing follows

The Drum Parade

I love making music with kids. On the very first day, I lead a parade–I play the hand drum, and we march around the room. I stop playing and freeze; almost every child freezes, too. I play a fast tempo for jogging, then a slow tempo for creeping. I rub the drum with my fingers and circle in place; I tap it and tiptoe. I hold the drum high over my head and play strongly. I lead the group into a circle where we slowly sit down to the sound of the drum getting softer and softer. Not a word has been said (except for the child who whispers, "Can't she talk?"); but we have communicated–music has spoken, music is what happens in this room.

I am currently the music teacher in a small public elementary school. I do the Drum Parade with each class–kindergarten through fifth grade. The "big kids" still love it; with them I include meter and complex rhythms. After the parade, we pass the drum and tap the rhythm of our names, or play simple beat patterns. The musical conversation develops as I get to know their skills and preferences. More hand drums are passed out–I've made sure that I have over 30 hand drums; there is nothing worse than having to wait to play a drum! Most kids love the drums–as the days go by, we drum the rhythms of nursery rhymes and poems, we accompany stories, we add other percussion instruments such as shakers, triangles, cabasas, and guiros. On xylophones, we play the rhythms we've explored, creating melodies and more accompaniments.

I learned the Drum Parade at a workshop years ago, when I was a preschool teacher. I tried it out right away–what fun the three-year-olds had! We marched all over the school, inside, outside, and back to the Circle Time rug. Later in the morning, two little boys went to the instrument box, took out a tambourine and a shaker, and marched around the room again. Another child picked up a couple of blocks she was building with and matched their beat–oh dear, things got way too noisy. We teachers let it go on a few minutes and then told the children it was too loud; take the instruments outside!

The Drum Parade is fun for me, too. I like improvising and exploring all the sounds one drum can make. I like surprising children by varying the rhythms, the tempo, and the dynamics. I like a parade's simplicity–you don't really have to follow the beat; but it feels good when you do. I believe the Drum Parade encourages a sense of belonging. A few of my current students find school safer and more dependable than their own home situations. I like to imagine that for them, the Drum Parade and other communal music activities manifest positive family relationships. For all of us, music communicates heart, emotions, and sharing in something bigger than each of us alone.

–Joan Bell Dakin

a developmental process. It begins when a baby's rhythmic or melodic babbling is acknowledged as music. A toddler who hums, chants, sings little songs, or follows along when someone else sings he is developing his group singing skills. A two-year-old in a group sing usually shows interest, follows, sings or talks at times, and is silent at others. Even when she follows along, she may not be able to keep up with the adult, but she will probably be making music with her body–moving to the beat.

Rhythm bands allow children to become actively involved in music.
fatihhoca/Getty Images

I remember lots of circle time songfests with three-year-olds newly enrolled in preschool. These memories stand out because I did a lot of solo performances for them, even though it was my intention to create a group sing. At the beginning of the year, the kids just sat there. If I had hand movements or body movements, I could get some participation, but not a lot of them sang. After a few weeks, some began to join me. Eventually, some would sing the chorus and maybe one or two parts of the verses–not always right along with me.

By the time those three-year-olds became four-year-olds, however, their willingness and ability to sing in a group improved. Not all always joined in, but more did than when they were three. They knew the words, and some would sing along. I would occasionally hear them singing on their own–throughout the day, in the block corner, or at snack. Some knew a surprising number of lyrics and even got the tune right. The ones who did not eventually learned.

Children five and six years old have skills to stay together in a group sing, and they get better at it as they get older. The adult can introduce a larger variety of songs because their pitch range increases and they can sing both higher and lower. Older children can also learn simple rounds. Singing rounds leads to harmonizing and, eventually, accompanying singing with simple melodic instruments. A round is "a piece of music sung by at least three people in the same melody but each person begins at a different time to create a harmony. A common example is 'Row, Row, Row, Your Boat' with which the first child would sing this line, the next would continue with 'gently down the stream', and so on in harmony."[2]

Language and literacy are prominent aspects of a music curriculum. Children develop language skills when they are encouraged to think up new words to old songs or to create their own songs. Writing down the words connects the sounds

POINTS OF VIEW

Music as Background?

Some adults enjoy having music as a background all the time. Here is one person's perspective: "I love having music in my life. When there is no music, I feel something is missing. I always turn on the radio or CD the minute I walk into a room. Music is important to me. It helps my mood, animates me, gives me something to respond and react to. I think children need music all the time, everywhere. I know I do. By changing the beat and the kind of music, I can change the atmosphere of a room. I love to *create* change when it is needed. For example, when things slow down and kids start looking bored, I put on some lively music and things perk up again. Or at nap time, I put on classical music. It puts them to sleep right away."

Some adults, however, believe that music should be heard, not tuned out, as happens when it is played constantly. Here is another person's perspective: "I love music too, but I like to really focus on it and concentrate. When music is playing constantly in the background, people unconsciously tune it out. It takes energy to tune out sounds. Why waste energy like that? Constant music adds one more noise to a usually bustling room full of young children. I like silence as much as music. Although I don't get much of it around young children, when it comes it's so wonderful! Why fill up those few precious moments of silence? I think music should be played with purpose; by telling children when music is about to begin, they are prepared to listen and appreciate it. And when it's over, turn it off! Otherwise it's like a constantly running TV. Oh yes, about classical music at nap time. When you use music like that, it's a form of training. When children hear classical music, they're conditioned to go to sleep. Think what's going to happen when they are older and go to a concert!"

to symbols and is one of the many ways that demonstrates the utility of print to children.

Group Time: Instruments

In addition to singing, introducing, playing, and making instruments is a holistic form of play that encourages learning through art and social experiences. One preschool teacher–who has a master's in music and is a professional gamelan player–has created a whole gamelan music program for children. (Gamelan is an Indonesian form of orchestra–primarily from the islands of Java and Bali–that is predominantly composed of percussion instruments, such as metal xylophones, drums, and gongs.) In this program, the children are allowed to explore and experiment with rhythm, sound, and movement. They do not perform, but they do learn to appreciate playing music interactively as a group.

Of course, most programs will not have a full array of gamelan instruments at their disposal, but they can offer an assortment of rhythm instruments–sand blocks, rhythm sticks, triangles, tone blocks, shakers, and drums. Rhythm bands allow children to get actively involved in music. They also tend to get a greater response at group time than singing; almost all children, no matter what age, love banging away on instruments.

In some programs, children create the instruments they play. Instrument creation and playing can give children a chance to learn about music from different cultures as they research what instruments are used and how to make them as well as listen to music made by others. Instrument making is a way of connecting music to art, social studies, science, language, and emergent literacy.

Creative Movement

creative movement An activity, usually involving music, where a child is encouraged to move in creative ways, often with the direction or suggestion of the teacher.

Creative movement to music often occurs spontaneously throughout the day or can be planned as a group time activity. Sometimes all that is needed is to set the stage a little, put on some music, and let the children go at it. At other times, they may need more from you, such as modeling, suggesting, or playing some movement or rhythm games. Sometimes simply describing what the children are doing, instead of telling them what to do, prompts more creativity and spontaneity. The *Points of View* box on page 425 illustrates two perspectives on this subject.

Social Studies

NAEYC Program Standards
Program Standards 2 and 3
Curriculum and Teaching

Although the topic of "social studies" per se has not been directly discussed in this book, in some ways, the whole book is about social studies; the emphasis throughout has been on helping children gain social competence. Think of all the examples that illustrate how adults help children interact and relate in positive ways to others.

Social Studies Begins with Self

Social studies learning entails knowledge, values and attitudes, and skills. It begins with the study of the self. When children know, like, and respect themselves, they are better equipped to know, like, and respect others. When children are able to empathize with others, they are more likely to interact in cooperative ways. Power is also a social studies issue. Unless children gain a sense of their own personal power, they have difficulty interacting in a constructive way with others who exhibit power.

Understanding and appreciating others does not happen right away. There is a progression from being able to see the world from one's own point of view (egocentrism) to being able to take multiple perspectives. The progression is a result of both cognitive and social-emotional development. It is important to understand this progression to avoid thinking of young children as selfish. Children, particularly the younger ones, are inherently self-centered at this development stage and will gradually become less so over time as their brain further develops and with proper guidance from nurturing adults. Adults must broaden their view and understand that they cannot hurry the process. They can train a child to say "I'm sorry," but they cannot make the child feel the emotion that goes with a genuine apology until he or she is developmentally ready. Children develop in their own time, and we have to be patient and understanding with children. This, above all, requires educators to have the appropriate expectations about their abilities to empathize.

The child's first years are about acquiring **self-knowledge.** We have lots of terms to describe this body of knowledge: "self-concept," "body awareness," "self-image," "self-worth," and "self-esteem." Children develop this self-knowledge by constructing it from the input they receive from others. Children regard the way people react to them as mirrors that reflect who they are. They do not know that mirrors can be distorted. They take in the reflected images as truth.

self-knowledge
A term that relates to individuals learning about themselves. Words such as self-concept, body awareness, self-image, self-worth, and self-esteem all fall under the heading of self-knowledge.

Children note the messages they receive, both verbal and nonverbal, about their gender, race, ethnicity, class, appearance, and ability. As such, we need to be very careful about the messages we send children. If an adult unconsciously believes that girls are the weaker gender and need extra protection, that belief may manifest itself in his or her behavior, which, in turn, sends messages that may be incorporated into a child's self-image. Many times children live up to what they perceive adults expect from them; in this case, a young girl may believe she needs to act helpless to receive approval from the adult. Similar adult expectations influence children of different races, cultures, and abilities. Adult expectations can have powerful effects. We must be careful to allow children to be who they are. We must not let our visions of them limit them and get in the way of their development.

It is important that we become aware of our expectations and the messages we are giving children, which is not easy to do. We all have hidden biases, false assumptions, and stereotypical images, yet we must not pass them on to children. When we do, we risk giving them negative views of themselves and, possibly, other people. Just as children engage in self-study, so must their caregivers, teachers, and providers. Self-study is an interactive process no matter what a person's age. To help you reflect on how your attitudes come out in your actions, get feedback from coworkers or other people with whom you interact, and help them in return.

From Self to Others

As children grow in their self-awareness, they begin to expand in their knowledge of others. We help them by pointing out other people's perspectives whenever the opportunity arises: "She doesn't like it when you grab the book from her." "See how he's crying? It hurt when you pushed him down." We can also help children help others see another perspective: "Tell him how you feel about what he did to your drawing."

Along with self-knowledge and knowledge of others, children also need values to develop prosocial attitudes: "I won't let you hurt her. If you are angry with her, tell her how you feel, but no hitting." To develop these values and attitudes, they need skills.

Many social skills have already been discussed in this book. One that has not is how to join another child or a group of children who are playing, or entering

How do children skillfully enter the play of others? Observe a master player, and you will find out.
Tim Davis/Science Source

play. If you do not know how to help a child learn this skill, watch children who are good at it. You will not have to observe long in a free-play situation to see children cleverly work their way into, for example, a block construction or pretend play. Watch and you will see that they seldom ask outright, "Can I play with you?" Maybe they have already experienced a "no." Being told no is a form of rejection, and children who feel rejected are not likely to do something negative that results in further rejection. Rather than asking, experienced children will begin playing without interfering or taking over the play in progress.

Let us look at two contrasting examples. First, let us watch Jasmine enter the play of two children who are standing by the play stove in the housekeeping area. One is the mommy and one is the daddy, and they are in disagreement about what to do about their dog. Jasmine listens to what they are saying as she goes into the area quietly. She goes to the cupboard near the stove and takes out an eggbeater. Then, she does something a little out of the ordinary that catches the other children's attention. She holds the eggbeater upside down and turns the beaters in the air.

"What are you doing?" asks the boy. Jasmine picks up the thread of the two children's play argument and says, "This is my dog's helicopter. That's how we get him to the doctor when he's sick." "Oh," says the boy. "Our dog is sick too. That's why he is throwing up all over the place." "Yeah," says the girl. "We have to keep cleaning up." She gets down on her knees and starts scrubbing the floor. The new player has meshed her play theme with that of the other two, and the three come together without skipping a beat.

A less-skilled player might have just butted in and tried to take over the play, directing the other two, reassigning roles, or even changing the theme: "Pretend this is a rocket ship and we're going to Mars." The couple trying to deal with their sick dog might not have taken this interruption well. They probably would have responded by rejecting the new player.

Less-skilled players who start out with "Can I play?" often run to the teacher when told no. "Teacher . . . they won't let me play with them!" What can the teacher do besides trying to talk the playing children into accepting a new player, or forcing the issue? Let us see how that scene might play out when the teacher uses a strategy similar to the one Jasmine demonstrated.

Lacy has just asked to play with the three children in the dramatic play area who are now dealing with sick dogs and veterinarians. Upon being rejected, she runs to the teacher for help. The teacher moves over to the dramatic play area, with Lacy by the hand. Both of them stand at the edge of the area, trying to get a sense of the action. The "mommy" and "daddy" have stopped focusing on their dog and are now talking about how sick their baby is. Jasmine is holding their baby in her arms.

The teacher, joining their pretend play, suggests that maybe they need to call 911 for an ambulance. They go for that idea. The boy scrambles around to find the toy phone. When he finds it, he punches in the numbers and waits. The teacher answers the call with "This is the ambulance dispatcher." She listens as all three tell her about the emergency. She says in a serious play voice, "I'll send an ambulance right away. Lacy is the driver's name. Watch for her to arrive." She turns to Lacy beside her and asks, "Are you ready to go on a call?" Lacy happily makes a siren noise and moves into the dramatic play area.

"Here's the ambulance," she says cheerfully. "Oh, I'm glad you're here," says the distraught mother. "I think my baby needs an operation." The teacher hangs around a bit longer to see if Lacy will continue to need her support. She moves away as all four children take out their imaginary surgical instruments and begin operating on the doll Jasmine has laid on the stove.

From Self and Others to Community

Social studies learning involves more than just developing self-knowledge and increasing interactive skills–though these are the foundational concepts. Learning interactive skills leads to developing a knowledge of social systems, first in the classroom or family child care home and then in the greater community. An educator's job is to help children come to know the community in which they live as well as the resources and services it provides. Through learning about community, children begin to get a sense of the past (history), understand notions of place (geography), and explore the concepts of producers and consumers (economics). Some of this knowledge they acquire in the classroom or family child care home in their day-to-day experiences, but they can also learn it from taking field trips.

A field trip can entail a simple walk down the street to a park or fire station or a more elaborate behind-the-scenes tour of a library or bakery. A prime requirement for field trips is that children have hands-on experiences that are appropriate for their age. Simple is usually best. A visit to an apple farm, for example, is very meaningful for young children. In some programs, children visit parents at their jobs.

It is important that field trips not occur out of the blue. They should be planned so that they are meaningful to the children and connected to their project work.

holistic approach An approach or a curriculum that focuses on the whole child rather than addressing mind, body, and emotions separately. For example, though an activity may be planned with cognitive (intellectual) goals, the teacher also pays attention to how the child responds physically and emotionally.

How the Early Childhood Professional Weaves the Fabric of the Program

Chapter by chapter, this text dissected the early childhood profession into its most elemental components, from theory and documentation to math, science, art. Early childhood education is unlike the pedagogy typically found in primary school settings. A terrific early childhood education (ECE) educator teaches with a fabric that seamlessly weaves the tools and themes discussed prior into a curriculum that disguises learning through experiences and nurtures the development of a child's mind, body, and heart. ECE professionals calls this the **holistic approach**. This integrated, developmentally appropriate, approach helps children form the foundation that underpins their sense of place, confidence to tackle new challenges, and the self-regulation skills required to move forward in the face of difficulty.

Teaching the Whole Child

NAEYC Program Standards
Program Standards 2 and 3 Curriculum and Teaching

As an early childhood professional, your responsibility–either alone or in collaboration with others–is to create a holistic curriculum. It is your job as caregiver/teacher/provider to manufacture a seamless program fabric. In fact, the curriculum may be so seamless that a parent might wonder if her child is learning anything. She observes day after day and sees lots of playing and what looks like one art project after another. She wonders if her child is learning math or getting the foundations for reading. If she asks, you will have plenty of documentation (in the form of portfolios, journals, and anecdotal records) and can show the parent samples of math work or emergent literacy that were embedded in a project or play activity. (Remember also that documentation and assessment are vital elements of emergent curriculum and project work; they allow children to revisit ideas and apply them to new areas of exploration.) Documenting learning is also part of your public relations role to help articulate program goals and outcomes and to illustrate to parents, supervisors, boards, and funders that the curriculum is balanced and that the children are gaining knowledge, skills, and dispositions (such as curiosity, resourcefulness, enthusiasm, and creativity) through their play and project work.

Instead of waiting to be questioned by anxious parents, it is a good idea to explain up front how the program works and then keep families informed as you go along. An orientation illustrated with slides or videos helps parents understand what you are doing *before* they get worried. But remember that the orientation should not be a one-way presentation. Getting input from parents and building bridges between the home and the program are also important aspects of the orientation and of ongoing communication in general. A truly seamless program fabric includes threads from a child's home life.

As you plan for learning and development and implement your plans, you become a facilitator, helping children construct knowledge of all sorts. In the role of facilitator, you observe children, pick up on their interests, set up the environment accordingly, interact appropriately, ask provocative questions, and observe some more until further interaction is called for.

When children slow down in their exploration, it is your job to facilitate further inquiry. When necessary, provide resources and change the environment in order to help children follow new trails of interest. While doing this, help children get along with each other and resolve social dilemmas.

You may be in the background much of the time. Maintaining a hands-off approach is as important as knowing when to intervene. Of course, by observing, you can make sensitive decisions about when help is needed. Some forms of intervention are very subtle. You might only ask a question, but it may be enough to keep the children moving when they were just about to give up on some problem. Sometimes just an open-ended statement like, "I wonder what would happen if . . ." may be enough to renew the children's curiosity to delve deeper or further.

Just remember that questions can make or break an activity. It is easy to disrupt an activity by asking inane questions such as, "What color is this?" If children already know the answer, why bother to ask? If they do not know the answer, then what is the point of asking a question, especially if it has nothing to do with the activity? Often, adults take a drill approach to questioning–to test children's knowledge of isolated labels or facts. Keep in mind that your job as facilitator is to help children make connections and see relationships.

By using a sensitive, responsive approach, you show that you take children's intellect seriously. Rather than seeking to entertain children with cute, fun activities, you look instead for projects that provoke thought, satisfy intellectual thirst, and, at the same time, integrate the whole curriculum–combining science and art together, for example, or integrating math into play. As an early childhood professional, your job is not to teach subjects; the job is to teach each child in a culturally consistent manner to facilitate development and learning in each child–the whole child. Your job is also to bring individual children into the group and help them create a community.

A Story to End With

Back in the distant past, when I was a preschool teacher and mother of a bunch of little kids, I bought a set of unit blocks for home. I thought long and hard before I invested in what was then a tremendous expense. But I finally shelled out half a fortune and ended up with a big box of smooth, clean, wooden blocks for my children. Those blocks were used over the years first by my children and then by my grandchildren. One son used them far beyond the early years when as a skateboarder, he would create skate parks and use "fingerboards"–small homemade devices he and his friends used to figure out their newest skateboard tricks before trying them out on concrete. My granddaughters used them as walls of the rooms in houses as they practiced their architectual and family-life skills. The original blocks are mixed in with the newer ones I have bought over the years as I had more money to spend. But those first blocks stick out from the rest. They show their age: the wood has darkened and mellowed, and they are covered with the scratches, dents, and nicks that come from more than a quarter century of continual use. But they are intact—strong and sturdy as ever. I do not imagine they will ever wear out.

I have not been back to my preschool for years, but I am sure the same blocks that graced the block corner when I was a teacher are still there. I think about all the ways those blocks—mine and the preschool's—have been used over the years.

They have been stacked thousands of times. I remember how many children I watched build a tower and then knock it over. I also remember a child's grief when someone else knocked over the newly built tower before the builder got a chance to knock it over him- or herself. There are some emotional moments in block building, but children eventually learn to get along with each other and begin to collaborate. They also learn lessons in physics and at the same time they're honing their social skills.

Then there are the social studies lessons; children build houses, forts, farms, cities—you name it, I have seen kids build it. Populated by small figures like people, farm animals, and dinosaurs, the play goes even further and deeper, especially when followed up with projects and field trips.

I did not used to think of block building as art when I was a teacher, but I do now. I never considered drawing, photographing, or making displays of block structures. Recently I saw a bulletin board of children's block creations carefully documented by photographs and the children's comments. What the photographs froze in time were the unique artistic perspectives each child built into his or her structure—before it could be knocked down or cleaned up.

Then there is the drama in the block area. I remember the blocks being used to make battlefields where good guys fought bad guys day after day. Even though we may disapprove as adults, children find ways to play out the theme of good versus evil under many guises and as skillfully as movie directors. Good-versus-evil themes are nothing new. Great works of literature throughout the ages have been based on the same themes that children dream up in the block corner.

Block play often involves metaphor. I watched one boy go through a long period of trap building.

Early childhood education is like a good set of blocks. It is not only a building tool—it is also an investment for the future that will last!
Susan Woog-Wagner/Science Source

Whatever figures were available in the block area became the victims of the traps he was constantly inventing. Traps were important to him. To this day, I am still trying to figure out the meaning those traps had for him.

As I came to the end of this book, I thought long and hard about how to write the last story. I decided to pick a topic that does not end. I imagine that the blocks I invested in long ago will outlive me and that future generations will use them to replay the themes that their predecessors thought they invented.

Early childhood education is like a good set of blocks. It is an investment for the future that lasts. Our contributions to early childhood education will outlive all of us, and each succeeding generation, influenced by us, will reinvent the field.

SUMMARY

Unlike higher education, early childhood education cannot be divided into distinct subject areas. Every learning aspect of the early childhood program is interconnected. Art, music, and social studies–as well as language, emergent literacy, math, and science–all relate to one another.

In the early childhood program, art is an open-ended creative process that helps children sharpen their perceptions, organize their thoughts and actions, express feelings, and get to know their world better. The early childhood educator facilitates the art process by having age-appropriate expectations, valuing children's art, talking constructively to children about their artwork, providing a variety of materials as well as the freedom and time to explore different mediums on a deep level, and providing opportunities for both individual and collaborative art projects.

Like art, music should be taught in an open-ended fashion, not as formal lessons. All children–whether musically talented or not–should be encouraged to make music with their voices, bodies, and instruments in creative ways. To enhance children's music experiences, adults should appreciate children's spontaneous efforts to create music, encourage them to respond to music with movement, show interest in what they are doing without judging, and set up a music center with a variety of instruments, apparatus, and props to encourage the creative, visual, and auditory aspects of music and movement. Group time is another opportunity for incorporating music experiences, such as group singing, creative movement, and instrument playing.

Social studies learning is woven throughout the entire early childhood curriculum, which emphasizes helping children gain social competence. Social studies learning–which entails knowledge, values and attitudes, and skills–begins with the study of the self (self-concept, body awareness, self-image, self-worth, and self-esteem) and progresses on to the study of others and, ultimately, to the study of the community. To develop constructive social skills, adults need to be careful about the messages (both verbal and nonverbal) they send young children about gender, race, ethnicity, class, appearance, and ability. They also need to help children develop empathy and prosocial attitudes and skills. To teach children about community, adults need to give children a sense of the past and help them understand notions of place and the concepts of producers and consumers.

As mentioned earlier, it is impossible to divide early childhood education into separate pieces and label them according to traditional areas of study. It is the early childhood educator's job to weave the different parts into the whole fabric of the program–including the family. Such a sensitive, responsive approach to education takes the child's intellect seriously. Early childhood education is not about teaching subjects; it is about facilitating development and learning in each child–the whole child.

REFLECTION QUESTIONS

1. Are you a "creative person?" What are your experiences with discovering your own creativity? How do those experiences relate to young children discovering their creativity?
2. What is your personal experience of and reaction to art? Do you have some ideas about how to promote creativity in children through art?
3. What is your personal experience of and reaction to music? Do you have some ideas about how to promote creativity in children through music?
4. What do you think children need in order to gain a sense of community? Do you belong to a community? How does your experience of community relate to what you want children to learn?

TERMS TO KNOW

Can you use the following words in a sentence? Do you know what they mean?

body tracing 422
creative movement 426
self-knowledge 427
holistic approach 430

FOR FURTHER READING

Armstrong, L. J. (2012). *Family Child Care Homes: Creative Spaces for Children to Learn.* St. Paul, MN: Redleaf.

Brillante, P., and Mankiw, S. (2015). Preschool through grade 3: A sense of place: Human geography in the early childhood classroom. *YC Young Children, 70* (3), 16–23.

Duncan, S. (2011). Breaking the code–Changing our thinking about children's environments. *Exchange, 33* (4), 13-17.

Furmanek, D. (2014). Preschool through grade 2: Classroom choreography: Enhancing learning through movement. *YC Young Children*, 69 (4), 80–85.

Geist, K., Geist, E., and Kuznik, K. (2012). The patterns of music: Young children learning mathematics through beat, rhythm, and melody. *YC Young Children, 67* (1), 74–79.

Helm, J. H. and Katz, L. (2016). *Young Investigators: The Project Approach in the Early Years* (3rd ed.). New York: Teachers College Press.

Katz, L. (2012). Developing professional insight. In G. Perry, B. Henderson, and D. R. Meier (Eds.). *Our Inquiry, Our Practice: Undertaking, Supporting, and Learning from Early Childhood Teacher Researcher(ers)* (pp. 27–132). Washington D.C.: National Association for the Education of Young Children.

McLennan, D., and Bombardier, J. (2015). Kindergarten: "Paper shoes aren't for dancing!" Children's explorations of music and movement through inquiry. *YC Young Children, 70* (3), 70–75.

Mindes, G. (2015). Preschool through grade 3: Pushing up the social studies from early childhood education to the world. *YC Young Children, 70* (3), 10–15.

Moyles, J. R. (2012). *A–Z of play in early childhood.* Maidenhead, Berkshire, UK: McGraw Hill Open University Press.

Stacey, S. (2011). *The Unscripted Classroom: Emergent Curriculum in Action.* St. Paul, MN: Redleaf.

Weissman, P., and Hendrick, J. (2014). *The whole child: Developmental education for the early years.* Boston, MA: Pearson.

Design credits: Tips and Techniques: ©Ingram Publishing; Focus on Diversity: ©Pixelic/Getty Images

Endnotes

Chapter 1

1. Stephanie Feeney and Nancy Freeman, *Ethics and the Early Childhood Educator. 2005 Code Edition* (Washington, D.C.: NAEYC, 2005).
2. Examples of NAEYC position statements can be found on its website, www.NAEYC.org
3. This table reflects only part of Piaget's stages of cognitive development.
4. This table reflects only part of Freud's psychosexual stages.
5. This table reflects only part of Erikson's psychosocial stages.
6. C. Eggers-Pierla, *Connections and Commitments: Reflecting Latino Values in Early Childhood Programs* (Portsmouth, NH: Heinemann, 2005).
7. Robert A. LeVine, "Child Rearing as Cultural Adaption," in *Culture and Infancy: Variations in the Human Experience,* eds. P. Herbert Leiderman, Steven R. Tulkin, and Anne Rosenfeld (New York: Academic Press, 1977).
8. Jack P. Shonkoff and Deborah A. Phillips, eds., *From Neurons to Neighborhoods: The Science of Early Childhood Development* (Washington, D.C.: National Academy Press, 2000).
9. Feeney and Freeman, *Ethics and the Early Childhood Educator.*
10. *National Association for the Education of Young Children Code of Ethical Conduct,* prepared by the Ethics Commission of the National Association for the Education of Young Children (Stephanie Feeney, Chair) (Washington, D.C.: NAEYC, 1992).

Chapter 2

1. RIE is a program through which students can work with parents and their infants in a playroom setting.
2. WestEd Center for Child and Family Studies is an educational research and development organization that has created a training program for infant and toddler caregivers that is used across the United States.
3. *Focus on Diversity* was compiled from the following sources: Lou Matheson. "If You Are Not an Indian, How Do You Treat an Indian?" in *Cross-Cultural Training for Mental Health Professionals,* eds. H. Lefley and P. Pedersen (Springfield, IL: Thomas, 1986): 124; Maria Root, Christine Ho, and Stanley Sue, "Issues in the Training of Counselors for Asian Americans," in *Cross-Cultural Training for Mental Health Professionals,* eds. H. Lefley and P. Pedersen (Springfield, IL: Thomas, 1986): 202; Jeffrey W. Trawick-Smith, *Early Childhood Development: A Multicultural Perspective* (Columbus, OH: Merrill, 1997): 349; V. H. Young, "Family and Childhood in a Southern Georgia Community," *American Anthropologist 72* (1970): 269–288.
4. Personal conversation with Anna Tardos, Pikler's daughter and present director of the Pikler Institute.
5. *Caring for Our Children. National Health and Safety Performance Standards: Guidelines for Out-of-Home Child Care Programs* (2nd ed.) (Washington, DC: American Public Health Association and American Academy of Pediatrics, 2002).

6. Janet Gonzalez-Mena, *The Program for Infant-Toddler Caregivers: A Guide to Routines* (2nd ed.) (Sacramento: Far West Laboratory for Educational Research and Development and the California Department of Education, 2000).

CHAPTER 3

1. M. R. Gunnar, J. Bruce, and S. E. Hickman, "Salivary Cortisol Response to Stress in Children," in T. Theorell, ed., *Everyday Biological Stress Mechanisms* (Basel, Switzerland: Karger, 2001): 52–60.
2. Rima Shore, *Rethinking the Brain: New Insights in Early Development* (New York: Families and Work Institute, 1997). J. Ronald Lally, "Brain Research, Infant Learning, and Child Care Curriculum." *Child Care Information Exchange* (May 1998): 46–48.
3. Louise Derman-Sparks discusses this point in her book *Antibias Curriculum: Tools for Empowering Young Children* (Washington, D.C.: NAEYC, 1989). A revision of that classic book is coauthored with Julie Olsen Edwards and came out in 2010. Published by NAEYC, it's called *Antibias Education.*
4. Anne Wilson Schaef, *Beyond Therapy, Beyond Science* (San Francisco: Harper San Francisco, 1992).
5. I know of one teacher who had a student ask, "What's happening with the guinea pig we buried?" The teacher, who was comfortable with the subject of death, responded by digging up the animal's remains for the class to examine. The question ended up prompting an ongoing science lesson; the class unburied and reburied the body regularly throughout the year. Of course, such an approach to science might upset children and parents in another program. I'm not recommending it, merely reporting it.

   Another way to observe transformation would be to watch a carved pumpkin rot slowly over a period of time and record its various changes–an easy activity if your program celebrates Halloween. Or a neater and more traditional transformation project might be to watch silkworms change from eggs, to caterpillars, to cocoons, to moths that lay eggs. They too will die in the end, but that's less upsetting to most children.
6. For a detailed discussion of this subject, see Derman-Sparks, *Antibias Curriculum.*
7. Duong Thanh Binh, *A Handbook for Teachers of Vietnamese Students: Hints for Dealing with Cultural Differences in Schools* (Arlington, VA: Center for Applied Linguistics, 1975): 18.
8. Jerome Kagan, *The Nature of the Child* (New York: Basic Books, 1984): 244.

CHAPTER 4

1. Edward T. Hall, *The Dance of Life: The Other Dimension of Time* (New York: Anchor, 1983): 169.
2. Gaye Gronlund, "Bringing the DAP Messages to Kindergarten and Primary Teachers," *Young Children* (July 1995): 9–13.
3. Linda Gillespie, "It's Never 'Just Play'!" *Young Children* 71.3 (July 2016): 92–94.
4. J. Van Horn, P. Monighan Nourot, B. Scales, and K. Rodriguez Alward, *Play at the Center of the Curriculum* (Columbus, OH: Merrill, 2007).
5. Patricia Monihan-Nourot, Barbara Scales, Judith van Hoorn, and Milly Almy, *Looking at Children's Play: A Bridge Between Theory and Practice* (New York: Teachers College Press, 1987): 23.
6. Vivian Paley, *Boys and Girls, Superheroes in the Doll Corner* (Chicago: University of Chicago Press, 1994).
7. Monihan-Nourot et al., *Looking*, 26.
8. Jayanthi Mistry, "Culture and Learning in Infancy: Implications for Caregiving," in *Program for Infant Toddler Caregivers: A Guide to Culturally Sensitive Care,* ed. Peter Mangione (Sacramento: California Department of Education, 1995): 20.

9. Aaron Hochanadel and Dora Finamore, "Fixed and Growth Mindset in Education and How Grit Helps Students Persist in the Face of Adversity," *Journal of International Education Research* 11.1 (2015): 47-50.

Chapter 5

1. Erik Erikson, *Childhood and Society* (2nd ed.) (New York: Norton, 1963).
2. Erikson, *Childhood and Society*.
3. Jean Piaget, *The Construction of Reality in the Child* (New York: Basic Books, 1954).
4. Lawrence Kohlberg, "Moral Stages and Moralization: The Cognitive-Developmental Approach," in *Moral Development and Behavior,* ed. T. Lickona (New York: Holt, 1976); William Damon, *Moral Child: Nurturing Children's Natural Moral Growth* (New York: Free Press, 1988).
5. Ruth Chao, "Beyond Parental Control and Authoritarian Parenting Style: Understanding Chinese Parenting Through the Cultural Notion of Training," *Child Development* 65 (1994): 1111-1119.
6. M. Lindner Gunnoe, "Associations between Parenting Style, Physical Discipline, and Adjustment in Adolescents' Reports," *Psychological Reports* 112.3 (2013): 933.
7. U.S. Department of Education, "Civil Rights Data Collection for the 2013-14 School Year" (2108). Retrieved from www2.ed/gov/about.offices/list/ocr/docs/crdc-2013-14.html.
8. J. F. Casas, S. M. Weigel, N. R. Crick, J. M. Ostrov, K. E. Woods, E. A. J. Yeh, and C. A. Huddleston-Casas, "Early Parenting and Children's Relational and Physical Aggression in the Preschool and Home Contexts," *Journal of Applied Developmental Psychology* 27 (2006): 209-227. https://doi.org/10.1016/j.appdev.2006.02.003
9. Cynthia Ballenger, "Because You Like Us: The Language of Control," *Harvard Educational Review* 62.2 (Summer 1992): 199-208; Cynthia Ballenger, *Teaching Other People's Children: Literacy and Learning in a Bilingual Classroom* (New York: Teachers College Press, 1999).
10. Jerome Kagan, *The Nature of the Child* (New York: Basic Books, 1984): 244-245.
11. U.S. Department of Education, "Civil Rights Data Collection: 2013-14 State and National Estimations," Washington, D.C., accessed September 2018, https://ocrdata.ed.gov/StateNationalEstimations/Estimations_2013_14; Yale University Child Study Center, www.edweek.org/links.
12. Malik, Rasheed, "4 Disturbing Facts About Preschool Suspension," Center for American Progress, March 30, 2017, https://www.americanprogress.org/issues/early-childhood/news/2017/03/30/429552/4-disturbing-facts-preschool-suspension/.
13. Yale University Child Study Center, https://medicine.yale.edu/childstudy/zigler/publications/Preschool%20Implicit%20Bias%20Policy%20Brief_final_9_26_276766_5379_v1.pdf.
14. B. Kaiser and J. S. Rasminsky, *Challenging Behavior in Young Children* (Upper Saddle River, NJ: Pearson, 2012).
15. Division for Early Childhood of the Council for Exceptional Children, "Position Statement on Challenging Behavior and Young Children–Executive Summary," 2017, https://www.decdocs.org/executive-summary-challenging-behavior.
16. Stanley Greenspan, "Emotional Development in Infants and Toddlers and the Role of the Caregiver," in *A Caregiver's Guide to Social Emotional Growth and Socialization,* ed. J. Ronald Lally (Sacramento: California Department of Education, 1990).

Chapter 6

1. As I wrote these words, it occurred to me that, by necessity, child care workers and preschool teachers often work against their own mandates to share; what belongs to and is used by adults in a program setting is often inappropriate for

sharing with children. Sharp scissors, for example, have to be kept away from little hands, and teachers don't usually let children play with their personal possessions. It might be worth thinking about how to create a real-life sharing lesson, when you purposely share something of your own periodically. Left to chance, you probably won't tend to be a good model for sharing.

2. Lilian G. Katz, *Talks with Teachers* (Washington, D.C.: NAEYC, 1977).
3. Some teachers complain that they don't have the luxury of spending a lot of time with one problem when the rest of the children require supervision. This complaint is at the very center of the argument for better adult–child ratios and more financial support for child care and early childhood programs. For all of us who are concerned about eliminating adult violence through the education of young children, we must have the resources—both financial and human—that allow us to teach young children nonviolent ways. The scene you just read about is one of the most effective—though time-consuming—ways to teach these lessons.
4. I first saw this approach when Magda Gerber, a child therapist and infant expert, demonstrated it in a program that eventually became the present-day Resources for Infant Educarers (RIE). I've used the approach ever since and continually marvel at how it really works! If you talk children through their problem (it's kind of like sports announcing without making judgments), they eventually either get tired of the problem or find solutions. Even toddlers manage!
5. Stanley Coopersmith, *The Antecedents of Self-Esteem* (San Francisco: Freeman, 1967). Other researchers share Coopersmith's ideas but employ different words. Susan Harter, for example, uses the words *acceptance, power and control, moral virtue,* and competence in "Developmental Perspectives on the Self-System," in *Handbook of Child Psychology,* ed. E. Mavis Hetherington, 4th ed., Vol. 1 (New York: Wiley, 1983): 275–386.
6. Coopersmith, *The Antecedents of Self-Esteem.*
7. Coopersmith, *The Antecedents of Self-Esteem.*
8. Unfortunately, because of society's attitudes, taking pride in one's profession is not always an easy task for the early childhood educator. When child care professionals are paid poorly, talked down to, and exploited, they have to make a special effort to bolster their feelings of significance. They have to remind themselves that they are part of a valuable team of early childhood professionals. As Lynn Graham (a reviewer of this book and an assistant professor at Iowa State University) says, we have "the future sitting in our laps." If all early childhood educators remind themselves of this fact regularly, then self-respect and feelings of significance should follow naturally.
9. Aaron Hochanadel and Dora Finamore, "Fixed and Growth Mindset in Education an How Grit Helps Students Persist in the Face of Adversity," *Journal of International Education Research,* 11.1 (2015): 47–50.
10. However, in the latter example, he may also be protecting himself from allegations of child abuse. Unfortunately, men are more vulnerable to suspicion of wrongdoing than women, and some now refuse to change diapers or help with toileting to protect themselves.
11. Louise Derman-Sparks and Julie Olsen Edwards, *Antibias Education for Young Children and Ourselves* (Washington, D.C.: National Association for the Education of Young Children, 2010). and the ABC Task (Washington, D.C.: NAEYC, 1989).
12. Southern Poverty Law Center, "Testing Yourself for Hidden Bias" (2018). Retrieved from https://www.tolerance.org/professional-development/test-yourself-for-hidden-bias.
13. Seymour Papert, *Mindstorms: Children, Computers, and Powerful Ideas* (New York: Basic Books, 1980): vi–viii.
14. Elizabeth Jones and John Nimmo, *Emergent Curriculum* (Washington, D.C.: NAEYC, 1994).
15. Adapted from Janet Gonzalez-Mena, *From a Parent's Perspective* (Salem, WI: Sheffield, 1994).

CHAPTER 7

1. G. Music, *Nurturing Natures: Attachment and Children's Emotional, Sociocultural and Brain Development* (Abingdon, Oxon, and New York: Routledge, 2017).
2. Daphna Bassok, Jenna E. Finch, RaeHyuck Lee, Sean F. Reardon, & Jane Waldfogel. (2016). Socioeconomic Gaps in Early Childhood Experiences. AERA Open, Vol 2 (2016). https://doi.org/10/1177/2332858416653924
3. Lilian G. Katz, *Talks with Teachers* (Washington, DC: NAEYC, 1977).
4. Janet Gonzalez-Mena and Anne Stonehouse, "In the Child's Best Interests," *Child Care Information Exchange* (November 1995): 17–20.
5. Janet Gonzalez-Mena, "Taking a Culturally Sensitive Approach in Infant-Toddler Programs," *Young Children* 47.2 (January 1992): 4–9.
6. Alicia Lieberman, "Approaches to Infant Mental Health: Working with Infants and Their Families," address given at a conference held at the University of Victoria, 20 October. 1995.
7. Alicia Lieberman, "Approaches to Infant Mental Health: Working with Infants and Their Families," address given at a conference held at the University of Victoria, 20 October. 1995.

CHAPTER 8

1. Some of the great theorists and practitioners of early childhood education have written about the importance of the environment. Loris Malaguzzi, of Reggio Emilia fame, regarded the environment as a teacher. Maria Montessori also stressed the teaching qualities of the environment. Jean Piaget regarded the environment as important for presenting opportunities for children to interact with objects and with each other. Emmi Pikler put a lot of thought into what went into the environment at the Pikler Institute–not only the furniture, which was specially designed for children of each age group, but also the play objects to encourage manipulation.
2. California Department of Social Services (2018). "Child Care Regulations," 2018. Retrieved from http://www.cdss.ca.gov/inforesources/Letters-Regulations/Legislation-and-Regulations/Community-Care-Licensing-Regulations/Child-Care.
3. Elizabeth Prescott, "Is Day Care as Good as a Good Home?" *Young Children* (January 1978).
4. The child care centers of Reggio Emilia in Italy are noted for their ateliers, which are like combination workshops/artist studios.
5. At Pacific Oaks College and Children's School in Pasadena, California, the outdoors is the major work and play area. The attitude is that if something can be learned indoors, it can probably be learned outdoors as well.
6. As the seasons permit, children sleep and eat outdoors in Lóczy, the orphanage in Hungary where Magda Gerber, a famous infant specialist, devised her RIE philosophy. RIE stands for Resources for Infant Educarers.
7. Elizabeth Jones, *Dimensions of Teaching-Learning Environments* (Pasadena, CA: Pacific Oaks, 1978); E. Prescott, "The Physical Environment–A Powerful Regulator of Experience," *Exchange* (November 1994): 9–15.
8. Louis Torelli and Charles Durrett, *Landscapes for Learning* (Berkeley, CA: Spaces for Children, 1995).
9. Sybil Kritchevsky, Elizabeth Prescott, and Lee Walling, *Planning Environments for Young Children: Physical Space* (Washington, D.C.: NAEYC, 1969).
10. Torelli and Durrett, *Landscapes.*
11. Kritchevsky, Prescott, and Walling, *Planning.*
12. As mentioned, not everyone believes strongly in independence and individuality. If you believe in developing children's self-help skills and set up the environment to promote independence, it is important that you discuss this subject with parents. If there is disagreement (between a provider/staff member and a parent or among staff members), open up a dialogue and discuss your differences. (See Chapter 7 for a review of dialoguing.)

13. American Academy of Pediatrics, American Public Health Association, National Resource Center for Health and Safety in Child Care and Early Education, *Caring for Our Children: National Health and Safety Performance Standards; Guidelines for Early Care and Education Programs* (4th Ed.). (Elk Grove Village, IL: American Academy of Pediatrics; Washington, D.C.: American Public Health Association, 2019).
14. Ibid.
15. George Forman and Fleet Hill's book on constructive play has many examples of materials and equipment that promote cooperation; George Forman and Fleet Hill, Constructive Play: *Applying Piaget in the Preschool* (Menlo Park, CA: Addison, 1984).
16. Lyn Fasoli and Janet Gonzalez-Mena, "Let's Be Real: Authenticity in Child Care," *Child Care Information Exchange* 114 (March 1997): 35–40.
17. S. R. Kellert, "Experiencing Nature: Affective, Cognitive, and Evaluative Development," *Children and Nature: Psychological, Sociocultural, and Evolutionary Investigations,* eds. S. Clayton and S. Opotow (Cambridge, MA: MIT Press, 2002); R. Moore, Countering Children's Sedentary Lifestyles by Design. Natural Learning Initiative. Accessed on June 12, 2004, from www. naturalearning .org; Robert Pyle, "Eden in a Vacant Lot: Special Places, Species and Kids in Community of Life," in *Children and Nature: Psychological, Sociocultural and Evolutionary Investigations,* eds. S. Clayton and S. Opotow (Cambridge, MA: MIT Press, 2002).
18. I have yet to observe a good example of an early childhood program that emphasizes human interaction over exploration. In my experience, placing groups of children in sparse environments creates a situation where the adults must be good entertainers or use crowd-control techniques. Indeed, neither of these situations seems to promote the value of human interaction.
19. Rebecca New, "Excellent Early Education: A City in Italy Has It." *Young Children* 45.6 (1990): 4–11. Indeed, anyone who has seen the Reggio Emilia traveling display can't fail to be impressed by the aesthetic quality of the display itself and the documentation pictures. The children's work is obviously handled with great care and respect; nowhere is there a hint that someone might take a piece of children's art, scribble a name across the bottom, and stick it up randomly on an empty space of wall with two pieces of torn-off masking tape.
20. New, "Excellent Early Education."

CHAPTER 9

1. J. Ho and S. Funk. (March 2018). "Promoting Young Children's Social and Emotional Health," *Young Children,* 73.1 (March 2018). Retrieved from https://www.naeyc.org/resources/pubs/yc/mar2018/promoting-social-and-emotional-health.
2. Ronald J. Lally, "The Impact of Child Care Policies and Practices on Infant/Toddler Identity Formation," *Young Children* 51.1 (November 1995): 58–67.
3. Lally, J. R., & Signer, S. *Working with caregivers feelings: Scenario based on Marisabel.* [Handout from Module IV, Protective Urges: Parent-Care Teacher Relations training session]. Unpublished document. Sausalito, CA: The Program for Infant/Toddler Care.
4. S. Workman, *Where Do Your Child Care Dollars Go? Understanding the True Cost of Quality Early Childhood Education* (Washington, D.C.: Center for American Progress, February 14, 2018). Retrieved from www.americanprogress.org/issues/early-childhood/reports/2018/02/14/446330/child-care-dollar-go/
5. U.S. Department of Education, *Fact Sheet: Troubling Pay Gap for Early Childhood Teachers* (June 14, 2016). Retrieved from www.ed.gov/news/press-releases/fact-sheet-troubling-pay-gap-early-childhood-teachers
6. S. Workman, *Where Do Your Child Care Dollars Go? Understanding the True Cost of Quality Early Childhood Education* (Washington, D.C.: Center

for American Progress, February 14, 2018). Retrieved from www.americanprogress.org/issues/early-childhood/reports/2018/02/14/446330/child-care-dollar-go/

7. Linda Brault, "Inclusion of Children with Disabilities and Other Special Needs," panel presentation at the WestEd Program for Infant-Toddler Caregivers, Community College Special Seminar, Riverside, California, 2003. Linda directs a program to promote full inclusion of children with special needs in infant-toddler care and education programs. The program, *Beginning Together,* is part of the California Institute for Human Services.
8. Joseph J. Tobin, David Y. H. Wu, and Dana H. Davidson, video companion to *Preschool in Three Cultures,* University of New Hampshire, 1990.
9. Louise Derman-Sparks and the ABC Task Force, *Antibias Curriculum: Tools for Empowering Young Children* (Washington, DC: NAEYC, 1989).
10. Sue Bredekamp and Carol Copple, eds., *Developmentally Appropriate Practice in Early Childhood Education Programs* (Washington, DC: NAEYC, 2009).
11. A project carried on by WestEd in California for a number of years is called *Bridging Cultures,* which looks closely at the issues of independence and interdependence using a framework of individualism and collectivism. A number of books and articles related to this project are included in the references. The latest focus of this ongoing project is early care and education. This project resulted in a training module. See M. Zepeda, J. Gonzalez-Mena, C. Rothstein-Fisch, and E. Trumbell, *Bridging Cultures in Early Care and Education* (Mahwah, NJ: Erlbaum, 2006).
12. This information on oppression comes from a personal communication from Intisar Shareef, Ed.D.
13. Lally, "Impact of Child Care."
14. Barbara Rogoff, *The Cultural Nature of Human Development* (New York: Oxford University Press, 2003).
15. Carol Brunson Phillips, "Culture: A Process That Empowers," *Program for Infant/Toddler Caregiving: A Guide to Culturally Sensitive Care,* ed. Peter Mangione (Sacramento: Far West Laboratory and California Department of Education, 1995).
16. Nisha Patel, Thomas G. Power, and Navaz Peshotan Bhavnagri, "Socialization Values and Practices of Indian Immigrant Parents: Correlates of Modernity and Acculturation," *Child Development* 67 (1996): 302-313.

Chapter 10

1. NAEYC, *NAEYC Early Childhood Program Standards and Accreditation Criteria* (Washington, DC: National Association for the Education of Young Children, 2005): 9. NAEYC's program standard 1 for accreditation is "Relationships." The rationale for that standard says that warm, sensitive, and responsive interactions foster children's sense of security and a positive sense of self. Relationships also help children develop personal responsibility, self- regulation and the skills to get along with others.
2. In a 1997 visit to Napa Valley College, Amelia Gambetti (of the Reggio Emilia schools) stated emphatically how important group time is to help children feel connected to each other and see themselves as belonging to a group.
3. Louise Derman-Sparks and the ABC Task Force, *Antibias Curriculum: Tools for Empowering Young Children* (Washington, D.C.: NAEYC, 1989).

Chapter 11

1. Susan Bredekamp and Carol Copple, eds., *Developmentally Appropriate Practice in Early Childhood Education Programs* (Washington, D.C.: NAEYC, 2009); Carol Copple and Susan Bredekamp, *Basics of Developmentally Appropriate Practice: An Introduction for Teachers of Children 3 to 6.*

(Washington, D.C.: National Association for the Education of Young Children, 2006); Carol Copple, Susan Bredekamp, and Janet Gonzalez-Mena, *Basics of Developmentally Appropriate Practice: An Introduction for Teachers of Infants and toddlers.* (Washington, D.C.: National Association for the Education of Young Children.

2. This example is loosely built on my own cross-cultural experience and from information in Carolyn P. Edwards and Lella Gandini's "Teacher's Expectations About the Timing of Developmental Skills: A Cross-Cultural Study," *Young Children* (May 1989): 15-19.
3. Edwards and Gandini, "Teacher's Expectations."
4. L.A. Werner, "Infant Auditory Capabilities," *Current Opinion in Otolaryngology and Head and Neck Surgery, 10*.5 (2002): 398-402. DOI: 10.1097/00020840200210000000013
5. P. Kuhl., F. Tsao, H. Liu., Y. Zhang, and B. Boer, (2001). "Language/Culture/Mind/Brain," *Anna1ls of the New York Academy of Sciences, 1* (2001): 136. doi:10.1111/j.1749-6632.2001.tb03478.x
6. D. F. Bjorklund and K. B. Causey, *Children's Thinking: Cognitive Development and Individual Differences* (6th ed.). (Thousand Oaks, CA: Sage, 2018).
7. Magda Gerber, *Manual for Resources for Infant Educarers* (Los Angeles: Resources for Infant Educarers, 1991).
8. J. J. Beaty and L. Pratt, *Early Literacy in Preschool and Kindergarten: A Multicultural Perspective* (4th ed.). (Upper Saddle River, NJ: Merrill/Prentice Hall, 2014).

CHAPTER 12

1. V. Buysse, D. C. Castro, T. West, and M. L. Skinner, *Addressing the Needs of Latino Children: A National Survey of State Administrators of Early Childhood Programs* (executive summary) (Chapel Hill: The University of North Carolina: FPG Child Development Institute, 2004).
2. NAEYC, *NAEYC Early Childhood Program Standards and Accreditation Criteria* (Washington, D.C.: National Association for the Education of Young Children, 2005): 10.
3. Lilian Katz, keynote address, Southeast Alaska AEYC Conference, Sitka, Alaska, February 1995.

CHAPTER 13

1. R. Chetty, J. N. Friedman, N. Hilger, E. Saez, D. Whitmore Schanzenbach, and D. Yagan, "How Does Your Kindergarten Classroom Affect Your Earnings? Evidence from Project Star," *Quarterly Journal of Economics* 4 (2011): 1593, https://doi.org/10.1093/qje/qjr04.
2. Lilian Katz uses this phrase when discussing the subject of language acquisition. It brings home the idea that young children do not learn language through formal lessons.
3. Fred Genesee, "Myths About Early Childhood Bilingualism," *Canadian Psychology/Psychologie Canadienne* 56 (2015): 6-15. 10.1037/a0038599.
4. R. Barac, E. Bialystok, D. C. Castro, and M. Sanchez, "Review: The Cognitive Development of Young Dual Language Learners: A Critical Review," *Early Childhood Research Quarterly* 29 (2014): 699-714, https://doi.org/10.1016/j.ecresq.2014.02.003.
5 E. Bialystok, F. I. Craik, and G. Luk, "Bilingualism: Consequences for Mind and Brain," *Trends in Cognitive Sciences,* 16.4 (2012): 240-250.
6. It is important that children not acquire English at the cost of losing their home language. English-only programs pose the risk of inadequate proficiency in both English and the home language and contribute greatly to school failure, especially among low-income children. V. Buysse, D. C. Castro, T. West, and M. L. Skinner, *Addressing the Needs of Latino Children: A National Survey of State Administrators of Early Childhood Programs* (executive summary) (Chapel Hill: University

of North Carolina: FPG Child Development Institute, 2004); Diane August and Eugene Garcia, *Language Minority Education in the United States* (Springfield, IL: Thomas, 1988); Lily Wong Fillmore, "When Learning a Second Language Means Losing the First," *Early Childhood Research Quarterly* 6.3 (1991): 323–346.

7. Lourdes Diaz Soto, *Language, Culture, and Power* (Albany, NY: State University of New York Press, 1997).
8. The Annie E. Casey Foundation, (2018, January 11). "The Number of Bilingual Kids in America Continues to Rise," January 11, 2018, https://datacenter.kidscount.org/updates/show/184-the-number-of-bilingual-kids-in-america-continues-to-rise.
9. "NAEYC Position Statement: Responding to Linguistic and Cultural Diversity–Recommendations for Effective Early Childhood Education," *Young Children* 51.2 (1996): 4–12.
10. Laurie Makin, Julie Campbell, and Criss Jones Diaz, *One Childhood, Many Languages* (Pymble, NSW, Australia: HarperEducational, 1995): 69.
11. Lev Vygotsky, *Thought and Language* (Cambridge, MA: MIT Press, 1962); Lev Vygotsky, "Thinking and Speech," in *The Collected Works of L. S. Vygotsky,* eds. Robert Reiber and Arron Carton (New York: Plenum, 1987).
12. Lilian Katz, "What We Can Learn From Reggio Emilia," in *The Hundred Languages of Children,* eds. Carolyn Edwards, Lella Gandini, and George Forman (Norwood, NJ: Ablex, 1994): 29.
13. Peter Mangione, *Program for Infant Toddler Caregivers: A Guide to Language Development and Communication* (Sacramento: California Department of Education, 1992).
14. W. Caudill and H. Winstein, "Maternal Care and Infant Behavior in Japan and America," in *Japanese Culture and Behavior,* eds. Takie Sugiyam Lebra and William P. Lebra (Honolulu, HI: University Press of Hawaii, 1974); Patricia M. Clancy, "The Acquisition of Communicative Style in Japanese," in *Language Socialization Across Cultures,* eds. Bambi B. Schieffelin and Elinor Ochs (Cambridge, MA: Cambridge University Press, 1986).
15. This example was inspired by the observation of Tammy Todd, a student at Napa Valley College.
16. Philip R. Harris and Robert T. Moran, *Managing Cultural Differences* (Houston, TX: Gulf, 1987): 37.
17. Edward T. Hall, *Beyond Culture* (New York: Anchor, 1989).
18. Laura Berk, *Infants and Children: Prenatal Through Middle Childhood* (5th ed.) (Boston, MA: Allyn and Bacon, 2005).
19. "Literacy Development and Pre-First Grade: A Joint Statement of Concerns About Present Practices in Pre-First Grade Reading Instruction and Recommendations for Improvement," statement written jointly by the Association for Childhood Education International, the Association for Supervision and Curriculum Development, the International Reading Association–National Association for the Education of Young Children, the National Association of Elementary School Principals, and the National Council of Teachers of English (issued by the International Reading Association, 1985).

    The new position statement "Learning to Read and Write" (1998) was endorsed by American Speech-Language-Hearing Association, Association for Childhood Education International, Association of Teacher Educators, Council for Early Childhood Professional Recognition, Division for Early Childhood/Council for Exceptional Children, National Association of Early Childhood Specialists in State Departments of Education, National Association of Early Childhood Teacher Educators, National Association of Elementary School Principals, National Association of State Directors of Special Education, National Council of Teachers of

English, Zero to Three/National Center for Infants, Toddlers, & Families. The concepts are supported by American Academy of Pediatrics, American Association of School Administrators, American Educational Research Association, and the National Head Start Assocation.

20. J. Gonzalez-Mena, "Caregiving Routines and Literacy," in *Learning to Read the World: Language and Literacy in the First Three Years,* eds. S. E. Rosenkoetter and J. Knapp-Philo (Washington, D.C.: Zero to Three, 2006).
21. This is also a useful technique if you are trying to limit the number of children in the space. If the reasonable limit is, say, four children in the area at a time, print no more than four tags. Children without a tag must then wait their turn.
22. Elizabeth Jones and John Nimmo, *Emergent Curriculum* (Washington, D.C.: NAEYC, 1994): 112.

CHAPTER 14

1. Jean Piaget, "Piaget's Theory," in *Carmichael's Manual of Child Psychology,* ed. Paul Mussen, 3rd ed., vol. 1 (New York: Wiley, 1970).
2. Peter H. Johnston, *Opening Minds: Using Language to change Lives* (Portland, ME: Stenhouse, 2012).
3. George E. Forman and David S. Kuschner, *The Child's Construction of Knowledge: Piaget for Teaching Children* (Washington, D.C.: NAEYC, 1984).
4. Christine Chaille and Lory Britain, The Young Child as Scientist: *A Constructivist Approach to Early Childhood Science Education* (New York: Longman, 1997).
5. Constance Kamii and Rheta DeVries, *Physical Knowledge in Preschool Education: Implications of Piaget's Theory* (Englewood Cliffs, NJ: Prentice Hall, 1993).
6. Kamii and DeVries, *Physical Knowledge.*
7. The expression "the hundred languages of children" is the name of a book and a traveling exhibit about Reggio Emilia schools and their educational philosophy. It is also a metaphor for the many avenues of communication children use to symbolize their ideas, perceptions, and experiences.

CHAPTER 15

1. Elizabeth Jones and Georgina Villarino, "What Goes Up on the Classroom Walls–and Why?" *Young Children,* 49.2 (January 1994): 38-40.
2. David Johnson, "Round," *The New Grove Dictionary of Music and Musicians,* (2nd ed.), eds. Stanley Sadie and John Tyrrell (London: Macmillan Publishers, 2001).

Glossary

ACEI The Association for Childhood Education International is an early childhood professional organization that publishes a journal, holds international conferences, and guides and supports professionals in the field. Founded in the late 1800s as a kindergarten organization, the ACEI broadened its focus in the 1930s to include preschools and elementary schools.

ages and stages A catch phrase that relates to childhood developmental features and behaviors that tend to correlate with specific ages. Each stage describes a particular period of development that differs qualitatively from the stages that precede and follow it. The sequence of stages never varies.

aggressiveness The quality of dominating power that results in pushing forward (sometimes in hostile, harmful, attacks) without regard to the welfare of the other person or persons.

Americans with Disabilities Act A 1992 law (Public Law 101-336), also called ADA, that defines disability, prohibits discrimination, and requires employers, transportation, and other public agencies to provide access to the disabled in places of employment, public facilities, and transportation services.

anecdotal record A documentation method that briefly describes an activity, a snatch of conversation, a chant, and so on. Anecdotal records can be based on reflection or written on the spot.

antibias focus An activist approach to valuing diversity and promoting equity by teaching children to accept, respect, and celebrate diversity as it relates to gender, race, culture, language, ability, and so on.

assertiveness The quality of standing up for one's own needs and wants in ways that recognize and respect what other people need and want.

assisted performance A concept described by Russian researcher Lev Vygotsky that suggests that children cannot perform as well on their own in some cases as they can when they receive a bit of help from a more skilled person.

associative play A form of play in which children use the same materials, interact with each other, and carry on conversations. It is not as organized as cooperative play, in which children take on differentiated roles.

attachment An enduring affectionate bond between a child and a person who cares for the child, giving the child a feeling of safety or security. Building a trusting secure attachment through consistency, responsiveness, and predictability shows children they can trust the caregiver to meet their needs and frees them to explore their environment.

authentic assessment A method of assessing children according to what they know, can do, and are interested in, which can then be applied to ongoing curriculum planning. Authentic assessment avoids comparing children to a norm or grading them. It also avoids standardized testing, which measures isolated skills and bits of knowledge out of context.

behavior contagion A phenomenon that occurs when children are influenced by each other's behavior. It is most noticeable in its negative form, when children are doing something they are not supposed to do.

behaviorism The scientific study of behaviors that can be seen and measured. Behaviorism, also called "learning theory," attributes all developmental change to environmental influences.

behavior modification A form of systematic training that attempts to change unacceptable behavior patterns. It involves reinforcing acceptable behavior rather than paying attention to and, thus, rewarding unacceptable behavior.

body tracing An example of an activity that requires cooperation. The child lies on a piece of paper on the floor that is at least as long as he or she is tall and someone traces an outline of the child's body.

both-and thinking An approach to decision making in which the early-childhood educator considers what is developmentally, individually, and culturally appropriate in all situations; it involves coming up with a solution that may incorporate all the conflicting elements. Both-and thinking contrasts with either-or thinking, in which the choice is between one solution or the other.

CDF The Children's Defense Fund is an organization that advocates for children, particularly those in poverty and/or of color.

center-based program An early childhood program, usually child care, that operates in a building other than a person's home.

cephalo-caudal development The developmental pattern of human beings that proceeds in a head-to-foot direction.

child-centered curriculum An educational philosophy created by John Dewey that emphasizes designing curriculum according to the interests of the children rather than specific subject matter.

child-centered (or child-directed) learning A teaching–learning process in which the child learns from interacting with the environment, other children, and adults. This type of learning contrasts with a classroom in which the educator's main role is to teach specific subject matter or formal lessons.

child development The study of how children change as they grow from a qualitative rather than a merely quantitative standpoint.

cognitive stages A set of stages described and named by Jean Piaget that focuses on intellectual development.

concrete operations A cognitive stage described by Jean Piaget during which young children (ages seven through eleven) can use what Piaget calls "mental operations" to reason about the concrete world. At this stage, children are not yet capable of purely abstract thinking that involves taking variables into account or coming up with propositions.

constructivist approach A view based on Jean Piaget's work that suggests that children do not passively receive knowledge through being taught but rather actively construct it themselves.

cooperative play A form of play that involves a significant degree of organization. Interactive role playing and creating a joint sculpture are two examples of cooperative play.

creative movement An activity, usually involving music, where a child is encouraged to move in creative ways, often with the direction or suggestion of the teacher.

curriculum A plan for learning. Curriculum can be both written (an official plan in the form of, for example, an outline or web) and unwritten (that is, unconscious learning that occurs through the adult–child relationship).

descriptive feedback A form of nonjudgmental commentary. Adults use descriptive feedback to put children's actions and feelings to words to convey recognition, acceptance, and support: "You're putting a lot of work into that drawing" or "Looks like you don't like him to touch your painting." Descriptive feedback should be used to facilitate rather than disrupt.

developmental checklist A method of documenting and assessing a child's development. A developmental checklist might be broken down into specific categories, such as physical, psychomotor, cognitive, social-emotional, and language.

developmentally appropriate practice (DAP) A set of practices that directly relates to a child's stage of development as defined by such theorists as Piaget and Erikson.

dialoguing An approach to conflict whose goal is to reach agreement and solve problems. Unlike arguing, whose aims are to persuade and win, dialoguing involves gathering information and understanding multiple viewpoints in order to find the best solution for all parties concerned.

distraction A device to keep a child from continuing an action or behavior. Distraction can also be used to take children's minds off a strong feeling. Distraction works but has side effects as children learn that their energy or feelings are not acceptable to adults who distract them. Distraction is sometimes confused with redirection and may look similar, but distraction is aimed at stopping the energy behind the behavior or the feeling, whereas redirection moves it in a more acceptable direction.

double bind A kind of mixed message that causes confusion. For example, a mother embraces a child and says, "Why don't you go play with the other children?" Her body language says "Stay here with me," but her actual words say the opposite.

dual focus A method of supervision that allows the adult to focus on a child or small group of children while still being aware of what else is going on in the environment at large.

early childhood culture The culture (largely unrecognized) that results from early childhood training. Related to the dominant culture of the society but not exactly like it.

emergent curriculum A curriculum that grows out of children's interests and activities and takes shape over time. Although emergent curriculum often has a spontaneous aspect and is child-centered, it is also facilitated and, thus, planned for by adults.

emergent literacy The ongoing, holistic process of becoming literate–that is, learning to read and write. Emergent literacy contrasts with a reading-readiness approach, which emphasizes teaching isolated skills rather than allowing literacy to naturally unfold in a print-rich environment.

empowerment Helping someone experience his or her sense of personal power. For example, an adult can empower a child by giving him or her the opportunity to make some decisions rather than being told what to do.

environmental checklist A method of documenting the setup and/or use of the environment in an early childhood program. An environmental checklist can be used to assess a specific child's use of the environment, or it can be used to assess the effectiveness of the setup itself.

event sampling *See* incident reports.

expressive language Language that is produced to convey ideas, feelings, thoughts, and so on. Expressive language develops later than its counterpart, receptive language.

family child care program An early childhood program that provides child care services in the home of the provider.

feedforward A guidance tool that helps children understand beforehand what consequences might result from certain behavior (often unacceptable behavior). It is only feedforward if it is presented in a neutral tone and is neither judgmental nor threatening.

full inclusion A concept that goes beyond simply including children with special needs into whatever setting is the natural environment of their typically developing peers. Full inclusion means that such children, regardless of their disability or challenge, are always integrated into a natural environment and that services are as culturally normative as possible.

gross-motor spaces Indoor and outdoor areas specifically designed for gross-motor skill building and/or vigorous play involving large-muscle activities, such as running, stretching, climbing, jumping, rolling, swinging, ball throwing, and (in the case of older children) game playing.

group time A period during which children come together to participate in some specially planned activities, such as singing, storytelling, movement activities, or discussions. The content, duration, and frequency of group time (sometimes called "circle time") vary according to the children's age and developmental needs.

guidance Nonpunishing methods of leading children's behavior in positive directions so that children learn to control themselves, develop a healthy conscience, and preserve their self-esteem.

Head Start A comprehensive, federally funded program that provides education, health screening, and social services to help low-income families give their children–from birth to age five–the start they need to succeed in public school. There are some state-supported versions of Head Start as well.

high-context culture A culture that depends more on context than on spoken or written language to get messages across.

holism A view that considers the whole as more important than an analysis of the parts. A holistic view of the child, for example, integrates the mind, body, feelings, and personal context into an inseparable unit. A holistic curriculum is an integrated approach to a plan for learning in which the teaching-learning process occurs throughout the day rather than being broken down into separate subjects.

holistic approach An approach or a curriculum that focuses on the whole child rather than addressing mind, body, and feelings separately. For example, though an activity may be planned with cognitive (intellectual) goals, the teacher also pays attention to how the child responds physically and emotionally.

holistic listening A form of listening that goes beyond merely hearing. Holistic listening involves the whole body and uses all the senses in order to pick up subtle cues that are not put into words or otherwise readily apparent.

home culture The family life of the child, which encompasses cultural beliefs, goals, and values–including how they play out in child-rearing practices.

home language The language spoken at home. For many children that language is English, but for many others, it is a language other than English. The term can also be used for a particular way of speaking English that differs from what is called "standard English."

IDEA An acronym that stands for Individuals with Disabilities Education Act.

IEP An acronym that stands for individualized education program and is the result of a process carried out by a team of multidisciplinary specialists who come together to work in collaboration with a family on appropriate and meaningful goals and objectives for the family and child with special needs.

IFSP An acronym that stands for individualized family service plan and is a written plan for early intervention generated by a team of infant experts and the parents for infants and toddlers with identified special needs and their families.

impression management A nonconstructive way of talking to children that discounts their feelings and their sense of reality. For example, a child says, "I don't like that sandwich." In response, the adult says, "Yes, you do." Impression management teaches children to mistrust their senses.

incidents reports (or event sampling) A method of documenting a particular type of repeated occurrence from beginning to finish. Sometimes called "event sampling," incidents reports focus on one of a variety of behaviors, such as aggressive incidents or parent-child separations.

incongruence A type of mixed message that causes confusion. For example, a person's body language might convey anger while the words contradict the emotion.

integration The incorporation of children with special needs into programs with their typically developing peers and giving them the support they need so they really belong. Part of integration is giving attention to the interactions between the two groups of children. The goal is for all children to participate in the program to the greatest degree possible.

interest centers In the early childhood setting, the floor space, equipment, and materials for play, interaction, and exploration. Examples of interest centers include dramatic play, block, science, art, and music centers.

intrinsic motivation Inner rewards that drive a child to accomplish something. Intrinsic motivation contrasts with extrinsic motivation, in which rewards are given to the child in the form of praise, tokens, stickers, stars, privileges, and so forth.

invented spelling The way children spell when they first begin to write, going by the sounds of the language more than by conventional spelling rules. In other words, they invent their own spelling.

language immersion programs The purpose of these programs is learning a language that is not the child's home language. When children who are at risk for losing their home language are put into English language immersion programs, the result is often the replacement of their home language with English. An approach that has had more positive results in the United States is called two-way language immersion programs or dual language immersion programs. In this approach, half the children have English as their home language and the other half come from a different language group–such as Spanish. In this situation, each group learns the language of the other as instruction occurs in both. Children are more likely to end up bilingual in two-way language immersion programs.

learning theory A theory that focuses on the scientific study of behaviors that can be seen and measured. Learning theory, also called "behaviorism," attributes all change to environmental influences.

limits Boundaries placed on children's behavior. They can be physical boundaries or verbal boundaries.

logico-mathematical knowledge One of three kinds of knowledge described by Jean Piaget. Logico-mathematical knowledge comes from physical knowledge and involves an understanding of relationships between objects through the use of comparison and seriation.

low-context culture A low-context culture is one that depends a great deal on words to convey messages rather than emphasizing context.

mainstreaming A term that means placing children with special needs into programs that serve children who are typically developing. In some such programs support for the children with special needs may be minimal, so those who can't handle the mainstream may never feel they belong there.

mapping A method of documenting how a specific child functions in the early childhood environment. Using a map of a room or area, the recorder plots the path of the child and records such activities as interactions with other children or with adults. Start and end points are notated as well as the duration of the observation. Mappings can also be used to assess the use and effectiveness of the environment itself.

meaning-making A practice through which children construct knowledge by finding meaning in their experience.

modeling A teaching device and guidance tool in which an adult's attitude or behavior becomes an example the child consciously or unconsciously imitates.

Montessori A particular approach to education founded by Maria Montessori (Italy's first woman physician) that emphasizes the active involvement of children in the learning process and promotes the concept of a prepared environment.

multiculturalism An approach to education that accepts and respects cultural differences and supports the vision of a pluralistic society.

NAEYC The National Association for the Education of Young Children is the largest and best-known early childhood professional organization. In addition to publishing a journal and holding conferences, the NAEYC sets standards for the early childhood field and advocates for young children and their families through its position statements.

natural environment A setting (such as a home or early care and education program) where children with disabilities will find their typically developing peers. A natural environment can be defined by the fact that it will continue to exist whether or not children with disabilities are there.

nature-nurture question The question that asks, "What causes children to turn out the way they do?" In other words, is a child's development influenced more by his or her heredity (nature) or by his or her environment (nurture)? This question is sometimes called a "controversy," because nature proponents insist that genetics plays a stronger role in influencing development, while nurture proponents make the same claim about environmental experiences.

normalization A term which means that services such as early care and education programs provided to those with special needs are based on circumstances that are as culturally normative as possible.

object permanence A cognitive milestone described by Jean Piaget that occurs in later infancy, when babies develop the understanding that objects and people continue to exist when they can't be seen.

observation The act of watching carefully and objectively. The usual goal involves paying close attention to details for the purpose of understanding behavior.

one-to-one correspondence The understanding that counting involves assigning one number to each object or person being counted. This form of counting differs from reciting numbers by rote.

parallel play A form of play in which two or several children are playing by themselves, but within close proximity of each other. Each child's play may be influenced by what another child is doing or saying, but there is no direct interaction or acknowledgment of the other child.

parent cooperative preschool A program designed to educate parents while serving their children. This type of program, sometimes called a "parent participation nursery school," operates on a cooperative basis and is often run by parents under the auspices of public school systems.

physical-care centers Areas of the early childhood environment that are designated and equipped for cooking, eating, cleaning up, hand washing, diapering and toileting, and napping.

physical knowledge One of three kinds of knowledge described by Jean Piaget. Physical knowledge involves an understanding–in concrete rather than abstract terms–of how objects and materials behave in the physical world.

physical milestones of development Events that mark progress in the path of physical development, such as the first time a baby rolls over, sits up, or takes a first step. These milestones, which were introduced by Arnold Gesell, are based on norms that come from the scientific study of children's physical behavior.

portfolio One of the tools of authentic assessment. Portfolios are collections of samples of children's work; they assess both process and product. Teachers, children, and parents can all contribute to portfolios in order to broaden the assessment to reflect developmental progress in the home as well as the early childhood setting.

primary caregiving system A caregiving system in which infants are divided up and assigned (in groups of three or four) to specific primary caregivers who are responsible for meeting their needs and record keeping. The goal of this approach is to promote closeness and attachment, but not exclusivity. An important aspect of this system is for each child to know and relate to other caregivers as well.

professionalism A set of attitudes, theories, and standards that guides the early childhood professional. An early childhood professional is someone who is (1) trained in the principles and practices of the education of young children between birth and age eight; (2) knows about the ethics, standards, and legal responsibilities of the profession; and (3) conducts himself or herself accordingly.

project-based approach An in-depth teaching-learning process that emerges from an idea–thought up by either a child or an adult–and is carried out over days or weeks. Unlike free play, project work emphasizes product as well as process. Documentation of the process (during and upon completion) is an important element of the project approach.

proximal-distal development The developmental pattern of human beings that progresses from the middle of the body out to the extremities.

psychosexual stages A set of stages described and named by Sigmund Freud that focuses on sexual development.

psychosocial stages A set of developmental stages described and named by Erik Erikson that focuses on successive social crises.

real-world math Math activities that relate directly to problems in the child's own world–as compared to theoretical problems that have nothing to do with the child's reality. Real-world math is sometimes called "authentic math."

receptive language Language that can be understood, though perhaps not spoken. Receptive language develops earlier than its counterpart, expressive language.

redirection A form of early childhood guidance that diverts a child from unacceptable behavior to acceptable behavior without stopping the energy flow. Ideally, redirection involves giving the child a choice to lead him or her toward a constructive behavior or activity.

reflective practice A way of examining one's own experience–both past and present–in order to understand it, learn from it, and grow. Reflective thinking is a useful exercise for examining personal reactions to certain situations or people that may get in the way of developing and maintaining relationships or effectively facilitating the teaching-learning process.

register A particular style of language or way of speaking that varies according to the circumstances and the role a person is filling at the moment.

RERUN An acronym that lists all the elements needed to resolve a conflict through dialoguing: reflect, explain, reason, understand, and negotiate. RERUN is a holistic process, and, as such, it is not a series of steps that must always occur in the same order; but as the acronym suggests, the process can be repeated as often as necessary.

RIE stands for Resources for Infant Educarers a group dedicated to improving the quality of infant care and education through teaching, mentoring, and supporting parents and professionals. RIE was started by Magda Gerber, A Hungarian infant expert.

round A piece of music sung by at least three people in the same melody but each person begins at a different time to create a harmony. A common example is "Row, Row, Row Your Boat" with which the first child would sing this line, the next would continue with "gently down the stream," and so on in harmony.

running record observation A method of documenting that gives a blow-by-blow, objective, written description of what is happening while it is happening. A running record can include adult interpretations about the meaning of the observed behaviors, but it must separate objective data from subjective comments.

scaffolding A form of assistance that supports and furthers understanding and performance in a learner.

self-esteem A realistic assessment of one's worth that results in feelings of confidence and satisfaction.

self-knowledge A term that relates to individuals learning about themselves. Words such as self-concept, body awareness, self-image, self-worth, and self-esteem all fall under the heading of self-knowledge.

sensorimotor cognition A cognitive stage described by Jean Piaget that occurs from birth to age two. During this stage, children's cognition develops through movement and sensory exploration of the physical world.

sensorimotor play A form of play that involves exploring, manipulating, using movement, and experiencing the senses. It is sometimes called "practice play" or "functional play." In sensori-motor play, the child interacts with his or her environment using both objects and other people.

social knowledge One of three kinds of knowledge described by Jean Piaget. Social knowledge relates to knowledge about the world that can only be transmitted socially, such as labels for objects.

social learning theory A branch of behaviorism that focuses on the significance of modeling and imitation in a child's development.

sociocultural theory A theory developed by Lev Vygotsky that focuses on the effect of cultural context on development.

solitary play A form of play in which a child plays alone even though other children may be present.

stage theorist A theorist who believes that children develop according to specific, sequential stages of development.

surround care Child care that extends beyond the regular daily program. It may be offered in a child care center for infants, toddlers, and preschoolers during the early morning and evening, when there are fewer children present. Surround care is also offered in some programs for school-age children who attend before and after school.

symbolic play A form of play that uses one thing to stand for another and shows the person's ability to create mental images. Three types of symbolic play are dramatic play, constructive play, and playing games with rules.

synchronous interaction A coordinated interaction in which one person responds to the other in a timely way so that one response influences the next in a kind of rhythmic chain reaction that creates connections.

third space The process or ability to move beyond a dualistic or exclusive mindset ("me versus you") to an inclusive perspective that focuses on the complementary aspects of opposing values, behaviors, and beliefs ("sum is greater than its parts"). Without making concessions, both parties try to reduce emphasis on the differences while validating each others' language, values, and ideas. By doing so both parties can identify and integrate the strengths of each others' views to come to a satisfactory agreement.

time-out A nonviolent alternative to punishment that removes a child from a situation in which he or she is behaving in an unacceptable way. Time-out is an effective guidance measure when the child is truly out of control and needs to be removed to settle down. Used as a punishing device by controlling adults, however, it has side effects–as does any punishment–including undermining self-esteem.

time sample A documentation technique that involves collecting samples of targeted behaviors of small groups of children within a specific time frame as a way to learn about individual and group patterns.

transformation and representation Two processes that distinguish the constructivist approach from other teaching-learning approaches. Transformation involves processes of change, while representation portrays change in the form of traces. Activities of transformation and representation facilitate children's symbolic thinking.

transition The passages between one place and another or one activity and another. Examples of transitions include arrivals and departures, cleanup time, and going outside. Transitions occur as often as children change activities either as a group or as individuals.

trickle-down effect The result of expectations and approaches appropriate for older children appearing in programs for younger ones in the name of getting them ready for what is to come.

zone of proximal development (ZPD) According to Lev Vygotsky, the gap between a child's current performance and his or her potential performance if helped by a more competent person—child or adult.

References

Abbott, L., and Langston, A. (2005). *Birth to Three Matters: Supporting the Framework of Effective Practice.* London: Open University Press.

Adams, E. J. (2011). Teaching children to name their feelings. *Young Children, 66* (3), 66–67.

Adams, S. K., and Baronberg, J. (2005). *Promoting Positive Behavior.* Saddle River, NJ: Prentice Hall.

Adamson, S. (2005). Making the calendar meaningful. *Young Children, 60* (5), 42.

Akbar, N. (1985). *The Community of Self.* Tallahassee, FL: Mind Productions.

Alati, S. (2005). What about our passions as teachers? *Young Children, 60* (6), 86–89.

Allen, K. E., and Marotz, L. R. (1989). *Developmental Profiles: Birth to Six.* Albany, NY: Delmar.

Allison, J. (2005). Building literacy curriculum using the project approach. In B. Neugebauer (Ed.), *Literacy: A Beginnings WorkshopBook* (pp. 22–28). Redmond, WA: Exchange.

Althouse, R., Johnson, M.H., and Mitchell, S.T. (2003). *The Color of Learning: Integrating the Visual Arts into the Early Childhood Curriculum.* New York: Teachers College Press.

American Academy of Pediatrics, American Public Health Association, National Resource Center for Health and Safety in Child Care and Early Education. (2019). *Caring for Our Children: National Health and Safety Performance Standards; Guidelines for Early Care and Education Programs* (4th ed.). Elk Grove Village, IL: American Academy of Pediatrics; Washington, D.C.: American Public Health Association.

American Academy of Pediatrics Council on Environmental Health. (2012). *American Academy of Pediatrics.* Chicago, IL: American Academy of Pediatrics.

Andrews, A. G., and Trafton, P. R. (2002). *Little Kids–Powerful Problem Solvers: Math Stories from a Kinder-garten Classroom.* Portsmouth, NH: Heinemann.

Aronson, S., and Shope, T. (2005). *Managing Infectious Diseases in Child Care and Schools.* Chicago, IL: American Academy of Pediatrics.

Aronson, S., and Spahr, P. M. (2002). *Healthy Young Children: A Manual for Programs.* Washington, D.C.: National Association for the Education of Young Children.

Balaban, N. (2011). Easing the separation process for infants, toddlers, and families. In D. Koralek and L. G. Gillespie (Eds.) (pp. 14–19), *Spotlight on Infants and Toddlers.* Washington, D.C.: National Association for the Education of Young Children.

Balaban, N. (2006). *Everyday Goodbyes: Starting School and Early Care: A Guide to the Separation Process.* New York: Teachers College Press.

Ball, J. (2004). *Early Childhood Care and Development Program as Hook and Hub for Community Development: Promising Practices in First Nations.* Victoria, B., Canada: University of Victoria.

Ballenger, C. (1999). *Teaching Other People's Children: Literacy and Learning in a Bilingual Classroom.* New York: Teachers College Press.

Bandtec Network for Diversity Training. (2003). *Reaching for Answers: A Workbook on Diversity in Early Childhood Education.* Oakland, CA: Bandtec Network for Diversity Training.

Bandura, A. (1977). *Social Learning Theory.* Englewood Cliffs, NJ: Prentice-Hall.

Barac, R., Bialystok, E., Castro, D. C., and Sanchez, M. (2014). Review: The cognitive development of young dual language learners: A critical review. *Early Childhood Research Quarterly, 29,* 699–714. https://doi.org/10.1016/j.ecresq.2014.02.003

Bardige, B. S., and Segal, M. M. (2005). *Poems to Learn Read By: Building Literacy with Love.* Washington, D.C.: Zero to Three.

Barrera, I., and Corso, R. (2003). *Skilled Dialogue: Strategies for Responding to Cultural Diversity in Early Childhood.* Baltimore: Brookes.

Barrera, I., Kramer, L., and Macpherson, T. D. (2012). *Skilled Dialogue: Strategies for Responding to Cultural Diversity in Early Childhood* (2nd ed.). Baltimore: Brookes.

Basso, K. (2007). To give up on words: Silence in western Apache culture. In L. Monaghan and J. E. Goodman (Eds.), *A Cultural Approach to Interpersonal Communication* (pp. 77-87). Malden, MA: Blackwell.

Bassok, D., Finch, J. E., Lee, R. H., Reardon, S. F., and Waldfogel, J. (2016). Socioeconomic gaps in early childhood experiences. *AERA Open, 2* (2016). https://doi.org/10.1177/2332858416653924

Bates, C. (2014). Profile: Digital portfolios: Using technology to involve families. *YC Young Children,* 69 (4), 56-57.

Beal, S. M., and Finch, C. F. (1993). An overview of retrospective case control slides investigating the relationship between prone sleep positions and SIDS. *Journal of Pediatrics and Child Health, 27,* 334-339.

Beaty, J. J., and Pratt, L. (2014). *Early Literacy in Preschool and Kindergarten: A Multicultural Perspective* (4th ed.). Upper Saddle River, NJ: Merrill/Prentice Hall.

Bell, S. H., Carr, V. W., Denno, D., and Johnson, L. J. (2004). *Challenging Behaviors in Early Childhood Settings: Creating a Place for All Children.* Baltimore: Brookes.

Benjet, C., and Kazdin, A. E. (2003). Spanking children: The controversies, findings and new directions. *Clinical Psychology Review, 23,* 197-224.

Bennett, T. (2006). Future teachers forge family connections. *Young Children, 61* (1), 22-27.

Bennett, T. (2007). *Mapping Family Resources and Support: Spotlight on Young Children and Families.* Washington, D.C.: National Association for the Education of Young Children.

Bernard, B., and Quiett, D. (2002). *Nurturing the Nurturer: The Importance of Sound Relationships in Early Childhood Intervention.* San Francisco: WestEd.

Bernhard, J. K. (2012). *Stand Together or Fall Apart: Professionals Working with Immigrant Families.* Winnepeg: CA, Fernwood.

Bernhard, J. K., and Gonzalez-Mena, J. (2000). The cultural context of infant and toddler care. In D. Cryer and T. Harms (Eds.), *Infants and Toddlers in Out-of-Home Care* (pp. 237-267). Baltimore: Brookes.

Bernhard, J. K., Nirdosh, S., Freire, M., and Torres, F. (1997). Latin Americans in a Canadian primary school: Perspectives of parents, teachers and children on cultural identity and academic achievement. *Canadian Journal of Regional Science, 20,* 217-236.

Bialystok, E. (2001). *Bilingualism in Development: Language, Literacy, and Cognition.* Cambridge: Cambridge University Press.

Bialystok, E., Craik, F. I., & Luk, G. (2012). Bilingualism: consequences for mind and brain. *Trends in cognitive sciences,* 16 (4), 240-250.

Biggar, H. (2005). *NAEYC* recommendations on screening and assessment of young English-language learners. *Young Childen, 60* (6), 44-46.

Bisson, J. (2008). Holiday lessons learned in an early childhood classroom. In A. Pelo (Ed.), *Rethinking Early Childhood Education* (pp. 165-170). Milwaukee, WI: Rethinking Schools.

Bixler, R. D., Floyd, M. E., and Hammutt, W. E. (2002). Environmental socialization: Qualitative tests of the childhood play hypothesis. *Environment and Behavior, 34* (6), 795-818.

Bjorklund, D. F., and Causey, K. B. (2018). *Children's Thinking: Cognitive Development and Individual Differences* (6th ed.). Thousand Oaks, CA: Sage.

Blimes, J. (2004). *Beyond Behavior Management: The Six Life Skills Children Need to Thrive in Today's World.* St. Paul, MN: Redleaf Press.

Bloom, P. J., Eisenberg, P., and Eisenberg, E. (2003, Spring/Summer). Reshaping early childhood programs to be more family responsive. *America's Family Support Magazine,* 36-38.

Bodrova, E., and Leong, D. J. (2007). *Tools of the Mind: The Vygotsky Approach to Early Childhood Education* (2nd ed.). Upper Saddle River, NJ: Pearson/Merrill Prentice Hall.

Bodrova, E., and Leong, D. J. (2012). Chopsticks and counting chips: Do play and foundational skills need to compete for teacher's attention in an early childhood classroom? In C. Coppel, (Ed.), *Growing Minds: Building Strong Cognitive Foundations in Early Childhood* (pp. 67-74). Washington, D.C.: National Association for the Education of Young Children.

Bohart, H., and Procopio, R. (2017). *Spotlight on Young Children: Social and Emotional Development.* Washington, D.C.: National Association for the Education of Young Children.

Bowman, B. (2003). *Love to Read: Essays in Developing and Enhancing Early Literacy Skills of African American Children.* Washington, D.C.: National Black Child Development Institute.

Bowman, B. T., and Stott, F. M. (1994). Understanding development in a cultural context: The challenge for teachers. In B. L. Mallory and R. S. New (Eds.), *Diversity and Developmentally Appropriate Practices: Challenges for Early Childhood Education* (pp. 119-133). New York: Teachers College Press.

Bradley, J., and Kibera, P. (2006). Closing the gap: Culture and the promotion of inclusion in child care. *Young Children, 61* (1), 34-41.

Brault, L., and Brault, T. (2005). *Children with Challenging Behavior.* Phoenix, AZ: CPG Publishing.

Brault, L. M. V. (2007). *Making Inclusion Work: Strategies to Promote Belonging for Children with Special Needs in Child Care Settings.* Sacramento, CA: California Department of Education.

Bravo, E. (2008). It's all of our business: What fighting for family-friendly policies could mean for early childhood educators. In A. Pelo (Ed.), *Rethinking Early Childhood Education* (pp. 197–200). Milwaukee, WI: Rethinking Schools.

Bredekamp, S., and Copple, C. (Eds.). (2009). *Developmentally Appropriate Practice in Early Childhood Programs.* Washington, D.C.: National Association for the Education of Young Children.

Breslin, D. (2005). Children's capacity to develop resiliency: How to nurture it. *Young Children, 60* (1), 47–52.

Brillante, P., and Mankiw, S. (2015). Preschool through grade 3: A sense of place: Human geography in the early childhood classroom. *YC Young Children, 70* (3), 16–23.

Briody, J., and McGarry, K. (2005). Using social stories to ease children's transitions. *Young Children, 60* (5), 38–42.

Brock, C. (2001). Serving English language learners. *Language Arts, 78,* 467–475.

Bronfenbrenner, U. (2004). *Making Human Beings Human: Bioecological Perspectives on Human Development.* London: Sage.

Bronfenbrenner, U. (2005). *The Ecology of Human Development: Experiments by Nature and Design.* Cambridge, MA: Harvard University Press.

Bronson, M. (2012). Recognizing and supporting the development of self-regulation in young children. In C. Coppel, (Ed.), *Growing Minds: Building Strong Cognitive Foundations in Early Childhood,* (pp. 97–104), Washington, D.C.: National Association for the Education of Young Children.

Bronson, M. (2000). Research in review: Recognizing and supporting the development of self-regulation in young children. *Young Children, 55,* 32–37.

Bruce, T. (2004). *Developing Learning in Early Childhood.* Thousand Oaks, CA: Sage.

Bruno, H. E. (2005, September). At the end of the day: Policies, procedures and practices to ensure smooth transitions. *Exchange, 165,* 66–69.

Burningham, L. M., and Dever, M. T. (2005). An interactive model for fostering family literacy. *Young Children, 60* (5), 87–94.

Buysse, V., Castro, D. C., West, T., and Skinner, M. L. (2004). *Addressing the Needs of Latino Children: A National Survey of State Administrators of Early Childhood Programs* (executive summary). Chapel Hill: The University of North Carolina: FPG Child Development Institute.

Butler, A. M., and Ostrosky, M. M. (2018, September). Reducing challenging behaviors during transitions: Strategies for early childhood educators to share with parents. *Young Children, 73* (4). https://www.naeyc.org/resources/pubs/yc/sept2018/reducing-challenging-behaviors-during-transitions

Buysse, V., Wesley, P., Coleman, M. R., Snyder, P., and Winton, P. (Eds.). (2006). *Evidence-Based Practice in the Early Childhood Field.* Washington, D.C.: Zero to Three.

California Department of Social Services. (2018). *Child Care Regulations.* http://www.cdss.ca.gov/inforesources/Letters-Regulations/Legislation-and-Regulations/Community-Care-Licensing-Regulations/Child-Care

Carlebach, D., and Tate, B. (2002). *Creating Caring Children.* Miami: Peace Education Foundation.

Carlson, F. M. (2010). *Big Body Play: Why Boisterous, Vigorous, and Very Physical Play Is Essential to Children's Development and Learning.* Washington, D.C.: National Association for the Education of Young Children.

Carlsson-Paige, N., and Levin, D. E. (2005). When push comes to shove: Reconsidering children's conflicts. In B. Neugebauer (Ed.), *Behavior: A Beginnings WorkshopBook* (pp. 39–41). Redmond, WA: Exchange.

Carr, M. (2001). *Assessment in Early Childhood Settings: Learning Stories.* Thousand Oaks, CA: Sage.

Carson, R. (1965). *The Sense of Wonder.* New York: Harper & Row.

Carter, D. R., Norman, R., and Tredwell, C. (2011). Program-wide positive behavior support in preschool: Lessons for getting started. *Early Childhood Education Journal, 38* (5): 349–355.

Carter, M., and Curtis, D. (2003). *Designs for Living and Learning: Transforming Early Childhood Environments.* St. Paul. MN: Redleaf.

Casas, J. F., Weigel, S. M., Crick, N. R., Ostrov, J. M., Woods, K. E., Yeh, E. A. J., and Huddleston-Casas, C. A. (2006). Early parenting and children's relational and physical aggression in the preschool and home contexts. *Journal of Applied Developmental Psychology,* 27, 209–227. https://doi.org/10.1016/j.appdev.2006.02.003

Caspe, M., Seltzer, A., Kennedy, J. Cappio, M., and DeLorenzo, C. (2013). Infants, toddlers, and preschool: Engaging families in the child assessment process. *YC Young Children, 68* (3), 8–15.

Casper, V. (2003). Very young children in lesbian- and gay-headed families: Moving beyond acceptance. *Zero to Three, 23* (3), 18–26.

Chaille, C., and Britain, L. (1997). *The Young Child as Scientist: A Constructivist Approach to Early Childhood Science Education.* New York: Longman.

Chalufour, I., and Worth, K. (2004). *Building Structures with Young Children.* St. Paul, MN: Redleaf.

Chang, H. (2005). *Getting Ready for Quality: The Critical Importance of Developing and Supporting a Skilled, Ethnically and Linguistically Diverse Early Childhood Workforce.* Oakland, CA: California Tomorrow.

Charmian, K. (2007). Childhood bilingualism: Research on infancy through school age. *Literacy, 41* (2), 110–111.

Chenfeld, M. B. (2004). Education is a moving experience: Get movin'! *Young Children, 59* (4), 56–57.

Chetty, R., Friedman, J. N., Hilger, N., Saez, E., Whitmore Schanzenbach, D., and Yagan, D. (2011). How does your kindergarten classroom affect your earnings? Evidence from Project Star. *Quarterly Journal of Economics, 4,* 1593. https://doi.org/10.1093/qje/qjr04

Christian, L. G. (2006). Understanding families: Applying family systems theory to early childhood practice. *Young Children, 61* (1), 12–21.

Christie, J. (1998). Play: A medium for literacy development. In D. Fromberg and D. Bergen (Eds.), *Play from Birth to Twelve and Beyond: Contexts, Perspectives, and Meanings* (pp. 50–55). New York: Garland.

Cianciolo, S., Trueblood-Noll, R., and Allingham, P. (2004). Health consultation in early childhood settings. *Young Children, 59* (2), 56–61.

Clarke-Stewart, A., and Alhusen, V. D. (2005). *What We Know about Childcare.* Cambridge, MA: Harvard University Press.

Clay, J. W. (2004). Creating safe, just places to learn for children of lesbian and gay parents: The NAEYC Code of Ethics in action. *Young Children, 59* (6), 34–38.

Cohen, D., Stern, V., and Balaban, N. (2008). *Observing and Recording the Behavior of Young Children* (5th ed.). New York: Teachers College Press.

Colkeer, L. J. (2005). *The Cooking Book: Fostering Young Children's Learning and Delight.* Washington, D.C.: National Association for the Education of Young Children.

Collins, M. C. (2010). ELL preschoolers' English vocabulary acquisition from storybook reading. *Early Childhood Research Quarterly, 25* (1), 84–97.

Columbo, M. (2005). Reflections from teachers of culturally diverse children. *Young Children, 60* (6).

Cooper, R., and Jones, E. (2005, October). Enjoying diversity. *Exchange,* 6–9.

Coopersmith, S. (1967). *The Antecedents of Self-Esteem.* San Francisco: Freeman.

Copley, J. V. (2010). *The Young Child and Mathematics* (2nd ed.). Washington, D.C.: National Association for the Education of Young Children.

Coppel, C., Ed. (2012). *Growing Minds: Building Strong Cognitive Foundations in Early Childhood.* Washington, D.C.: National Association for the Education of Young Children.

Copple, Carol (Ed.). (2003). *A World of Difference: Readings on Teaching Young Children in a Diverse Society.* Washington, D.C.: National Association for the Education of Young Children.

Copple, C., and Bredekamp, S. (2006). *Basics of Developmentally Appropriate Practice: An Introduction for Teachers of Children 3 to 6.* Washington, D.C.: National Association for the Education of Young Children.

Copple, C., and Bredekamp, S. (2009). *Developmentally Appropriate Practice in Early Childhood Programs Serving Children Birth through Age 8* (3rd ed.). Washington, D.C.: National Association for the Education of Young Children.

Copple, C., Bredekamp, S., and Gonzalez-Mena, J. (2011). *Basics of Developmentally Appropriate Practice: An Introduction for Teachers of Infants and Toddlers.* Washington, D.C.: National Association for Education of Young Children.

Corsaro, W. A., and Molinari, L. (2005). *I Compagni: Understanding Children's Transition from Preschool to Elementary School.* New York: Teachers College Press.

Council on Interracial Books for Children. (2008). 10 quick ways to analyze children's books for racism and sexism. In A. Pelo (Ed.), *Rethinking Early Childhood Education* (pp. 211–213). Milwaukee, WI: Rethinking Schools.

Courtney, A. M., and Montano, M. (2006). Teaching comprehension from the start: One first grade classroom. *Young Children, 61* (2), 68–74.

Crawford, L. (2004). *Lively Learning: Using the Arts to Teach the K-8 Curriculum.* Greenfield, MA: Northeast Foundation for Children.

Crawford, M. J., and Weber, B. (2016). *Autism Intervention Every Day: Embedding Activities in Daily Routines for Young Children and Their Families.* Baltimore: Brookes Publishing.

Cuffaro, H. (1995). *Experimenting with the World: John Dewey and the Early Childhood Classroom.* New York: Teachers College Press.

Curtis, D., and Carter, M. (2012). *The Art of Awareness.* St. Paul, MN: Redleaf.

Cushner, K., McClelland, A., and Stafford, P. (2003). *Human Diversity in Education: An Integrative Approach* (4th ed.). Boston: McGraw-Hill.

Cutler, K. M., Gilkerson, D., Parrott, S., and Bowne, M.T. (2003, January). Developing math games based on children's literature. *Young Children, 58,* 22–27.

D'Addesio, J., Grob, B., and Furman, L. (2005). Social studies: Learning about the world around us. *Young Children, 60* (5), 50–57.

Daniel, J., and Friedman, S. (2005). Preparing teachers to work with culturally and linguistically diverse children. *Young Children, 60* (6).

Darling-Hammond, L., French, J., and Garcia-Lopez, S. P. (2002). *Learning to Teach for Social Justice.* New York: Teachers College Press.

Darragh, J. (2008, July–August). Access and inclusion: Ensuring engagement in EC environments. *Exchange* (182), 20–23.

David, J., Onchonga, O., Drew, R., Grass, R., Stechuk, R., and Burns, M. S. (2005). Head start embraces language diversity. *Young Children, 60* (6), 40–43.

David, M., and Appell, G. (2001). Lóczy: An unusual approach to mothering. In J. M. Clark, revised translation by J. Falk (Ed.), *Lóczy ou Le Maternage Insolite.* Budapest: Association Pikler-Lóczy for Young Children.

Davis, C., and Yang, A. (2005). *Parents and Teachers Working Together.* Turners Falls, MA: Northeast Foundation for Children.

Day, M., and Parlakian, R. (2004). *How Culture Shapes Social-Emotional Development: Implications for Practice in Infant-Family Programs.* Washington, D.C.: Zero to Three.

DeJong, L. (2003, March). Using Erikson to work more effectively with teenage parents. *Young Children, 58,* 87–95.

Delpit, L., and Dowdy, J. K. (2002). *The Skin That We Speak: Thoughts on Language and Culture in the Classroom.* New York: New Press.

Derman-Sparks, L. (2011, July/August). Anti-bias education. *Exchange, 33* (4), 55–58.

Derman-Sparks, L., and Edwards, J. O. (2010). *Anti-bias Education for Young Children and Ourselves.* Washington, D.C.: National Association for the Education of Young Children.

Derman-Sparks, L., LeeKeenan, E., and Nimmo, J. (2015). *Leading Anti-bias Early Childhood Programs: A Guide for Change.* New York: Teachers College, Columbia University.

Derman-Sparks, L. (1989). *Antibias Curriculum: Tools for Empowering Young Children.* Washington, D.C.: National Association for the Education of Young Children.

Derman-Sparks, L. (2008). Why an anti-bias curriculum? In A. Pelo (Ed.), *Rethinking Early Childhood Education* (pp. 7–12). Milwaukee, WI: Rethinking Schools.

Derman-Sparks, L., and Olsen, E. J. (2010). *Anti-bias Education for Ourselves and Others.* Washington, D.C.: National Association for the Education of Young Children.

Derman-Sparks, L., and Ramsey, P. G. (2005). What if all the children in my class are white? *Young Children, 60* (6), 20–27.

Derman-Sparks, L., and Ramsey, P. G. (2006). *What If All the Kids Are White: Antibias Multicultural Education with Young Children and Families.* New York: Teachers College Press.

DeVries, R., Zan, B., Hildebrandt, C., Edmiaston R., and Sales, C. (2002). *Developing Constructivist Early Childhood Curriculum: Practical Principles and Activities.* New York: Teachers College Press.

DeWeese-Parkinson, C. (2008). Talking the talk: Integrating indigenous languages into a Head Start classroom. In A. Pelo (Ed.), *Rethinking Early Childhood Education* (pp. 175–176). Milwaukee, WI: Rethinking Schools.

Diaz Soto, L. (1997). *Language, Culture, and Power.* Albany: State University of New York Press.

Diffily, D., and Sassman, C. (2002). *Project-Based Learning with Young Children.* Portsmouth, NH: Heinemann.

Dischler, P. A. (2010). *Teaching the 3 Cs: Creativity, Curiosity, and Courtesy.* Thousand Oaks, CA: Corwin.

Division for Early Childhood of the Council for Exceptional Children. (2017). *Position Statement on Challenging Behavior and Young Children–Executive Summary.* https://www.decdocs.org/executive-summary-challenging-behav

Division of Research to Practice, Office of Special Education Programs. (2001). *Synthesis on the Use of Assistive Technology with Infants and Toddlers.* Washington, D.C.: U.S. Department of Education.

Dombro, A. L., Jablon, J. R., and Stetson, C. (2011). Powerful interactions. *Young Children, 66* (1), 12–19.

Dombro, A. L., and Lerner, C. (2006). Sharing the care of infants and toddlers. *Young Children, 61* (1), 29–33.

Dowhey, M. (2008). Heather's moms got married. In A. Pelo (Ed.), *Rethinking Early Childhood Education* (pp. 177–179). Milwaukee, WI: Rethinking Schools.

Dragon, P. B. (2003). *Everything You Need to Know to Teach First Grade.* Portsmouth, NH: Heinemann.

Dragon, P. B. (2006). *A How-To Guide for Teaching English Language Learners in the Primary Classroom.* Washington, D.C.: National Association for the Education of Young Children.

Dreikurs, R., and Grey, L. (1990). *Logical Consequences: A New Approach to Discipline.* New York: Dutton.

Drew, W. F., and Rankin, B. (2004). Promoting creativity for life using open-ended materials. *Young Children, 59* (4), 38–45.

Dubosarky, M., Murphy, B., Roehrig, G., Frost, L. C., Jones, J., and Bement, J. (2011, September). Animal tracks on the playground, minnows in the sensory table: Incorporating cultural themes to promote preschoolers' critical thinking in American Indian head start classrooms. *Young Children, 66* (5), 20–29.

Duckworth, E. (2001). *"Tell Me More": Listening to Learners.* New York: Teachers College Press.

Duke, N. K. (2003, March). Reading to learn from the very beginning: Information books in early childhood. *Young Children, 58,* 14–20.

Duncan, S. (2011). Breaking the code–Changing our thinking about children's environments. *Exchange 33* (4), 13–17.

Dunn, J. (2004). *Children's Friendships: The Beginnings of Intimacy.* Malden, MA: Blackwell.

Dweck, C. S. (2006). *Mindset: The New Psychology of Success.* New York: Random House.

Eberly, J. L., and Golbeck, S. L. (2004). Blocks, building and mathematics: Influences of task format and gender of play partners among preschoolers. In S. Reifel and M. H. Brown (Eds.), *Advances in Early Education and Day Care: Social Contexts of Early Education, and Reconceptualizing Play (II)* (2nd ed., Vol. 13, pp. 39–54). Greenwich, CT: Jai Press.

Edelman, L. (2004). A relationship-based approach to early intervention. *Resources and Connections, 3* (2), 2–10.

Editors of Teaching Young Children. (2014). *Exploring Math and Science in Preschool.* Washington, D.C.: National Association for the Education of Young Children.

Edwards, C. P., Gandini, L., and Forman, G. (Eds.). (1993). *The Hundred Languages of Children: The Reggio Emilia Approach to Early Childhood Education.* Norwood, NJ: Ablex.

Eggers-Pierola, C. (2005). *Connections and Commitments: Reflecting Latino Values in Early Childhood Programs.* Portsmouth, NH: Heinemann.

Eisaguirre, L. (2007). *Stop Pissing Me Off!* Cincinnati, OH: Adams Media.

Elicker, J., and McMullen, M. (2013). Infants and toddlers: Appropriate and meaningful assessment in family centered programs. *YC Young Children,* 68 (3), 22–27.

Elkind, D. (2006). Work, chores, and play. Setting a healthy balance. *Exchange*, 39–41.

Elkind, D. (2007). *The Power of Play: How Spontaneous Imaginative Activities Lead to Happier, Healthier Children.* Cambridge, MA: Da Capo.

Elliot, E., and Gonzalez-Mena, J. (2011). Babies' self-regulation: Taking a broad perspective. *Young Children, 66* (1), 28–33.

Epstein, A. S. (2012). How planning and reflection develop young children's thinking skills. In C. Coppel, (Ed.), *Growing Minds: Building Strong Cognitive Foundations in Early Childhood,* (pp. 111–118). Washington, D.C.: National Association for the Education of Young Children.

Erikson, E. (1950). *Childhood and Society.* New York: Norton.

Erikson, E., and Coles, R. (2000). *The Erik Erikson Reader.* New York: W. W. Norton.

Espinosa, L. M. (2010). *Getting it Right for Young Children from Diverse Backgrounds: Applying Research to Improve Practice with a focus on Dual Language Learners (2nd ed.).* Upper Saddle River, NJ: Pearson.

Estok, V. (2005). One district's study on the propriety of transition-grade classrooms. *Young Children, 60* (2), 28–31.

Fadiman, A. (1997). *The Spirit Catches You and You Fall Down: A Hmong Child, Her American Doctors, and the Collision of Two Cultures.* New York: Noonday Press.

Falk, B. (Ed.) (2012). *Defending Childhood: Keeping the Promise of Early Education.* New York: Teachers College Press.

Feeney, S. (2012). *Professionalism in Early Childhood Education: Doing Our Best for Young Children.* Upper Saddle River, NJ: Pearson.

Feeney, S., and Freeman, N. K. (2012). *Ethics and the Early Childhood Educator: Using the NAEYC Code* (3rd ed.). Washington, D.C.: National Association for the Education of Young Children.

Feeney, S., Galper, A. and Seefeldt C. (Eds.) (2009). *Continuing Issues in Early Childhood Education.* Upper Saddle River, NJ: Pearson/Merril.

Fenion, A. (2005). Collaborative steps: Paving the way to kindergarten for young children with disabilities. *Young Children, 60* (2), 32–37.

Ferguson, C., Green, S., and Marchel, C. (2013). Kindergarten and primary grades: Teacher-made assessments show children's growth. *YC Young Children, 68* (3), 28–37.

Fernandez, M. T., and Marfo, K. (2005). Enhancing infant-toddler adjustment during transitions to care. *Zero to Three, 26,* 41–48.

Fernea, E. W. (1995). *Children in the Muslim Middle East.* Austin: University of Texas Press.

Fillmore, L. W. (1991). When learning a second language means losing the first. *Early Childhood Research Quarterly, 6* (3), 323–346.

Fillmore, L. W. (2000). Loss of family languages: Should educators be concerned? *Theory into Practice, 39,* 203–210.

Fitzgerald, D. (2004). *Parent Partnership in the Early Years.* London: Continuum.

Flicker, E. S., and Hoffman, J. A. (2002, September). Developmental discipline in the early childhood classroom. *Young Children, 78,* 82–88.

Florez, I. R. (2011). Developing young children's self-regulation through everyday experiences. *Young Children, 66* (4), 46–51.

Freiburg, S. (1959). *The Magic Years.* New York: Scribner's.

Freire, P. (1998). *Teachers as Cultural Workers: Letters to Those Who Dare Teach.* Boulder, CO: Westview Press.

French, K. (2004). Supporting a child with special health care needs. *Young Children, 59* (2), 62–63.

French, K., and Cain, H. M. (2006). Including a young child with spina bifida. *Young Children, 61* (3), 78–85.

Friend, M. D., and Bursuck W. D. (2019). *Getting It Right for Young Children from Diverse Backgrounds: Applying Research to Improve Practice with a Focus on Dual Language Learners* (2nd ed.). Upper Saddle River, NJ: Pearson.

Friend, M., and Cook, L. (2003). *Interactions: Collaboration Skills for School Professionals* (4th ed.). Boston: Allyn & Bacon.

Fromberg, D. P. (2002). *Play and Meaning in Early Childhood Education.* Boston: Allyn & Bacon.

Frost, J., Wortham, S., and Reifel, S. (2001). *Play and Child Development.* Upper Saddle River, NJ: Merrill/Prentice Hall.

Furmanek, D. (2014). Preschool through grade 2: Classroom choreography: Enhancing learning through movement. *YC Young Children, 69* (4), 80–85.

Frost, J. L., Brown, P. -S., Sutterby, J. A., and Thornton, C. D. (2004). *The Developmental Benefits of Playgrounds.* Olney, MD: Association of Childhood Education International.

Frost, J. L., Wortham, S. C., and Reifel, S. (2005). *Play and Child Development* (2nd ed.). Upper Saddle River, NJ: Merrill/Prentice Hall.

Galinsky, E. (2010). *Mind in the Making: The Seven Essential Life Skills Every Child Needs.* New York: Harper Studios.

Gandini, L., and Edwards, C.P. (Eds.). (2000). *Bambini: The Italian Approach to Infant-Toddler Care.* New York: Teachers College Press.

Gandini, L., Hill, L., and Schwall, C. (Eds.). (2005). *In the Spirit of the Studio: Learning from the Atelier of Reggio Emilia.* New York: Teachers College Press.

Gantley, M., Davies, D. P., and Murcett, A. (1993). Sudden infant death syndrome: Links with infant care practices. *British Medical Journal, 306,* 16–20.

García, E. E., and McLaughlin, B. (Eds.). (1995). *Meeting the Challenge of Linguistic and Cultural Diversity in Early Childhood Education.* New York: Teachers College Press.

Gardner, H. (1983). *Frames of Mind.* New York: Basic Books.

Gardner, H. (1993). *Frames of Mind: The Theory of Multiple Intelligences.* New York: Basic Books.

Gardner, H. (2000). *Intelligence Reframed: Multiple Intelligences for the 21st Century.* New York: Basic Books.

Gardner, H. (2000). *Multiple Intelligences: The Theory in Practice.* New York: Basic Books.

Garner, A. (2004). *Families Like Mine: Children of Gay Parents Tell It Like It Is.* New York: Harper Collins Publishers.

Gartell, D. (2011). Aggression, the prequel: Preventing the need. *Young Children, 66* (6), 62–64.

Gartrell, D. (2006). Guidance matters: Boys and men teachers. *Young Children, 61* (3), 92–93.

Gauvian, M. (2001). *The Social Context of Cognitive Development.* New York: Guilford.

Geist, E. (2003, January). Infants and toddlers exploring mathematics. *Young Children, 58,* 10–12.

Geist, K., Geist, E., and Kuznik, K. (2012). The patterns of music: Young children learning mathematics through beat, rhythm, and melody. *YC Young Children, 67* (1), 74–79.

Gelman, R., and Brenneman, K. (2004). Science learning pathways for young children. *Early Childhood Research Quarterly, 19* (1), 150–159.

Gelnaw, A. (2005). Belonging: Including children of gay and lesbian parents–and all children–in your program. *Exchange,* 42–44.

Gelnaw, A., Brickley, M., Marsh, H., and Ryan, D. (2004). *Opening Doors: Lesbian and Gay Parents and Schools.* Washington, D.C.: Family Pride Coalition.

Genesee, F. (2015). Myths about early childhood bilingualism. *Canadian Psychology/Psychologie Canadienne,* 56, 6–15. https://doi.org/10.1037/a0038599

Genesse, F., Paradis, J., and Crago, M. B. (2004). *Dual Language Development and Disorders: A Handbook on Bilingualism and Second Language Learning.* Baltimore: Brookes.

Genishi, C. (2002, July). Research in review: Young English language learners. *Young Children, 57,* 66–71.

Genishi, C., and Haas Dyson, A. (2005). Ways of talking: Respecting differences. In B. Neugebauer (Ed.), *Literacy: A Beginnings WorkshopBook* (pp. 32–25). Redmond, WA: Exchange.

Gerber, M. (1985). Modifying the environment to respond to the changing needs of the child. *Educaring, 6* (1), 1–2.

Gerber, M. (Ed.). (forthcoming). *Manual for Resources for Infant Educarers* (2nd ed.). Los Angeles: Resources for Infant Educarers.

Gerber, M. (1998). *Dear Parent: Caring for Infants with Respect.* Los Angeles: Resources for Infant Educarers.

Gerber, M. (2005). RIE principles and practices. In S. Petrie and S. Owen (Eds.), *Authentic Relationships in Group Care for Infants and Toddlers: Resources for Infant Educarers (RIE) Principles into Practice* (pp. 35-49). London and Philadelphia: Jessica Kingsley Publishers.

Gillanders, C., and Castro, C. (2011). Storybook reading for young dual language learners. *Young Children, 66* (1), 91-95.

Gillespie, L. (2016, July). It's never 'just play'! *Young Children, 71* (3), 92-94.

Gilliam, W. S., Maupin, A. N., Reyes, C. R., Accavitti, M., and Shic, F. (2016, September 28). Do early educators' implicit biases regarding sex and race relate to behavior expectations and recommendations of preschool expulsions and suspensions? *Yale Child Study Center.* https://medicine.yale.edu/childstudy/zigler/publications/Preschool%20Implicit%20Bias%20Policy%20Brief_final_9_26_276766_5379_v1.pdf

Gillespie. L. (2016, July). It's Never "Just Play"! *YC Young Children,* (3), 92.

Ginott, H. (1956). *Between Parent and Child.* New York: Macmillan.

Goldberg, R. J., Haugen, K., Sivanathan, A., and Spakota, R. D. (2011, September/October). World forum working group report on inclusion. *Exchange, 33* (55, Serial No. 201), 56.

Goleman, D. (1997). *Emotional Intelligence.* New York: Bantam.

Goleman, D. (2000). *Working with Emotional Intelligence.* New York: Bantam.

Gonzalez, M., Moll, L. C., and Amanti, C. (2005). *Funds of Knowledge: Theorizing Practices in Households, Communities, and Classrooms.* Mahwah, NJ: Erlbaum.

Gonzalez-Mena, J. (2012). *Child, Family and Community: Family-Centered Early Care and Education.* Upper Saddle River, NJ: Pearson.

Gonzalez-Mena, J. (2011). Culture and communication in the child care setting. In P. Mangione and D. Greenwald, (Eds.), *A Guide to Language Development and Communication* (2nd ed.), (pp. 51-60). Sacramento, CA: California Department of Education.

Gonzalez-Mena, J. (2010a). Cultural responsiveness and routines: When center and home don't match. *Exchange, 42* (4), 42-44.

Gonzalez-Mena, J. (2010b). Compassionate roots Begin with babies. *Exchange, 32* (3), 46-49.

Gonzalez-Mena, J. (1986). Praise: Motivator or manipulator? *Educaring, 7* (1), 1-4.

Gonzalez-Mena, J. (1990). *Program for Infant-Toddler Caregivers: A Guide to Routines.* Sacramento: California Department of Education.

Gonzalez-Mena, J. (1992). Taking a culturally sensitive approach in infant-toddler programs. *Young Children, 47* (2), 4-9.

Gonzalez-Mena, J. (1997). *Multicultural Issues in Child Care.* Mountain View, CA: Mayfield.

Gonzalez-Mena, J. (1997). Independence or interdependence? Understanding the parent's perspective. *Child Care Information Exchange, 117,* 61-63.

Gonzalez-Mena, J. (2002). *Infant/Toddler Caregiving: A Guide to Routines* (2nd ed.). Sacramento: California Department of Education with WestEd.

Gonzalez-Mena, J. (2002, September). Working with cultural differences: Individualism and collectivism. *The First Years: Nga TuaTuatahi, 4,* 13-15.

Gonzalez-Mena, J. (2004). What can an orphanage teach us? Lessons from Budapest. *Young Children, 59* (5), 26-30.

Gonzalez-Mena, J. (2012). *50 Strategies for Working and Communicating with Parents.* Columbus, OH: Merrill.

Gonzalez-Mena, J. (2006). Caregiving routines and literacy. In S. E. Rosenkoetter and J. Knapp-Philo (Eds.), *Learning to Read the World: Language and Literacy in the First Three Years.* Washington, D.C.: Zero to Three.

Gonzalez-Mena, J. (2007, May-June). Thinking about thinking: How can I get inside your head?" *Exchange* (175), 50-52.

Gonzalez-Mena, J. (2008). *Diversity in Early Care and Education: Honoring Differences* (5th ed.). New York: McGraw-Hill.

Gonzalez-Mena, J. (2012). *Child, Family and Community: Family Centered Early Care and Education* (5th ed.). Upper Saddle River, NJ: Pearson/Merrill.

Gonzalez-Mena, J., and Briley, L. (2011). Improving infant mental health in orphanages. *The Signal: Newsletter of the World Association for Infant Mental Health, 19* (4), 14-17.

Gonzalez-Mena, J., Chahin, E., and Briley, L. (2005, November/December). The Pikler Institute: A unique approach to caring for children. *Exchange,* 49-51.

Gonzalez-Mena, J., and Eyer, D. W. (2013). *Infants, Toddlers, and Caregivers.* Mountain View, CA: Mayfield.

Gonzalez-Mena, J., and Shareef, I. (2005). Discussing diverse perspectives on guidance. *Young Children, 60* (6), 34-38.

Gonzalez-Mena, J., and Stonehouse, A. (1995). In the child's best interests. *Child Care Information Exchange, 106,* 17-18, 20.

Gonzalez-Mena, J., and Stonehouse, A. (2008). *Making Links: A Collaborative Approach to Planning and Practice in Early Childhood Programs* (American ed.). New York: Teachers College Press.

Gopnik, A., Meltzoff, A. N., and Kuhl, D. K. (1999). *The Scientist in the Crib.* New York: HarperCollins.

Gordon, T. (1970). *Parent Effectiveness Training: The No-Lose Program for Raising Responsible Children.* New York: P. H. Wyden.

Graue, E. (2001). Research in review: What's going on in the children's garden? Kindergarten today. *Young Children, 56,* 67–73.

Gray, H. (2004). You go away and you come back. *Young Children, 59* (5), 100–107.

Greenfield, P. M., Quiroz, B., Rothstein-Fisch, C., and Trumbull, E. (2001). *Bridging Cultures between Home and School.* Mahwah, NJ: Lawrence Erlbaum.

Greenman, J. (1989). Living in the real world: Diversity and conflict: The world will never sing in perfect harmony. *Child Care Information Exchange, 69,* 11–13.

Greenman, J. (2006). *What Happened to My World? Helping Children Cope with Natural Disasters and Castastrophe.* Washington, D.C.: National Association for the Education of Young Children.

Greenman, J., and Stonehouse, A. (1996). *Prime Times: A Handbook for Excellence in Infant and Toddler Care.* St. Paul, MN: Redleaf Press.

Greenman, M. (2011, March/April). The family partnership. *Exchange, 33* (2), 46–48.

Greenough, W., Emde, R. N., Gunnar, M., Massinga, R., and Shonkoff, J. P. (2001, April/May). The impact of the caregiving environment on young children's development: Different ways of knowing. *Zero to Three,* 21, 16–24.

Greenspan, S. I. (1999). *Building Healthy Minds.* Cambridge, MA: Perseus.

Greenspan, S. I. (with Benderly, B. L.). (1997). *The Growth of the Mind: And the Endangered Origins of Intelligence.* Reading, MA: Addison-Wesley.

Greenspan, S. I. (2004). *The First Idea: How Symbols, Language, and Intelligence Evolved from Our Primate Ancestors to Modern Humans.* Cambridge, MA: Da Capo Press.

Greenspan, S. I., and Greenspan, N. T. (1985). *First Feelings: Milestones in the Emotional Development of Your Baby and Child.* New York: Viking.

Greenspan, S. I., and Wieder, S. (1998). *The Child with Special Needs: Encouraging Intellectual and Emotional Growth.* Reading, MA: Perseus.

Grefsrud, S. (2011, March/April). Room at the table: Parent engagement in head start. *Exchange, 33* (2, Serial No. 198), 57–59.

Grisham-Brown, J., Hemmeter, M. L., and Pretti-Frontczak, K. (2005). *Blended Practices for Teaching Young Children in Inclusive Settings.* Baltimore: Brookes.

Gronlund, G. (2006). *Make Early Learning Standards Come Alive: Connecting Your Practice and Curriculum to State Guidelines.* St. Paul, MN: Redleaf Press.

Gullo, D. F. (2005). *Understanding Assessment and Evaluation in Early Childhood Education* (2nd ed.). New York: Teachers College Press.

Gullo, D. F. (Ed.). (2006). *K Today: Teaching and Learning in the Kindergarten Year.* Washington, D.C.: National Association for the Education of Young Children.

Gunnar, M. R., Bruce, J., and Hickman, S. E. (2001). Salivary cortisol response to stress in children. In T. Theorell (Ed.), *Everyday Biological Stress Mechanisms* (pp. 52–60). Basel, Switzerland: Karger.

Guralnick, M. J. (Ed.). (2001). *Early Childhood Inclusion: Focus on Change.* Baltimore: Brookes.

Guss, S., Horm, D., Lang, E., Krehbiel, S., Petty, J., Austin, K., Bergren, C., Brown, A., and Holloway, S. (2013). Toddlers: Using classroom quality assessments to inform teacher decisions. *YC Young Children, 68* (3), 16–21.

Hale, J. E. (1986). *Black Children: Their Roots, Culture and Learning Styles.* Baltimore: Johns Hopkins University Press.

Hall, E. T. (1989). *Beyond Culture.* New York: Anchor.

Hammond, R. A. (2001, Fall). Preparing for literacy: Communication comes first. Educating, 22, 1–5.

Hancock, C., and Carter, D. (2016). Preschool: Building environments that encourage positive behavior: The preschool behavior support self-assessment. *YC Young Children, 71* (1), 66–73.

Hanson, M. J., and Lynch, E. W. (2004). *Understanding Families: Approaches to Diversity, Disability, and Risk.* Baltimore: Brookes.

Harris, P. R., and Moran, R. T. (1987). *Managing Cultural Differences.* Houston, TX: Gulf.

Harwood, R. L., Miller, J. G., and Irizzary, N. L. (1995). *Culture and Attachment.* New York: Guilford Press.

Hatch, J. A. (2005). *Teaching in the New Kindergarten.* Clifton Park, NY: Thomson Delmar Learning.

Hatcher, B., and Petty, K. (2004). Seeing is believing: Visible thought in dramatic play. *Young Children, 59* (6), 79–82.

Haynes-Lawrence, D. (2009, January–February). Crisis nurseries: Emergency services for children and families in need. *Exchange* (185), 16–20.

Healy, J. (2008). Cybertots: Technology and the preschool child. In A. Pelo (Ed.), *Rethinking Early Childhood Education* (pp. 75–84). Milwaukee, WI: Rethinking Schools.

Healy, J. M. (2011, March/April). Brain readiness: Impacting readiness–Nature and nurture. *Exchange, 33* (2, Serial No. 198), 18–21.

Heffron, M. C., Ivins, B., and Weston, D. (2005). Finding an authentic voice. Use of self: Essential learning process for relationship-based work. *Infants and Young Children, 18* (4), 323–336.

Helm, J. H., and Beneke, S. (Eds.). (2003). *The Power of Projects: Meeting Contemporary Challenges in Early Childhood Classrooms–Strategies and Solutions.* New York: Teachers College Press.

Helm, J. H., and Katz, L. (2016). *Young Investigators: The Project Approach in the Early Years* (3rd ed.). New York: Teachers College Press and Washington, D.C.: National Association for the Education of Young Children.

Hensen, R. (2005). Real super-hero play. *Young Children, 60* (5), 37.

Hernandez, L., and Smith, C. J. (2009, January-February). Disarming cantankerous people: Coping with difficult personalities in the ECE work setting. *Exchange* (185), 12–14.

Heroman, C. (2016). *Making and Tinkering with STEM: Solving Design Challenges with Young Children.* Washington, D.C.: National Association for the Education of Young Chilldren.

Hewitt, K. (2001, January). Blocks as a tool for learning. *Young Children, 56,* 6–12.

Hill-Clarke, K. Y., and Robinson, N. R. (2004). It's as easy as A-B-C and Do-Re-Mi: Music, rhythm, and rhyme enhance children's literacy skills. *Young Children, 59* (5), 91–95.

Hillman, C. B. (2011, November/December). Home visits: Building relationships by revisiting home visits. *Exchange, 33* (6, Serial No. 202), 80–85.

Hoffman, E. (2004). *Magic Capes, Amazing Powers: Transforming Superhero Play in the Classroom.* St. Paul, MN: Redleaf Press.

Ho, J., and Funk, S. (2018, March). Promoting young children's social and emotional health. *Young Children, 73* (1). https://www.naeyc.org/resources/pubs/yc/mar2018/promoting-social-and-emotional-health

Hochanadel, A., and Finamore, D. (2015). Fixed and growth mindset in education and how grit helps students persist in the face of adversity. *Journal of International Education Research, 11* (1), 47–50.

Honig, A. S. (2005). The language of lullabies. *Young Children, 60* (5), 30–36.

Howes, C. (2010). *Culture and Child Development in Early Childhood Programs: Practices for Quality Education and Care.* New York: Teachers College Press.

Howes, C., and Ritchie, S. (2002). *A Matter of Trust.* New York: Teachers College Press.

Hudson, R. A. (2007). *Speech communities.* In L. Monaghan and J. E. Goodman (Eds.). *A Cultural Approach to Interpersonal Communication* (pp. 212–217). Malden, MA: Blackwell.

Hullinger-Sirken, H., and Staley, L. (2016). Preschool through grade 3: Understanding writing development: Catie's continuum. *YC Young Children, 71* (5), 74–78.

Humpal, M. E., and Wolf, J. (2003, March). Music in the inclusive environment. *Young Children, 58,* 103–107.

Hymes, J. L., Jr. (1981). *Teaching the Child under Six* (3rd ed.). Columbus, OH: Merrill.

Hymes, J. L., Jr. (1998). A child development point of view. *Young Children, 53* (3), 49–51.

Hynes-Berry, M. (2011). *Don't Leave the Story in the Book: Using Literature to Guide Inquiry in Early Childhood Classrooms.* New York: Teachers College Press.

Hyson, M. (2004). *The Emotional Development of Young Children* (2nd ed.). New York: Teachers College Press.

Isbell, C., and Isbell, R. (2005). *The Inclusive Learning Center Book for Preschool Children with Special Needs.* Beltsville, MD: Gryphon House.

Israel, M. S. (2004). Ethical dilemmas for early child hood educators: The ethics of being accountable. *Young Children, 59* (6), 24–32.

Jablon, J. R., Dombro, A. L., and Dichtelmiller, M. L. (1999). *The Power of Observation.* Washington, D.C.: Teaching Strategies.

Jacobs, G., and Crowley, K. (2011). *Reaching Standards and Beyond in Kindergarten: Nurturing Children's Sense of Wonder and Joy in Learning.* Thousand Oaks, CA: Corwin.

Jacobson, T. (2003). *Confronting Our Discomfort: Clearing the Way for Anti-Bias in Early Childhood.* Portsmith, NH: Heinemann.

Jacobson, T. (2008). *Don't Get So Upset: Helping Young Children Manage Their Feelings by Understanding Your Own.* St. Paul, MN: Redleaf.

Jalongo, M. R. (2008). *Learning to Listen, Listening to Learn: Building Essential Skills in Young Children.* Washington, D.C.: National Association for the Education of Young Children.

Johnson, D. (2001). "Round." *The New Grove Dictionary of Music and Musicians* (2nd ed.), Stanley Sadie and John Tyrrell (Eds.). London: Macmillan Publishers.

Johnston, P. H. (2004). *Choice Words: How Our Language Affects Children's Learning.* Portland, ME: Stenhouse.

Jones, E. (2003, May). Playing to get smart. *Young Children, 58,* 32–33.

Jones, E., and Cooper, R. (2006). *Playing to Get Smart.* New York: Teachers College Press.

Jones, E., and Nimmo, J. (1994). *Emergent Curriculum.* Washington, D.C.: National Association for the Education of Young Children.

Jones, E., and Reynolds, G. (2011). *The Play's the Thing.* New York: Teachers College Press.

Jones, N. P. (2005). Big jobs: Planning for competence. *Young Children, 60* (2), 86–93.

Kaiser, B., and Rasminsky, J. S. (2012). *Challenging Behavior in Young Children: Understanding, Preventing, and Responding Effectively.* (4th ed.). Upper Saddle River, NJ: Pearson.

Kaiser, B., and Rasminsky, J. S. (2005, July). Including children with challenging behavior in your child care community. *Exchange,* 32–34.

Kamii, C. (with Housman, L. B.). (2000). *Young Children Reinvent Arithmetic: Implications of Piaget's Theory* (2nd ed.). New York: Teachers College Press.

Katz, L., and Schery, T. K. (2006). Including children with hearing loss in early childhood programs. *Young Children, 61* (1), 86–95.

Katz, L. G., and Chard, S. (2000). *Engaging Children's Minds: The Project Approach* (2nd ed.). Stamford, CT: Ablex.

Keeler, R. (2009, January–February). A spring playscape project: Building a tree circle. *Exchange* (185), 70–71.

Kemple, K. M. (2003). *Let's Be Friends.* New York: Teachers College Press.

Kemple, K. M., Batey, J. J., and Hartle, L. C. (2004). Music play: Creating centers for musical play and exploration. *Young Children, 59* (4), 30–37.

Kersey, K., and Malley, C. R. (2005). Helping children develop resiliency: Providing supportive relationships. *Young Children, 60* (1), 53–58.

Klein, M. D., and Chen, D. (2001). *Working with Children from Culturally Diverse Backgrounds.* New York: Delmar.

Klug, B. (2011). Daring to teach: Challenging the Western narrative of the American Indians in the Classroom. In J. G. Landsman, and C. W. Lewis (Eds.). White Teachers/Diverse Classrooms (2nd ed.). Sterling, VA: Stylus.

Koc, K., and Buzzelli, C. A. (2004). The moral of the story is . . . using children's literature in moral education. *Young Children, 59* (1), 92–97.

Kohl, M. F. (2005). *Primary Art: It's the Process, Not the Product.* Beltsville, MD: Gryphon House.

Koplow, L. (2007). *Unsmiling Faces: How Preschool Can Heal.* New York: Teachers College Press.

Koralek, D. (Ed.). (2006). *Spotlight on Young Children and Social Studies.* Washington, D.C.: National Association for the Education of Young Children.

Kordt-Thomas, C., and Lee, I. M. (2006). Floor time: Rethinking play in the classroom. *Young Children, 61* (3), 86–90.

Korte, K. M., Fielden, L. J., and Agnew, J. C. (2005). To run, stomp, or study: Hissing cockroaches in the classroom. *Young Children, 60* (2), 12–18.

Kostelnik, M., E. Onaga, Rohde, B., and Whiren, A. (2002). *Children with Special Needs: Lessons for Early Childhood Professionals.* New York: Teachers College Press.

Kristal, J. (2005). *The Temperament Perspective: Working with Children's Behavioral Styles.* Baltimore: Brookes.

Kroeger, J. (2008). Doing the difficult: Schools and lesbian, gay, bisexual, transgendered, and queer families. In T. Turner-Vorbeck and M. M. Marsh (Eds.), *Other Kinds of Families: Embracing Diversity in Schools.* (pp. 121–138). New York: Teachers College Press.

Kuby, C. R. (2011, September). Humpty Dumpty and Rosa Parks: Making space for critical dialogue with 5- and 6-year-olds. *Young Children, 66* (5), 36–43.

Kuhl, P., Tsao, F., Liu, H., Zhang, Y., and Boer, B. (2001). Language/culture/mind/brain. *Annals of the New York Academy of Sciences, 1,* 136. doi:10.1111/j.1749-6632.2001.tb03478.x

Kyger, C. S., Ramming, P., and Thompson, S. D. (2006). A new bit on toddler biting: The influence of food, oral motor development, and sensory activities. *Young Children, 61* (2), 17–23.

Kyttä, M. (2004). The extent of children's independent mobility and the number of actualized affordances as criteria for child-friendly environments. *Journal of Environmental Psychology, 24* (2), 179–198.

Lally, J. R. (1995). The impact of child care policies and practices on infant/toddler identity formation. *Young Children, 51* (1), 58–67.

Lally, J. R. (1998, May). Brain Research, infant learning, and child care curriculum. *Child Care Information Exchange,* 46–48.

Lally, J. R., Griffin, A., Fenichel, E., Segal, M., Szanton, E., and Weissbourd, B. (2004). *Caring for Infants and Toddlers in Groups: Developmentally Appropriate Practice* (2003 ed.). Washington, D.C.: Zero to Three.

Lally, J. R., and Mangione, P. (2011). The uniqueness of infancy demands a responsive approach to care. In D. Koralek and L. G. Gillespie (Eds.) (pp. 7–13), *Spotlight on Infants and Toddlers.* Washington, D.C.: National Association for the Education of Young Children.

Laski, E. (2013). Preschool and kindergarten: Portfolio picks: An approach for developing children's metacognition. *YC Young Children, 68* (3), 38–43.

Laurion, J., and Schmiedicke, C. (2005). *Creating Connection: How to Lead Family Child Care Support Groups.* St. Paul, MN: Redleaf.

Lee, L. (2006). *Stronger Together: Family Support and Early Childhood Education* (2nd ed.). San Rafael, CA: Parent Services Project.

Leong, D. J., and Bodrova, E. (2012). Assessing and scaffolding: Make-believe play. *Young Children, 67* (1), 28–35.

Lesser, L. K., Burt, T., and Glenaw, A. (2005). *Making Room in the Circle: Lesbian, Gay, Bisexual and Transgender Families in Early Childhood Settings.* San Rafael, CA: Parent Services Project.

Levin, D. E. (2003). *Teaching Young Children in Violent Times: Building a Peaceable Classroom* (2nd ed.). Washington, D.C.: National Association for the Education of Young Children.

Levin, D. E., and Carlsson-Paige, N. (2005). *The War Play Dilemma.* New York: Teachers College Press.

LeVine, R. A. (1977). Child rearing as cultural adaption. In P. H. Leiderman, S. R. Tulkin, and A. Rosenfeld (Eds.), *Culture and Infancy: Variations in the Human Experience.* New York: Academic.

Lewin-Benham, A. (2006). One teacher, 20 preschoolers, and a goldfish: Environmental awareness, emergent curriculum, and documentation. *Young Children, 61* (1), 28–34.

Lewin-Benham, A. (2006). *Possible Schools: The Reggio Approach to Urban Education.* New York: Teachers College Press.

Lieberman, A. F. (1993). *The Emotional Life of the Toddler.* New York: Free Press.

Liewra, C., Reeble, T., and Rosenow, N. (2011). *Growing with Nature: Supporting Whole-Child Learning in Outdoor Classrooms.* Lincoln, NE: Arbor Day Foundation.

Lindner Gunnoe, M. (2013). Associations between parenting style, physical discipline, and adjustment in adolescents' reports. *Psychological Reports, 112* (3), 933.

Logue, M. E. (2006). Teachers observe to learn—differences in social behavior of toddlers and preschoolers in same-age and multiage groupings. *Young Children, 61* (3), 70–77.

Logue, M. E., Shelton, H., Cronkite, D., and Austin, J. (2007, March). Family ties: Strengthening partnerships with families through toddlers' stories. *Young Children, 62* (2), 85–87.

Loomis, C., and Wagner, J. (2005). A different look at challenging behavior. *Young Children, 60* (2), 94–99.

Lopéz, E. J., Salas, L., and Flores, J. P. (2005). Hispanic preschool children: What about assessment and intervention? *Young Children, 60* (6), 48–54.

Louv, R. (2005). *Last Child in the Woods: Saving Our Children from Nature Deficit Disorder.* New York: Workman.

Louv, R. (2008). Don't know much about natural history: Education as a barrier to nature. In A. Pelo (Ed.), *Rethinking Early Childhood Education.* Milwaukee, WI: Rethinking Schools.

Luckenbill, J. (2012). Getting the picture: Using the digital camera as a tool to support reflective practice and responsive care. *Young Children, 67* (2), 28–36.

Lutton, A. (Ed.) (2012). *Advancing the Early Childhood Profession: NAEYC Standards and Guidelines for Professional Development.* Washington, D.C.: National Association for the Education of Young Children.

Lynn, L. L., and Kieff, J. (2002, May). Including *everyone* in outdoor play. *Young Children, 57,* 20–26.

MacDonald, S. (2004). *Sanity Savers for Early Childhood Teachers: 200 Quick Fixes for Everything from Big Messes to Small Budgets.* Beltsville, MD: Gryphon House.

MacDonald, B. (2005). Purposeful work: A Montessori approach to everyday challenging behaviors. *Exchange,* 51–54.

MacDonald, B. (2005). Purposeful work: A Montessori approach to everyday challenging behaviors. In B. Neugebauer (Ed.), *Behavior: A Beginnings WorkshopBook* (pp. 30–32). Redmond, WA: Exchange.

MacNaughton, G. (2000). *Rethinking Gender in Early Childhood Education.* St. Leonards, NSW, Australia: Allen and Unwin.

Malik, Rasheed, "4 Disturbing Facts About Preschool Suspension," *Center for American Progress,* 30 March 2017, https://www.americanprogress.org/issues/early-childhood/news/2017/03/30/429552/4-disturbing-facts-preschool-suspension/

Mallory, B. L., and Rous, B. (2008). Educating young children with developmental differences: Principles of inclusive practice. In S. Feeney, A. Galper, and C. Seefeldt (Eds.), *Continuing Issues in Early Childhood Education* (pp. 278–302). Upper Saddle River, NJ: Pearson.

Malone, K., and Tranter, P. (2003). Children's environmental learning and the use, design and management of schoolgrounds. *Children, Youth and Environments, 13* (2). Retrieved June 9, 2004, from cye.colorado.edu

Maltz, D. N., and Borker, R. A. (2007). A cultural approach to male-female miscommunication. In L. Monaghan and J. E. Goodman (Eds.), *A Cultural Approach to Interpersonal Communication* (pp. 77–87). Malden, MA: Blackwell.

Mardell, B., Rivard, M., and Krechevsky, M. (2012). Visible learning, visible learners: The power of the group in a kindergarten classroom. *Young Children, 67* (1), 12–19.

Marion, M. (2018). *Guidance for Young Children* (10th ed.). Upper Saddle River, NJ: Pearson

Marshall, E., and Sensoy, O. (2011). *Rethinking Popular Culture and Media.* Milwaukee, WI: Rethinking Schools.

Martin, J. (2009, January-February). Using the principles of intentional teaching to communicate effectively with parents. *Exchange* (185), 53-56.

Mary, M. (2002, February/March). How mothers in four American cultural groups shape infant learning during mealtimes. *Zero to Three, 22,* 14-20.

Maschinot, B. (2008). *The Changing Face of the United States: The Influence of Culture on Child Development.* Washington, D.C.: Zero to Three.

Maslow, A. (1968). *Toward a Psychology of Being.* New York: Van Nostrand.

Matlock, R., and Hornstein, J. (2004). Sometimes a smudge is just a smudge, and sometimes it's a saber-toothed tiger: Learning and the arts through the ages. *Young Children, 59* (4), 12-17.

McDaniel, G. L., Isaac, M. Y., Brooks, H. M., and Hatch, A. (2005). Confronting K-3 teaching challenges in an era of accountability. *Young Children, 60* (2), 20-26.

McDermont, L.B. (2011, September). Play school: Where children and families learn and grow together. *Young Children, 66* (5), 81-86.

McGinnis, M.H., Getskow, V., and Dicker, B. S. (2012, March/April). Parental rights and authorization: Parental rights and release authorization. *Exchange, 34* (2, Serial No. 204), 16-18.

McKay, F. (1988). Discipline. In A. Stonehouse (Ed.), *Trusting Toddlers: Programming for One- to Three-Year-Olds in Child Care Centers* (pp. 65-78). Canberra, Australia: Australian Early Childhood Association.

McLennan, D., and Bombardier, J. (2015). Kindergarten: "Paper shoes aren't for dancing!": Children's explorations of music and movement through inquiry. *YC Young Children, 70* (3), 70-75.

McWilliams, S. M., Maldonado-Mancebo, T., Szczpaniak, P. S., and Jones, J. (2011, November). Supporting Native Indian preschoolers and their families: Family-school-community partnerships. *Young Children, 66* (6), 34-41.

Meisels, S. J., and Atkins-Burnett, S. (2005). *Developmental Screening in Early Childhood.* Washington, D.C.: National Association for the Education of Young Children.

Meyer, J. (2003). *Kids Talking: Learning Relationships and Culture with Children.* Lanham, MD: Rowman & Littlefield Publishers.

Michael-Luna, S. (2015). What parents have to teach us about their dual language children.*YC Young Children, 70* (5), 42-49.

Miller, K. (2005). Developmental issues that affect behavior. In B. Neugebauer (Ed.), *Behavior: A Beginnings WorkshopBook* (pp. 12-13). Redmond, WA: Exchange.

Miller, K. (2005). *Simple Transitions for Infants and Toddlers.* Beltsville, MD: Gryphon House.

Mills, H., O'Keefe, T., and Jennings, L. B. (2004). *Looking Closely and Listening Carefully: Learning Literacy through Inquiry.* Urbana, IL: National Council of Teachers of English.

Mindes, G. (2005). Social studies in today's early childhood curricula. *Young Children, 60* (5), 12-18.

Mindes, G. (2015). Preschool through grade 3: Pushing up the social studies from early childhood education to the world. *YC Young Children, 70* (3), 10-15.

Mitchell, L. C. (2004). Making the MOST of creativity in activities for young children with disabilities. *Young Children, 59* (4), 46-49.

Moerman, M. (2007). Talking culture: ethnography and conversation analysis. In L. Monaghan and J.E. Goodman (Eds.), *A Cultural Approach to Interpersonal Communication* (119-124). Malden, MA: Blackwell.

Monaghan, L. (2007). Conversations: The link between words and the world. In L. Monaghan and J.E. Goodman (Eds.), *A Cultural Approach to Interpersonal Communication* (pp. 145-149). Malden, MA: Blackwell.

Monaghan, L., and Goodman, J. E. (Eds.). (2007). *A Cultural Approach to Interpersonal Communication.* Malden, MA: Blackwell.

Morelli, G., Rogoff, B., and Oppenheim, D. (1992). Cultural Variation in Infants' Sleeping Arrangements: Questions of Independence. *Developmental Psychology, 28,* 604-619.

Morgan, G. (2005). Is education separate from care? *Exchange,* 6-10.

Morse, A. (2003). *Language Access: Helping Non-English Speakers Navigate Health and Human Services.* Washington, D.C.: National Conference of State Legislature Children's Policy Initiative.

Moutray, C. L., and Snell, C. A. (2003, March). Three teachers' quest: Providing daily writing activities for kindergartners. *Young Children, 58,* 24-28.

Moyles, J. R. (2012). *A-Z of Play in Early Childhood.* Maidenhead, Berkshire, UK: McGraw-Hill Open University Press.

Mueller, S. (2005). *Everyday Literacy: Environmental Print Activities for Young Children Ages 3 to 8.* Beltsville, MD: Gryphon House.

Murray, A. (2001). Ideas on manipulative math for young children. *Young Children, 56,* 28-29.

Musatti, T., and Mayer, S. (2001). Knowing and learning in an educational context: A study in the infant-toddler centers of the city of pistoia. In L. Gandini and C. P., Edwards (Eds.), *Bambini: The Italian Approach to Infant/ Toddler Care*. New York: Teachers College Press.

Music, G. (2017). *Nurturing Natures: Attachment and Children's Emotional, Sociocultural and Brain Development.* Abingdon, Oxon; New York: Routledge

NAEYC and NCTM (National Association for the Education of Young Children and National Council of Teachers of Mathematics). (2002, April). Early childhood mathematics: Promoting good beginnings. Joint Position Statement. Washington, D.C.: NAEYC and Reston, VA: NCTM. Online at www.naeyc.org/resources/position-statements/psmath.htm

NAEYC and NCTM (National Association for the Education of Young Children and National Council of Teachers of Mathematics). (2003, January). Learning paths and teaching strategies in early mathematics. *Young Children, 58,* 41-43.

NAEYC position statement: Responding to linguistic and cultural diversity: Recommendations for effective early childhood education. (1996). *Young Children, 51* (2), 4-12.

NAEYC standards for early childhood professional preparation: Initial licensure standards. Online at www.naeyc.org/profdev/prep_review/2001.pdf

National Association for the Education of Young Children. (2005). NAEYC Code of Ethical Conduct. In *NAEYC Early Childhood Program Standards and Accreditation Criteria*. Washington, D.C.: National Association *Childhood Program Standards and Accreditation Criteria* for the Education of Young Children.

National Association for the Education of Young Children. (2005). *NAEYC Early Learning Program Accreditation Standards and Assessment Items.* Washington, D.C.: National Association for the Education of Young Children.

National Research Council and Institute of Medicine. (2000). *From Neurons to Neighborhoods: The Science of Early Childhood Development.* Board on Children, Youth, and Families, Commission on Behavioral and Social Sciences and Education. Washington, D.C.: National Academy Press.

Neelly, L. P. (2001, May). Developmentally appropriate music practice: Children learn what they live. *Young Children, 56,* 32-36.

Neelly, L. P. (2002, July). Practical ways to improve singing in early childhood classrooms. *Young Children, 57,* 80-83.

Negri-Pool, L. L. (2008). Welcoming Kalenna: Making our students feel at home. In A. Pelo (Ed.), *Rethinking Early Childhood Education* (pp. 161-169). Milwaukee, WI: Rethinking Schools.

Nemeth, K. N. (2012). *Basics of Supporting Dual Language Learners: An Introduction for Educators of Children from Birth through Age 8.* Washington, D.C.: National Association for the Education of Young Children.

Nemeth, K. N. (2009). *Many Languages, One Classroom: Teaching Dual and English Language Learners.* Beltsville, MD: Gryphon House.

Neugebauer, B. (Ed.). (2005). *Behavior: A Beginnings WorkshopBook.* Redmond, WA: Exchange.

Neubauer, B. (2009, July-August). Celebrating Mother Nature. *Exchange 182,* 18-19.

Neuman, S. B., Copple, C., and Bredekamp, S. (2001). *Learning to Read and Write: Developmentally Appropriate Practice for Young Children.* Washington, D.C.: National Association for the Education of Young Children.

New, R. S. (2000). Reggio Emilia: An approach or an attitude. In J. L. Rupermarine and D. E. Johnson (Eds.), *Approaches to Early Childhood Education* (3rd ed., pp. 341-358). Upper Saddle River, NJ: Merrill.

New, R. S., and Beneke, M. (2008). Negotiating diversity in early childhood education: Rethinking notions of expertise. In S. Feeney, A. Galper, and C Seefeldt (Eds.), *Continuing Issues in Early Childhood Education* (pp. 303-324). Upper Saddle River, NJ: Pearson.

Newburger, A., and Vaughan, E. (2006). *Teaching Numeracy, Language, and Literacy with Blocks.* St. Paul, MN: Gryphon House.

NICHD Early Child Care Research Network (Ed.). (2006). *Child Care and Child Development: Results from the NICHD Study of Early Child Care and Youth Development.* New York: Guilford.

Nicolopoulou, A. (2001). Peer-group culture and narrative development. In S. Blum-Kulka and C. Snow (Eds.), *Talking with Adults.* Mahwah, NJ: Erlbaum.

Nieto, S. (Ed.). (2005). *Why We Teach.* New York: Teachers College Press.

North, M., Durekas, T., Siegel, B., and Sisbarro, A. (2009, January-February). The ins and outs of transporting children on field trips. *Exchange* (185). 84-85.

Norton-Meier, L., and Whitmore, K. (2015). Toddlers through grade 2: Developmental moments: Teacher decision making to support young writers. *YC Young Children, 70* (4), 76-83.

Ohman-Rodriguez, J. (2004). Music from inside out: Promoting emergent composition with young children. *Young Children, 59* (4), 50-55.

Okagaki, L., and Diamond, K. E. (2000). Responding to cultural and linguistic differences in the beliefs and practices of families with young children. *Young Children, 55,* 74–78.

Olsen, L., Bhattacharya, J., and Scharf, A. (2005). *Ready or Not: School Readiness and Immigrant Communities.* Oakland, CA: California Tomorrow.

Olson, M. (2007, March). Strengthening families: Community strategies that work. *Young children, 62* (2), 26–32.

Olson, M., and Hyson, M. (2005). Supporting teachers, strengthening families' initiative adds a national leadership program for early childhood professionals. *Young Children, 60* (1), 44–45.

Ordonez-Jasis, R., and Ortiz, R. W. (2006). Reading their worlds: Working with diverse families to enhance children's early literacy development. *Young Children, 61* (1), 41–48.

Oremland, J., Flynn, L., and Kieff, J. E. (2002, Spring). Merry-go-round: Using interpersonal influence to keep inclusion spinning smoothly. *Childhood Education, 78,* 153–159.

Orenstein, P. (2011). *Cinderella Ate My Daughter.* New York: HarperCollins.

Paley, V. G. (2004). *A Child's Work: The Importance of Fantasy Play.* Chicago: University of Chicago Press.

Pate, R. R., Pfeiffer, K. A., Trost, S. G., Ziegler, P., and Dowda, M. (2004). Physical activity among children attending preschools. *Pediatrics, 114* (5), 1258–1263.

Patterson, K., Grenny, J., McMillan, R., and Switzler, A. (2002). *Crucial Conversations: Tools for Talking When Stakes Are High.* New York: McGraw-Hill.

Pelo, A. (2008). A pedagogy for ecology. In A. Pelo (Ed.), *Rethinking Early Childhood Education.* (pp. 123–130). Milwaukee, WI: Rethinking Schools.

Pelo, A. (2008). Bringing the lives of lesbian and gay people into our programs. In A. Pelo (Ed.), *Rethinking Early Childhood Education* (pp. 180–182). Milwaukee, WI: Rethinking Schools.

Pelo, A. (Ed.). (2008). *Rethinking Early Childhood Education.* Milwaukee, WI: Rethinking Schools.

Perry, G., Henderson, B., and Meier (Eds.) (2012). *Our Inquiry, Our Practice: Undertaking, Supporting, and Learning from Early Childhood Teacher Research(ers).* Washington, D.C.: National Association for the Education of Young Children.

Petrie, S., and Owen, S. (Eds.). (2005). *Authentic Relationships in Group Care for Infants and Toddlers: Resources for Infant Educarers (RIE) Principles into Practice.* London and Philadelphia: Jessica Kingsley.

Piaget, J. (1936/1952). *The Origins of Intelligence in Children.* (M. Cook, Trans.). New York: Norton.

Piaget, J. (1965/1972). *Science of Education and the Psychology of the Child* (rev. ed.). New York: Viking.

Pica, R. (2006). Physical fitness and the early childhood Curriculum. *Young Children, 61* (3), 12–19.

Pikler, E. (1971). Learning of motor skills on the basis of self-induced movements. In J. Hellmuth (Ed.), *Exceptional Infant* (Vol. 2, pp. 54–89). New York: Bruner/Mazel.

Pikler, E. (1973). Some contributions to the study of gross motor development of children. In A. Sandovsky (Ed.), *Child and Adolescent Development* (pp. 52–64.). New York: Free Press.

Pikler, E. (1979). *A quarter of a century of observing infants in a residential center.* In M. Gerber (Ed.), RIE Manual (pp. 90–92). Los Angeles: Resources for Infant Educarers.

Pikler, E. (1979). Can infant-child care centers promote optimal development? In M. Gerber (Ed.), *RIE Manual* (pp. 93–102). Los Angeles: Resources for Infant Educarers.

Pikler, E., and Tardos, A. (1968). Some contributions to the study of infants' gross motor activities. In *Proceedings of the 16th International Congress of Applied Psychology.* Amsterdam: ICAP.

Pilonieta, P., Shue, P., and Kissel, B. (2014). Preschool: Reading books, writing books: Reading and writing come together in a dual language classroom. *YC Young Children, 69* (3), 14–21.

Pinto, C. (2001, Spring). Supporting competence in a child with special needs: One child's story. *Educaring, 22,* 1–6.

Poussaint, A. F. (2006). Understanding and involving African American parents. *Young Children, 61* (1), 48.

Powers, J. (2005). *Parent-Friendly Early Learning: Tips and Stategies for Working Well with Families.* St. Paul, MN: Redleaf.

Powers, J. (2006). Six fundamentals for creating relationships with families. *Young Children, 61* (1), 28.

Prieto, H. V. (2009, January). One language, two languages, three languages . . . more? *Young Children, 64* (1), 52–53.

Project Zero. (2003). *Making Teaching Visible: Documenting Individual and Group Learning as Professional Development.* Cambridge, MA: Project Zero.

Quann, V., and Wein, C. A. (2011). The visible empathy of infants and toddlers. In D. Koralek and L. G. Gillespie, Eds. (pp. 21–28), *Spotlight on Infants and Toddlers.* Washington, D.C.: National Association for the Education of Young Children.

Quintero, E. P. (2005). Multicultural literature: A source of meaningful content for kindergartners. *Young Children, 60* (6), 28–32.

Raeff, C., Greenfield, P. M., and Quiroz, B. (2000). Conceptualizing interpersonal relationships in the cultural contexts of individualism and collectivism. In S. Harkness, C. Raeff, and C. M. Super (Eds.), *New Directions for Child and Adolescent Development* (pp. 59-74). San Francisco: Jossey-Bass.

Ramirez, A. Y. (2008). Immigrant families and schools. In A. Pelo (Ed.), *Rethinking Early Childhood Education* (pp. 171-174). Milwaukee, WI: Rethinking Schools.

Ramsey, P. G. (2004). *Teaching and Learning in a Diverse World* (3rd ed.). New York: Teachers College Press.

Ranck, E. R., and Anderson, C. (2010). Blocks: A versatile learning tool for yesterday, today, and tomorrow. *Young Children, 65* (2), 54-56.

Ranson Jacobs, L. (2006, January). The value of real work with children exhibiting challenging behavior. *Exchange,* 36-38.

Reifel, S., and Sutterby, J. (2008). Play theory and practice in contemporary classrooms. In S. Feeney, A. Galper, and C. Seefeldt (Eds.), *Continuing Issues in Early Childhood Education* (pp. 238-257). Upper Saddle River, NJ: Pearson.

Reynolds, G. (2008). Observations are essential in supporting children's play. In B. Neugebauer (Ed.), *Professionalism* (pp.92-95). Redmond, WA: Exchange Press.

Rieger, L. (2008). A welcoming tone in the classroom: Developing the potential of diverse students and their families. In T. Turner-Vorbeck and M. M. Marsh (Eds.), *Other Kinds of Families: Embracing Diversity in Schools* (pp. 64-69). New York: Teachers College Press.

Rinehart, N. M. (2006). The curriculum belongs to the community: Curriculum planning and development for tlingit and haida young children. *Zero to Three, 26* (4), 46-48.

Riojas-Cortez, M. (2011). Culture, play, and family: Supporting young children on the autism spectrum. *Young Children, 66* (5), 94-99.

Rishel, T. J. (2008). From the principal's desk: Making the school environment more inclusive. In T. Turner-Vorbeck and M.M. Marsh (Eds.), *Other Kinds of Families: Embracing Diversity in Schools* (pp. 46-63). New York: Teachers College Press.

Ritter, J. (2007, March). Tips for starting a successful community partnership. *Young Children, 62* (2), 38.

Roberts, L. C., and Hill, H. T. (2003, March). Come and listen to a story about a girl named rex: Using children's literature to debunk gender stereotypes. *Young Children, 58,* 39-42.

Rofrano, F. (2002, January). "I care for you": A reflection on caring as infant curriculum. *Young Children, 57,* 49-51.

Rogoff, B. (2011). *Developing Destinies: Child Development in Cultural Context.* New York: Oxford University Press.

Rogoff, B. (2003). *The Cultural Nature of Human Development.* Oxford and New York: Oxford University Press.

Rogoff, B., Stott, F., and Bowman, B. (1996). Child development knowledge: A slippery base for practice. *Early Childhood Research Quarterly, 11* (2), 1169-1184.

Rosenkoetter, S., and Barton, L. (2002, February/ March). Bridges to literacy: Early routines that promote later school success. *Zero to Three, 22,* 33-38.

Rosenkoetter, S.E., and Knapp-Philo, J. (Eds.). (2006). *Learning to Read the World: Language and Literacy in the First Three Years.* Washington, D.C.: Zero to Three.

Rosenow, N. (2005). The impact of sensory integration on behavior: Discovering our best selves. In B. Neugebauer (Ed.), *Behavior: A Beginnings WorkshopBook* (pp. 33-35). Redmond, WA: Exchange.

Rosenow, N. (2011). Planning intentionally for children's outdoor environments. *Exchange, 33* (4), 46-40.

Roskos, K., and Christie, J. (2001). On not pushing too hard: A few cautionary remarks about linking literacy and play. *Young Children, 56,* 64-66.

Roskos, K. A., Christie, J. F., and Richgels, D. J. (2003, March). The essentials of early literacy instruction. *Young Children, 58,* 52-59.

Rothstein-Fisch, C, (2003). *Bridging Cultures: Teacher Education Module.* Mahwah, NJ: Erlbaum.

Russell, G. M. (2004). Surviving and thriving in the midst of anti-gay politics. *The Policy Journal of the Institute for Gay and Lesbian Strategic Studies, 7* (20), 1-7.

Ruzzo, K., and Sacco, M. A. (2004). *Significant Studies for Second Grade: Reading and Writing Investigations for Children.* Portsmouth, NH: Heinemann.

Ryan, S., and Grieshaber, S. (2004). It's more than child development: Critical theories, research, and teaching young children. *Young Children, 59* (6), 44-52.

Sandall, S.R. (2003, May). Play modifications for children with disabilities. *Young Children, 58,* 54-55.

Sandall, S.R., and Schwartz, I.S. (2002). *Building Blocks for Teaching Preschoolers with Special Needs.* Baltimore: Brookes.

Scarlett, W. G., Naudeau, S. C., Salonius-Pasternak, D., and Ponte, I. C. (2004). *Children's Play.* Thousand Oaks, CA: Sage.

Schaef, A. W. (1992). *Beyond Therapy, Beyond Science: A New Model for Healing the Whole Person.* San Francisco: HarperSanFrancisco.

Schall, J. (2000, Winter). Rites of passage. *Educaring, 23,* 1-5.

Schein, D., and Rivkin, M. (2014). *The Great Outdoors: Advocating for Natural Spaces for young Children* (rev. ed.). Washington, D.C.: National Association for the Education of Young Children.

Schickedanz, J. A. (1999). *Much More than the ABC's: The Early Stages of Reading and Writing.* Washington, D.C.

Schickedanz, J. A. (2005). Setting the stage for literacy events in the classroom. In B. Neugebauer (Ed.), *Literacy: A Beginnings WorkshopBook* (pp. 17-21). Redmond, WA: Exchange.

Schickedanz, J. A., and Casberque, R. M. (2004). *Writing in Preschool: Learning to Orchestrate Meaning and Marks.* Newark, DE: International Reading Association.

Scott, D. M. (2005). The pathway to leadership takes many roads: A personal journey. *Young Children, 60* (1), 42.

Seefeldt, C., and Galper, A. (2007). *Active Experiences for Active Children: Science.* Upper Saddle River, NJ: Pearson/Merrill/Prentice Hall.

Segal, M., Masi, W., and Leiderman, R. (2001). *In Time and with Love: Caring for Infants and Toddlers with Special Needs* (2nd ed.). New York: New Market Press.

Seo, K.-H. (2003, January). What children's play tells us about teaching mathematics. *Young Children, 58,* 28-33.

Seplocha, H. (2004). Partnerships of learning: Conferencing with families. *Young Children, 59* (5), 96-99.

Shabazian, A. (2016). Birth to grade 3: The role of documentation in fostering learning. *YC Young Children, 71* (3), 73-79.

Shonkoff, J. P., and Phillips, D. A. (2000). *From Neurons to Neighborhoods: The Science of Early Childhood Development.* Washington, D.C.: National Academy Press.

Shonkoff, J. P., and Phillips, D. (2001, April/May). From neurons to neighborhoods: The science of early childhood development—an introduction. *Zero to Three, 21,* 4-7.

Shore, R., and Strasser, J. (2006). Music for their minds. *Young Children, 61* (2), 62-67.

Simon, F. (2011, September/October). Social media: Everyone is doing it! Managing social media in the early childhood ecosystem. *Exchange, 33* (5, Serial No. 201), 12-16.

Simons, K. A., and Curtis, P. A. (2007, March). Connecting with communities: Four successful schools. *Young Children, 62* (2) 12-20.

Siraj-Blatchford, I., and Clarke, P. (2000). *Supporting Identity, Diversity and Language in the Early Years.* Philadelphia: Open University Press.

Sluss, D. J. (2004). *Supporting Play: Birth through Age Eight.* Clifton Park, NY: Thompson Delmar Learning.

Sorte, J. M., and Daeschel, I. (2006). Health in action: A program approach to fighting obesity in young children. *Young Children, 61* (3), 40-49.

Soto, L. D. (2002). Making a difference in the lives of bilingual/bicultural children. In J. L. Kincheloe and S. R. Steinberg (Eds.), *Counterpoints: Studies in the Post Modern Theory of Education* (Vol. 134). New York: Peter Lang.

Souto-Manning, M. (2010, March). Family involvement: Challenges to consider, strengths to build on. *Young Children, 65* (2), 82-89.

Sprung, B., Froschi, M., and Hinitz, B. (2005). *The Anti-Bullying and Teasing Book for Preschool Classrooms.* Beltsville, MD: Gryphon House.

Stacey, S. (2011). *The Unscripted Classroom: Emergent Curriculum in Action.* St. Paul, MN: Redleaf.

Stacy, S. (2008). *Emergent Curriculum in Early Childhood Settings: From Theory to Practice.* St. Paul, MN: Redleaf.

Stegelin, D. A. (2005). Making the case for play policy: Research-based reasons to support play-based environments. *Young Children, 60* (2), 76-85.

Stephens, K. (2010, July/August). Parent relationships: Building relationships—What parents can teach us about their children. *Exchange, 32* (4, Serial No. 194), 38-40.

Stephens, K. (2005, May). Meaningful family engagement. *Exchange,* 18-25.

Stephens, K. (2005). Responding professionally and compassionately to challenging behavior. In B. Neugebauer (Ed.), *Behavior: A Beginnings WorkshopBook* (pp. 7-11). Redmond, WA: Exchange.

Stoecklin, K. L. (2005, July). Creating environments that sustain children, staff, and our planet. *Exchange,* 39-48.

Stonehouse, A. (2011, March/April). Moving from family participation to partnerships: Not always easy; always worth the effort. *Exchange, 33* (2), 48-51.

Stonehouse, A., and Gonzalez-Mena, J. (2004). *Making Links: A Collaborative Approach to Planning and Practice in Early Childhood.* Sydney, Australia: Pademelon Press.

Stroll, J., Hamilton, A., Oxley, E., Mitroff Eastman, A., and Brent, R. (March, 2012). Young thinkers in motion: Problem solving and physics in preschool. *Young Children, 67* (2), 20-26.

Sturm, L. (2004). Temperament in early development. *Zero to Three* (March), 56.

Sullivan, D. R.-E. (2010). *Learning to Lead.* St. Paul, MN: Redleaf Press.

Swartz, M. I. (2005). Playdough: What's standard about it? *Young Children, 60* (2), 100-109.

Swick, K. J. (2004). *Empowering Parents, Families, Schools and Communities during the Early Childhood Years.* Champaign, IL: Stipes.

Swim, T. J., and Freeman, R. (2004). Time to reflect: Using food in early childhood classrooms. *Young Children, 59* (6), 18-22.

Szanton, E. S. (2001, January). For America's infants and toddlers, are important values threatened by our zeal to "teach"? *Young Children, 56,* 15–21.

Tan, A. L. (2004). *Chinese American Children and Families: A Guide for Educators and Service Providers.* Olney, MD: Association for Childhood Education International.

Tannen, D. (2007). Conversational signals and devices. In L. Monaghan and J. E. Goodman (Eds.). *A Cultural Approach to Interpersonal Communication* (pp. 150–160). Malden, MA: Blackwell.

Tardos, A. (1985). Facilitating the play of children at lóczy. *Educaring, 6* (3), 1–2.

Tardos, A. (1986). Patterns of play observed at lóczy. *Educaring, 7* (2), 1–7.

Thatcher, D. H. (2001). Reading in math class: Selecting and using picture books for math investigations. *Young Children, 56,* 20–26.

The Early Math Collaborative–Erickson Institute. (2014). *Big Ideas of Early Mathematics: What Teachers of Young Children Need to Know (Practical Resources in ECE).* London: Pearson.

Thelen, P., and Klifman, T. (2011). Using daily transition strategies to support all children. *Young Children, 66* (4), 92–98.

The Southern Poverty Law Center. (2018). *Testing Yourself for Hidden Bias.* https://www.tolerance.org/professional-development/test-yourself-for-hidden-bias

Thoennes, T. (2008). Emerging faces of homelessness: Young children, their families, and schooling. In T. Turner-Vorbeck and M. M. Marsh (Eds.), *Other Kinds of Families: Embracing Diversity in Schools* (pp. 162–176). New York: Teachers College Press.

Thompson, N. L., and Hare, R. D. (2006). Early education for American Indian and Alaska native children in rural America. *Zero to Three, 26* (4), 43–45.

Thornberg, R. (2006). The situated nature of preschool children's conflict strategies. *Educational Psychology, 26* (1), 109–112.

Tobin, J., Hsueh, Y., and Karasawa, M. (2011). *Preschool in Three Cultures Revisited: China, Japan, and the United States.* Chicago: University of Chicago Press.

Tobin, J., Wu, D., and Davidson, D. (1989). *Preschool in Three Cultures: Japan, China and the United States.* New Haven, CT: Yale University Press.

Tortora, S. (2005). *The Dancing Dialogue: Using the Communicative Power of Movement with Young Children.* Baltimore: Brookes.

Trepanier-Street, M. (2000). Multiple forms of representation in long-term projects: The garden project. *Childhood Education, 77,* 8–25.

Trumbull, E., and Farr, B. (2005). *Language and Learning: What Teachers Need to Know.* Norwood, MA: Christopher Gordon.

Trumbull, E., Rothstein-Fisch, C., and Greenfield, P. M. (2000). *Bridging Cultures in Our Schools: New Approaches That Work.* San Francisco: WestEd.

Turnbull, A., and Turnbull, R. (2001). *Families, Professionals, and Exceptionality.* Upper Saddle River, NJ: Merrill and Prentice Hall.

Turner-Vorbeck, T., and Marsh, M. M. (2008). *Other Kinds of Families: Embracing Diversity in Schools.* New York: Teachers College Press.

Udell, T., and Glasenapp, G. (2005). Managing challenging behaviors: Adult communication as a prevention and teaching tool. In B. Neugebauer (Ed.), *Behavior: A Beginnings WorkshopBook* (pp. 26–29). Redmond, WA: Exchange.

Ulmen, M. C. (2005). Somebody read to me!: Ten easy ways to include reading every day. *Young Children, 60* (6), 96–97.

U.S. Department of Education, *Civil Rights Data Collection: 2013-14 State and National Estimations,* Washington, D.C., accessed September 2018. https://ocrdata.ed.gov/StateNationalEstimations/Estimations_2013_14

U.S. Department of Education. (2016, June 14). *Fact Sheet: Troubling Pay Gap for Early Childhood Teachers.* https://www.ed.gov/news/press-releases/fact-sheet-troubling-pay-gap-early-childhood-teachers

U.S. Department of Education. (n.d.). *Sec. 300.323, When IEPs Must Be in Effect.* https://sites.ed.gov/idea/regs/b/d/300.323

Vandell, D. L. (2004). Early child care: The known and the unknown. *Merrill-Palmer Quarterly, 50* (3), 387–414.

Vandenbroeck, M. (2006). *Globalization and Privatization: The Impact on Child Care Policy and Practice* (Working Paper No. 38). The Hague, The Netherlands: Bernard van Leer Foundation.

Vartuli, S. (2005). Beliefs: The heart of teaching. *Young Children, 60* (5), 76–86.

Vesely C. K., and Ginsberg, M. R. (2011). Strategies and practices for working with immigrant families in early education programs. *Young Children, 66* (1), 84–89.

Villa, J., and Colker, L. (2006). A personal story: Making inclusions work. *Young Children, 61* (1), 96–100.

Villegas, M., Neugebauer, S. R., and Venegas, K. R. (2008). *Indigenous Knowledge and Education: Sites of struggle, Strength, and Survivance.* Cambridge, MA: Harvard Education Press.

Vinson, B. M. (2001, January). Fishing and Vygotsky's concept of effective education. *Young Children, 56* (1), 88–89.

Virmani, E. A., and Ontal, L. L. (2010). Supervision and training in child care: Does reflective supervision foster caregiver insightfulness? *Infant Mental Health Journal, 31* (1): 16–32.

Volk, D., and Long, S. (2005). Challenging myths of the deficit perspective: Honoring children's literacy resources. *Young Children, 60* (6), 12–19.

Vygotsky, L. S. (1962). *Thought and Language* (E. Hanfmann and G. Vakar, Trans. and Eds.). Cambridge, MA: MIT Press.

Vygotsky, L. S. (1978). *Mind in Society: The Development of Higher Psychological Processes* (M. Cole, Ed.). Cambridge, MA: Harvard University Press.

Vygotsky, L. S. (1987). Thinking and speech. In R. Reiber and A. Carton (Eds.), *The Collected Works of L. S. Vygotsky.* New York: Plenum.

Ward, G., and Dahlmeier. (2011). Rediscovering joyfulness. *Young Children, 66* (6), 94–98.

Washington, V. (2005). Sharing leadership: A case study of diversity in our profession. *Young Children, 60* (1), 23–31.

Wassermann, S. (2004). *This Teaching Life: How I Taught Myself to Teach.* New York: Teachers College Press.

Weatherson, D., Weigand, R. F., and Weigand, B. (2010). Reflective supervision: Supporting reflection as a cornerstone for competency. *Zero to Three, 31* (2), 22–40.

Weissman, P., and Hendrick, J. (2014). *The Whole Child: Developmental Education for the Early Years. Boston: Pearson.*

Wellhousen, K., and Kieff, J. (2001). *A Constructivist Approach to Block Play in Early Childhood.* Albany, NY: Delmar.

Werner, L. A. (2002). Infant auditory capabilities. *Current Opinions in Otolaryngology and Head and Neck Surgery, 10* (5), 398–402. 10.1097/00020840200210000000013

Wessels, S., and Trainin, G. (2014). Kindergarten and grade 1: Bringing literacy home: Latino families supporting children's literacy learning. *YC Young Children, 71* (5), 74–78.

Westervelt, G., Sibley, A., and Schaack, D. (2008). Quality in early childhood programs. In S. Feeney, A. Galper, and C. Seefeldt (Eds.), *Continuing Issues in Early Childhood Education* (pp. 83–99). Upper Saddle River, NJ: Pearson.

White, R. (2004.) Young children's relationship with nature: Its importance to children's development and the earth's future. Retrieved in 2004, from www.white-hutchinson.com/children/articles/childrennature.shtml

Whitebook, M., and Sakai, L. (2004). *By a Thread: How Child Care Centers Hold on to Teachers, How Teachers Build Lasting Careers.* Kalamazoo, MI: W. E. Upjohn Institute for Employment Research.

Whitelaw Drogue, P. (2006, January/February). Stop refereeing and start building communication skills. *Exchange,* 6–8.

Whitin, D. J. (2005). Pairing books for children's mathematical learning. *Young Children, 60* (2), 42–48.

Whitin, P., and Whitin, D. J. (2003, January). Developing mathematical understanding along the yellow brick road. *Young Children, 58,* 36–40.

Widerstrom, A. H. (2005). *Achieving Learning Goals through Play: Teaching Young Children with Special Needs* (2nd ed.). Baltimore: Brookes.

Wien, C. A. (2004). *Negotiating Standards in the Primary Classroom: The Teacher's Dilemma.* New York: Teachers College Press.

Wien, C. A. (2014). *The Power of Emergent Curriculum: Stories from Early Childhood Settings.* Washington, D.C.: National Association for the Education of Young Children.

Wien, C. A. (2008). *Emergent Curriculum in the Primary Classroom: Interpreting the Reggio Emilia Approach in Schools.* New York: Teacher's College Press.

Williams, B., Cunningham, D., and Lubawy, J. (2005). *Preschool Math.* Beltsville, MD: Gryphon House.

Williams, K. C., and Cooney, M. H. (2006). Young children and social justice. *Young Children, 61* (2), 75–82.

Williamson, G. G., and Anzalone, M. (2001). *Sensory Integration and Self-Regulation in Infants and Toddlers: Helping Very Young Children Interact with Their Environment.* Washington, D.C.: Zero to Three.

Wirth, S., and Rosenow, N. (2012). Supporting whole-child learning in nature-filled outdoor classrooms. *Young Children, 67* (1), 42–48.

Wolpert, E. (2005). *Start Seeing Diversity: The Basic Guide to an Anti-Bias Classroom.* St. Paul, MN: Redleaf.

Wood, K., and Youcha, V. (2009). *The ABC's of the ADA: Your Early Childhood Program Guide to the American's with Disabilities Act.* Baltimore: Brookes.

Woodard, C., Haskins, G., Schaefer, G., and Smolen, L. (2004). Let's talk: A different approach to oral language development. *Young Children, 59* (4), 92–95.

Workman, S. (2018, February 14). *Where Do Your Child Care Dollars Go? Understanding the True Cost of Quality Early Childhood Education.* Washington, D.C.: Center for American Progress. https://www.americanprogress.org/issues/early-childhood/reports/2018/02/14/446330/child-care-dollar-go/

Worsley, M., Beneke, S., and Helm, J. H. (2003, January). The pizza project: Planning and integrating math standards in project work. *Young Children, 58,* 44–49.

Worth, K., and Grollman, S. (2003). *Worms, Shadows and Whirlpools: Science in the Early Childhood Classroom.* Portsmouth, NH: Heinemann.

Woyke, P. P. (2004). Hopping frogs and trail walks: Connecting young children and nature. *Young Children, 59* (1), 82–85.

Wright, K., Stegelin, D. A., and Hartle, L. (2007). *Building Family, School, and Community Partnership* (3rd Ed.). Upper Saddle River, NJ: Pearson/Merrill/Prentice Hall.

Wurm, J. (2005). *Working in the Reggio Way: A Beginner's Guide for American Teachers.* St. Paul, MN: Redleaf.

Wylie, S., and Fenning, K. (2012). *Observing Young Children: Transforming Early Learning Through Reflective Practice* (4th ed.). Toronto, Ontario: Nelson.

Xu, S. H., and Rutledge, A. L. (2003, March). Chicken starts with *Ch!* Kindergartners learn through environmental print. *Young Children, 58,* 44–51.

Yoshida, H. (2008, November). The cognitive consequences of early bilingualism. *Zero to Three, 29* (2) 26–30.

Young Children. (2004). Resources for early childhood education around the globe. *Young Children, 59* (5), 82–83.

Young Children. (2004). Resources for exploring the creative arts with young children. *Young Children, 59* (4), 58–59.

Young Children. (2004). Resources for exploring the ethical dimensions of the early childhood profession. *Young Children, 59* (6), 40–42.

Young Children. (2004). Resources on health and safety for early childhood educators. *Young Children, 59* (2), 64–66.

Young Children. (2005). Resources on embracing diversity in early childhood settings. *Young Children, 60* (6), 55–59.

Young Children. (2005). Resources on kindergarten and beyond. *Young Children, 60* (2), 59–62.

Young Children. (2005). Resources on leadership in early childhood education. *Young Children, 60* (1), 46.

Youngquist, J. (2004). From medicine to microbes: A project investigation of health. *Young Children, 59* (2), 28–32.

Zepeda, M., Gonzalez-Mena, J., Rothstein-Fisch, C., and Trumbell, E. (2006). *Bridging Cultures in Early Care and Education.* Mahwah, NJ: Erlbaum.

Zigler, E., Gillliam, W. S., and Barnett, W. S. (Eds.). (2011). *The Pre-k Debates: Current Controversies and Issues.* Baltimore: Brookes.

Zigler, E., Gilliam, W. S., and Jones, S. M. (2006). *A Vision for Universal Preschool Education.* Cambridge, MA: Cambridge University Press.

Zigler, E. F., Singer, D. G., and Bishop-Josef, S. J. (Eds.). (2004). *Children's Play: The Roots of Reading.* Washington, D.C.: Zero to Three.

Index

Note: Page numbers followed by "f" & "t" indicate figures & tables respectively.

E

H

I

J

N

O

T